Classroom Companion: Economics

The Classroom Companion series in Economics includes undergraduate and graduate textbooks alike. It welcomes fundamental textbooks aimed at introducing students to the core concepts, empirical methods, theories and tools of the field, as well as advanced textbooks written for students at the Master and PhD level seeking a deeper understanding of economic theory, mathematical tools and quantitative methods.

Sergey Khrushchev

Linear Algebra with Applications to Economics

 Springer

Sergey Khrushchev
Department of Management
and Mathematical Economics
Satbayev University
Almaty, Kazakhstan

ISSN 2662-2882   ISSN 2662-2890
Classroom Companion: Economics
ISBN 978-3-031-68681-8   ISBN 978-3-031-68682-5   (eBook)
https://doi.org/10.1007/978-3-031-68682-5

Mathematics Subject Classification (2020): 15, 91, 39, 97, 92

This Springer imprint is published by the registered company Springer Nature Switzerland AG
The registered company address is: Gewerbestrasse 11, 6330 Cham, Switzerland

If disposing of this product, please recycle the paper.

Foreword

Dear Reader,

It is my great pleasure to introduce you to this remarkable textbook on Linear Algebra. Whether you are an eager student about to embark on your first journey into this fundamental branch of mathematics, or an instructor searching for the ideal resource to guide your students, I am confident that the pages ahead will exceed your expectations.

If you're an instructor searching for the perfect linear algebra textbook for your economics, finance or management science students, your quest ends here. Professor Khrushchev has crafted a text that brilliantly fuses rigorous linear algebra with fascinating economic applications. Your students will thank you for choosing a book that brings abstract concepts to life and equips them with invaluable analytical tools.

And if you're a student, get ready for a thrilling intellectual adventure! You'll be in the capable hands of a world-class guide who knows exactly how to demystify linear algebra and reveal its hidden powers. Professor Khrushchev's warm and engaging style will have you enjoying every page, while absorbing knowledge that will pay huge dividends in your future studies and career.

Linear algebra is the cornerstone of modern mathematics and its applications. Its tenets and techniques permeate nearly every scientific field - from computer science and data analysis to physics, engineering, and economics. A solid grasp of linear algebra is therefore essential for any student who aspires to truly understand the mathematical underpinnings of our world. Yet for many, the abstract concepts and notations of vectors, matrices, and linear transformations can seem daunting at first encounter.

This is precisely where this textbook shines. With unparalleled clarity and an engaging style, its author takes the reader by the hand and guides them through the core concepts and methods of linear algebra. Starting from the basic task of solving systems of linear equations, an early linear algebra problem that has once occupied mathematicians for centuries, the book gradually builds up the essential tools of the trade - matrices, vector spaces, inner products, eigenvalues, and more.

Along the way, the author never loses sight of what makes mathematics truly exciting - the thrill of discovery, the satisfaction of grasping a deep insight, and the empowerment that comes with mastering a new technique. Each chapter is sprinkled with well-chosen examples, visualizations, and applications that illuminate the theory and reveal the hidden structure underlying everyday problems. Careful explanations and detailed proofs, presented with a light touch, ensure that no reader is left behind.

The author masterfully weaves together rigorous mathematical proofs with concrete examples drawn from various fields, demonstrating the far-reaching impact of linear algebra. From the organic introduction to the Fredholm Alternative in Chapter 2, to the thought-provoking exploration of Markov chains in Chapter 5, to the elegant derivation of the Gram-Schmidt algorithm in Chapter 6 (just to name a few results herein), the book consistently showcases the relevance of linear algebra in solving real-world problems.

Professor Khrushchev has a knack for breaking down complex concepts into digestible pieces, making the most challenging topics accessible to students. The lucid explanations of eigenvectors and eigenvalues in Chapter 5, for instance, demystify these pivotal concepts and provide a solid foundation for understanding their applications in diagonalization and quadratic forms. The book's prose is crisp and lively, keeping readers engaged and motivated throughout their learning journey.

The author also recognizes the importance of computational tools in modern linear algebra and provides plenty of examples and exercises that encourage students to use software such as MATLAB. This emphasis on computation not only prepares students for the demands of the modern workplace but also deepens their understanding of the underlying mathematics.

In short, this book is much more than a mere catalog of definitions and results. It is an invitation to active learning, to grappling with ideas, to mathematical exploration. Numerous problems and exercises, strategically placed and varying in difficulty, challenge the reader to put their newly acquired knowledge into practice. Detailed solutions to the examples in text are provided, but the greatest reward comes from the struggle to solve a tricky problem on one's own. This is how true learning happens, and this book is the perfect companion for that journey.

For the instructor, this textbook is a dream come true. Its thoughtful structure and clear presentations make it an ideal basis for a semester course. Students will appreciate the lucid explanations and worked examples, while instructors will find ample material for problem sets and exams. The inclusion of relatively advanced topics, such as the Jordan normal form, means that this book can be also used for more advanced courses. It is truly a one-stop shop for all your linear algebra teaching needs.

In an age where information is freely available at our fingertips, one might wonder about the role of a traditional textbook. But information is not the same as knowledge, and knowledge is not the same as understanding. This book provides all three, in a coherent package that has been carefully designed and refined to optimize the learning experience. It is the product of its author's deep expertise and his empathy for the challenges faced by learners of mathematics.

So whether you are a student taking your first steps into the abstract world of mathematics, or an instructor looking to inspire a new generation of problem solvers and critical thinkers, this book is for you. Open its pages, sharpen your pencil, and get ready for an exhilarating intellectual journey. Linear algebra awaits, and with this book as your guide, you are well equipped for the adventure ahead.

Happy reading, and happy learning!

Berkeley, June 2024   *Olga Holtz*

Preface

With the proliferation of Big Data Analysis, there has been an increased demand for education in Linear Algebra among economists. Unlike Calculus, Linear Algebra courses typically do not employ Descartes' method of using Geometry for conceptual understanding and Algebra for computations, except in Analytic Geometry concerning lines and planes in space. I have observed that students often struggle with this aspect, as well as with grasping abstract concepts like vector space, bases, and subspaces. Therefore, this book adopts a traditional approach, treating Linear Algebra as a theory for solving finite systems of linear equations in a finite number of unknowns. Since every linear system can be represented by an augmented matrix, we focus on the study of matrices, – which being mere tables, are generally more accessible to students. However, this book does not follow the texts by Kolman B. and Hill D. (2007) or the top-level Linear Algebra course by Axler S. (2015).

Nowadays, students have access to tools like Wolfram's Mathematica, Microsoft Math, and others. Additionally, ChatGPT Plus and Microsoft Copilot have emerged. As a result, numerical exercises such as calculating the reduced row echelon form of a matrix or the determinant of a square matrix or something else purely technical have diminished in value for grading student's works. Such problems should be replaced with conceptual questions. This book suggests several ways to do this, offering, in particular, many solved economical problems. Some illustrate theoretical results, while others serve as exercises. In the era of ChatGPT, there is actually no need to include many problems in a book, as any problem can be sent to the bot. While the bot might initially solve it with mistakes, this can be rectified by asking follow-up questions. When the bot solves the problem, one can also request it to generate similar problems.

Chapter 1 introduces the Gauss-Jordan method of elimination. It reduces to the construction of the reduced row echelon form for the augmented matrix of the system of linear equations considered. In what follows, it is important to prove that any matrix has a unique reduced row echelon form. In its own turn any matrix $\mathbf{A}$ in reduced row echelon form determines a unique basis for the null-space $N(\mathbf{A})$ of $\mathbf{A}$. I call it the Gauss basis for the subspace $N(\mathbf{A})$. It is clear that every subspace in $\mathbb{R}^n$ is a linear span of a finite number of vectors in it and therefore can be considered as the null-space of some matrix. Hence, every subspace of $\mathbb{R}^n$ has a unique Gauss basis. For the zero $n \times n$ matrix the Gauss basis is nothing but the standard basis in $\mathbb{R}^n$. Chapter 1 concludes with a section on Portfolio theory, presenting some concepts of risk-less and arbitrage portfolios. This section also includes the non-arbitrage theorem.

In Chapter 2, I introduce a new and straightforward method of solving linear systems using the so-called Gauss matrices associated with the augmented matrices. I hope that these methodological innovations will make this book valuable for both educators and students. The last section of Chapter 2 is an elementary introduction to the Linear Programming.

Chapter 3 deals with a simultaneous elimination, which is only possible for square coefficient matrices of linear systems. This elimination immediately leads to formulas for solutions of Cramer's type. To reveal this relationship, the notion of the determinant of a square matrix is introduced. In this chapter, I prove the rank factorization theorem and use it to demonstrate that for any subspace V of $\mathbb{R}^n$ with dimension r and any subspace W of dimension $n - r$, there exists a matrix with $N(\mathbf{A}) = V$ and with the range $R(\mathbf{A}) = W$. The rank factorization theorem plays a crucial role in construction of SGI (Strong Generalized Inverse) for a given matrix. The economical section completes Chapter 3. It is devoted to Leontief' s theory.

Chapter 4 deals with vector spaces. I provide the formal definition of a vector space, primarily to show that ten axioms of a vector space are essentially the usual rules of arithmetics for the set of real numbers $\mathbb{R}$. Then, given a set X, I consider the set $\mathfrak{F}(X,\mathbb{R})$ of all functions from X to $\mathbb{R}$. If $X = \{1,2,\ldots,n\}$, then $\mathfrak{F}(X,\mathbb{R})$ is a vector space since all ten axioms should be checked pointwise, while $\mathbb{R}$ satisfies them as it is known from elementary school. Similarly, if V is any vector space, then $\mathfrak{F}(X,V)$ is also a vector space for similar reasons. An elementary theorem on subspaces of a vector space completes this schema. To illustrate the usefulness of vector spaces, I construct complex numbers as a subspace of special 2×2 matrices with real entries. In fact, vector spaces V appear in this book as subspaces of $\mathbb{R}^n$. For instance, if some $n \times n$ matrix $\mathbf{A}$ maps the subspace V into itself, then by using the Gauss basis for V, one can establish correspondence between V and $\mathbb{R}^r$, where $r = \dim(V)$, by mapping the standard basis of $\mathbb{R}^r$ onto the Gauss bases of V. This process can be repeated in every concrete problem, but it is more reasonable instead to extend theorems already proved for matrices to the setting of vector spaces.

In Chapter 5, a very important topic of diagonalization of square matrices is considered. It includes standard theorems on a possibility of diagonalization and also a short theory of Jordan normal form. In particular, it is proved that any square matrix $\mathbf{A}$ is similar to its transpose matrix $\mathbf{A}^T$. Although school mathematics covers arithmetic and geometric progressions, it completely ignores arithmetic-geometric sequences. These are described by the recurrences

$$y_n = q y_{n-1} + d, \ y_0 = c, \ n = 1,2,\ldots \ .$$

This elementary equation encompasses practically all services that ordinary people can find in any bank. Therefore, it is surprising that this theory is not studied in elementary schools. I do not develop this subject too far but solve a number of practical problems and show that the diagonalization of special 2×2 matrices is a way to solve them. Applications to solutions of linear recurrence systems, the Cob-Web model, stability problems of national economics, and Markov's theory are considered at the end of this chapter.

Chapter 6 is devoted to inner product spaces, orthogonal diagonalization and spectral theorem for symmetric matrices. Applications considered in this chapter include the theory of quadratic forms. Lagrange's method is applied to prove Silvester's theorem. At the end of the chapter, I consider two applications to the optimization of profit in several variables.

Chapter 7 begins with the pioneering paper on regression by Chebyshev. It includes a standard theory of normal systems and the theory of weak and strong generalized inverses matrices.

Regarding other books, which may be helpful, I mention an outstanding introduction in Mathematical Economics by Anthony M. and Biggs N. (2024) as well as a very nice course of Linear Algebra by Anthony M. and Harvey M. (2012).

Almaty,   *Sergey Khrushchev*
June 2024   *Satbayev University*

Acknowledgements

I am grateful to Iskander Beisembetov, the former rector of the Kazakh-British Technical University, and Zoya Tuiebakhova, the former vice-rector, for attracting me to work at the International School of Economics (ISE) at this university. The efforts made by both of them to promote the ISE to a leading role among the centers affiliated with the University of London deserve careful consideration for dissemination in other international projects of Kazakhstan. My work at ISE has had a significant influence on the content of this textbook.

I am also deeply appreciative of Meiram Begentayev, the rector of Satbayev University, and the leadership of the Project Management Institute and the Department of Management and Mathematical Economics. Their unwavering support has been crucial in advancing my research on the integration of Microsoft Copilot and Chat-GPT within Microsoft 365 Education, as detailed in Khrushchev (2023). This pioneering work has significantly contributed to the developing of this book.

Contents

Chapter 1
Gauss-Jordan Elimination

1.1 Introduction

When considering a system of two linear equations in two unknowns, such as the system

$$\begin{cases} x + y & = 3 \\ x - y & = 1 \end{cases},$$

(1.1)

we can solve it by eliminating the unknown y. The most straightforward approach is to replace the first equation with the sum of the two equations, yielding $2x = 4$, from which $x = 2$. Substituting this value into the second equation gives $y = x - 1 = 1$. The aforementioned operations can be organized as a sequence of equivalent systems:

$$\begin{cases} x + y & = 3 \\ x - y & = 1 \end{cases} \Leftrightarrow \begin{cases} 2x & = 4 \\ x - y & = 1 \end{cases} \Leftrightarrow \begin{cases} x & = 2 \\ x - y & = 1 \end{cases} \Leftrightarrow \begin{cases} x & = 2 \\ -y & = -1 \end{cases} \Leftrightarrow \begin{cases} x & = 2 \\ y & = 1 \end{cases}.$$

In this sequence, the second system is obtained from the first by adding the second equation to the first one. Subtracting the second equation from the new first equation reverts us to the initial system. The same is true for other operations that follow. They are all reversible, which implies that the solution set does not change. The third system arises from the second via a reversible operation, specifically by dividing the first equation by 2. The fourth system is obtained from the third by substraction the first equation from the second. Finally, the last system in the chain derives from its predecessor by multiplying the second equation by -1.

The variables x and y in these operations are redundant if we represent systems in the tabular form known as matrices:

$$\begin{pmatrix} 1 & 1 & | & 3 \\ 1 & -1 & | & 1 \end{pmatrix}$$

(1.2)

S. Khrushchev, *Linear Algebra with Applications to Economics*, Classroom Companion: Economics, https://doi.org/10.1007/978-3-031-68682-5_1

To retrieve the system (1.1) from the matrix (1.2), we multiply the first column of (1.2) by x, then multiply the second column of (1.2) by y, sum them, and finally replace the separator | by the equality symbol =. Clearly, these operations are reversible. In matrix notation (1.1) appears as:

$$\begin{pmatrix} 1 & 1 & | & 3 \\ 1 & -1 & | & 1 \end{pmatrix} \sim \begin{pmatrix} 2 & 0 & | & 4 \\ 1 & -1 & | & 1 \end{pmatrix} \sim \begin{pmatrix} 1 & 0 & | & 2 \\ 1 & -1 & | & 1 \end{pmatrix} \sim \begin{pmatrix} 1 & 0 & | & 2 \\ 0 & -1 & | & -1 \end{pmatrix} \sim \begin{pmatrix} 1 & 0 & | & 2 \\ 0 & 1 & | & 1 \end{pmatrix}. \quad (1.3)$$

In Section 1.2 we discuss the matrix method for solutions of systems of linear equations in more detail, with concrete examples of textual problems of economical character.

Row operations described in (1.3) are the subject of Section 1.3. Problem 1.3 deserves special attention since it opens a perspective for students to use calculators, MATLAB, Wolfram's Mathematica, and Microsoft Math Solver for solutions of systems involving parameters.

By (1.1) and (1.2) the system can be rewritten as a matrix product $\mathbf{AX} = \mathbf{b}$, where

$$\mathbf{A} = \begin{pmatrix} 1 & 1 \\ 1 & -1 \end{pmatrix}, \mathbf{X} = \begin{pmatrix} x \\ y \end{pmatrix}, \mathbf{b} = \begin{pmatrix} 3 \\ 1 \end{pmatrix}.$$

So, the product of a matrix by a column $\mathbf{X}$ is the column $\mathbf{b}$ of the same size. If it is necessary to solve simultaneously a system for several data columns $\mathbf{b}$, then this task can be written in the form of matrix products. These questions are studied in Section 1.4.

Any row operation on a matrix $\mathbf{A}$ can be represented as $\mathbf{EA}$, where $\mathbf{E}$ is an elementary matrix. This fact is mainly of a theoretical importance, and the theory of elementary matrices is present in Section 1.5

Every matrix determines the row and column spaces. In Section 1.6 we consider methods to find them explicitly.

Analytic geometry of lines and planes in space considered in Section 1.7 is a simplified version of Descartes approach to Calculus. Vector spaces considered in Linear Algebra are homogeneous in structure. Since usually in Linear Algebra one can consider problems on examples including two or three vectors, it is crucial to understand the geometry behind these constructions. For example, with formulas (1.1) and (1.3) I explained the Gauss-Jordan elimination for 2×3 matrices, i.e. for two dimensional case. Although higher dimensions involve more theory, the case of dimension two is very illustrative.

In Section 1.8, we study the reduced row echelon forms. I present a straightforward proof that every matrix has a unique reduced row echelon form. It is crucial for students to immediately realize which matrices are in a reduced row echelon form and which are not. To achieve this understanding, there are two prevalent strategies. The conventional method involves requiring students to find reduced row echelon form by hand. Personally, I do not view this approach as particularly effective, even though I often demonstrate the process in my book for the sake of completeness. A more efficient strategy is to give students problems similar to Problem 1.24, which cannot be solved using calculators or programs like Wolfram's Mathematica. In such a case, one can allow calculators even during exams and quizzes without compromising the student's comprehension of the topic.

An important topic of constructing solutions to linear systems using the reduced row echelon form of the augmented matrix of this system is discussed in Section 1.9.

Section 1.10 presents the theory of the matrix rank.

Finally, in Section 1.11, I consider The Linear Algebra aspects of the Portfolio theory. In this section, all methods developed in Chapter 1 are applied to interesting financial problems of Mathematical Economics.

1.2 Systems of Linear Equations and Matrices

Historically Linear Algebra appeared as the theory for systematized solutions of systems of linear equations in a finite number of unknowns. Here is a typical practical problem of this sort.

Problem 1.1 A company produces three types of smart phones: type A, type B, and type C. There are three departments in this company: Preparatory, Production, and Packing departments. These departments have available 310, 150, 100 work hours per week. It takes 6 minutes to prepare, 2 minutes to produce, and 1 minute to pack one phone of type A. The corresponding production times for phones of type B are 4 minutes, 3 minutes, and 2 minutes. Each phone of type C consumes 5 minutes of preparatory work, 3 minutes of production, and 3 minutes for packing. Determine the weekly levels of production of each type of phone so that the company would be operated to full capacity.

Solution: To analyze the data given in the problem, we first arrange them in a table. The columns of Table 1.1 are labeled by the phones' types, whereas its rows correspond to the names of the departments.

Let a, b, c be the numbers of phones of types A, B, C produced per week. Then the equations of operating in full capacity are given by

$$6a + 4b + 5c = 310 \times 60 \quad \longleftarrow \textbf{Preparatory department}$$
$$2a + 3b + 3c = 150 \times 60 \quad \longleftarrow \textbf{Production department}$$
$$a + 2b + 3c = 100 \times 60 \quad \longleftarrow \textbf{Packing department}$$

To simplify calculations we introduce new variables x, y, z:

$$a = 600 \cdot x, \quad b = 600 \cdot y, \quad c = 600 \cdot z. \tag{1.4}$$

We obtain the system:

$$\begin{cases} 6x + 4y + 5z = 31 \\ 2x + 3y + 3z = 15 \\ x + 2y + 3z = 10 \end{cases} \tag{1.5}$$

We swap the first and the third equations and obtain an equivalent system:

$$\begin{cases} x + 2y + 3z & = 10 \\ 2x + 3y + 3z & = 15 \\ 6x + 4y + 5z & = 31 \end{cases}.$$

To eliminate x from the second and third equations, we subtract from them the

	Type A	Type B	Type C
Preparatory department	6 min	4 min	5 min
Production department	2 min	3 min	3 min
Packing department	1 min	2 min	3 min

Table 1.1: The Table of the Problem.

multiples of the first equation:

$$\begin{cases} x + 2y + 3z & = 10 \\ 0 \cdot x - y - 3z & = -5 \\ 0 \cdot x - 8y - 13z & = -29 \end{cases} \Leftrightarrow \begin{cases} x + 2y + 3z & = 10 \\ y + 3z & = 5 \\ 8y + 13z & = 29 \end{cases}.$$

To eliminate y from the third equation, we subtract from it the multiple of the second equation:

$$\begin{cases} x + 2y + 3z & = 10 \\ y + 3z & = 5 \\ -11z & = -11 \end{cases} \Leftrightarrow \begin{cases} x + 2y + 3z & = 10 \\ y + 3z & = 5 \\ z & = 1 \end{cases} \qquad (1.6)$$

Moving from the bottom of the system to the top, we find that

$$\begin{cases} x & = 3 \\ y & = 2 \\ z & = 1 \end{cases} \Rightarrow \begin{cases} a & = 3 \cdot 600 = 1800 \\ b & = 2 \cdot 600 = 1200 \\ c & = 1 \cdot 600 = 600 \end{cases} \quad \square$$

Definition 1.1 A system of m **linear equations in n unknowns** $x_1, x_2, \ldots, x_n$ is a set of m equations of the form

$$\begin{cases} a_{11}x_1 + a_{12}x_2 + \cdots + a_{1n}x_n & = b_1 \\ a_{21}x_1 + a_{22}x_2 + \cdots + a_{2n}x_n & = b_2 \\ \vdots \quad \vdots \quad \vdots \quad \vdots \quad \vdots & \quad \vdots \\ a_{m1}x_1 + a_{m2}x_2 + \cdots + a_{mn}x_n & = b_m \end{cases}. \qquad (1.7)$$

The numbers a_{ij} are called the **coefficients** of the system. The numbers b_i are called the **constant terms**.

For example, system (1.5) is a system of 3 linear equations in 3 unknowns.

Definition 1.2 We say that the ordered set $s_1, s_2, \ldots, s_n$ is a **solution** of system (1.7) if the substitution

$$x_1 = s_1, \ x_2 = s_2, \ \ldots \ x_n = s_n,$$

makes **all** m equations in (1.7) hold true.

For example, system (1.5) has a unique solution: $x = 3$, $y = 2$, $z = 1$.

Definition 1.3 Given a system (1.7) the matrices $\mathbf{A}$ and $\mathbf{b}$

$$\mathbf{A} = \begin{pmatrix} a_{11} & a_{12} & \cdots & a_{1n} \\ a_{21} & a_{22} & \cdots & a_{2n} \\ \vdots & \vdots & \ddots & \vdots \\ a_{m1} & a_{m2} & \cdots & a_{mn} \end{pmatrix}, \quad \mathbf{b} = \begin{pmatrix} b_1 \\ b_2 \\ \vdots \\ b_m \end{pmatrix}, \tag{1.8}$$

are called the **coefficient matrix** of the system (1.7) and the **data column**. The number a_{ij} in the ith row and jth column of $\mathbf{A}$ is called the **entry** of $\mathbf{A}$. To save space the matrix with entries a_{ij} is often denoted by (a_{ij}). If $\mathbf{A}$ has m rows and n columns as in (1.8), then we say that the **size** of $\mathbf{A}$ is $m \times n$.

The first matrix below

$$\mathbf{A} = \begin{pmatrix} 6 & 4 & 5 \\ 2 & 3 & 3 \\ 1 & 2 & 3 \end{pmatrix}, \quad \mathbf{b} = \begin{pmatrix} 31 \\ 15 \\ 10 \end{pmatrix}$$

is the coefficient matrix for the system of linear equations (1.5), whereas the second is the data column in (1.5).

If the coefficient matrix $\mathbf{A} = (a_{ij})$ and the data column $\mathbf{b}$ are given, then one can rewrite (1.7) in matrix form as follows. In (1.9), for every j, $1 \le j \le n$, the unknown x_j is placed above the jth column of $\mathbf{A}$. We multiply each entry of the jth column by x_j,

$$
\begin{array}{c}
\quad x_1 \ x_2 \ \cdots \ x_n \\
\quad \downarrow \ \downarrow \ \cdots \ \downarrow \\
\begin{array}{c} + \\ + \\ \vdots \\ + \end{array}
\begin{pmatrix}
a_{11} & a_{12} & \cdots & a_{1n} \\
a_{21} & a_{22} & \cdots & a_{2n} \\
\vdots & \vdots & \ddots & \vdots \\
a_{m1} & a_{m2} & \cdots & a_{mn}
\end{pmatrix}
\end{array}
\tag{1.9}
$$

and, finally, add all products obtained in each row separately. These operations define the **product** of an $m \times n$ matrix $\mathbf{A}$ by the column $\mathbf{x}$ of n unknowns:

$$
\begin{pmatrix}
a_{11} & a_{12} & \cdots & a_{1n} \\
a_{21} & a_{22} & \cdots & a_{2n} \\
\vdots & \vdots & \ddots & \vdots \\
a_{m1} & a_{m2} & \cdots & a_{mn}
\end{pmatrix}
\begin{pmatrix}
x_1 \\ x_2 \\ \vdots \\ x_n
\end{pmatrix}
=
\begin{pmatrix}
a_{11}x_1 + a_{12}x_2 + \cdots + a_{1n}x_n \\
a_{21}x_1 + a_{22}x_2 + \cdots + a_{2n}x_n \\
\vdots \ \ \vdots \\
a_{m1}x_1 + a_{m2}x_2 + \cdots + a_{mn}x_n
\end{pmatrix}
\tag{1.10}
$$

In Problem 1.1 the unknowns a, b, c correspond to the labels **Type A**, **Type B**, **Type C** of Table 1.1. In other words, they correspond to its columns. Similarly, variables x, y, z correspond to the columns of (1.5).

Notice that the column of unknowns is an $n \times 1$ matrix $\mathbf{x}$. If $\mathbf{b}$ is the $m \times 1$ matrix which is a data column, then the linear system of m equations in n unknowns can be written in the **matrix form**:

$$
\mathbf{Ax} = \mathbf{b}.
\tag{1.11}
$$

The ith equation of the system (1.11) can be written as follows:

$$
\text{row}_i(\mathbf{A}) \cdot \mathbf{x} = a_{i1}x_1 + a_{i2}x_2 + \cdots + a_{in}x_n = b_i,
\tag{1.12}
$$

where $\text{row}_i(\mathbf{A})$ is the ith row of the matrix $\mathbf{A}$. In what follows we often drop the sign of multiplication. The system (1.5) can be written in the form (1.11) with

$$
\mathbf{A} = \begin{pmatrix} 6 & 4 & 5 \\ 2 & 3 & 3 \\ 1 & 2 & 3 \end{pmatrix}, \quad
\mathbf{x} = \begin{pmatrix} x \\ y \\ z \end{pmatrix}, \quad
\mathbf{b} = \begin{pmatrix} 31 \\ 15 \\ 10 \end{pmatrix}.
\tag{1.13}
$$

Let us analyze the solution to Problem 1.1. To save time on rewriting signs + and =, as well as the unknowns, we may agree to drop them all, since they can be easily recovered from matrices obtained by (1.10) any time we wish. In these notations, we associate with any system (1.7) the **augmented matrix**:

$$
(\mathbf{A}|\mathbf{b}) =
\left(
\begin{array}{cccc|c}
a_{11} & a_{12} & \cdots & a_{1n} & b_1 \\
a_{21} & a_{22} & \cdots & a_{2n} & b_2 \\
\vdots & \vdots & \ddots & \vdots & \vdots \\
a_{m1} & a_{m2} & \cdots & a_{mn} & b_m
\end{array}
\right).
$$

The augmented matrix for the system (1.5) is given by:

$$\mathbf{B} = (\mathbf{A}|\mathbf{b}) = \begin{pmatrix} 6 & 4 & 5 & | & 31 \\ 2 & 3 & 3 & | & 15 \\ 1 & 2 & 3 & | & 10 \end{pmatrix}. \tag{1.14}$$

The rows of the matrix (1.14) correspond to the linear equations of the system (1.5), whereas its columns to the coefficients at x, y, z, and the numerical data given on the right-hand part of the equations.

Problem 1.2 In the former year, a farmer invested 40 million KZT (monetary unit of Kazakhstan) in the production of grain, meat and vegetables. Currently, the farmer invests 88 million KZT by doubling the investments in the production of grain, tripling the investments in the production of meat, and keeping the investments in the production of vegetables on the same level. Last year, the farmer got profit of 3 120 000 KZT. It is known that grain makes 10% profit, meat makes 8%, and vegetables 6%. What will be farmer's profit in the current year?

Solution: Let x_1, x_2, x_3 be the investments (in millions of KZT) which the farmer made in grain, meat, and vegetables last year.

	Grain	Meat	Vegetables
Former Year	x_1	x_2	x_3
Current Year	$2x_1$	$3x_2$	x_3
Profit	$0.1x_1$	$0.08x_2$	$0.06x_3$

Then we obtain the system of equations:

$$\begin{cases} x_1 + x_2 + x_3 & = 40 \\ 2x_1 + 3x_2 + x_3 & = 88 \\ 0.1x_1 + 0.08x_2 + 0.06x_3 & = 3.12 \end{cases}$$

The matrix form of the system and the augmented matrix $\mathbf{B} = (\mathbf{A}|\mathbf{b})$ are given below:

$$\begin{pmatrix} 1 & 1 & 1 \\ 2 & 3 & 1 \\ 0.1 & 0.08 & 0.06 \end{pmatrix} \begin{pmatrix} x_1 \\ x_2 \\ x_3 \end{pmatrix} = \begin{pmatrix} 40 \\ 88 \\ 3.12 \end{pmatrix}, \quad (\mathbf{A}|\mathbf{b}) = \begin{pmatrix} 1 & 1 & 1 & | & 40 \\ 2 & 3 & 1 & | & 88 \\ 0.1 & 0.08 & 0.06 & | & 3.12 \end{pmatrix}. \tag{1.15}$$

The symbol $\sim$ between matrices $\mathbf{A}$ and $\mathbf{B}$ indicates that the matrix $\mathbf{B}$ is obtained from the matrix $\mathbf{A}$ by a row operation. The operation, if necessary, is shown above/below $\sim$. For instance, $r_3 := 100r_3$ shows that the 3rd row of $\mathbf{B}$ equals hundred times the 3rd row of $\mathbf{A}$. Formula $r_2 := r_2 - 2r_1$ shows that the 2nd row of $\mathbf{B}$ equals the second row of $\mathbf{A}$ minus the double of its first row.

$$\begin{pmatrix} 1 & 1 & 1 & 40 \\ 2 & 3 & 1 & 88 \\ 0.1 & 0.08 & 0.06 & 3.12 \end{pmatrix} \underset{\sim}{r_3:=100r_3} \begin{pmatrix} 1 & 1 & 1 & 40 \\ 2 & 3 & 1 & 88 \\ 10 & 8 & 6 & 312 \end{pmatrix} \underset{r_3:=r_3-10r_1}{\overset{r_2:=r_2-2r_1}{\underset{\sim}{}}} \begin{pmatrix} 1 & 1 & 1 & 40 \\ 0 & 1 & -1 & 8 \\ 0 & -2 & -4 & -88 \end{pmatrix}$$

$$\underset{r_3:=-r_3/2}{\overset{}{\underset{\sim}{}}} \begin{pmatrix} 1 & 1 & 1 & 40 \\ 0 & 1 & -1 & 8 \\ 0 & 1 & 2 & 44 \end{pmatrix} \underset{\sim}{r_3:=r_3-r_2} \begin{pmatrix} 1 & 0 & 2 & 32 \\ 0 & 1 & -1 & 8 \\ 0 & 0 & 3 & 36 \end{pmatrix} \underset{\sim}{r_3:=r_3/3} \begin{pmatrix} 1 & 0 & 2 & 32 \\ 0 & 1 & -1 & 8 \\ 0 & 0 & 1 & 12 \end{pmatrix}$$

$$\underset{r_2:=r_2+r_3}{\overset{r_2:=r_2-2r_3}{\underset{\sim}{}}} \begin{pmatrix} 1 & 0 & 0 & 8 \\ 0 & 1 & 0 & 20 \\ 0 & 0 & 1 & 12 \end{pmatrix} \Rightarrow \begin{cases} x_1 & = 8 \\ x_2 & = 20 \\ x_3 & = 12 \end{cases}.$$

Hence the profit in the current year equals

$$\Pi = 0.1 \cdot (2x_1) + 0.08 \cdot (3x_2) + 0.06 \cdot x_3 = 1.6 + 4.8 + 0.72 = \mathbf{7.12} \ \mathbf{million \ KZT}. \quad \square$$

Problems

Prob. 1 — An investor owns three types of stocks: A, B, and C. The closing prices of each stock as well as their total values are given in the table:

Day	Stock A	Stock B	Stock C	Total
Monday	$8	$30	$25	$83 500
Tuesday	$10	$25	$30	$90 000
Wednesday	$12	$10	$30	$79 000
Thursday	$10	$15	$25	$72 500
Friday	$15	$20	$25	$87 500

How many shares of each stock does the investor own?

Prob. 2 — A businessman owns three stores: A, B, and C. The sales figures for each store over four consecutive days are as follows:

- Day 1: Store A: $5000, Store B: $3000, Store C: $2000
- Day 2: Store A: $4000, Store B: $4000, Store C: $2000
- Day 3: Store A: $4000, Store B: $2000, Store C: $4000
- Day 4: Store A: $3000, Store B: $4000, Store C: $3000

Despite the variation in daily sales, the total sales value for the business remained unchanged at $150,000 at the end of each day. How many units does each store sell, if we assume that each unit sells for the same price across all stores?

Prob. 3 — An art collector has works by three artists: A, B, and C. Prices for these works for a three-day auction period are:

- Day 1: Artist A:$2000, Artist B: $1500, Artist C: $1200
- Day 2: Artist A: $1800, Artist B: $2000, Artist C: $1100
- Day 3: Artist A: $2200, Artist B: $1300, Artist C: $1500

Despite fluctuations in art prices, the total value of the collectible works remained constant at \$95,000 at the end of each day. How many pieces from each artist does the collector own?

Prob. 4 — A man owns three properties: A, B, and C. The rental prices of the properties on three consecutive months are as follows:

- Month 1: Property A: \$1000, Property B: \$1200, Property C: \$1300
- Month 2: Property A: \$1200, Property B: \$1000, Property C: \$1400
- Month 3: Property A: \$1300, Property B: \$1100, Property C: \$1300

Despite the fluctuations in rental prices, the total rental income remained unchanged at \$26,000 at the end of each of these three months. How many units are there in each property?

Prob. 5 — Write the systems in the matrix form, construct the augmented matrices, and using elementary row operations, solve the systems:

$$\begin{cases} 2x_1 - x_2 - x_3 & = 4 \\ 3x_1 + 4x_2 - 2x_3 & = 11 \\ 3x_1 - 2x_2 + 4x_3 & = 11 \end{cases} \qquad \begin{cases} x_1 + x_2 + 2x_3 + 3x_4 & = 1 \\ 3x_1 - x_2 - x_3 - 2x_4 & = -4 \\ 2x_1 + 3x_2 - x_3 - x_4 & = -6 \\ x_1 + 2x_2 + 3x_3 - x_4 & = -4 \end{cases}$$

Prob. 6 — Solve the systems:

$$\begin{cases} 2x_1 - x_2 + 3x_3 + 2x_4 & = 4 \\ 3x_1 + 3x_2 + 3x_3 + 2x_4 & = 6 \\ 3x_1 - x_2 - x_3 + 2x_4 & = 6 \\ 3x_1 - x_2 + 3x_3 - x_4 & = 6 \end{cases}$$

Prob. 7 — It is known that the system of equations

$$\begin{cases} ay + bx & = c \\ cx + az & = b \\ bz + cy & = a \end{cases}$$

has a single solution. Determine which of the following conditions is necessary and sufficient for this: $abc \neq 0$, $abc \neq 1$, $abc \neq -1$, $a \neq 0$, $b \neq 0$?

Prob. 8 — A hockey team owns an arena that has a seating capacity of 20 000 fans. With the ticket price set at \$15, average attendance at recent games has been 3000. A market survey indicates that for each dollar the ticket price is lowered, the average attendance increases by 1000. Determine the ticket price which maximizes the revenue.

1.3 Row Operations

Our solutions of the systems with the augmented matrices (1.14) and (1.15) are reduced to a finite number of simple steps: swapping couples of equations, multiplying
equations by non-zero numbers, and adding a multiple of one equation to another
equations. In matrix notations these operations look as:

RO1 multiplying a row by a non-zero constant: $\text{row}_i(\mathbf{B}) \to c \cdot \text{row}_i(\mathbf{B})$;

RO2 swapping couples of rows: $\text{row}_i(\mathbf{B}) \leftrightarrow \text{row}_j(\mathbf{B})$:

RO3 adding a multiple of one row to another row:

$$\text{row}_i(\mathbf{B}) \to \text{row}_i(\mathbf{B}) + c \cdot \text{row}_j(\mathbf{B}).$$

The operations **RO1 − 3** are called the **elementary row operations**.

Lemma 1.1 *The elementary row operation* **RO2** *can be obtained as a sequence of
four row operations of types* **RO1** *and* **RO3**.

Proof. To save space we illustrate this on the example of 3×3 matrices:

$$\begin{pmatrix} a_{11} & a_{12} & a_{13} \\ a_{21} & a_{22} & a_{23} \\ a_{31} & a_{32} & a_{33} \end{pmatrix} \underset{\sim}{r_3 := r_3 + r_1} \begin{pmatrix} a_{11} & a_{12} & a_{13} \\ a_{21} & a_{22} & a_{23} \\ a_{31} + a_{11} & a_{32} + a_{12} & a_{33} + a_{13} \end{pmatrix} \underset{\sim}{r_1 := r_1 - r_3}$$

$$\begin{pmatrix} -a_{31} & -a_{32} & -a_{33} \\ a_{21} & a_{22} & a_{23} \\ a_{31} + a_{11} & a_{32} + a_{12} & a_{33} + a_{13} \end{pmatrix} \underset{\sim}{r_3 := r_3 + r_1} \begin{pmatrix} -a_{31} & -a_{32} & -a_{33} \\ a_{21} & a_{22} & a_{23} \\ a_{11} & a_{12} & a_{13} \end{pmatrix}$$

$$\underset{\sim}{r_1 := -r_1} \begin{pmatrix} a_{31} & a_{32} & a_{33} \\ a_{21} & a_{22} & a_{23} \\ a_{11} & a_{12} & a_{13} \end{pmatrix}. \qquad \square$$

Since operation **RO2** is direct and simple, it is considered as an elementary row
operation as well. However, in theoretical questions it is convenient to consider only
RO1 and **RO3**. Moreover, we may assume that $c = 1$ in **RO3** (see Theorem 1.3).

Definition 1.4 An $m \times n$ matrix $\mathbf{A}$ is said to be row equivalent to an $m \times n$
matrix $\mathbf{B}$ if $\mathbf{B}$ can be obtained from $\mathbf{A}$ by a finite number of elementary row
operations. In this case we write $\mathbf{B} \sim \mathbf{A}$.

It is clear that the relation $\sim$ satisfies the following **equivalence** conditions:

1. $\mathbf{A} \sim \mathbf{A}$,
2. $\mathbf{A} \sim \mathbf{B} \Rightarrow \mathbf{B} \sim \mathbf{A}$,
3. $\mathbf{A} \sim \mathbf{B}$ and $\mathbf{B} \sim \mathbf{C} \Rightarrow \mathbf{A} \sim \mathbf{C}$.

Theorem 1.1 *Let* $\mathbf{Ax} = \mathbf{c}$ *and* $\mathbf{Bx} = \mathbf{d}$ *be two linear systems of m equations in n unknown. If the augmented matrices of these equations are row equivalent, then both systems have the same solutions.*

Proof. Any elementary row operation on $(\mathbf{A}|\mathbf{c})$ is reversible. Therefore, it keeps invariant the set of solutions of the corresponding linear system. Since $(\mathbf{B}|\mathbf{d})$ is obtained from $(\mathbf{A}|\mathbf{c})$ by a finite number of row operations, we conclude that the solutions to the corresponding systems coincide. □

Corollary 1.1 *If* $\mathbf{A}$ *and* $\mathbf{C}$ *are row equivalent, then the linear systems* $\mathbf{Ax} = \mathbf{0}$ *and* $\mathbf{Cx} = \mathbf{0}$ *have the same solutions.*

Proof. Notice that elementary row operations on $(\mathbf{A}|\mathbf{0})$ keep zeros in the last column of the augmented matrix. □

Therefore, in case of a **homogeneous system** $\mathbf{Ax} = \mathbf{0}$ one usually makes row operations with the coefficient matrix $\mathbf{A}$ rather than with the augmented matrix. The reason is that its last column is zero and cannot be changed by elementary row operations.

Theorem 1.1 has the following obvious extension.

Theorem 1.2 *Let* $\mathbf{A}$ *and* $\mathbf{B}$ *be* $m \times n$ *matrices and* $\mathbf{C}, \mathbf{D}$ *be* $m \times k$ *matrices. Suppose that* $m \times (n+k)$ *matrix* $(\mathbf{A}|\mathbf{C})$ *is row equivalent to* $m \times (n+k)$ *matrix* $(\mathbf{B}|\mathbf{D})$. *Then the systems* $\mathbf{Ax} = \mathbf{Cb}$ *and* $\mathbf{Bx} = \mathbf{Db}$ *are equivalent systems of linear equations.*

Proof. If $(\mathbf{A}|\mathbf{C}) \sim (\mathbf{B}|\mathbf{D})$, then in terms of systems the vertical line between matrices is transformed into equalities, whereas the formal variables $(\mathbf{x}|\mathbf{b})$ are considered as unknowns. Then the system $\mathbf{Ax} = \mathbf{Cb}$ is transformed by **invertible** row operations into the system $\mathbf{Bx} = \mathbf{Db}$. Hence they are equivalent. □

Problem 1.3 Solve the system

$$\begin{cases} 2x_1 + 3x_2 + x_3 & = 4b_1 + 2b_2 \\ x_1 + x_2 + 2x_3 & = 2b_1 + 2b_2 + 4b_3 \\ x_1 + 2x_2 + x_3 & = 4b_1 + 2b_2 + 2b_3 \end{cases},$$

where b_1, b_2, b_3 are arbitrary real numbers.

Solution: We apply Theorem 1.2 with $\mathbf{A}$ and $\mathbf{C}$ equal the coefficient matrices of the left-hand and right-hand parts of the system correspondingly. Then

$$
\begin{pmatrix} 2 & 3 & 1 & 4 & 2 & 0 \\ 1 & 1 & 2 & 2 & 2 & 4 \\ 1 & 2 & 1 & 4 & 2 & 2 \end{pmatrix}
\underset{\sim}{r_1 := r_1 - r_2 - r_3}
\begin{pmatrix} 0 & 0 & -2 & -2 & -2 & -6 \\ 1 & 1 & 2 & 2 & 2 & 4 \\ 1 & 2 & 1 & 4 & 2 & 2 \end{pmatrix}
\begin{array}{l} r_1 := -r_1/2 \\ r_3 := r_3 - r_2 \\ \sim \end{array}
$$

$$
\begin{pmatrix} 0 & 0 & 1 & 1 & 1 & 3 \\ 1 & 1 & 2 & 2 & 2 & 4 \\ 0 & 1 & -1 & 2 & 0 & -2 \end{pmatrix}
\underset{\sim}{r_2 := r_2 - r_3}
\begin{pmatrix} 0 & 0 & 1 & 1 & 1 & 3 \\ 1 & 0 & 3 & 0 & 2 & 6 \\ 0 & 1 & -1 & 2 & 0 & -2 \end{pmatrix}
\begin{array}{l} r_2 := r_2 - 3r_1 \\ r_3 := r_3 + r_1 \\ \sim \end{array}
$$

$$
\begin{pmatrix} 0 & 0 & 1 & 1 & 1 & 3 \\ 1 & 0 & 0 & -3 & -1 & -3 \\ 0 & 1 & 0 & 3 & 1 & 1 \end{pmatrix}
\sim
\begin{pmatrix} 1 & 0 & 0 & -3 & -1 & -3 \\ 0 & 1 & 0 & 3 & 1 & 1 \\ 0 & 0 & 1 & 1 & 1 & 3 \end{pmatrix}
\Rightarrow
\begin{cases} x_1 & = -3b_1 - b_2 - 3b_3 \\ x_2 & = 3b_1 + b_2 + b_3 \\ x_3 & = b_1 + b_2 + 3b_3 \end{cases}
\quad \square
$$

Theorem 1.3 *Let $\mathbf{A}$ and $\mathbf{B}$ be two $m \times n$ row equivalent matrices. Then $\mathbf{B}$ can be obtained from $\mathbf{A}$ by a finite number of elementary row operations of type* **RO1** *or* **RO3** *with $c = 1$.*

Proof. By Lemma 1.1 only **RO1** or **RO3** can be used. If $c \neq 1$ then the row operation

$$\mathrm{row}_i(\mathbf{A}) \rightarrow \mathrm{row}_i(\mathbf{A}) + c \cdot \mathrm{row}_j(\mathbf{A})$$

is a combination of two row operations:

$$\mathrm{row}_j(\mathbf{A}) \rightarrow c \cdot \mathrm{row}_j(\mathbf{A}), \ \mathrm{row}_i(\mathbf{A}) \rightarrow \mathrm{row}_i(\mathbf{A}) + \mathrm{row}_j(\mathbf{A}). \qquad \square$$

Now, a solution to Problem 1.1 is compressed in a few lines:

$$
\begin{pmatrix} 6 & 4 & 5 & 31 \\ 2 & 3 & 3 & 15 \\ 1 & 2 & 3 & 10 \end{pmatrix}
\underset{}{r_1 \leftrightarrow r_3}
\begin{pmatrix} 1 & 2 & 3 & 10 \\ 2 & 3 & 3 & 15 \\ 6 & 4 & 5 & 31 \end{pmatrix}
\begin{array}{l} r_2 := r_2 - 2r_1 \\ r_3 := r_3 - 6r_1 \\ \sim \end{array}
\begin{pmatrix} 1 & 2 & 3 & 10 \\ 0 & -1 & -3 & -5 \\ 0 & -8 & -13 & -29 \end{pmatrix}
$$

$$
\begin{array}{l} r_2 := -r_2 \\ r_3 := -r_3 \\ \sim \end{array}
\begin{pmatrix} 1 & 2 & 3 & 10 \\ 0 & 1 & 3 & 5 \\ 0 & 8 & 13 & 29 \end{pmatrix}
\begin{array}{l} r_1 := r_1 - 2r_2 \\ r_3 := r_3 - 8r_2 \\ \sim \end{array}
\begin{pmatrix} 1 & 0 & -3 & 0 \\ 0 & 1 & 3 & 5 \\ 0 & 0 & -11 & -11 \end{pmatrix}
\underset{\sim}{r_3 := -r_3/11}
$$

$$\begin{pmatrix} 1 & 0 & -3 & | & 0 \\ 0 & 1 & 3 & | & 5 \\ 0 & 0 & 1 & | & 1 \end{pmatrix} \overset{r_1 := r_1 + 3r_3}{\underset{\sim}{r_2 := r_2 - 3r_3}} \begin{pmatrix} 1 & 0 & 0 & | & 3 \\ 0 & 1 & 0 & | & 2 \\ 0 & 0 & 1 & | & 1 \end{pmatrix} \Rightarrow \begin{cases} x & = 3 \\ y & = 2 \\ z & = 1 \end{cases}. \quad \square \qquad (1.16)$$

A system of linear equations may have no solutions.

Problem 1.4 Solve the following system

$$\begin{cases} 6x + 4y + 5z & = 31 \\ 2x + 3y + 3z & = 15 \\ x + 2y + 1.9z & = 10 \end{cases}. \qquad (1.17)$$

Solution: Using row operations one can check that

$$\begin{pmatrix} 6 & 4 & 5 & | & 31 \\ 2 & 3 & 3 & | & 15 \\ 1 & 2 & 1.9 & | & 10 \end{pmatrix} \sim \begin{pmatrix} 1 & 0 & 0.3 & | & 0 \\ 0 & 1 & 0.8 & | & 0 \\ 0 & 0 & 0 & | & 1 \end{pmatrix}.$$

It follows that if system (1.18) has a solution x, y, z, then

$$0 \cdot x + 0 \cdot y + 0 \cdot z = 1,$$

which is impossible, since $0 \neq 1$. Hence system (1.18) has no solutions. $\square$
A system of linear equations may have infinitely many solutions.

Problem 1.5 Solve the following system

$$\begin{cases} 6x + 4y + 5z & = 31 \\ 2x + 3y + 3z & = 15 \\ x + 2y + 8.9z & = 10 \end{cases}. \qquad (1.18)$$

Solution: Using row operations one can check that

$$\begin{pmatrix} 6 & 4 & 5 & | & 31 \\ 2 & 3 & 3 & | & 15 \\ 1 & 2 & 8.9 & | & 10 \end{pmatrix} \sim \begin{pmatrix} 1 & 0 & 0.3 & | & 3.3 \\ 0 & 1 & 0.8 & | & 2.8 \\ 0 & 0 & 0 & | & 0 \end{pmatrix}$$

If $z = s$, then $x = 3.3 - 0.3s$, $y = 2.8 - 0.8t$. We obtain a formula for all solutions as follows:

$$\begin{pmatrix} x \\ y \\ z \end{pmatrix} = \begin{pmatrix} 3.3 - 0.3s \\ 2.8 - 0.8s \\ s \end{pmatrix} = \begin{pmatrix} 3.3 \\ 2.8 \\ 0 \end{pmatrix} + s \begin{pmatrix} -0.3 \\ -0.8 \\ 1 \end{pmatrix} \quad \square$$

These examples show that a system of linear equations may have a unique solution, no solutions, or infinitely many solutions.

Problems

Prob. 9 — Using appropriate row operations solve the following systems of linear equations.

$$\begin{cases} 2x + 3y + 4z & = 3 \\ 3x + 2y + z & = 2 \\ 9x + 5y + z & = 5 \end{cases} \qquad \begin{cases} 2x + 3y + 7z & = 5 \\ 3x + 2y + 8z & = 5 \\ 9x + 5y + 23z & = 14 \end{cases}$$

Prob. 10 —

$$\begin{cases} 2x + 3y + 7z & = 5 \\ 3x + 2y + 8z & = 5 \\ 9x + 5y + 23z & = 14 \end{cases} \qquad \begin{cases} x - 4y + z & = -12 \\ 6x - 31y + 9z & = 11 \\ 3x - 17y + 5z & = -30 \end{cases}$$

Prob. 11 —

$$\begin{cases} 2x_1 + 4x_2 + x_3 + 2x_4 & = 4 \\ x_1 + 2x_2 + x_3 + x_4 & = 4 \\ x_1 + 2x_2 + x_4 & = 2 \\ x_1 + 2x_2 + x_3 + x_4 & = 3 \end{cases} \qquad \begin{cases} 2x_1 + 4x_2 + x_3 + 2x_4 & = 4 \\ x_1 + 2x_2 + x_3 + x_4 & = 3 \\ x_1 + 2x_2 + x_4 & = 1 \\ x_1 + 2x_2 + x_3 + x_4 & = 3 \end{cases}$$

Prob. 12 — Let $\mathbf{A}$ be an $m \times n$ matrix and $\mathbf{B} \sim \mathbf{A}$. Let $\mathbf{A}_j$ be the matrix of size $m \times (n - 1)$ obtained from the matrix $\mathbf{A}$ by deleting a column $\text{col}_j(\mathbf{A})$. Let $\mathbf{B}_j$ be the matrix obtained from the matrix $\mathbf{B}$ by deleting a column $\text{col}_j(\mathbf{B})$. Is it true that $\mathbf{B}_j \sim \mathbf{A}_j$? Justify your answer.

Prob. 13 — Using the properties of row operations prove that

$$\begin{pmatrix} 6 & 4 & 5 & | & 1 & 0 & 0 \\ 2 & 3 & 3 & | & 0 & 1 & 0 \\ 1 & 2 & 3 & | & 0 & 0 & 1 \end{pmatrix} \sim \begin{pmatrix} 1 & 0 & 0 & | & 3/11 & -2/11 & -3/11 \\ 0 & 1 & 0 & | & -3/11 & 13/11 & -8/11 \\ 0 & 0 & 1 & | & 1/11 & -8/11 & 10/11 \end{pmatrix}.$$

Deduce that the system

$$\begin{pmatrix} 6 & 4 & 5 \\ 2 & 3 & 3 \\ 1 & 2 & 3 \end{pmatrix} \begin{pmatrix} x \\ y \\ z \end{pmatrix} = \begin{pmatrix} 1 & 0 & 0 \\ 0 & 1 & 0 \\ 0 & 0 & 1 \end{pmatrix} \begin{pmatrix} 31 \\ 15 \\ 10 \end{pmatrix}$$

and the system

$$\begin{pmatrix} 1 & 0 & 0 \\ 0 & 0 & 1 \\ 0 & 0 & 1 \end{pmatrix} \begin{pmatrix} x \\ y \\ z \end{pmatrix} = \begin{pmatrix} 3/11 & -2/11 & -3/11 \\ -3/11 & 13/11 & -8/11 \\ 1/11 & -8/11 & 10/11 \end{pmatrix} \begin{pmatrix} 31 \\ 15 \\ 10 \end{pmatrix}$$

are equivalent.

Prob. 14 — A Trust Fund plans to invest three hundred thousand dollars in governmental bonds, in deposit accounts, and in shares. The annual interest rates for bonds, deposits, and shares are 2%, 3%, and 6%. The rules of the Trust state that the investment in bonds must be equal to that in shares plus the triple of one in deposits. Determine the maximal possible yield of the Trust at the end of the year.

Prob. 15 — A factory makes furniture: tables, chairs, and cupboards. Each piece of furniture requires three operations: cutting materials, assembling, and finishing. Each operation requires the number of hours given in the table:

Furniture	Table	Chair	cupboard
Cutting	3	1	2
Assembling	2	2	1
Finishing	2	3	1

Monthly labor resources in the factory are allocated as follows: 60 hours of cutting, 50 hours of assembling, and 70 hours of finishing. Determine the optimal number of tables, chairs, and cupboards to be produced in order to fully utilize all available labor-hours.

Prob. 16 — Check whether the two matrices below are row-equivalent:

$$\mathbf{A} = \begin{pmatrix} 1 & 1 & 2 \\ 0 & -1 & 2 \\ 3 & 1 & 2 \end{pmatrix}, \quad \mathbf{B} = \begin{pmatrix} 1 & 2 & 3 \\ 4 & 3 & 6 \\ 5 & 5 & 10 \end{pmatrix}.$$

1.4 Matrix Algebra

Any $m \times n$ matrix $\mathbf{A}$ can be written in the **entry** notations $\mathbf{A} = (a_{ij})$. These notations are used if individual entries of $\mathbf{A}$ are important. A matrix $\mathbf{A}$ can be also written in the **row** notations

$$\mathbf{A} = \begin{pmatrix} \mathrm{row}_1(\mathbf{A}) \\ \mathrm{row}_2(\mathbf{A}) \\ \vdots \\ \mathrm{row}_m(\mathbf{A}) \end{pmatrix}, \tag{1.19}$$

where

$$\mathrm{row}_i(\mathbf{A}) = \begin{pmatrix} a_{i1} & a_{i2} & \cdots & a_{in} \end{pmatrix}$$

is the ith row of $\mathbf{A}$. Any row $\mathrm{row}_i(\mathbf{A})$ is a $1 \times n$ matrix. Finally, $\mathbf{A}$ can be expressed in the **column** notations

$$\mathbf{A} = \begin{pmatrix} \mathrm{col}_1(\mathbf{A}) & \mathrm{col}_2(\mathbf{A}) & \cdots & \mathrm{col}_n(\mathbf{A}) \end{pmatrix}, \tag{1.20}$$

where

$$\text{col}_j(\mathbf{A}) = \begin{pmatrix} a_{1j} \\ a_{2j} \\ \vdots \\ a_{mj} \end{pmatrix}$$

is the jth column of $\mathbf{A}$. Any $\text{col}_j(\mathbf{A})$ is an $m \times 1$ matrix. For example, the matrix $\mathbf{A}$ in (1.13) is given in the entry notations. It can also be written in the row notations

$$\mathbf{A} = \begin{pmatrix} \text{row}_1(\mathbf{A}) \\ \text{row}_2(\mathbf{A}) \\ \text{row}_3(\mathbf{A}) \end{pmatrix}.$$

Here $\text{row}_2(\mathbf{A}) = \begin{pmatrix} 2 & 3 & 3 \end{pmatrix}$.

The row notations (1.19) are space consuming compared with (1.20). To convert rows into columns we introduce a special operation on matrices. With any row $\begin{pmatrix} a_{i1} & a_{i2} & \cdots & a_{in} \end{pmatrix}$ we associate the column

$$\begin{pmatrix} a_{i1} & a_{i2} & \cdots & a_{in} \end{pmatrix}^T \overset{def}{=} \begin{pmatrix} a_{i1} \\ a_{i2} \\ \vdots \\ a_{in} \end{pmatrix}.$$

Given an $m \times n$ matrix $\mathbf{A}$ written in the row notations (1.19) we define the **transpose** of $\mathbf{A}$ as the $n \times m$ matrix $\mathbf{A}^T$:

$$\mathbf{A}^T = \begin{pmatrix} \text{row}_1(\mathbf{A})^T & \text{row}_2(\mathbf{A})^T & \cdots & \text{row}_m(\mathbf{A})^T \end{pmatrix}. \tag{1.21}$$

Equivalently,

$$\text{row}_i(\mathbf{A})^T = \text{col}_i(\mathbf{A}^T), \quad i = 1, 2, \ldots, m. \tag{1.22}$$

For example, if

$$\mathbf{A} = \begin{pmatrix} 1 & 0 & 0 & 2 \\ 0 & 0 & 1 & 1 \\ 0 & 0 & 0 & 0 \end{pmatrix} \quad \text{then} \quad \mathbf{A}^T = \begin{pmatrix} 1 & 0 & 0 \\ 0 & 0 & 0 \\ 0 & 1 & 0 \\ 2 & 1 & 0 \end{pmatrix}. \tag{1.23}$$

Definition 1.5 Given two matrices $\mathbf{A} = (a_{ij})$ and $\mathbf{B} = (b_{ij})$ of the same size $m \times n$ we define $\mathbf{A} + \mathbf{B}$ and $\lambda \mathbf{A}$, where λ is a real number, in the entry notations as follows

$$\mathbf{A} + \mathbf{B} = \begin{pmatrix} a_{ij} + b_{ij} \end{pmatrix}, \quad \lambda \mathbf{A} = (\lambda a_{ij}). \tag{1.24}$$

For every positive integer n we denote by $\mathbf{I}_n$ the **identity matrix**:

$$\mathbf{I}_1 = \begin{pmatrix} 1 \end{pmatrix}, \ \mathbf{I}_2 = \begin{pmatrix} 1 & 0 \\ 0 & 1 \end{pmatrix}, \ \mathbf{I}_3 = \begin{pmatrix} 1 & 0 & 0 \\ 0 & 1 & 0 \\ 0 & 0 & 1 \end{pmatrix}, \ \cdots, \ \mathbf{I}_n = \begin{pmatrix} 1 & 0 & 0 & \cdots & 0 \\ 0 & 1 & 0 & \cdots & 0 \\ 0 & 0 & 1 & \cdots & 0 \\ \vdots & \vdots & \vdots & \ddots & \vdots \\ 0 & 0 & 0 & \cdots & 1 \end{pmatrix}.$$

> **Definition 1.6** An $n \times n$ matrix is called the **identity matrix** if its entries a_{ij} satisfy
> $$a_{ij} = \begin{cases} 1 & \text{if } i = j \\ 0 & \text{if } i \neq j \end{cases}.$$

The identity matrix is the coefficient matrix of the linear system

$$\begin{cases} x_1 & +0 \cdot x_2 + \cdots +0 \cdot x_n = b_1 \\ 0 \cdot x_1 & +x_2 \ + \cdots +0 \cdot x_n = b_2 \\ \vdots & \vdots \ \cdots \ \vdots \ \ \vdots \\ 0 \cdot x_1 & +0 \cdot x_2 + \cdots \ +x_n = b_n \end{cases} \Rightarrow \begin{cases} x_1 & = b_1 \\ x_2 & = b_2 \\ \vdots & \vdots \\ x_n & = b_n \end{cases},$$

which has an obvious solution.

There is another series of important matrices:

$$\mathbf{0}_{1 \times 1} = \begin{pmatrix} 0 \end{pmatrix}, \ \mathbf{0}_{2 \times 2} = \begin{pmatrix} 0 & 0 \\ 0 & 0 \end{pmatrix}, \ \cdots, \ \mathbf{0}_{n \times n} = \begin{pmatrix} 0 & 0 & 0 & \cdots & 0 \\ 0 & 0 & 0 & \cdots & 0 \\ 0 & 0 & 0 & \cdots & 0 \\ \vdots & \vdots & \vdots & \ddots & \vdots \\ 0 & 0 & 0 & \cdots & 0 \end{pmatrix}.$$

We denote by $\mathbf{0}_{m \times n}$ the $m \times n$ matrix with zero entries. In general we write $\mathbf{0}$ if the size of this zero matrix is clear from the context. Often $\mathbf{0}_n$ denotes either the zero column or row of length n.

If $\mathbf{A} = \mathbf{I}_n$ we denote by ${}^n\mathbb{R} = \mathbf{row}(\mathbf{I}_n)$ the set of all possible rows of length n. The rows

$$_n\mathbf{f}_i = \mathbf{f}_i = \mathrm{row}_i(\mathbf{I}_n)$$

are used to represent uniquely any row $\mathbf{y} = \begin{pmatrix} y_1 & y_2 & \cdots & y_n \end{pmatrix}$ of length n as the sum

$$\mathbf{y} = y_1 \mathbf{f}_1 + y_2 \mathbf{f}_2 + \cdots + y_n \mathbf{f}_n. \tag{1.25}$$

We denote by $\mathbb{R}^n = \mathbf{col}(\mathbf{I}_n)$ the set of all columns of length n. Let

$$_n\mathbf{e}_j = \mathbf{e}_j = \mathrm{col}_j(\mathbf{I}_n)$$

be the j-th column of $\mathbf{I}_n$. We drop the first index n if it is clear that we consider columns of $\mathbf{I}_n$. Any column $\mathbf{x} = \begin{pmatrix} x_1 & x_2 & \cdots & x_n \end{pmatrix}^T$ of length n can be uniquely represented as the sum

$$\mathbf{x} = x_1\mathbf{e}_1 + x_2\mathbf{e}_2 + \cdots + x_n\mathbf{e}_n. \tag{1.26}$$

Both (1.25) and (1.26) are partial cases of matrix addition and multiplication by a constant, see (1.24). Since two columns of equal length are equal if and only if their entries are the same, we see that the representation (1.26) is unique. The set $\{\mathbf{e}_1,\ldots,\mathbf{e}_n\}$ is called the **standard basis** for $\mathbf{col}(\mathbf{I}_n)$. The standard basis has a nice property. For every $m \times n$ matrix $\mathbf{A}$ and every index $j = 1,\ldots,n$

$$\boxed{\mathbf{A}\mathbf{e}_j = \mathrm{col}_j(\mathbf{A})}. \tag{1.27}$$

Lemma 1.2 *Two $m \times n$ matrices $\mathbf{A}_1$ and $\mathbf{A}_2$ are equal if and only if*

$$\mathbf{A}_1\mathbf{x} = \mathbf{A}_2\mathbf{x} \tag{1.28}$$

for every vector $\mathbf{x}$ in $\mathbb{R}^n$.

Proof. If (1.28) holds for every $\mathbf{x}$ in $\mathbb{R}^n$ then it holds for $\mathbf{x} = \mathbf{e}_j$, $j = 1,\ldots,n$. By (1.27) the jth column of $\mathbf{A}_1$ equals the jth column of $\mathbf{A}_1$ for $j = 1,\ldots,n$. It follows that the matrices $\mathbf{A}_1$ and $\mathbf{A}_2$ have equal entries. Therefore, $\mathbf{A}_1 = \mathbf{A}_2$. If $\mathbf{A}_1 = \mathbf{A}_2$ then (1.28) obviously holds for every $\mathbf{x}$ in $\mathbb{R}^n$. $\square$

Definition 1.7 Following (1.10), given an $m \times n$ matrix $\mathbf{A}$ and an $n \times p$ matrix $\mathbf{B}$, we define the matrix product $\mathbf{AB}$ in the entry notations by

$$\mathbf{AB} = \left(\mathrm{row}_i(\mathbf{A})\mathrm{col}_j(\mathbf{B})\right). \tag{1.29}$$

Notice that the $1 \times n$ matrix $\mathrm{row}_i(\mathbf{A})$ is multiplied by the column $\mathrm{col}_j(\mathbf{B})$ of length n, so that

$$\mathrm{row}_i(\mathbf{A})\mathrm{col}_j(\mathbf{B}) = a_{i1}b_{1j} + a_{i2}b_{2j} + \cdots + a_{in}b_{nj}.$$

In the row and column notations the product of two matrices can be written as follows:

$$\mathbf{AB} = \begin{pmatrix} \mathbf{A}\mathrm{col}_1(\mathbf{B}) & \cdots & \mathbf{A}\mathrm{col}_p(\mathbf{B}) \end{pmatrix} = \begin{pmatrix} \mathrm{row}_1(\mathbf{A})\mathbf{B} \\ \mathrm{row}_2(\mathbf{A})\mathbf{B} \\ \vdots \\ \mathrm{row}_m(\mathbf{A})\mathbf{B} \end{pmatrix}. \tag{1.30}$$

Theorem 1.4 *Let* **A**, **B**, **C** *be matrices such that one of two products* **(AB)C** *and* **A(BC)** *makes sense. Then the second product makes sense too and*

$$(\mathbf{AB})\mathbf{C} = \mathbf{A}(\mathbf{BC}). \tag{1.31}$$

Proof. If **(AB)C** makes sense, then the number p of rows of **C** (equally, the length of the columns of **C**) must be equal the number of columns of **AB**, which is nothing but the number of columns of **B**, see (1.30). It follows that **C** is a $p \times q$ matrix and **B** is $n \times p$ matrix. Since **AB** makes sense, formulas (1.30) show that **A** is an $m \times n$ matrix. Hence the product **A(BC)** exists. The second case is considered similarly.

To prove (1.31) we consider the matrices **A, B, C** as functions

$$f_{\mathbf{A}} : \mathbb{R}^n \to \mathbb{R}^m, \;\; f_{\mathbf{A}}(\mathbf{z}) = \mathbf{Az},$$
$$f_{\mathbf{B}} : \mathbb{R}^p \to \mathbb{R}^n, \;\; f_{\mathbf{B}}(\mathbf{y}) = \mathbf{By},$$
$$f_{\mathbf{C}} : \mathbb{R}^q \to \mathbb{R}^p, \;\; f_{\mathbf{C}}(\mathbf{x}) = \mathbf{Cx},$$

Then by the definition of the superposition ∘ of functions we obtain

$$(\mathbf{AB})\mathbf{Cx} = (f_{\mathbf{A}} \circ f_{\mathbf{B}}) (f_{\mathbf{C}}(\mathbf{x})) = f_{\mathbf{A}} (f_{\mathbf{B}} (f_{\mathbf{C}}(\mathbf{x}))).$$

Similarly,

$$\mathbf{A}(\mathbf{BC})\mathbf{x} = f_{\mathbf{A}} \circ (f_{\mathbf{B}} \circ f_{\mathbf{C}}) (\mathbf{x}) = f_{\mathbf{A}} (f_{\mathbf{B}} (f_{\mathbf{C}}(\mathbf{x}))),$$

implying that

$$(\mathbf{AB})\mathbf{Cx} = \mathbf{A}(\mathbf{BC})\mathbf{x}$$

for every **x** in $\mathbb{R}^q$. By Lemma 1.2 this implies (1.31). □

This proof looks a little bit tricky for beginners. Therefore, we present also another proof showing that the matrix multiplication is **associative**. It is based on (1.30).

Proof. By the second formula in (1.30) we see that $\mathrm{row}_i(\mathbf{AB}) = \mathrm{row}_i(\mathbf{A})\mathbf{B}$. Then by (1.29)

$$(\mathbf{AB})\mathbf{C} = \left(\mathrm{row}_i(\mathbf{AB})\mathrm{col}_j(\mathbf{C})\right) = \left((\mathrm{row}_i(\mathbf{A})\mathbf{B})\mathrm{col}_j(\mathbf{C})\right).$$

Similarly, by the first formula of (1.29)

$$A(BC) = \left(\text{row}_i(A)\text{col}_j(BC)\right) = \left(\text{row}_i(A)(B\text{col}_j(C))\right).$$

These formulas show that it is sufficient to prove that

$$(yB)\,x = y\,(Bx) \tag{1.32}$$

for any $n \times p$ matrix B, any row y of length n, and any column x of length p. Since any column x is a linear combination of the columns $_p e_j$, and any row y is the linear combination of the columns $_n f_i$, see (1.26), it is sufficient to establish (1.32) for $y = {_n}f_i$ and $x = {_p}e_j$. We have

$$({_n}f_i B)\,{_p}e_j = \text{row}_i(B)\,{_p}e_j = b_{ij}.$$

Similarly,

$${_n}f_i\left(B\,{_p}e_j\right) = {_n}f_i\text{col}_j(B) = b_{ij},$$

which proves the theorem.   □

Problem 1.6 Let A, B, C be the 3×3 matrices defined by

$$A = \begin{pmatrix} 1 & -1 & 0 \\ 0 & 1 & -1 \\ 1 & 1 & 1 \end{pmatrix}, \quad B = \begin{pmatrix} 1 & -2 & 1 \\ -1 & 0 & -2 \\ 2 & 1 & 0 \end{pmatrix}, \quad C = \begin{pmatrix} 1 & -1 & 1 \\ 0 & 1 & -1 \\ 1 & 1 & 1 \end{pmatrix}.$$

Show that $AB = BA$ but $AC \neq CA$.

Solution: By the definition of the product we have

$$AB = \begin{pmatrix} 2 & -2 & 3 \\ -3 & -1 & -2 \\ 2 & -1 & -1 \end{pmatrix} = BA$$

$$AC = \begin{pmatrix} 1 & -2 & 2 \\ -1 & 0 & -2 \\ 2 & 1 & 1 \end{pmatrix} \neq \begin{pmatrix} 2 & -1 & 2 \\ -1 & 0 & -2 \\ 2 & 1 & 0 \end{pmatrix} = CA. \quad □$$

In Problem 1.6 the matrix B is the square A^2 of the matrix A which explains why $AB = BA$, see Theorem 1.4. The matrix C differs from A only at the entry $(1,3)$.

Problem 1.7 A company manufactures three types of perfume - floral (F), woody (W), and oriental (O), in four factories in different locations A, B, C, D. The distributions of costs in producing a single piece of perfume are given in the first table below. The number of pieces of perfume produced in one month at the four locations are given in the second table.

	F	W	O
material	$1	$4	$6
labor	$2	$3	$5
overheads	$2	$1	$4

	A	B	C	D
F	3000	2000	1000	2000
W	600	1000	3000	1000
O	300	800	1000	3000

Find the total monthly costs of materials, labor and overhead at each factory.

Solution: Let $\mathbf{C}$ be the cost matrix and $\mathbf{D}$ be the 'data' matrix:

$$\mathbf{C} = \begin{pmatrix} 1 & 4 & 6 \\ 2 & 3 & 5 \\ 2 & 1 & 4 \end{pmatrix}, \quad \mathbf{D} = 100 \begin{pmatrix} 30 & 20 & 10 & 20 \\ 6 & 10 & 30 & 10 \\ 3 & 8 & 10 & 30 \end{pmatrix}$$

Then

$$\mathbf{CD} = 100 \begin{pmatrix} 1 & 4 & 6 \\ 2 & 3 & 5 \\ 2 & 1 & 4 \end{pmatrix} \begin{pmatrix} 30 & 20 & 10 & 20 \\ 6 & 10 & 30 & 10 \\ 3 & 8 & 10 & 30 \end{pmatrix} = 100 \begin{pmatrix} 72 & 108 & 190 & 240 \\ 93 & 110 & 160 & 220 \\ 78 & 82 & 90 & 170 \end{pmatrix} =$$

$$\begin{pmatrix} 7200 & 10800 & 19000 & 24000 \\ 9300 & 11000 & 16000 & 22000 \\ 7800 & 8200 & 9000 & 17000 \end{pmatrix}.$$

The corresponding monthly costs, therefore, can be seen in the table:

	A	B	C	D
materials	$7200	$10800	$19000	$24000
labor	$9300	$11000	$16000	$22000
overheads	$7800	$8200	$9000	$17000

$\square$

Theorem 1.5 *Let $\mathbf{A}$ be an $m \times k$ matrix and $\mathbf{B}$ a $k \times n$ matrix. Then*

$$(\mathbf{AB})^T = \mathbf{B}^T \mathbf{A}^T.$$

Proof. By (1.29) we obtain that the (i,j) entry of $\mathbf{B}^T \mathbf{A}^T$

$$\text{row}_i(\mathbf{B}^T)\text{col}_j(\mathbf{A}^T) = \text{col}_i(\mathbf{B})\text{row}_j(\mathbf{A}) = \text{row}_j(\mathbf{A})\text{col}_i(\mathbf{B})$$

equals the (i,j) entry of $(\mathbf{AB})^T$. $\square$

If $\mathbf{A}$ and $\mathbf{B}$ are of the same size and λ is a real constant we obtain by the definition that

$$(\mathbf{A} + \mathbf{B})^T = \mathbf{A}^T + \mathbf{B}^T, \quad (\lambda \mathbf{A})^T = \lambda \mathbf{A}^T.$$

There are two useful formulas for matrix products:

$$\left(\alpha_1 \; \alpha_2 \; \cdots \; \alpha_m \right) \begin{pmatrix} \mathrm{row}_1(\mathbf{A}) \\ \mathrm{row}_2(\mathbf{A}) \\ \vdots \\ \mathrm{row}_m(\mathbf{A}) \end{pmatrix} = \alpha_1 \mathrm{row}_1(\mathbf{A}) + \cdots + \alpha_m \mathrm{row}_m(\mathbf{A}), \qquad (1.33)$$

$$\left(\mathrm{col}_1(\mathbf{A}) \; \cdots \; \mathrm{col}_n(\mathbf{A}) \right) \begin{pmatrix} \alpha_1 \\ \alpha_2 \\ \vdots \\ \alpha_n \end{pmatrix} = \alpha_1 \mathrm{col}_1(\mathbf{A}) + \cdots + \alpha_n \mathrm{col}_n(\mathbf{A}). \qquad (1.34)$$

Problem 1.8 A rent a car company owns 1000 cars. Each day 30% of cars available are rented, and 60% of all rented cars are returned. At present, 500 cars are available for rent. How many cars will be in the garage in two days?

Solution: Let

$$\begin{cases} x_t & = \text{\bf the number of cars available for rent in } t \text{ \bf days,} \\ y_t & = \text{\bf the number of cars at rent in } t \text{ \bf days,} \end{cases}$$

be the number of cars in different states by 12 A.M. of the tth day. The total number x_{t+1} of the cars in the garage is made by the number $0.7x_t$ of cars remained in the garage for the next day, plus 60% of all y_t cars at rent returned back:

$$x_{t+1} = 0.7x_t + 0.6y_t . \qquad (1.35)$$

The number y_{t+1} of cars at rent equals the number $0.3x_t$ of cars rented previously, plus the number $0.4y_t$ of cars at rent:

$$y_{t+1} = 0.3x_t + 0.4y_t . \qquad (1.36)$$

We obtain the system

$$\begin{cases} x_{t+1} & = 0.7x_t + 0.6y_t \\ y_{t+1} & = 0.3x_t + 0.4y_t \end{cases}$$

Let

$$\mathbf{A} = \begin{pmatrix} 0.7 & 0.6 \\ 0.3 & 0.4 \end{pmatrix}, \quad \mathbf{x}_t = \begin{pmatrix} x_t \\ y_t \end{pmatrix}$$

Then

$$\mathbf{x}_2 = \mathbf{A}^2 \mathbf{x}_0 = \begin{pmatrix} 0.7 & 0.6 \\ 0.3 & 0.4 \end{pmatrix} \begin{pmatrix} 0.7 & 0.6 \\ 0.3 & 0.4 \end{pmatrix} \begin{pmatrix} 500 \\ 500 \end{pmatrix} = \begin{pmatrix} 0.67 & 0.66 \\ 0.33 & 0.34 \end{pmatrix} \begin{pmatrix} 500 \\ 500 \end{pmatrix} = \begin{pmatrix} 665 \\ 335 \end{pmatrix} . \quad \square$$

Problem 1.9 A rent a car company owns 1000 cars. Customers can either rent a car with a fixed day of return (the scheduled rent) or can leave this free. The later rent is more expensive. Each day, customers rent 40% of the cars in the garage of which 10% are not fixed. The same day, 50% of cars with fixed date and 50% of free date cars return. At present, 500 cars wait for customers in the garage, whereas 300 cars are on a scheduled rent. How many cars will be in the garage, rented on schedule and free, in two days?

Solution: Let

$$\begin{cases} x_t & = \text{\textbf{the number of cars in the garage at the } } t\text{\textbf{th day}} \\ y_t & = \text{\textbf{the number of cars on scheduled rent at the } } t\text{\textbf{th day}} \\ z_t & = \text{\textbf{the number of cars with free rent at the } } t\text{\textbf{th day}} \end{cases}$$

be the number of cars in different states by 12 A.M. of the tth day.

The number x_{t+1} of the cars in the garage is the sum of $0.6x_t$ not rented cars at the tth day, of $0.5y_t$ cars returned to the garage, and of $0.5z_t$ cars:

$$x_{t+1} = 0.6x_t + 0.5y_t + 0.5z_t.$$

The number y_{t+1} of the cars on scheduled rent is the sum of $0.3x_t$ rented cars at the tth day and $0.5y_t$ cars, which remain on lease:

$$y_{t+1} = 0.3x_t + 0.5y_t.$$

The number z_{t+1} of free leased cars equals the sum of $0.1x_t$ plus the sum of $0.5z_t$:

$$z_{t+1} = 0.1x_t + 0.5z_t.$$

We obtain the system

$$\begin{cases} x_{t+1} & = 0.6x_t + 0.5y_t + 0.5z_t \\ y_{t+1} & = 0.3x_t + 0.5y_t \\ z_{t+1} & = 0.1x_t + 0.5z_t \end{cases}$$

Let

$$A = \begin{pmatrix} 0.6 & 0.5 & 0.5 \\ 0.3 & 0.5 & 0 \\ 0.1 & 0 & 0.5 \end{pmatrix}, \quad x_t = \begin{pmatrix} x_t \\ y_t \\ z_t \end{pmatrix}$$

Then $x_2 =$

$$A^2 x_0 = \begin{pmatrix} 0.6 & 0.5 & 0.5 \\ 0.3 & 0.5 & 0 \\ 0.1 & 0 & 0.5 \end{pmatrix} \begin{pmatrix} 0.6 & 0.5 & 0.5 \\ 0.3 & 0.5 & 0 \\ 0.1 & 0 & 0.5 \end{pmatrix} \begin{pmatrix} 500 \\ 300 \\ 200 \end{pmatrix} = \begin{pmatrix} 0.56 & 0.55 & 0.55 \\ 0.33 & 0.4 & 0.15 \\ 0.11 & 0.05 & 0.3 \end{pmatrix} \begin{pmatrix} 500 \\ 300 \\ 200 \end{pmatrix} = \begin{pmatrix} 555 \\ 315 \\ 130 \end{pmatrix}.$$

Notice that $x_2 + y_2 + z_3 = 555 + 315 + 130 = 1000$. $\qquad\square$

Problems

Prob. 17 — Let $\mathbf{A}$ be the matrix defined in (1.23). Evaluate the matrix products if possible: $\mathbf{A}\mathbf{A}^T$, $\mathbf{A}^T\mathbf{A}$, $\mathbf{A}\mathbf{A}^T\mathbf{A}$, $\mathbf{A}^T\mathbf{A}\mathbf{A}^T$, $\mathbf{I} - \mathbf{A}^T\mathbf{A}$.

Prob. 18 — Evaluate the matrix products

$$
\begin{pmatrix} 1 & 0 & 0 & 0 \\ 0 & 0 & 0 & 1 \\ 0 & 0 & 1 & 0 \\ 0 & 1 & 0 & 0 \end{pmatrix}
\begin{pmatrix} 2 & 7 & 3 & 8 \\ 7 & 4 & 9 & 2 \\ 9 & 4 & 3 & 8 \\ 7 & 5 & 4 & 1 \end{pmatrix}
\quad \text{and} \quad
\begin{pmatrix} 1 & 0 & 0 & 1 \\ 0 & 1 & 0 & 0 \\ 0 & 0 & 1 & 0 \\ 0 & 0 & 0 & 1 \end{pmatrix}
\begin{pmatrix} 2 & 7 & 3 & 8 \\ 7 & 4 & 9 & 2 \\ 9 & 4 & 3 & 8 \\ 7 & 5 & 4 & 1 \end{pmatrix}.
$$

Prob. 19 — Let

$$
\mathbf{A} = \begin{pmatrix} 0 & 1 & 0 & 0 \\ 0 & 0 & 1 & 0 \\ 0 & 0 & 0 & 1 \\ 0 & 0 & 0 & 0 \end{pmatrix}, \quad
\mathbf{B} = \begin{pmatrix} 0 & 0 & 0 & 0 \\ a & 0 & 0 & 0 \\ 0 & b & 0 & 0 \\ 0 & 0 & c & 0 \end{pmatrix}.
$$

Evaluate the products $\mathbf{AB}$ and $\mathbf{BA}$.

Prob. 20 — Prove that the formulas

$$
\mathbf{A}(\mathbf{B} + \mathbf{C}) = \mathbf{AB} + \mathbf{AC},
$$
$$
(\mathbf{B} + \mathbf{C})\mathbf{A} = \mathbf{BA} + \mathbf{CA},
$$

are true as soon as their matrix products are defined.

Prob. 21 — Let $\mathbf{A}$ and $\mathbf{B}$ be 2×2 matrices such that $\mathbf{AB} = \mathbf{I}_2$. Show that then $\mathbf{BA} = \mathbf{I}_2$.

Prob. 22 — Let

$$
\mathbf{A} = \begin{pmatrix} 1 & 0 & 2 & 3 & -1 \\ 1 & 2 & -1 & 0 & 1 \end{pmatrix}, \quad
\mathbf{B} = \begin{pmatrix} 1 & -1 & 1 \\ 1 & 3 & -1 \end{pmatrix}.
$$

Evaluate, when it is possible, $\mathrm{row}_1(\mathbf{B}) \cdot \mathbf{B}^T \cdot \mathbf{A} \cdot \mathrm{row}_2(\mathbf{A})^T$, $\mathbf{B}^T\mathbf{B}$, $\mathbf{BB}^T$, $\mathbf{ABB}^T$, $\mathbf{BB}^T\mathbf{A}$, $2\mathbf{A} + \mathbf{BB}^T\mathbf{A}$, $\mathbf{B} - \mathbf{BB}^T\mathbf{A}$, $\mathbf{AA}^T - \mathbf{BB}^T$.

Prob. 23 — Let $\mathbf{x}$ be a column of length m and $\mathbf{y}$ be the column of length n. Write down the entries of the matrix $\mathbf{A} = \mathbf{xy}^T$ and determine its size. Show that any $m \times n$ matrix can be uniquely represented in the form

$$
\mathbf{A} = \mathrm{col}_1(\mathbf{A})\mathbf{e_1}^T + \mathrm{col}_2(\mathbf{A})\mathbf{e_2}^T + \cdots + \mathrm{col}_n(\mathbf{A})\mathbf{e_n}^T.
$$

Prob. 24 — For an $n \times n$ matrix $\mathbf{A}$ its trace $\mathrm{tr}(\mathbf{A})$ is defined as the sum of the diagonal entries:

$$
\mathrm{tr}(\mathbf{A}) = a_{11} + a_{22} + \ldots + a_{nn}.
$$

Show that $\mathrm{tr}(\mathbf{AB}) = \mathrm{tr}(\mathbf{BA})$ for any two $n \times n$ matrices $\mathbf{A}$ and $\mathbf{B}$. Deduce from this that for $n \geq 2$ there are no $n \times n$ matrices $\mathbf{A}$ and $\mathbf{B}$ satisfying the matrix equation

$$\mathbf{AB} - \mathbf{BA} = \mathbf{I}_n.$$

Prob. 25 — Find all 2×2 matrices $\mathbf{A}$ satisfying the equation $\mathbf{A}^2 = \mathbf{I}_2$.

Prob. 26 — Find all 2×2 matrices satisfying the equation $\mathbf{A}^2 = \mathbf{0}_{2 \times 2}$.

Prob. 27 — An $n \times n$ matrix $\mathbf{A}$ is called **upper triangular** if its entries a_{ij} are zeros for $i > j$. It is called **lower triangular** if $a_{ij} = 0$ for $j > i$. The **main diagonal** of $\mathbf{A}$ is made by the entries $a_{ii}, i = 1, \ldots, n$. Show that a square matrix is upper triangular if and only if all its entries below the main diagonal are zeros. Show that the product of two upper triangular matrices is upper triangular.

Prob. 28 — Let $p(X) = a_0 X^d + a_1 X^{d-1} + \cdots + a_d$ be a polynomial in variable X and let $\mathbf{A}$ be an $n \times n$ matrix. Then

$$p(\mathbf{A}) \stackrel{def}{=} a_0 \mathbf{A}^d + a_1 \mathbf{A}^{d-1} + \cdots + a_d \mathbf{I}_n.$$

Show that $p(\mathbf{A}) = \mathbf{0}_{3 \times 3}$ for $p(X) = -X^3 + 3X^2 - 4X + 3$ and for the matrix $\mathbf{A}$ defined in Problem 1.6. If $p(X) = -X^3 + 3X^2 - 3X + 2$ then $p(\mathbf{C}) = \mathbf{0}_{3 \times 3}$.

Prob. 29 — Let

$$\mathbf{A} = \begin{pmatrix} a & 1 \\ 0 & b \end{pmatrix}$$

be a 2×2 matrix, where a and b are real numbers. Find a formula for $\mathbf{A}^n$.

Prob. 30 — A car rental company owns five types of cars: A, B, C, D, E. The prices for per day rent for each type of the cars are given in dollard by the 1×5 matrix

$$\mathbf{p} = \begin{bmatrix} 10 & 15 & 20 & 25 & 30 \end{bmatrix}.$$

The care demand for five days of a week ahead is given by the matrix $\mathbf{A}$, and $\mathbf{u}$ being an auxiliary matrix:

$$\mathbf{A} = \begin{pmatrix} 3 & 2 & 5 & 5 & 2 \\ 2 & 4 & 2 & 3 & 5 \\ 1 & 10 & 9 & 5 & 4 \\ 3 & 2 & 1 & 2 & 1 \\ 2 & 3 & 2 & 1 & 2 \end{pmatrix}, \quad \mathbf{u} = \begin{pmatrix} 1 \\ 1 \\ 1 \\ 1 \\ 1 \end{pmatrix}.$$

Using matrix multiplication calculate the total income of the company for this week. Check Theorem 1.4.

1.5 Elementary Matrices

Definition 1.8 An $n \times n$ matrix $\mathbf{E}$ is called **elementary** if it is obtained from the identity matrix $\mathbf{I}_n$ by one and only one row operation.

For example, the first three matrices below

$$\begin{pmatrix} 1 & 0 & 0 \\ 0 & 2 & 0 \\ 0 & 0 & 1 \end{pmatrix}, \quad \begin{pmatrix} 0 & 0 & 1 \\ 0 & 1 & 0 \\ 1 & 0 & 0 \end{pmatrix}, \quad \begin{pmatrix} 1 & 0 & 0 \\ 0 & 1 & 0 \\ 0 & 2 & 1 \end{pmatrix}, \quad \begin{pmatrix} 1 & 0 & 0 \\ 2 & 0 & 1 \\ 0 & 1 & 0 \end{pmatrix},$$

are elementary matrices, whereas the forth matrix is not.

Theorem 1.6 *Suppose that the $m \times n$ matrix $\mathbf{B}$ is obtained from an $m \times n$ matrix $\mathbf{A}$ by exactly one row operation. Then $\mathbf{B} = \mathbf{EA}$, where $\mathbf{E}$ is the elementary matrix obtained from $\mathbf{I}_m$ by the same row operation.*

Proof. Since $\mathbf{A} = \mathbf{I}_m \mathbf{A}$ we obtain by (1.30) that

$$\text{row}_i(\mathbf{A}) = \text{row}_i(\mathbf{I}_m)\mathbf{A}.$$

Therefore, any row operation on the rows of $\mathbf{A}$ is reflected by the same operation on the rows of $\mathbf{I}_m$. $\square$

Corollary 1.2 *Let $\mathbf{E}$ be an $m \times m$ elementary matrix and $\mathbf{E}^{-1}$ be the $m \times m$ elementary matrix corresponding to the opposite elementary row operation. Then $\mathbf{E}^{-1}\mathbf{E} = \mathbf{I}_m$.*

Theorem 1.7 *If $\mathbf{A} \sim \mathbf{B}$, then $\text{row}(\mathbf{A}) = \text{row}(\mathbf{B})$.*

Proof. By Lemma 1.1 it is sufficient to prove this theorem when $\mathbf{B}$ is obtained from $\mathbf{A}$ by only one elementary row operation of type **RO1** or **RO3**. By Theorem 1.6, $\mathbf{B} = \mathbf{EA}$, where $\mathbf{E}$ is the corresponding elementary matrix. Formulas (1.37) show that every row in $\text{row}(\mathbf{B})$ is a row in $\text{row}(\mathbf{A})$. By Corollary 1.2, $\mathbf{E}^{-1}\mathbf{E} = \mathbf{I}_m$, implying that

$$\mathbf{E}^{-1}\mathbf{B} = \mathbf{B} = \mathbf{E}^{-1}(\mathbf{EA}) = \left(\mathbf{E}^{-1}\mathbf{E}\right)\mathbf{A} = \mathbf{I}_m\mathbf{A} = \mathbf{A}.$$

It follows that $\mathbf{A}$ is obtained from $\mathbf{B}$ by an elementary row operation defined by $\mathbf{E}^{-1}$. Hence any row in $\mathbf{row}(\mathbf{A})$ is a row in $\mathbf{row}(\mathbf{B})$. □

Problem 1.10 Write the matrix

$$\mathbf{B} = \begin{pmatrix} 1 & 2 & 4 \\ 1 & 3 & 6 \\ -1 & 0 & 1 \end{pmatrix}$$

as a product of elementary matrices.

Solution: We apply row operations to obtain the identity matrix. A new trick here is that we not only write direct elementary row operations above symbols $\sim$, but also indicate in boxes the corresponding inverse operations below symbols $\sim$:

$$\begin{pmatrix} 1 & 2 & 4 \\ 1 & 3 & 6 \\ -1 & 0 & 1 \end{pmatrix} \overset{r_1:=r_1+r_3}{\underset{\boxed{r_1 := r_1 - r_3}}{\sim}} \begin{pmatrix} 0 & 2 & 5 \\ 1 & 3 & 6 \\ -1 & 0 & 1 \end{pmatrix} \overset{r_2:=r_2+r_3}{\underset{\boxed{r_2 := r_2 - r_3}}{\sim}} \begin{pmatrix} 0 & 2 & 5 \\ 0 & 3 & 7 \\ -1 & 0 & 1 \end{pmatrix} \overset{r_3:=-r_3}{\underset{\boxed{r_3 := -r_3}}{\sim}}$$

$$\begin{pmatrix} 0 & 2 & 5 \\ 0 & 3 & 7 \\ 1 & 0 & -1 \end{pmatrix} \overset{r_1 \leftrightarrow r_3}{\underset{\boxed{r_1 \leftrightarrow r_3}}{\sim}} \begin{pmatrix} 1 & 0 & -1 \\ 0 & 3 & 7 \\ 0 & 2 & 5 \end{pmatrix} \overset{r_2:=r_2-r_3}{\underset{\boxed{r_2 := r_2 + r_3}}{\sim}} \begin{pmatrix} 1 & 0 & -1 \\ 0 & 1 & 2 \\ 0 & 2 & 5 \end{pmatrix} \overset{r_3:=r_3-2r_2}{\underset{\boxed{r_3 := r_3 + 2r_2}}{\sim}}$$

$$\begin{pmatrix} 1 & 0 & -1 \\ 0 & 1 & 2 \\ 0 & 0 & 1 \end{pmatrix} \overset{r_1:=r_1+r_3}{\underset{\boxed{r_1 := r_1 - r_3}}{\sim}} \begin{pmatrix} 1 & 0 & 0 \\ 0 & 1 & 2 \\ 0 & 0 & 1 \end{pmatrix} \overset{r_2:=r_2-2r_3}{\underset{\boxed{r_2 := r_2 + 2r_3}}{\sim}} \begin{pmatrix} 1 & 0 & 0 \\ 0 & 1 & 0 \\ 0 & 0 & 1 \end{pmatrix}$$

Now we write the result of applications of boxed operations to $\mathbf{I}_3$ in the same order as above. This results in the concrete chain of elementary matrices.

$$\begin{pmatrix} 1 & 2 & 4 \\ 1 & 3 & 6 \\ -1 & 0 & 1 \end{pmatrix} = \underset{r_1:=r_1-r_3}{\begin{pmatrix} 1 & 0 & -1 \\ 0 & 1 & 0 \\ 0 & 0 & 1 \end{pmatrix}} \underset{r_2:=r_2-r_3}{\begin{pmatrix} 1 & 0 & 0 \\ 0 & 1 & -1 \\ 0 & 0 & 1 \end{pmatrix}} \underset{r_3:=-r_3}{\begin{pmatrix} 1 & 0 & 0 \\ 0 & 1 & 0 \\ 0 & 0 & -1 \end{pmatrix}}$$

$$\underset{r_1 \leftrightarrow r_3}{\begin{pmatrix} 0 & 0 & 1 \\ 0 & 1 & 0 \\ 1 & 0 & 0 \end{pmatrix}} \underset{r_3:=r_2+r_3}{\begin{pmatrix} 1 & 0 & 0 \\ 0 & 1 & 1 \\ 0 & 0 & 1 \end{pmatrix}} \underset{r_3:=r_3+2r_2}{\begin{pmatrix} 1 & 0 & 0 \\ 0 & 1 & 0 \\ 0 & 2 & 1 \end{pmatrix}} \underset{r_1:=r_1-r_3}{\begin{pmatrix} 1 & 0 & -1 \\ 0 & 1 & 0 \\ 0 & 0 & 1 \end{pmatrix}} \underset{r_2:=r_2+2r_3}{\begin{pmatrix} 1 & 0 & 0 \\ 0 & 1 & 2 \\ 0 & 0 & 1 \end{pmatrix}}$$

Check the above formula by matrix multiplication. □

Problems

Prob. 31 — Find elementary 3×3 matrices corresponding to the following elementary row operations:

$$r_1 \leftrightarrow r_3, r_2 := r_2 - 2r_3, r_3 = 2r_3, r_1 := r_1 - 3r_3,$$
$$\mathbf{E}_1, \qquad \mathbf{E}_2, \qquad \mathbf{E}_3, \qquad \mathbf{E}_4.$$

Prob. 32 — Find the inverse to each of the above elementary row operations and the corresponding elementary matrices $\mathbf{E}_i$, $i = 1, 2, 3, 4$.

Prob. 33 — Evaluate the products:

$$\mathbf{E}_1\mathbf{E}_2\mathbf{E}_3\mathbf{E}_4\mathbf{I}_3, \ \mathbf{E}_1\mathbf{E}_2^{-1}\mathbf{E}_3\mathbf{E}_4^{-1}\mathbf{I}_3, \ \mathbf{E}_1^5\mathbf{E}_2^{-4}\mathbf{E}_3^6\mathbf{E}_4^{-7}\mathbf{I}_3, \ \mathbf{E}_1^{-1}\mathbf{E}_2^{-1}\mathbf{E}_3^{-1}\mathbf{E}_4^{-1}\mathbf{I}_3.$$

Prob. 34 — Determine which of the matrices below are elementary and which are not:

$$\begin{pmatrix} 0 & 0 & 0 \\ 0 & 2 & 0 \\ 0 & 0 & 1 \end{pmatrix}, \ \begin{pmatrix} 0 & 0 & 1 \\ 1 & 0 & 0 \\ 1 & 0 & 0 \end{pmatrix}, \ \begin{pmatrix} 1 & 0 & 0 \\ 0 & 1 & 0 \\ 0 & 2 & 1 \end{pmatrix}, \ \begin{pmatrix} 1 & 0 & 0 \\ 0 & 0 & 1 \\ 0 & 1 & 0 \end{pmatrix}.$$

Prob. 35 — Let $\mathbf{A}$ be a 3×3 matrix with entries a_{ij} and $\mathbf{E}_i$ be the elementary matrices

$$\mathbf{E}_1 = \begin{pmatrix} 1 & 0 & 0 \\ 0 & 1 & 0 \\ -1 & 0 & 1 \end{pmatrix}, \ \mathbf{E}_2 = \begin{pmatrix} 1 & 0 & 0 \\ 0 & 1 & 0 \\ 1 & 0 & 1 \end{pmatrix}, \ \mathbf{E}_3 = \begin{pmatrix} 1 & 0 & 1 \\ 0 & 1 & 0 \\ 0 & 0 & 1 \end{pmatrix}.$$

Write down the products $\mathbf{E}_i\mathbf{A}$, $\mathbf{A}\mathbf{E}_i$, $i = 1, 2, 3$ in the entry notations. Write down $\mathbf{E}_i^{-1}$ in the entry notations.

Prob. 36 — Using Matrix Algebra and Elementary matrices, prove Lemma 1.1.

1.6 The Row and Column Spaces of a Matrix

In solutions to Problems 1.1, 1.3, 1.4, and 1.5 we used row operations. Therefore, it is useful to consider all possible rows, which can be obtained from the rows of a given matrix $\mathbf{A}$ by elementary row operations. Rows of equal lengths being matrices can be added and multiplied by constants. The set of all rows, which are obtained by multiplications of the rows of $\mathbf{A}$ by constants, and by addition of rows of $\mathbf{A}$:

$$c \cdot \text{row}_i(\mathbf{A}) = \left(c \cdot a_{i1} \ c \cdot a_{i2} \cdots c \cdot a_{in} \right),$$
$$\text{row}_i(\mathbf{A}) + \text{row}_k(\mathbf{A}) = \left(a_{i1} + a_{k1} \ a_{i2} + a_{k2} \cdots a_{in} + a_{kn} \right),$$

$$(1.37)$$

is called the **row space** of **A**.

> **Definition 1.9** Given an $m \times n$ matrix **A** we denote by **row(A)** the set of all rows in $^n\mathbb{R}$, which are **linear combinations** of rows of **A**:
>
> $$\alpha_1 \text{row}_1(\mathbf{A}) + \alpha_2 \text{row}_2(\mathbf{A}) + \cdots + \alpha_m \text{row}_m(\mathbf{A}), \qquad (1.38)$$
>
> where $\alpha_1, \alpha_2, \ldots, \alpha_m$ are real numbers.

Notice that the linear combination (1.38) can be obtained as the first row by the following sequence of elementary row operations on matrix **A**:

$$\text{row}_1(\mathbf{A}) := \alpha_1 \text{row}_1(\mathbf{A}), \quad \text{row}_1(\mathbf{A}) := \text{row}_1(\mathbf{A}) + \alpha_2 \text{row}_2(\mathbf{A}), \quad \ldots \ ,$$
$$\text{row}_1(\mathbf{A}) := \text{row}_1(\mathbf{A}) + \alpha_m \text{row}_m(\mathbf{A}).$$

For example, the row space of the matrix

$$\mathbf{A} = \begin{pmatrix} 1 & 2 & 1 \\ 1 & 1 & 2 \\ 2 & 3 & 3 \end{pmatrix} \qquad (1.39)$$

is a subset of **row(I₃)** $= {}^3\mathbb{R}$. Indeed, a general linear combination of the rows of **A** looks as follows

$$\alpha_1 \text{row}_1(\mathbf{A}) + \alpha_2 \text{row}_2(\mathbf{A}) + \alpha_3 \text{row}_3(\mathbf{A}) =$$
$$\alpha_1 \begin{pmatrix} 1 & 2 & 1 \end{pmatrix} + \alpha_2 \begin{pmatrix} 1 & 1 & 2 \end{pmatrix} + \alpha_3 \begin{pmatrix} 2 & 3 & 3 \end{pmatrix} =$$
$$\begin{pmatrix} \alpha_1 + \alpha_2 + 2\alpha_3 & 2\alpha_1 + \alpha_2 + 3\alpha_3 & \alpha_1 + 2\alpha_2 + 3\alpha_3 \end{pmatrix}.$$

Hence

$$\mathbf{row(A)} = \left\{ \begin{pmatrix} \alpha_1 + \alpha_2 + 2\alpha_3 & 2\alpha_1 + \alpha_2 + 3\alpha_3 & \alpha_1 + 2\alpha_2 + 3\alpha_3 \end{pmatrix} \right\}, \qquad (1.40)$$

where $\alpha_1, \alpha_2, \alpha_3$ are arbitrary real numbers. In (1.40) the number of independent parameters $\alpha_1, \alpha_2, \alpha_3$ used for the description of **row(A)** equals three.

Problem 1.11 Given **A** defined by

$$\mathbf{A} = \begin{pmatrix} 1 & 2 & 1 \\ 1 & 1 & 2 \\ 2 & 3 & 3 \end{pmatrix}$$

find such a description of **row(A)**, which requires a minimal number of rows not necessarily of **A**.

Solution: We have

$$\left(\alpha_1 + \alpha_2 + 2\alpha_3 \quad 2\alpha_1 + \alpha_2 + 3\alpha_3 \quad \alpha_1 + 2\alpha_2 + 3\alpha_3\right)^T =$$

$$\begin{pmatrix} \alpha_1 + \alpha_2 + 2\alpha_3 \\ 2\alpha_1 + \alpha_2 + 3\alpha_3 \\ \alpha_1 + 2\alpha_2 + 3\alpha_3 \end{pmatrix} = \begin{pmatrix} 1 & 1 & 2 \\ 2 & 1 & 3 \\ 1 & 2 & 3 \end{pmatrix} \begin{pmatrix} \alpha_1 \\ \alpha_2 \\ \alpha_3 \end{pmatrix} = \begin{pmatrix} b_1 \\ b_2 \\ b_3 \end{pmatrix}. \quad (1.41)$$

It follows that $\mathbf{row}(\mathbf{A})$ is the set of all rows $\mathbf{b}^T$ such that the linear system in (1.41) has a solution. To describe such columns $\mathbf{b}$ we consider the augmented matrix of the system (1.41)

$$\left(\mathbf{A}^T | \mathbf{b}\right) = \begin{pmatrix} 1 & 1 & 2 & | & b_1 \\ 2 & 1 & 3 & | & b_2 \\ 1 & 2 & 3 & | & b_3 \end{pmatrix} \sim \begin{pmatrix} 1 & 1 & 2 & | & b_1 \\ 0 & -1 & -1 & | & b_2 - 2b_1 \\ 0 & 1 & 1 & | & b_3 - b_1 \end{pmatrix} \sim$$

$$\begin{pmatrix} 1 & 0 & 1 & | & b_2 - b_1 \\ 0 & -1 & -1 & | & b_2 - 2b_1 \\ 0 & 0 & 0 & | & b_2 + b_3 - 3b_1 \end{pmatrix} \sim \begin{pmatrix} 1 & 0 & 1 & | & b_2 - b_1 \\ 0 & 1 & 1 & | & 2b_1 - b_2 \\ 0 & 0 & 0 & | & b_2 + b_3 - 3b_1 \end{pmatrix}$$

It follows that the system $\mathbf{A}^T \boldsymbol{\alpha} = \mathbf{b}$ has a solution if and only if

$$b_2 + b_3 - 3b_1 = 0 \Leftrightarrow \begin{cases} b_1 = s, \\ b_2 = t, \\ b_3 = 3s - t, \end{cases}$$

where s and t are arbitrary real numbers. So, we have

$$\mathbf{b} = \begin{pmatrix} b_1 \\ b_2 \\ b_3 \end{pmatrix} = \begin{pmatrix} s \\ t \\ 3s - t \end{pmatrix} = s \begin{pmatrix} 1 \\ 0 \\ 3 \end{pmatrix} + t \begin{pmatrix} 0 \\ 1 \\ -1 \end{pmatrix}.$$

It follows that $\mathbf{row}(\mathbf{A})$ equals the set of all linear combinations

$$s \begin{pmatrix} 1 & 0 & 3 \end{pmatrix} + t \begin{pmatrix} 0 & 1 & -1 \end{pmatrix} \quad (1.42)$$

of only two rows in $\mathbf{row}(\mathbf{A})$. This number cannot be reduced to one. Indeed, otherwise both rows in (1.42) would be proportional to some non-zero row, implying that these rows are proportional. However, this is not the case.   □

Definition 1.10 A nonempty subset V of rows in $^n\mathbb{R}$ is called a **subspace** if for any two rows $\mathbf{v}_1$ and $\mathbf{v}_2$ in V and any real number c both rows $c\mathbf{v}_1$ and $\mathbf{v}_1 + \mathbf{v}_2$ are also in V (see (1.37)).

By (1.37) the row space of any matrix is a subspace. Corollary 2.1 below shows that the converse statement is also true.

The set of all columns, which are obtained by multiplications of the columns of $\mathbf{A}$ by constants, and by addition of columns of $\mathbf{A}$:

$$c \cdot \text{col}_i(\mathbf{A}) = \left(c \cdot a_{1i} \quad c \cdot a_{2i} \quad \cdots \quad c \cdot a_{mi}\right)^T,$$

$$\text{col}_i(\mathbf{A}) + \text{col}_k(\mathbf{A}) = \left(a_{1i} + a_{1k} \quad a_{2i} + a_{2k} \quad \cdots \quad a_{mi} + a_{mk}\right)^T. \qquad (1.43)$$

is called the **column space** of $\mathbf{A}$.

Definition 1.11 Given an $m \times n$ matrix $\mathbf{A}$ we denote by $\text{col}(\mathbf{A})$ the set of all columns of $\mathbb{R}^m$, which are linear combinations of the columns of $\mathbf{A}$:

$$\alpha_1 \text{col}_1(\mathbf{A}) + \alpha_2 \text{col}_2(\mathbf{A}) + \cdots + \alpha_n \text{col}_n(\mathbf{A}),$$

where $\alpha_1, \alpha_2, \ldots, \alpha_n$ are real numbers.

The column space $\text{col}(\mathbf{A})$ for the matrix $\mathbf{A}$ in (1.39) consists of all columns of the form:

$$\text{col}(\mathbf{A}) = \left(\alpha_1 + 2\alpha_2 + \alpha_3 \quad \alpha_1 + \alpha_2 + 2\alpha_3 \quad 2\alpha_1 + 3\alpha_2 + 3\alpha_3\right)^T,$$

where $\alpha_1, \alpha_2, \alpha_3$ are arbitrary real numbers. The column space of $\mathbf{I}_n$ is the set of all columns

$$\left(\alpha_1 \quad \alpha_2 \quad \cdots \quad \alpha_n\right)^T, \qquad (1.44)$$

where $\alpha_1, \alpha_2, \ldots, \alpha_n$ are arbitrary real numbers. The column space $\text{col}(\mathbf{A})$ of an $m \times n$ matrix $\mathbf{A}$ is a subset of $\text{col}(\mathbf{I}_m) = \mathbb{R}^m$ of the identity matrix $\mathbf{I}_m$.

Definition 1.12 A nonempty subset V of columns in $\mathbb{R}^m$ is called a **subspace** if for any two columns $\mathbf{v}_1$ and $\mathbf{v}_2$ in V and any real number c both rows $c\mathbf{v}_1$ and $\mathbf{v}_1 + \mathbf{v}_2$ are also in V (see (1.43)).

By (1.43) the column space of any matrix is a subspace. It will be shown later that the converse statement is also true, see Corollary 2.1.

Problem 1.12 Given $\mathbf{A}$ defined by

$$\mathbf{A} = \begin{pmatrix} 1 & 2 & 1 \\ 1 & 1 & 2 \\ 2 & 3 & 3 \end{pmatrix}$$

find such a description of $\text{col}(\mathbf{A})$, which requires a minimal number of columns not necessarily of $\mathbf{A}$.

Solution: A column is in **col(A)** if and only if it can be represented as

$$
\left(\alpha_1 + 2\alpha_2 + \alpha_3 \ \ \alpha_1 + \alpha_2 + 2\alpha_3 \ \ 2\alpha_1 + 3\alpha_2 + 3\alpha_3\right)^T = \begin{pmatrix} 1 & 2 & 1 \\ 1 & 1 & 2 \\ 2 & 3 & 3 \end{pmatrix}\begin{pmatrix} \alpha_1 \\ \alpha_2 \\ \alpha_3 \end{pmatrix} = \begin{pmatrix} b_1 \\ b_2 \\ b_3 \end{pmatrix}
$$

for some real numbers $\alpha_1, \alpha_2, \alpha_3$. It follows that a column **b** is in **col(A)** if and only if the system above has a solution in $\alpha_1, \alpha_2, \alpha_3$. To find all **b** with this property we consider the augmented matrix of the system:

$$
(\mathbf{A}|\mathbf{b}) = \begin{pmatrix} 1 & 2 & 1 & b_1 \\ 1 & 1 & 2 & b_2 \\ 2 & 3 & 3 & b_3 \end{pmatrix} \sim \begin{pmatrix} 1 & 2 & 1 & b_1 \\ 0 & -1 & 1 & b_2 - b_1 \\ 0 & -1 & 1 & b_3 - 2b_1 \end{pmatrix} \sim
$$

$$
\begin{pmatrix} 1 & 2 & 1 & b_1 \\ 0 & 1 & -1 & b_1 - b_2 \\ 0 & 0 & 0 & b_3 - b_1 - b_2 \end{pmatrix} \sim \begin{pmatrix} 1 & 0 & 3 & 2b_2 - b_1 \\ 0 & 1 & -1 & b_1 - b_2 \\ 0 & 0 & 0 & b_3 - b_1 - b_2 \end{pmatrix}.
$$

It follows that **b** is in **col(A)** if and only if $b_3 = b_1 + b_2$. One can easily check that every column of **A** satisfies this condition. Therefore, if we put $b_1 = s$, $b_2 = t$, then

$$
\mathbf{b} = \begin{pmatrix} b_1 \\ b_2 \\ b_3 \end{pmatrix} = \begin{pmatrix} s \\ t \\ s + t \end{pmatrix} = s\begin{pmatrix} 1 \\ 0 \\ 1 \end{pmatrix} + t\begin{pmatrix} 0 \\ 1 \\ 1 \end{pmatrix}. \tag{1.45}
$$

Since the columns at s and t in (1.45) are not proportional, the minimal number of independent columns is two.   □

Definition 1.13 A finite set $\{\mathbf{v}_1, \mathbf{v}_2, \dots, \mathbf{v}_k\}$ of rows (columns) is called **linearly independent** if the only solution to the equation in α_j

$$
\alpha_1 \mathbf{v}_1 + \alpha_2 \mathbf{v}_2 + \cdots + \alpha_k \mathbf{v}_k = \mathbf{0}
$$

is the zero solution $\alpha_1 = \alpha_2 = \cdots = \alpha_k = 0$.

Since $\alpha\mathbf{0} = \mathbf{0}$ for every real number α, a **linearly independent set cannot include a zero row (column)**.

For any finite set of rows (columns) $\{\mathbf{v}_1, \mathbf{v}_2, \dots, \mathbf{v}_k\}$ in **row**$(\mathbf{I}_n)$ (**col**$(\mathbf{I}_n)$), we denote by

$$
\text{Lin} (\mathbf{v}_1, \mathbf{v}_2, \dots, \mathbf{v}_k)
$$

the set $\alpha_1 \mathbf{v}_1 + \alpha_2 \mathbf{v}_2 + \cdots + \alpha_k \mathbf{v}_k$ of all possible linear combinations of rows (columns) $\mathbf{v}_1, \mathbf{v}_2, \dots, \mathbf{v}_k$ with real coefficients $\alpha_1, \alpha_2, \dots, \alpha_k$. The set $\text{Lin} (\mathbf{v}_1, \mathbf{v}_2, \dots, \mathbf{v}_k)$ is called the **linear span** of rows (columns) $\mathbf{v}_1, \mathbf{v}_2, \dots, \mathbf{v}_k$. In particular, in these notations for every $m \times n$ matrix **A**:

$$\mathbf{row}(\mathbf{A}) = \mathrm{Lin}\,(\mathrm{row}_1(\mathbf{A}), \mathrm{row}_2(\mathbf{A}), \dots, \mathrm{row}_m(\mathbf{A})),$$

$$\mathbf{col}(\mathbf{A}) = \mathrm{Lin}\,(\mathrm{col}_1(\mathbf{A}), \mathrm{col}_2(\mathbf{A}), \dots, \mathrm{col}_n(\mathbf{A})).$$

In other words, the row space of any matrix is the linear span of its rows, whereas the column space is the linear span of its columns. If $\mathbf{A} = \begin{pmatrix} \mathbf{v}_1 & \mathbf{v}_2 & \cdots & \mathbf{v}_n \end{pmatrix}$ is a matrix of size $m \times n$ written in column notations, then the set $\mathrm{Lin}\,(\mathbf{v}_1, \mathbf{v}_2, \dots, \mathbf{v}_k)$ is the set of all columns $\mathbf{b}$ such that the system $\mathbf{Ax} = \mathbf{b}$ has a solution in $\mathbf{x}$. Indeed,

$$x_1\mathbf{v}_1 + x_2\mathbf{v}_2 + \cdots + x_n\mathbf{v}_n = \mathbf{Ax} = \mathbf{b}.$$

Theorem 1.8 *Let $\mathbf{A}$ be an $m \times n$ with linearly independent rows and let $\mathbf{B} \sim \mathbf{A}$. Then the rows of $\mathbf{B}$ are linearly independent.*

Proof. Since $\mathbf{B}$ is obtained from $\mathbf{A}$ by a finite chain of elementary row operations, it is sufficient to prove the Theorem for the case of one elementary row operation. By Theorem 1.6 in this case $\mathbf{B} = \mathbf{EA}$, where $\mathbf{E}$ is an elementary $m \times m$ matrix. By Theorem 1.9 it is sufficient to prove that

$$\mathbf{0} = \alpha\mathbf{B} = \alpha(\mathbf{EA}) = (\alpha\mathbf{E})\mathbf{A}$$

implies that $\alpha = \mathbf{0}$. Since the rows of $\mathbf{A}$ are linearly independent, we obtain that $\alpha\mathbf{E} = \mathbf{0}$. It follows that

$$\alpha = \alpha(\mathbf{EE}^{-1}) = (\alpha\mathbf{E})\mathbf{E}^{-1} = \mathbf{0E}^{-1} = \mathbf{0},$$

as stated.   □

Theorem 1.9 *Given a finite set of rows $\{\mathbf{v}_1, \mathbf{v}_2, \dots, \mathbf{v}_m\}$ in $^n\mathbb{R}$ let*

$$\mathbf{A} = \begin{pmatrix} \mathbf{v}_1 & \mathbf{v}_2 & \cdots & \mathbf{v}_m \end{pmatrix}^T$$

be an $m \times n$ matrix. Then the set $\{\mathbf{v}_1, \mathbf{v}_2, \dots, \mathbf{v}_m\}$ is linearly independent if and only if the system $\alpha\mathbf{A} = \mathbf{0}_n$ has only the trivial solution $\alpha = \mathbf{0}_m$.

Proof. Apply formula (1.33).   □

Theorem 1.10 *Given a finite set of columns* $\{v_1, v_2, \ldots, v_n\}$ *in* $\mathbb{R}^m$ *let*

$$\mathbf{A} = \begin{pmatrix} v_1 & v_2 & \ldots & v_n \end{pmatrix}$$

be an $m \times n$ *matrix. Then the set* $\{v_1, v_2, \ldots, v_n\}$ *is linearly independent if and only if the system* $\mathbf{A}\alpha = \mathbf{0}_m$ *has only the trivial solution* $\alpha = \mathbf{0}_n$.

Proof. Apply formula (1.34).   □

If $\{v_1, v_2, \ldots, v_m\}$ is a finite set of rows of length n then $\{v_1^T, v_2^T, \ldots, v_m^T\}$ is the set of columns. Applying Theorem 1.10 to these columns and solving the corresponding system of linear equations we can establish that the set of rows $\{v_1, v_2, \ldots, v_m\}$ is linearly independent. Equivalently, one can apply Theorem 1.9.

Corollary 1.3 *If* $\mathbf{A}$ *is an* $m \times n$ *matrix with linearly independent columns and a matrix* $\mathbf{B}$ *is row equivalent to* $\mathbf{A}$ *then the columns of* $\mathbf{B}$ *are linearly independent.*

Proof. By Theorem 1.10 the system $\mathbf{A}x = \mathbf{0}$ has only zero solution. Since $\mathbf{B} \sim \mathbf{A}$, the system $\mathbf{B}x = \mathbf{0}$ also has only zero solution. By Theorem 1.10 the columns of $\mathbf{B}$ are linearly independent.   □

Definition 1.14 A finite linearly independent set $\{v_1, v_2, \ldots, v_k\}$ of rows (columns) is called a **basis** for $\mathrm{Lin}(v_1, v_2, \ldots, v_k)$.

It is easy to see that if the set $\{v_1, v_2, \ldots, v_k\}$ is a basis, then every element of $\mathrm{Lin}(v_1, v_2, \ldots, v_k)$ has a unique representation as the linear combination

$$\alpha_1 v_1 + \alpha_2 v_2 + \cdots + \alpha_1 v_k.$$

The notion of a basis has a nice interpretation in terms of systems of linear equations.

Theorem 1.11 *Let* $S = \{v_1, v_2, \ldots, v_n\}$ *be a set of columns in* $\mathbb{R}^m$ *and let* $\mathbf{A}$ *be the matrix with columns* $\mathrm{col}_j(\mathbf{A}) = v_j$, $j = 1, \ldots, n$. *Then the system* $\mathbf{A}x = \mathbf{b}$ *has a unique solution if and only if* S *is a basis for* $\mathrm{Lin}(S)$ *and* $\mathbf{b}$ *is a column in* $\mathrm{Lin}(S)$.

Proof. The formula

$$\mathbf{Ax} = x_1\mathbf{v}_1 + x_2\mathbf{v}_2 + \cdots + x_n\mathbf{v}_n = \mathbf{b}$$

shows that the system $\mathbf{Ax} = \mathbf{b}$ has at least one solution if and only if $\mathbf{b}$ is a column in $\mathrm{Lin}(S)$. If for some $\mathbf{b}$ in $\mathrm{Lin}(S)$ the system has two different solutions $\mathbf{x} \neq \mathbf{y}$ then

$$\mathbf{0}_m = \mathbf{Ax} - \mathbf{Ay} = (x_1 - y_1)\mathbf{v}_1 + (x_2 - y_2)\mathbf{v}_2 + \cdots + (x_n - y_n)\mathbf{v}_n$$

is a non-trivial representation of the zero column, implying that S is linearly dependent.   □

Problem 1.13 For the matrix $\mathbf{A}$ considered in Problem 1.12 find a basis of its columns for $\mathbf{col}(\mathbf{A})$.

Solution: Using Theorem 1.10, we first determine if the columns of $\mathbf{A}$ are linearly independent or not. We have

$$\mathbf{0}_3 = \mathbf{A\alpha} \Leftrightarrow \begin{cases} \alpha_1 + 2\alpha_2 + \alpha_3 & = 0 \\ \alpha_1 + \alpha_2 + 2\alpha_3 & = 0 \\ 2\alpha_1 + 3\alpha_2 + 3\alpha_3 & = 0 \end{cases}.$$

Since

$$\begin{pmatrix} 1 & 2 & 1 & | & 0 \\ 1 & 1 & 2 & | & 0 \\ 2 & 3 & 3 & | & 0 \end{pmatrix} \sim \begin{pmatrix} 1 & 0 & 3 & | & 0 \\ 0 & 1 & -1 & | & 0 \\ 0 & 0 & 0 & | & 0 \end{pmatrix},$$

we conclude that

$$\begin{cases} \alpha_1 + 3\alpha_3 & = 0 \\ \alpha_2 - \alpha_3 & = 0 \end{cases} \Leftrightarrow \begin{cases} \alpha_1 & = -3t \\ \alpha_2 & = t \\ \alpha_3 & = t \end{cases},$$

where t is an arbitrary real number. It follows that the columns of $\mathbf{A}$ are linearly dependent. In particular, if $t = 1$, then

$$-3\mathrm{col}_1(\mathbf{A}) + \mathrm{col}_2(\mathbf{A}) + \mathrm{col}_3(\mathbf{A}) = \mathbf{0}_3 \Rightarrow \mathrm{col}_3(\mathbf{A}) = 3\mathrm{col}_1(\mathbf{A}) - \mathrm{col}_2(\mathbf{A}).$$

This means that $\mathrm{col}_3(\mathbf{A})$ is a linear combination of the first two columns. The first two columns are linear independent, since otherwise there is a non-trivial combination with $\alpha_3 = t = 0$, implying that $\alpha_1 = \alpha_2 = 0$, which is a contradiction. Hence, $\{\mathrm{col}_1(\mathbf{A}), \mathrm{col}_2(\mathbf{A})\}$ is a basis for $\mathbf{col}(\mathbf{A})$.   □

Problems

Prob. 37 — Given the matrices

$$A = \begin{pmatrix} 1 & 2 & 8 \\ 2 & 1 & 7 \\ 6 & 3 & 21 \end{pmatrix}, \; B = \begin{pmatrix} 1 & 2 & 8 \\ 2 & 1 & 7 \\ 3 & -1 & 3 \end{pmatrix},$$

investigate whether $\mathbf{row(A)} = \mathbf{row(B)}$ or not.

Prob. 38 — Given $\mathbf{A}$ defined by

$$\mathbf{A} = \begin{pmatrix} 1 & -1 & -4 \\ 2 & 1 & -2 \\ 1 & 2 & 2 \end{pmatrix}, \tag{1.46}$$

find rows of $\mathbf{A}$ which make a basis for $\mathbf{row(A)}$.

Prob. 39 — Given $\mathbf{A}$ defined by (1.46), find columns of $\mathbf{A}$ which make a basis for $\mathbf{col(A)}$.

Prob. 40 — A matrix $\mathbf{A}$ is row equivalent to a matrix with only one non-zero row. Show that every row of $\mathbf{A}$ is proportional to this row.

Prob. 41 — Determine whether the set of rows

$$\mathbf{u}_1 = \begin{pmatrix} 1 & 2 & -3 \end{pmatrix}, \; \mathbf{u}_2 = \begin{pmatrix} 3 & 5 & -7 \end{pmatrix}, \; \mathbf{u}_3 = \begin{pmatrix} 2 & 1 & 0 \end{pmatrix},$$

is linearly independent or not.

1.7 Analytic Geometry

Analytic Geometry says that points of the coordinate plane $\mathbb{R}^2$ with two perpendicular axes, see Fig 1.1, are in one-to-one correspondence with the pairs (x_1, x_2) of their coordinates. Similarly, points of the n-dimensional coordinate space $\mathbb{R}^n$ with n perpendicular coordinate axes are in one-to-one correspondence with the n-tuples $(x_1, x_2, \ldots, x_n)$ of their coordinates, see Fig 1.2 for $n = 3$. It follows that $\mathbb{R}^n$ can be naturally identified with either $\mathbf{row(I}_n)$ or with $\mathbf{col(I}_n)$. Since the matrix of unknowns $\mathbf{x}$ in (1.10) is a column of size $n \times 1$, we identify points of $\mathbb{R}^n$ with the **columns** of their coordinates:

$$\mathbb{R}^n \stackrel{\text{def}}{=} \mathbf{col(I}_n).$$

In case saving space is necessary we apply the transpose operation, which sends a column $\mathbf{v}$ to the row $\mathbf{v}^T$. Each such a column can also be identified with a **vector**.

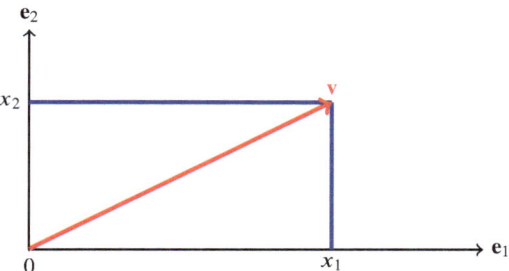

Fig. 1.1: Cartesian coordinates of a vector $\mathbf{v}$ in $\mathbb{R}^2$

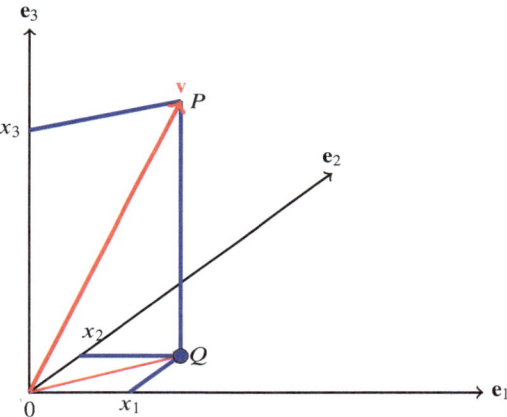

Fig. 1.2: Cartesian coordinates of a vector $\mathbf{v}$ in $\mathbb{R}^3$

Definition 1.15 A vector in $\mathbb{R}^n$ is the directed segment with its **tail** placed at the origin

$$\mathbf{0}_n = \underbrace{\left(0\ 0\ \cdots\ 0\right)}_{n}^{T}$$

and with the **head** at the point P with the coordinates

$$\left(\alpha_1\ \alpha_2\ \cdots\ \alpha_n\right)^{T}.$$

Since in concrete considerations vectors belong to $\mathbb{R}^n$ with fixed n, we often drop index n and write $\mathbf{0}_n = \mathbf{0}$. Fig 1.1 and 1.2 illustrate the definition of a vector in cases $n = 2$ and $n = 3$. Addition of vectors is the addition of columns, which is illustrated by Fig 1.3.

The vectors of the standard basis for $\mathbf{col}(\mathbf{I}_n) = \mathbb{R}^n$

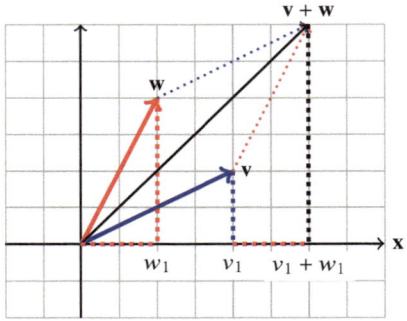

Fig. 1.3: Addition of vectors in $\mathbb{R}^2$

$$\mathbf{e}_1 = \mathrm{col}_1(\mathbf{I}_n), \ \mathbf{e}_2 = \mathrm{col}_2(\mathbf{I}_n), \ \ldots, \ \mathbf{e}_n = \mathrm{col}_n(\mathbf{I}_n), \tag{1.47}$$

are called the **coordinate vectors** in $\mathbb{R}^n$.

To evaluate the length $|\mathbf{v}|$ of vector $\mathbf{v}$ shown on Fig 1.2 we apply Pythagorean Theorem twice. In the right triangle $\triangle OPQ$

$$|\mathbf{v}|^2 = |OP|^2 = |PQ|^2 + |OQ|^2 = x_3^2 + |OQ|^2.$$

The Pythagorean Theorem for $\triangle OQx_1$ shows that

$$|OQ|^2 = x_1^2 + x_2^2 \Rightarrow |\mathbf{v}|^2 = x_1^2 + x_2^2 + x_3^2.$$

Similarly, the length $|\mathbf{x}|$ of a vector $\mathbf{x}$ in $\mathbb{R}^n$ satisfies $|\mathbf{x}|^2 = x_1^2 + x_2^2 + \cdots + x_n^2$.

Definition 1.16 Let $\mathbf{x}$ and $\mathbf{y}$ be two vectors in $\mathbb{R}^n$, i.e. two columns of size n. Following (1.12) we define the **Euclidean inner product** of $\mathbf{x}$ and $\mathbf{y}$, which is often called the **dot product**, by

$$\langle \mathbf{x}, \mathbf{y} \rangle = \mathbf{x}^T \mathbf{y} = x_1 y_1 + x_2 y_2 + \cdots + x_n y_n, \tag{1.48}$$

Theorem 1.12 *The Euclidean inner product satisfies the following properties:*

(a) *for every real α and β and vectors $\mathbf{x}$, $\mathbf{y}$, $\mathbf{z}$:*

$$\langle \alpha \mathbf{x} + \beta \mathbf{y}, \mathbf{z} \rangle = \alpha \langle \mathbf{x}, \mathbf{y} \rangle + \beta \langle \mathbf{y}, \mathbf{z} \rangle ;$$

(b) *for every $\mathbf{x}$ and $\mathbf{y}$*

$$\langle \mathbf{x}, \mathbf{y} \rangle = \langle \mathbf{y}, \mathbf{x} \rangle;$$

(c) *for every* $\mathbf{x}$, $\langle \mathbf{x}, \mathbf{x} \rangle \geq 0$, *and* $\langle \mathbf{x}, \mathbf{x} \rangle = 0$ *if and only if* $\mathbf{x} = \mathbf{0}$;
(d) *for every positive integers i and j not exceeding n*

$$\langle \mathbf{e}_i, \mathbf{e}_j \rangle = \begin{cases} 0 & \text{if } i \neq j \\ 1 & \text{if } i = j \end{cases}.$$

Proof. All properties easily follow from (1.48). □

The following formula shows that the Euclidean inner product is uniquely recovered by the length:

$$2 \cdot \langle \mathbf{x}, \mathbf{y} \rangle = |\mathbf{x}|^2 + |\mathbf{y}|^2 - |\mathbf{y} - \mathbf{x}|^2. \tag{1.49}$$

Formula (1.49) follows from simple identities for the coordinates:

$$2x_i y_i = y_i^2 + x_i^2 - (y_i - x_i)^2, \quad i = 1, 2, \ldots, n.$$

Lemma 1.3 *The Euclidean inner product of two non-zero vectors* $\mathbf{x}$ *and* $\mathbf{y}$ *is zero if and only if*

$$|\mathbf{y} - \mathbf{x}|^2 = |\mathbf{x}|^2 + |\mathbf{y}|^2. \tag{1.50}$$

Proof. See (1.49). □

Any non-zero vector is a segment of a line uniquely determined by its head and the tail. Any two non-zero and not proportional vectors $\mathbf{x}$, $\mathbf{y}$ are uniquely recovered by two heads and their common tail at $\mathbf{0}$. Elementary Geometry says that any three pairwise different points in the space (in our case the heads of $\mathbf{x}$ and $\mathbf{y}$, and their common tail) determine a plane. Therefore, we may define that two non-zero vectors $\mathbf{x}$ and $\mathbf{y}$ in $\mathbb{R}^n$ are **perpendicular** if their lines make a right angle at the point of their intersection at $\mathbf{0}$ inside their common plane.

Theorem 1.13 *Two non-zero vectors* $\mathbf{x}$ *and* $\mathbf{y}$ *in* $\mathbb{R}^n$ *are perpendicular if and only if their inner product is zero:* $\langle \mathbf{x}, \mathbf{y} \rangle = 0$.

Proof. Since $\mathbf{x}$ and $\mathbf{y}$ are in one plane, their difference $\mathbf{y} - \mathbf{x}$ is in this plane too. If we draw a vector parallel to $\mathbf{y} - \mathbf{x}$ with the tail at the head of $\mathbf{x}$, then its head will be at $\mathbf{y}$, see Fig 1.4. Then we obtain a triangle with vertices at $\mathbf{0}$ and at the heads of $\mathbf{x}$, $\mathbf{y}$. By Lemma 1.3 we obtain $\langle \mathbf{x}, \mathbf{y} \rangle = 0$ if and only if the lengths of sides of the triangle obtained satisfy (1.50). By the inverse Pythagorean theorem this means that the lines determined by $\mathbf{x}$ and $\mathbf{y}$ are perpendicular.

The conclusion is that two vectors are perpendicular in the sense of classical Geometry if and only if their inner product is zero.   □

In what follows we often call two non-zero vectors **x** and **y** **orthogonal** if $\langle \mathbf{x}, \mathbf{y} \rangle = 0$.

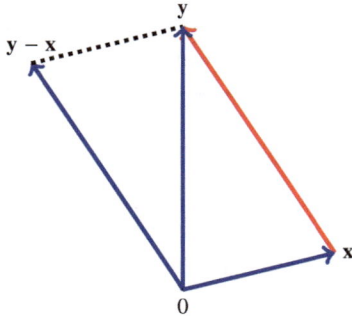

Fig. 1.4: $\langle \mathbf{x}, \mathbf{y} \rangle = 0 \Leftrightarrow \mathbf{y} \perp \mathbf{x}$

Any two non-zero and not proportional vectors **x** and **y** in $\mathbb{R}^n$ determine a unique plane passing through these vectors. Applying Elementary Geometry we may define an **angle** θ between these vectors.

Theorem 1.14 *For any two vectors* **x** *and* **y** *in* $\mathbb{R}^n$

$$\langle \mathbf{x}, \mathbf{y} \rangle = |\mathbf{x}| \cdot |\mathbf{y}| \cos \theta. \tag{1.51}$$

Proof. By the Law of Cosine

$$|\mathbf{x} - \mathbf{y}|^2 = |\mathbf{x}|^2 - 2|\mathbf{x}| \cdot |\mathbf{y}| \cdot \cos \theta + |\mathbf{y}|^2. \tag{1.52}$$

By Theorem 1.12

$$|\mathbf{x} - \mathbf{y}|^2 = \langle \mathbf{x} - \mathbf{y} \rangle = |\mathbf{x}|^2 - 2|\mathbf{x}| \cdot |\mathbf{y}| \cdot \langle \mathbf{x}, \mathbf{y} \rangle + |\mathbf{y}|^2. \tag{1.53}$$

Comparing (1.52) and (1.53) we obtain (1.51).   □

Theorem 1.15 *Let* **A** *be an* $n \times n$ *matrix with real entries. Then for any vectors* **x**, **y** *in* $\mathbb{R}^n$

$$\langle \mathbf{A}\mathbf{x}, \mathbf{y} \rangle = \langle \mathbf{x}, \mathbf{A}^T \mathbf{y} \rangle.$$

Proof.

$$\langle \mathbf{A}\mathbf{x}, \mathbf{y} \rangle = (\mathbf{A}\mathbf{x})^T \mathbf{y} = \mathbf{x}^T \mathbf{A}^T \mathbf{y} = \langle \mathbf{x}, \mathbf{A}^T \mathbf{y} \rangle. \qquad \square$$

Problem 1.14 Find the angle between the vectors

$$\mathbf{u} = \begin{pmatrix} 1 & 1 & -1 & 1 \end{pmatrix}^T, \quad \mathbf{v} = \begin{pmatrix} 1 & -1 & 1 & -1 \end{pmatrix}^T.$$

Solution:

$$\cos \theta = \frac{\langle \mathbf{u}, \mathbf{v} \rangle}{|\mathbf{u}| \cdot |\mathbf{v}|} = \frac{1 \cdot 1 + 1 \cdot (-1) + (-1) \cdot 1 + 1 \cdot (-1)}{\sqrt{4} \cdot \sqrt{4}} = \frac{1 - 1 - 1 - 1}{4} =$$

$$-\frac{2}{4} = -\frac{1}{2} \Rightarrow \theta = \arccos \left(-\frac{1}{2} \right) = \pi - \frac{\pi}{3} = \frac{2\pi}{3}.$$

Remark. Recall that the angle $\theta = \arccos x$, $-1 \le x \le 1$ is defined as the unique angle θ in the interval $[0, \pi]$ such that $\cos \theta = x$. This angle exists and is unique, since the function $x = \cos \theta$ is continuous and strictly decreasing on $[0, \pi]$.

Theorem 1.16 *Any* 3×3 *matrix* **A** *with linearly independent columns is row equivalent to the identity matrix* $\mathbf{I}_3$.

Proof. Since the columns of **A** are linearly independent, we see that $\mathrm{col}_1(\mathbf{A}) \ne$ **0**. It follows that there exists i such that $a_{i1} \ne 0$. Applying **RO1** to the ith row of **A** we make $a_{i1} = 1$. Subtracting multiples of the ith row from other two rows (see **RO3**), if necessary, we vanish all entries in the first column except for the ith row. Swapping the ith row with the first (see **RO2**), if necessary, we obtain a row equivalent matrix of the form

$$\mathbf{C} = \begin{pmatrix} 1 & c_{12} & c_{13} \\ 0 & c_{22} & c_{23} \\ 0 & c_{32} & c_{33} \end{pmatrix}.$$

By Corollary 1.3 the columns of $\mathbf{C}$ are linearly independent. It follows that $\mathrm{col}_2(\mathbf{C})$ is not proportional to $\mathrm{col}_1(\mathbf{C})$ implying that c_{22} and c_{32} cannot be zero both. Applying **RO2** to the second and third row, if necessary, we may assume that $c_{22} \neq 0$. It follows that $\mathbf{C}$ is row equivalent to the matrix

$$\mathbf{D} = \begin{pmatrix} 1 & 0 & d_{13} \\ 0 & 1 & d_{23} \\ 0 & 0 & d_{33} \end{pmatrix}.$$

The entry d_{33} of $\mathbf{D}$ cannot be zero, since otherwise the third column of $\mathbf{D}$ is a linear combination of the first two, which is impossible by Corollary 1.3. Then we obtain row equivalent matrices:

$$\begin{pmatrix} 1 & 0 & d_{13} \\ 0 & 1 & d_{23} \\ 0 & 0 & d_{33} \end{pmatrix} \sim \begin{pmatrix} 1 & 0 & d_{13} \\ 0 & 1 & d_{23} \\ 0 & 0 & 1 \end{pmatrix} \sim \mathbf{I}_3. \qquad \square$$

Corollary 1.4 *If $\{\mathbf{v}_1, \mathbf{v}_2, \mathbf{v}_3\}$ are any three linearly independent vectors in $\mathbb{R}^3$ then $\{\mathbf{v}_1, \mathbf{v}_2, \mathbf{v}_3\}$ is a basis for $\mathbb{R}^3$.*

Proof. We show that for any vector $\mathbf{b}$ in $\mathbb{R}^3$ there are unique real numbers x_1, x_2, x_3 such that

$$x_1\mathbf{v}_1 + x_2\mathbf{v}_2 + x_3\mathbf{v}_3 = \mathbf{b}. \tag{1.54}$$

Let

$$\mathbf{A} = \begin{pmatrix} \mathbf{v}_1 & \mathbf{v}_2 & \mathbf{v}_3 \end{pmatrix}, \quad \mathbf{x} = \begin{pmatrix} x_1 & x_2 & x_3 \end{pmatrix}^T \quad \mathbf{b} = \begin{pmatrix} b_1 & b_2 & b_3 \end{pmatrix}^T.$$

Then the system in (1.54) is equivalent to $\mathbf{Ax} = \mathbf{b}$. By Theorem 1.16 the augmented matrix of this system $(\mathbf{A}|\mathbf{b})$ is row equivalent to $(\mathbf{I}_3|\mathbf{x})$, implying that $\mathbf{Ax} = \mathbf{b}$ has a unique solution $\mathbf{x}$ for any $\mathbf{b}$ in $\mathbb{R}^3$. $\qquad \square$

The following theorem classifies subspaces of $\mathbb{R}^3$.

Theorem 1.17 *Let V be a subspace of $\mathbb{R}^3$. Then there are four possibilities:*

(1) $V = \{\mathbf{0}_3\}$;
(2) $V = \mathrm{Lin}(\mathbf{u})$ *for some non-zero vector* $\mathbf{u}$;
(3) $V = \mathrm{Lin}(\mathbf{u}, \mathbf{v})$ *for two non-zero not proportional vectors* $\mathbf{u}, \mathbf{v}$;
(4) $V = \mathbb{R}^3$.

Proof. If $V \neq \{\mathbf{0}_3\}$ then there is a non-zero vector $\mathbf{u}$ in V. If $V \neq \mathrm{Lin}(\mathbf{u})$, then there is vector $\mathbf{v}$ in V not proportional to $\mathbf{u}$. If $V \neq \mathrm{Lin}(\mathbf{u}, \mathbf{v})$, then there is $\mathbf{w}$ in V which is not in $\mathrm{Lin}(\mathbf{u}, \mathbf{v})$. The system $\{\mathbf{u}, \mathbf{v}, \mathbf{w}\}$ is linearly independent.

Indeed, if

$$a \cdot \mathbf{u} + b \cdot \mathbf{v} + c \cdot \mathbf{w} = \mathbf{0}$$

for some real numbers a, b, c, then $c = 0$, since $\mathbf{w}$ is not in $\mathrm{Lin}(\mathbf{u}, \mathbf{v})$. By Lemma 2.1 vectors $\mathbf{u}$ and $\mathbf{v}$ are linearly independent, since they are not proportional. It follows that $a = b = 0$.

By Corollary 1.4 the set $\{\mathbf{u}, \mathbf{v}, \mathbf{w}\}$ is a basis for $\mathbb{R}^3$ implying that $\mathrm{Lin}(\mathbf{u}, \mathbf{v}, \mathbf{w}) = \mathbb{R}^3$. Since V includes $\mathrm{Lin}(\mathbf{u}, \mathbf{v}, \mathbf{w})$, we obtain that $V = \mathbb{R}^3$. ☐

By Theorem 1.17 any non-zero proper subspace of $\mathbb{R}^3$ is either a plane or a line passing through the origin. By Theorem 1.17 any plane in $\mathbb{R}^3$ through its origin is uniquely determined by two not proportional vectors $\mathbf{u}$, $\mathbf{v}$. Any plane in $\mathbb{R}^3$ through its origin is uniquely determined by any non-zero vector $\mathbf{n}$ which is perpendicular to $\mathbf{u}$, $\mathbf{v}$. We can find such a vector $\mathbf{n} = \begin{pmatrix} n_1 & n_2 & n_3 \end{pmatrix}$ as a non-zero solution to the system

$$\begin{cases} \langle \mathbf{n}, \mathbf{u} \rangle = 0, \\ \langle \mathbf{n}, \mathbf{v} \rangle = 0, \end{cases} \Leftrightarrow \begin{cases} n_1 u_1 + n_2 u_2 + n_3 u_3 = 0, \\ n_1 v_1 + n_2 v_2 + n_3 v_3 = 0. \end{cases} \tag{1.55}$$

Problem 1.15 Find an equation of the plane through

$$\begin{pmatrix} 1 & 0 & 1 \end{pmatrix}^T, \quad \begin{pmatrix} 0 & 1 & 1 \end{pmatrix}^T. \tag{1.56}$$

Solution: Let $\mathbf{n} = \begin{pmatrix} a & b & c \end{pmatrix}^T$ be a vector orthogonal to both vectors in (1.56). Then

$$\begin{cases} a + 0 \cdot b + c = 0 \\ 0 \cdot a + b + c = 0 \end{cases} \Leftrightarrow \begin{pmatrix} 1 & 0 & 1 \\ 0 & 1 & 1 \end{pmatrix} \begin{pmatrix} a \\ b \\ c \end{pmatrix} = \mathbf{0}_2 \Leftrightarrow \begin{cases} a = -c \\ b = -c \end{cases}$$

It follows that the vector $\mathbf{n} = \begin{pmatrix} 1 & 1 & -1 \end{pmatrix}^T$ is perpendicular to both vectors in (1.56). The equation of the plane is: $x_1 + x_2 - x_3 = 0$. ☐

Any plane in $\mathbb{R}^3$ is obtained as a shift of some plane passing through $\mathbf{0}_n$. Since planes through $\mathbf{0}_n$ are described by Theorem 1.17 as linear spans $\mathrm{Lin}(\mathbf{u}, \mathbf{v})$, where $\mathbf{u}$ and $\mathbf{v}$ are non-zero not proportional vectors, we can easily obtain a **vector** formula for an arbitrary plane in $\mathbb{R}^3$:

$$\mathbf{x} = \mathbf{p} + s\mathbf{u} + t\mathbf{v}. \tag{1.57}$$

Definition 1.17 Formula (1.57) is called a **vector equation** of a plane.

Passing to the coordinates we obtain a **parametric** equation of a plane:

$$\begin{pmatrix} x_1 \\ x_2 \\ x_3 \end{pmatrix} = \begin{pmatrix} p_1 \\ p_2 \\ p_3 \end{pmatrix} + s \begin{pmatrix} u_1 \\ u_2 \\ u_3 \end{pmatrix} + t \begin{pmatrix} v_1 \\ v_2 \\ v_3 \end{pmatrix}.$$

If $\mathbf{n}$ is a vector perpendicular to the plane $\mathrm{Lin}(\mathbf{u}, \mathbf{v})$, then the equation of the plane through $\mathbf{p}$ is given by

$$\langle \mathbf{n}, \mathbf{x} \rangle = \langle \mathbf{n}, \mathbf{p} \rangle \Leftrightarrow \langle \mathbf{n}, \mathbf{x} - \mathbf{p} \rangle = 0 \Leftrightarrow$$
$$n_1(x_1 - p_1) + n_2(x - p_2) + n_3(x_3 - p_3) = 0 \Leftrightarrow$$
$$n_1 x_1 + n_2 x_3 + n_3 x_3 = n_1 p_1 + n_2 p_3 + n_3 p_3. \quad (1.58)$$

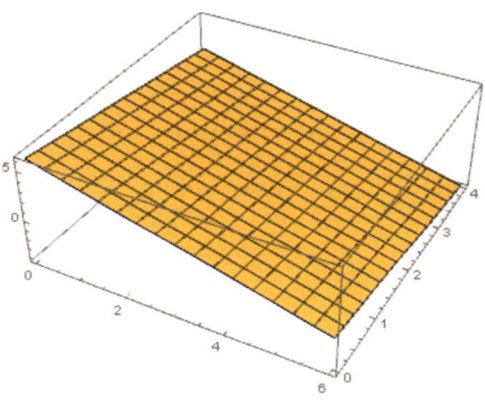

Fig. 1.5: $ax_1 + bx_2 + cx_3 = d$

Definition 1.18 The equation

$$ax_1 + bx_2 + cx_3 = d$$

is called a **Cartesian equation** of a plane, see Fig 1.5.

Problem 1.15 shows how one can obtain a Cartesian equation of a plane if a parametric equation is given.

Problem 1.16 A Cartesian equation of a plane is given by

$$2x_1 - 3x_2 + x_3 = 5. \quad (1.59)$$

Find vector and parametric equations of this plane.

Solution: The equation (1.59) is a system of one equation in three unknowns. Its augmented matrix is

$$\left(2 \;-3 \; 1 \; 5\right) \sim \left(1 \; -3/2 \; 1/2 \; 5/2\right).$$

Let $x_2 = 2s$, $x_3 = 2t$. Then $x_1 = 2.5 + 3s - t$. It follows that

$$\begin{pmatrix} x_1 \\ x_2 \\ x_3 \end{pmatrix} = \begin{pmatrix} 2.5 + 3s - t \\ 2s \\ 2t \end{pmatrix} = \begin{pmatrix} 2.5 \\ 0 \\ 0 \end{pmatrix} + s\begin{pmatrix} 3 \\ 2 \\ 0 \end{pmatrix} + t\begin{pmatrix} -1 \\ 0 \\ 2 \end{pmatrix} = \mathbf{p} + s\mathbf{u} + t\mathbf{v}. \quad \square$$

Problem 1.17 A parametric equation of a plane is

$$\begin{pmatrix} x_1 \\ x_2 \\ x_3 \end{pmatrix} = \begin{pmatrix} 2 \\ 1 \\ -1 \end{pmatrix} + s\begin{pmatrix} 1 \\ -1 \\ 1 \end{pmatrix} + t\begin{pmatrix} 1 \\ 1 \\ -1 \end{pmatrix}. \tag{1.60}$$

Find a Cartesian equation of this plane.

Solution: We consider (1.60) as a system of three linear equations in two unknowns s and t:

$$\begin{cases} s + t & = x_1 - 2, \\ -s + t & = x_2 - 1, \\ s - t & = x_3 + 1. \end{cases} \tag{1.61}$$

The augmented matrix of this system can be transformed with elementary row operations as follows:

$$\begin{pmatrix} 1 & 1 & x_1 - 2 \\ -1 & 1 & x_2 - 1 \\ 1 & -1 & x_3 + 1 \end{pmatrix} \sim \begin{pmatrix} 1 & 1 & x_1 - 2 \\ 0 & 2 & x_1 + x_2 - 3 \\ 0 & -2 & x_3 - x_1 + 3 \end{pmatrix} \sim \begin{pmatrix} 1 & 1 & x_1 - 2 \\ 0 & 2 & x_1 + x_2 - 3 \\ 0 & 0 & x_3 + x_2 \end{pmatrix} \sim$$

$$\begin{pmatrix} 1 & 1 & x_1 - 2 \\ 0 & 1 & 0.5x_1 + 0.5x_2 - 1.5 \\ 0 & 0 & x_3 + x_2 \end{pmatrix} \sim \begin{pmatrix} 1 & 0 & 0.5x_1 - 0.5x_2 - 0.5 \\ 0 & 1 & 0.5x_1 + 0.5x_2 - 1.5 \\ 0 & 0 & x_3 + x_2 \end{pmatrix}.$$

It follows that the system (1.61) has a solution in s and t if and only if $x_2 + x_3 = 0$. If $\mathbf{x}$ is a vector satisfying this condition, then

$$\begin{cases} s & = 0.5x_1 - 0.5x_2 - 0.5, \\ t & = 0.5x_1 + 0.5x_2 - 1.5. \end{cases}$$

A Cartesian equation of this plane is

$$x_2 + x_3 = 0. \quad \square$$

Problem 1.18 Find a Cartesian equation of the plane through three points P, Q, R with coordinates

$$P(-1,2,3), \quad Q(1,-1,2), \quad R(1,1,2).$$

Solution: A Cartesian equation of this plane is $ax_1 + bx_2 + cx_3 = d$. Since the points P, Q, R are on the plane, we obtain the system of equations

$$\begin{cases} -a + 2b + 3c - d &= 0 \\ a - b + 2c - d &= 0 \\ a + b + 2b - d &= 0 \end{cases} \tag{1.62}$$

The augmented matrix of the homogeneous system (1.62) is

$$\begin{pmatrix} -1 & 2 & 3 & -1 \\ 1 & -1 & 2 & -1 \\ 1 & 1 & 2 & -1 \end{pmatrix} \sim \begin{pmatrix} 1 & 0 & 0 & -1/5 \\ 0 & 1 & 0 & 0 \\ 0 & 0 & 1 & -2/5 \end{pmatrix} \Rightarrow \begin{cases} a &= 1 \\ b &= 0 \\ c &= 2 \\ d &= 5 \end{cases}.$$

A Cartesian equation is

$$x_1 + 2x_3 = 5. \quad \square$$

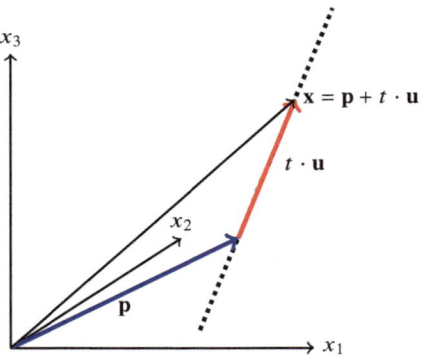

Fig. 1.6: A vector equation of a line

By Theorem 1.17 any line through $\mathbf{0}_3$ is described as $\mathrm{Lin}(\mathbf{u})$, where $\mathbf{u}$ is a non-zero vector. Any line in $\mathbb{R}^3$ is a shift of $\mathrm{Lin}(\mathbf{u})$. It follows that a **vector** equation of a line is given by

$$\mathbf{x} = \mathbf{p} + t\mathbf{u}, \tag{1.63}$$

see Fig 1.6. A non-zero vector $\mathbf{u}$ in (1.63) is called a **direction vector** of a line. Notice that two lines are parallel or coincide if and only if their directions vectors are proportional. A vector equation of a given line is not unique. As it is mentioned above **u** may be multiplied by any non-zero constant. Also, **p** may be any vector with its head on the line.

Problem 1.19 Check if the following vector equations

$$\mathbf{x} = \begin{pmatrix} 1 \\ -1 \\ 1 \end{pmatrix} + t \begin{pmatrix} 2 \\ 1 \\ -1 \end{pmatrix} \quad \text{and} \quad \mathbf{x} = \begin{pmatrix} 5 \\ 1 \\ -1 \end{pmatrix} + t \begin{pmatrix} -4 \\ -2 \\ 2 \end{pmatrix}$$

describe the same line.

Solution: We observe first that if these are vector equations of the same line then the direction vectors must be proportional. Since $-4 = (-2) \cdot 2$, the coefficient of proportionality must be -2. Inspecting the other coordinates we see that, indeed, the second direction vector is proportional to the first. We check if the vector $\begin{pmatrix} 1 & -1 & 1 \end{pmatrix}^T$ has its head on the second line:

$$\begin{pmatrix} 1 \\ -1 \\ 1 \end{pmatrix} = \begin{pmatrix} 5 \\ 1 \\ -1 \end{pmatrix} + t \begin{pmatrix} -4 \\ -2 \\ 2 \end{pmatrix} \Rightarrow t = 1.$$

So, the lines defined by these two equations have a common point and proportional direction vectors. Hence they coincide. □

To exclude the parameter t from a vector equation of a line we construct the system

$$\begin{cases} tu_1 &= x_1 - p_1 \\ tu_2 &= x_2 - p_2 \\ tu_3 &= x_3 - p_3 \end{cases} \Leftrightarrow \frac{x_1 - p_1}{u_1} = \frac{x_2 - p_2}{u_2} = \frac{x_3 - p_3}{u_3}. \tag{1.64}$$

The system in (1.64) is a **parametric** equation of a line, whereas the equivalent system of equalities is called a **Cartesian** equation of a line. Notice that it is assumed that all numbers in the denominators of a **Cartesian** equation are not equal to zero.

Theorem 1.18 *Let* **p** *be a vector in* $\mathbb{R}^3$ *and let*

$$ax_1 + bx_2 + cx_3 = d$$

be a Cartesian equation of a plane in $\mathbb{R}^3$. *Then the distance from the head of* **p** *to this plane is given by the following formula:*

$$\mathbf{dist} = \frac{|d - \langle \mathbf{n}, \mathbf{p} \rangle|}{|\mathbf{n}|},$$

where $\mathbf{n} = \begin{pmatrix} a & b & c \end{pmatrix}^T$ *is a normal vector to the plane.*

Proof. The closest point in the plane to the head of **p** can be obtained as the point of intersection of the perpendicular line to the plane passing through the head of **p**. This line is parallel to the normal **n**. Therefore its parametric equation is given by

$$\begin{cases} x_1 & = p_1 + at \\ x_2 & = p_2 + bt \\ x_3 & = p_3 + ct \end{cases}.$$

To find the point of intersection with the plane we substitute the formulas obtained above into the equation of the plane:

$$a(p_1+at)+b(p_2+bt)+c(p_3+ct) = d \Rightarrow t_0 = \frac{d - (ap_1 + bp_2 + cp_3)}{a^2 + b^2 + c^2} = \frac{d - \langle \mathbf{n}, \mathbf{p} \rangle}{|\mathbf{n}|}.$$

The coordinates of the point of intersection are

$$x_1 = p_1 + at_0, \quad x_2 = p_2 + bt_0, \quad x_3 = p_3 + ct_0.$$

It follows that

$$\mathbf{dist}^2 = (x_1 - p_1)^2 + (x_2 - p_2)^2 + (x_3 - p_3)^2 = (a^2 + b^2 + c^2)t_0^2 =$$

$$(a^2 + b^2 + c^2)\frac{|d - (ap_1 + bp_2 + cp_3)|^2}{(a^2 + b^2 + c^2)^2} \Rightarrow \mathbf{dist} = \frac{|d - \langle \mathbf{n}, \mathbf{p} \rangle|}{|\mathbf{n}|}. \qquad \square$$

Problem 1.20 Find vector, parametric, and Cartesian equations of the line through the points $P(1,-1,2)$ and $Q(2,0,1)$.

Solution: The direction of the line is determined by the vector

$$\mathbf{u} = \begin{pmatrix} 2 & 0 & 1 \end{pmatrix}^T - \begin{pmatrix} 1 & -1 & 2 \end{pmatrix}^T = \begin{pmatrix} 1 & 1 & -1 \end{pmatrix}^T.$$

It follows that vector, parametric, and Cartesian equations of this line are given by:

$$\mathbf{p} + t\mathbf{u} = \begin{pmatrix} 1 \\ -1 \\ 2 \end{pmatrix} + t \begin{pmatrix} 1 \\ 1 \\ -1 \end{pmatrix} \Leftrightarrow \begin{cases} x_1 & = 1 + t \\ x_2 & = -1 + t \\ x_3 & = 2 - t \end{cases} \Leftrightarrow$$

$$x_1 - 1 = x_2 + 1 = 2 - x_3. \qquad \square$$

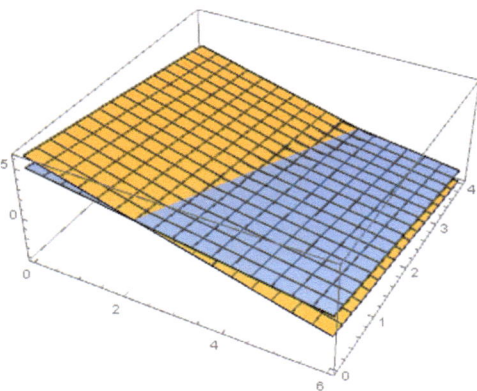

Fig. 1.7: A Geometrical Illustration to Problem 1.21

Problem 1.21 Find vector, parametric, and Cartesian equations for the line of intersection of the plane $2x + 3y + 3z = 15$ with the plane $6x + 4y + 5z = 31$, see Fig 1.7.

Solution: The coordinates x, y, z of the points on the line of intersection satisfy the linear system of two equations in three unknowns:

$$\begin{cases} 2x + 3y + 3z & = 15 \\ 6y + 4y + 5z & = 31. \end{cases}$$

We apply row operations to the augmented matrix of this system:

$$\left(\begin{array}{ccc|c} 2 & 3 & 3 & 15 \\ 6 & 4 & 5 & 31 \end{array}\right) \sim \left(\begin{array}{ccc|c} 1 & 0 & \frac{3}{10} & \frac{33}{10} \\ 0 & 1 & \frac{4}{5} & \frac{14}{5} \end{array}\right).$$

If $z = t$ then

$$x = \frac{33}{10} - \frac{3}{10}t, \quad y = \frac{14}{5} - \frac{4}{5}t, \quad z = t,$$

is a parametric equation of the line of intersection. Then

$$\mathbf{x} = \begin{pmatrix} x \\ y \\ z \end{pmatrix} = \begin{pmatrix} \frac{33}{10} \\ \frac{14}{5} \\ 0 \end{pmatrix} + t \begin{pmatrix} -\frac{3}{10} \\ -\frac{4}{5} \\ 1 \end{pmatrix} = \mathbf{p} + t\mathbf{u}$$

is a vector equation of the line. Excluding parameter t we obtain a Cartesian equation:

$$\frac{x - 3.3}{-0.3} = \frac{y - 2.8}{-0.8} = \frac{z}{1}. \quad \square$$

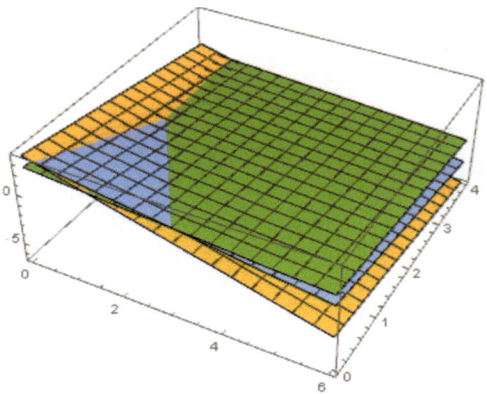

Fig. 1.8: A Geometrical Illustration to Problem 1.22

Problem 1.22 Find the coordinates of the point of intersection of three planes: $x + 2y + 3z = 8$, $2x + 3y + 3z = 11$, and $6x + 4y + 5z = 19$.

Solution: We have

$$\begin{pmatrix} 1 & 2 & 3 & | & 8 \\ 2 & 3 & 3 & | & 11 \\ 6 & 4 & 5 & | & 29 \end{pmatrix} \sim \begin{pmatrix} 1 & 0 & 0 & | & 1 \\ 0 & 1 & 0 & | & 2 \\ 0 & 0 & 1 & | & 1 \end{pmatrix} \Rightarrow \begin{cases} x & = 1, \\ y & = 2, \\ z & = 1. \end{cases} \quad \square$$

Problem 1.23 Find the distance from the point $P(8, 5, 12)$ to the plane

$$2x - y + 3z = 5.$$

Solution: The coefficients at x, y, z in the Cartesian equation of the plane are the coordinates of the following normal vector $\mathbf{n} = \begin{pmatrix} 2 & -1 & 3 \end{pmatrix}^T$ to the plane. Since the line perpendicular to a plane makes the shortest path from any of its points to the plane, we find the parametric equation of the line through P parallel to $\mathbf{n}$:

$$\begin{cases} x & = 8 + 2t, \\ y & = 5 - t, \\ z & = 12 + 3t. \end{cases}$$

To find the intersection point of this line with the plane we insert these formulas into the Cartesian equation of the plane:

$$5 = 2 - y + 3z = 2(8 + 2t) - (5 - t) + 3(12 + 3t) = 47 + 14t \Rightarrow t = -3.$$

It follows that the intersection point Q has the coordinates: $x = 8 - 6 = 2$, $y = 5 - (-3) = 8$, $z = 12 + 3(-3) = 3$. Hence

$$\text{dist}(P,Q) = \sqrt{(8-2)^2 + (5-3)^2 + (12-8)^2} = 2\sqrt{14}. \quad \square$$

Problems

Prob. 42 — Find a Cartesian equation of the subspace V of $\mathbb{R}^3$ defined as the set of solutions $\mathbf{v}$ to the linear system

$$\begin{cases} 5v_1 + 3v_2 + 13v_3 & = 0 \\ 6v_1 + 7v_2 + 19v_3 & = 0 \ . \\ 4v_1 + 4v_2 + 12v_3 & = 0 \end{cases} \tag{1.65}$$

Prob. 43 — Find Cartesian equations of lines determined by pairs of equations in (1.65).

Prob. 44 — Determine the angles between the lines in Problem 43.

Prob. 45 — Find the coordinates of the point of intersection of three planes: $x - y + 2z = -4$, $2x + y - z = 6$, and $x + 2y + z = 6$.

Prob. 46 — Find a general equation of the line through the point $P(1,3,1)$ which is parallel to the line

$$x + y + z + 2 = 0, \quad 2x + 3y + z = 0.$$

Prob. 47 — Find a general equation of the line through the points $P(1,3,1)$ and $Q(4,2,-1)$.

Prob. 48 — Given the parametric equation of the plane

$$x = 1 + u + v, \quad y = 2 - u + 2v, \quad z = -1 - u - 2v$$

find a general equation of the plane.

Prob. 49 — Given a general equation of the plane

$$3x + 2y + z - 1 = 0$$

find a parametric equation of the plane.

Prob. 50 — Given two planes

$$3x + y + z + 1 = 0 \text{ and } 2x + 3y + z + 2 = 0.$$

Determine whether they intersect, parallel or equal. Write a general equation of the line of intersection if it exists.

Prob. 51 — Find vector, parametric, and Cartesian equations for the line of intersection of the planes $x - y + z = 1$, $2x + y - 7z = 5$, $x + 2y - 8z = 4$.

Prob. 52 — Find vector, parametric, and Cartesian equations of the line through the points $P(2, -3, 4)$ and $Q(-5, 8, -7)$.

Prob. 53 — Three planes $x - y - z + 5 = 0$, $2x + y + 7z + 1 = 0$, $x + 2y + 8z - 4 = 0$ intersect at a line. Find the Cartesian equation of the plane passing through the point $P(1, 1, 1)$ and perpendicular to this line.

Prob. 54 — Find the distance from the point $P(1, 1, 2)$ to the plane $x - 2y + z = 5$.

Prob. 55 — Find vector and parametric equations of the plane $x + 2y + 3z = 6$.

Prob. 56 — The parametric equation of the plane is given by

$$\begin{cases} x & = -2 + s - 3t \\ y & = 2 + 3s - t \\ z & = 1 + 5s + 3t \end{cases}.$$

Find a Cartesian equation of this plane.

Prob. 57 — Find a cartesian equation of the subspace $V = \mathrm{Lin}(\mathbf{v}_1, \mathbf{v}_2, \mathbf{v}_3)$ of $\mathbb{R}^3$, where

$$\mathbf{v}_1 = \begin{pmatrix} 1 \\ -1 \\ -4 \end{pmatrix}, \ \mathbf{v}_2 = \begin{pmatrix} 2 \\ 1 \\ 7 \end{pmatrix}, \ \mathbf{v}_3 = \begin{pmatrix} 1 \\ 2 \\ 11 \end{pmatrix}.$$

1.8 Reduced Row Echelon Form

Since elementary row operations are revertible, they do not alter the set of solutions. In this section, we present a generalization of the methods, which were used to find the solution for the system (1.5). First, we define the target matrix for our row operations, see (1.6).

Definition 1.19 An $m \times n$ matrix $\mathbf{A}$ is said to be in **row echelon form** if it satisfies the following properties:

(a) the first left entry of every non-zero row is 1, which is called the **leading one**;

(b) the leading one in a lower row is placed in the column which is further to the right;

(c) zero rows are all at the bottom of the matrix.

In case of (1.6) this matrix is

$$\begin{pmatrix} 1\ 2\ 3 & 10 \\ 0\ 1\ 3 & 5 \\ 0\ 0\ 1 & 1 \end{pmatrix} \Rightarrow \begin{cases} x_1 & = 10 - 2x_2 - 3x_3 = 3 \\ x_2 & = 5 - 3x_3 = 2 \\ x_3 & = 1 \end{cases}$$

A disadvantage of a Row Echelon Form is that it does not give an immediate solution. The matrix, obtained by elementary row operations in (1.16),

$$\begin{pmatrix} 1\ 0\ 0 & 3 \\ 0\ 1\ 0 & 2 \\ 0\ 0\ 1 & 1 \end{pmatrix}$$

is better, since it allows one to write the solution directly, without any extra calculations.

Definition 1.20 An $m \times n$ matrix **A** in **row echelon form** is said to be in **reduced row echelon form** if it satisfies the following additional property:

(**d**) if a column contains a leading one of some row, then all other entries in this column are zeros;

Here are examples of matrices in reduced row echelon form

$$\mathbf{A} = \begin{pmatrix} 0\ 1\ 0\ 4 \\ 0\ 0\ 1\ 2 \\ 0\ 0\ 0\ 0 \end{pmatrix}, \quad \mathbf{B} = \begin{pmatrix} 1\ 9\ 0\ 0\ 8 \\ 0\ 0\ 1\ 0\ 4 \\ 0\ 0\ 0\ 1\ 0 \end{pmatrix}, \tag{1.66}$$

$$\mathbf{C} = \begin{pmatrix} 1\ 1\ 0\ 1\ 0 \\ 0\ 0\ 1\ 0\ 0 \\ 0\ 0\ 0\ 0\ 1 \\ 0\ 0\ 0\ 0\ 0 \\ 0\ 0\ 0\ 0\ 0 \end{pmatrix} \tag{1.67}$$

The following matrices are not in reduced echelon form

$$\mathbf{A} = \begin{pmatrix} 1\ 2\ 0\ 4 \\ 0\ 0\ 0\ 0 \\ 0\ 0\ 1\ 2 \end{pmatrix}, \quad \mathbf{B} = \begin{pmatrix} 1\ 0\ 4\ 4 \\ 0\ 2\ 0\ 4 \\ 0\ 0\ 0\ 0 \end{pmatrix},$$

$$\mathbf{C} = \begin{pmatrix} 0\ 1\ 0\ 1 \\ 0\ 0\ 0\ 0 \\ 0\ 0\ 1\ 0 \\ 0\ 0\ 0\ 0 \end{pmatrix} \quad \mathbf{D} = \begin{pmatrix} 1\ 1\ 1\ 1 \\ 0\ 1\ 1\ 1 \\ 0\ 0\ 1\ 1 \\ 0\ 0\ 0\ 0 \end{pmatrix}$$

Problem 1.24 A matrix

$$\mathbf{A} = \begin{pmatrix} * & * & 1 & * & 5 \\ * & * & 2 & * & 1 \\ * & * & * & * & 3 \end{pmatrix}$$

is in reduced row echelon form. Find $\mathbf{A}$.

Solution: Since $a_{23} = 2 \neq 1$, the leading one of $\text{row}_2(\mathbf{A})$ is either a_{21} or a_{22}. Similarly, the leading one of $\text{row}_1(\mathbf{A})$ is either a_{11} or a_{12}. If $\text{col}_1(\mathbf{A}) = \mathbf{0}_3$, then $a_{11} = 0$ implying that $a_{12} = 1$. In this case a_{12} is the leading one of $\text{row}_1(\mathbf{A})$, implying that $a_{21} = a_{22} = 0$, which says that $\text{row}_2(\mathbf{A})$ has no leading one at all. It follows that $\text{col}_1(\mathbf{A}) \neq \mathbf{0}_3$ and hence $a_{11} = 1$, $a_{21} = a_{31} = 0$.

Since $\text{row}_2(\mathbf{A})$ must have a leading one, we conclude that $a_{22} = 1$ and $a_{12} = a_{32} = 0$. It follows that

$$\mathbf{A} = \begin{pmatrix} 1 & 0 & 1 & * & 5 \\ 0 & 1 & 2 & * & 1 \\ 0 & 0 & * & * & 3 \end{pmatrix}.$$

Since $\text{col}_3(\mathbf{A})$ cannot contain a leading one, we conclude that $a_{33} = 0$. Since $a_{35} = 3 \neq 1$, the leading one of $\text{row}_3(\mathbf{A})$ is a_{34}, which implies that $a_{14} = a_{24} = 0$. It follows that

$$\mathbf{A} = \begin{pmatrix} 1 & 0 & 1 & 0 & 5 \\ 0 & 1 & 2 & 0 & 1 \\ 0 & 0 & 0 & 1 & 3 \end{pmatrix}. \quad \square$$

Theorem 1.19 *Every matrix is row equivalent to a matrix in row echelon form.*

Proof. We prove this theorem by induction on the number m of rows in $\mathbf{A}$. If $m = 1$, then

$$\mathbf{A} = \begin{pmatrix} a_{11} & a_{12} & \cdots & a_{1n} \end{pmatrix}.$$

If all entries a_{1j} of $\mathbf{A}$ are zeros, then $\mathbf{A}$ is already in row echelon form. If it is not the case, then there is the first index k such that $a_{1k} \neq 0$. It follows that $\mathbf{A}$ is row equivalent (see **RO1**) to the matrix

$$\mathbf{B} = \begin{pmatrix} 0 & \cdots & 1 & a_{1(k+1)}/a_{1k} & \cdots & a_{1n}/a_{1k} \end{pmatrix},$$

which is in row echelon form.

Suppose now that the Theorem is proved for all $m \times n$ matrices. Let $\mathbf{A}$ be an $(m + 1) \times n$ matrix. If all entries of this matrix are zeros, then it is already in row echelon form. Otherwise, there is k, $1 \leq k \leq m + 1$ such that $\text{row}_k(\mathbf{A})$ is a non-zero row of $\mathbf{A}$. Among all non-zero rows of $\mathbf{A}$ we chose a row k such that $a_{kj} \neq 0$ with the smallest possible j. Then all columns of $\mathbf{A}$ with the indexes smaller than j are zero columns. Using row operation **RO2** we may put $\mathrm{row}_k(\mathbf{A})$ to the place of the first row. It follows that $\mathbf{A}$ is row equivalent to the matrix

$$\mathbf{B} = \begin{pmatrix} 0 \cdots 0 & a_{1j} & \cdots & a_{1n} \\ 0 \cdots 0 & a_{2j} & \cdots & a_{2n} \\ \vdots \ddots \vdots & \vdots & \ddots & \vdots \\ 0 \cdots 0 & a_{(m+1)j} & \cdots & a_{(m+1)n} \end{pmatrix}.$$

Since $a_{1j} \neq 0$, we can apply a row operation of type **RO1**

$$\mathrm{row}_1(\mathbf{B}) := \frac{1}{a_{1j}}\mathrm{row}_1(\mathbf{B})$$

to obtain a leading one in the first row. Applying row operations **RO3**

$$\mathrm{row}_k(\mathbf{C}) := \mathrm{row}_k(\mathbf{B}) - a_{kj}\mathrm{row}_k(\mathbf{B}), \ 2 \leq k \leq m+1,$$

we obtain a row equivalent matrix $\mathbf{C}$ which entries in column j are all zero except for $c_{1j} = 1$.:

$$\mathbf{C} = \begin{pmatrix} 0 \cdots 0 \ 1 \cdots & c_{1n} \\ 0 \cdots 0 \ 0 \cdots & c_{2n} \\ \vdots \ddots \vdots \ \vdots \ddots & \vdots \\ 0 \cdots 0 \ 0 \cdots & c_{(m+1)n} \end{pmatrix} = \begin{pmatrix} 0 \cdots 0 \ 1 & \cdots & c_{1n} \\ \boxed{\begin{matrix} 0 \cdots 0 \ 0 \cdots & c_{2n} \\ \vdots \ddots \vdots \ \vdots \ddots & \vdots \\ 0 \cdots 0 \ 0 \cdots & c_{(m+1)n} \end{matrix}} \end{pmatrix}.$$

The boxed matrix $\mathbf{D}$ has dimension $m \times n$. Therefore, by the induction hypothesis it is row equivalent to a matrix in row echelon form. Since the leading one of the first row of $\mathbf{C}$ is placed above zero column of $\mathbf{D}$, this implies that $\mathbf{C}$ is row equivalent to a matrix in row echelon form, as stated. $\qquad\square$

Theorem 1.20 *Every matrix is row equivalent to a matrix in reduced row echelon form.*

Proof. By Theorem 1.19 any matrix is row equivalent to a matrix in row echelon form. Therefore, we may prove the Theorem only for matrices in row echelon form. We fix the first column which contains a leading one. Using elementary row operations of type **RO3** we vanish all entries of this column above the leading one. Applying similar operations to the second column with a leading one we continue up to the very last column. The obtained matrix is in reduced row echelon form. $\qquad\square$

Problem 1.25 Solve the system of four linear equations in four unknowns

$$\begin{cases} 2x_1 + x_2 + 2x_3 + x_4 = 4 \\ 3x_1 + 2x_2 + x_3 \qquad\;\; = 3 \\ 4x_1 + 2x_2 + x_3 + x_4 = 3 \\ 5x_1 + x_2 + 2x_3 \qquad\;\; = 8. \end{cases}$$

Solution: The augmented matrix of the system is

$$(\mathbf{A}|\mathbf{b}) = \begin{pmatrix} 2 & 1 & 2 & 1 & 4 \\ 3 & 2 & 1 & 0 & 3 \\ 4 & 2 & 1 & 1 & 3 \\ 5 & 1 & 2 & 0 & 8 \end{pmatrix}.$$

To find its reduced row echelon form we follow the **standard algorithm of reduced row echelon form** used in the proofs of Theorems 1.19 and 1.20. To find an echelon form of $\mathbf{A}$ we apply the following row operations:

$$\underset{r_1:=r_1/2}{\sim} \begin{pmatrix} 1 & 1/2 & 1 & 1/2 & 2 \\ 3 & 2 & 1 & 0 & 3 \\ 4 & 2 & 1 & 1 & 3 \\ 5 & 1 & 2 & 0 & 8 \end{pmatrix} \overset{\substack{r_2:=r_2-3r_1 \\ r_3:=r_3-4r_1 \\ r_4:=r_4-5r_1}}{\underset{\sim}{}} \begin{pmatrix} 1 & 1/2 & 1 & 1/2 & 2 \\ 0 & 1/2 & -2 & -3/2 & -3 \\ 0 & 0 & -3 & -1 & -5 \\ 0 & -3/2 & -3 & -5/2 & -2 \end{pmatrix}$$

$$\underset{r_2:=2r_2}{\sim} \begin{pmatrix} 1 & 1/2 & 1 & 1/2 & 2 \\ 0 & 1 & -4 & -3 & -6 \\ 0 & 0 & -3 & -1 & -5 \\ 0 & -3/2 & -3 & -5/2 & -2 \end{pmatrix} \underset{r_4:=r_4+3r_2/2}{\sim} \begin{pmatrix} 1 & 1/2 & 1 & 1/2 & 2 \\ 0 & 1 & -4 & -3 & -6 \\ 0 & 0 & -3 & -1 & -5 \\ 0 & 0 & -9 & -7 & -11 \end{pmatrix}$$

$$\underset{r_3:=-r_3/3}{\sim} \begin{pmatrix} 1 & 1/2 & 1 & 1/2 & 2 \\ 0 & 1 & -4 & -3 & -6 \\ 0 & 0 & 1 & 1/3 & 5/3 \\ 0 & 0 & -9 & -7 & -11 \end{pmatrix} \underset{r_4:=r_4+9r_3}{\sim} \begin{pmatrix} 1 & 1/2 & 1 & 1/2 & 2 \\ 0 & 1 & -4 & -3 & -6 \\ 0 & 0 & 1 & 1/3 & 5/3 \\ 0 & 0 & 0 & -4 & 4 \end{pmatrix}$$

$$\underset{r_4:=-r_4/4}{\sim} \begin{pmatrix} 1 & 1/2 & 1 & 1/2 & 2 \\ 0 & 1 & -4 & -3 & -6 \\ 0 & 0 & 1 & 1/3 & 5/3 \\ 0 & 0 & 0 & 1 & -1 \end{pmatrix} \underset{r_1:=r_1-r_2/2}{\sim} \begin{pmatrix} 1 & 0 & 3 & 2 & 5 \\ 0 & 1 & -4 & -3 & -6 \\ 0 & 0 & 1 & 1/3 & 5/3 \\ 0 & 0 & 0 & 1 & -1 \end{pmatrix} \overset{\substack{r_1:=r_1-3r_3 \\ r_2:=r_2+4r_3}}{\underset{\sim}{}}$$

$$\begin{pmatrix} 1 & 0 & 0 & 1 & 0 \\ 0 & 1 & 0 & -5/3 & 2/3 \\ 0 & 0 & 1 & 1/3 & 5/3 \\ 0 & 0 & 0 & 1 & -1 \end{pmatrix} \overset{\substack{r_1:=r_1-r_4 \\ r_2:=r_2+5r_4/3 \\ r_3:=r_3-r_4/3}}{\underset{\sim}{}} \begin{pmatrix} 1 & 0 & 0 & 0 & 1 \\ 0 & 1 & 0 & 0 & -1 \\ 0 & 0 & 1 & 0 & 2 \\ 0 & 0 & 0 & 1 & -1 \end{pmatrix} \Rightarrow \begin{cases} x_1 = 1 \\ x_2 = -1 \\ x_3 = 2 \\ x_4 = -1 \end{cases}.$$

These calculations can be simplified if one tries to conduct them in the set of integers until it is possible.

$$\begin{pmatrix} 2 & 1 & 2 & 1 & | & 4 \\ 3 & 2 & 1 & 0 & | & 3 \\ 4 & 2 & 1 & 1 & | & 3 \\ 5 & 1 & 2 & 0 & | & 8 \end{pmatrix} \begin{matrix} r_2 := r_2 - r_1 \\ r_3 := r_3 - 2r_1 \\ r_4 = r_4 - 2r_1 \\ \sim \end{matrix} \begin{pmatrix} 2 & 1 & 2 & 1 & | & 4 \\ 1 & 1 & -1 & -1 & | & -1 \\ 0 & 0 & -3 & -1 & | & -5 \\ 1 & -1 & -2 & -2 & | & 0 \end{pmatrix} \begin{matrix} r_1 := r_1 - 2r_2 \\ r_4 := r_4 - r_2 \\ r_3 = -r_3 \\ \sim \end{matrix}$$

$$\begin{pmatrix} 0 & -1 & 4 & 3 & | & 6 \\ 1 & 1 & -1 & -1 & | & -1 \\ 0 & 0 & 3 & 1 & | & 5 \\ 0 & -2 & -1 & -1 & | & 1 \end{pmatrix} \begin{matrix} r_2 := r_2 + r_1 \\ r_4 := r_4 - 2r_1 \\ \sim \end{matrix} \begin{pmatrix} 0 & -1 & 4 & 3 & | & 6 \\ 1 & 0 & 3 & 2 & | & 5 \\ 0 & 0 & 3 & 1 & | & 5 \\ 0 & 0 & -9 & -7 & | & -11 \end{pmatrix} \begin{matrix} r_1 := -r_1 \\ r_4 := r_4 + 3r_3 \\ \sim \end{matrix}$$

$$\begin{pmatrix} 0 & 1 & -4 & -3 & | & -6 \\ 1 & 0 & 3 & 2 & | & 5 \\ 0 & 0 & 3 & 1 & | & 5 \\ 0 & 0 & 0 & -4 & | & 4 \end{pmatrix} \begin{matrix} \textbf{RO2} \\ \sim \end{matrix} \begin{pmatrix} 1 & 0 & 3 & 2 & | & 5 \\ 0 & 1 & -4 & -3 & | & -6 \\ 0 & 0 & 3 & 1 & | & 5 \\ 0 & 0 & 0 & -4 & | & 4 \end{pmatrix} \begin{matrix} r_1 := r_1 - r_3 \\ r_2 := r_2 + r_3 \\ r_4 := -r_4/4 \\ \sim \end{matrix}$$

$$\begin{pmatrix} 1 & 0 & 0 & 1 & | & 0 \\ 0 & 1 & -1 & -2 & | & -1 \\ 0 & 0 & 3 & 1 & | & 5 \\ 0 & 0 & 0 & 1 & | & -1 \end{pmatrix} \begin{matrix} r_1 := r_1 - r_4 \\ r_2 := r_2 + 2r_4 \\ r_3 := r_3 - r_4 \\ \sim \end{matrix} \begin{pmatrix} 1 & 0 & 0 & 0 & | & 1 \\ 0 & 1 & -1 & 0 & | & -3 \\ 0 & 0 & 3 & 0 & | & 6 \\ 0 & 0 & 0 & 1 & | & -1 \end{pmatrix} \begin{matrix} r_3 := r_3/3 \\ \sim \end{matrix}$$

$$\begin{pmatrix} 1 & 0 & 0 & 0 & | & 1 \\ 0 & 1 & -1 & 0 & | & -3 \\ 0 & 0 & 1 & 0 & | & 2 \\ 0 & 0 & 0 & 1 & | & -1 \end{pmatrix} \begin{matrix} r_2 := r_2 + r_3 \\ \sim \end{matrix} \begin{pmatrix} 1 & 0 & 0 & 0 & | & 1 \\ 0 & 1 & 0 & 0 & | & -1 \\ 0 & 0 & 1 & 0 & | & 2 \\ 0 & 0 & 0 & 1 & | & -1 \end{pmatrix} \quad \square$$

Although the second approach is not algorithmic, it is more effective. To apply it successfully one needs some practice.

It turns out that every matrix has only one reduced row echelon form.

Theorem 1.21 *If* **A** *and* **B** *are two* $m \times n$ *matrices in reduced row echelon form with equal row spaces, then* **A** = **B**.

Proof. If $\mathbf{row(A)} = \mathbf{row(B)} = \{\mathbf{0}_n\}$, then both **A** and **B** are in reduced row echelon form with zero entries. Hence they are equal as stated.

In the general case we apply induction on the number m of rows and assume that $\mathbf{row(A)} = \mathbf{row(B)} \neq \{\mathbf{0}_n\}$. For $m = 1$

$$\mathbf{A} = \mathrm{row}_1(\mathbf{A}), \quad \mathbf{B} = \mathrm{row}_1(\mathbf{B}),$$

implying that

$$\mathbf{row}(\mathbf{A}) = \{c \cdot \mathrm{row}_1(\mathbf{A}) \mid c \text{ is real}\}, \quad \mathbf{row}(\mathbf{B}) = \{c \cdot \mathrm{row}_1(\mathbf{B}) \mid c \text{ is real}\}.$$

It follows that $\mathbf{row}(\mathbf{A}) = \mathbf{row}(\mathbf{B})$ if and only if $\mathrm{row}_1(\mathbf{A})$ and $\mathrm{row}_1(\mathbf{B})$ are proportional. Since $\mathbf{A} \neq \mathbf{0}$ and $\mathbf{B} \neq \mathbf{0}$, both rows begin with leading ones. Being proportional they must be equal.

Assuming that the Theorem holds for any pair of matrices of size $(m-1) \times n$, we prove that it is true for any pair $\mathbf{A}$ and $\mathbf{B}$ of matrices of size $m \times n$. Let $\mathbf{A}_1$ and $\mathbf{B}_1$ be the matrices obtained from $\mathbf{A}$ and $\mathbf{B}$ by deleting the first rows.

Since $\mathbf{A} \neq \mathbf{0}$ and $\mathbf{B} \neq \mathbf{0}$ are in reduced row echelon forms, their first non-zero entries in $\mathrm{row}_1(\mathbf{A})$ and $\mathrm{row}_1(\mathbf{B})$ are $a_{1r} = 1$ and $b_{1s} = 1$. If $r < s$, then $b_{ir} = 0$ for every i and, therefore, $\mathrm{row}_1(\mathbf{A})$ cannot be a linear combination of rows of $\mathbf{B}$. Similarly, the case $r > s$ is impossible. It follows that $r = s$. Since $\mathbf{row}(\mathbf{A}) = \mathbf{row}(\mathbf{B})$, every $\mathrm{row}_i(\mathbf{B})$ with $i \geq 2$ can be written as the sum

$$\mathrm{row}_i(\mathbf{B}) = \alpha_1 \mathrm{row}_1(\mathbf{A}) + \alpha_2 \mathrm{row}_2(\mathbf{A}) + \cdots + \alpha_m \mathrm{row}_m(\mathbf{A}).$$

Since $\mathbf{B}$ and $\mathbf{A}$ are in a reduced row echelon form, the entries of $\mathrm{row}_i(\mathbf{B})$ for $i \geq 2$, and of $\mathrm{row}_j(\mathbf{A})$ for $j \geq 2$ in the rth column are all zeros. It follows that $\alpha_1 = 0$. Hence, every $\mathrm{row}_i(\mathbf{B})$, $i \geq 2$, is in $\mathbf{row}(\mathbf{A}_1)$. Similarly, every $\mathrm{row}_i(\mathbf{A})$, $i \geq 2$, is in $\mathbf{row}(\mathbf{B}_1)$. Therefore, $\mathbf{row}(\mathbf{A}_1) = \mathbf{row}(\mathbf{B}_1)$. Since $\mathbf{A}_1$ and $\mathbf{B}_1$ are both in reduced row echelon form, we conclude by induction hypothesis that $\mathbf{A}_1 = \mathbf{B}_1$.

Since $\mathrm{row}_1(\mathbf{A})$ is a linear combination of rows of $\mathbf{B}$, there are real numbers $\alpha_1, \ldots, \alpha_m$ such that

$$\mathrm{row}_1(\mathbf{A}) = \alpha_1 \mathrm{row}_1(\mathbf{B}) + \alpha_2 \mathrm{row}_2(\mathbf{B}) + \cdots + \alpha_m \mathrm{row}_m(\mathbf{B}) \qquad (1.68)$$

Since $\mathbf{A}$ and $\mathbf{B}$ are also in reduced row echelon form, the entries of both $\mathrm{row}_1(\mathbf{A})$ and $\mathrm{row}_1(\mathbf{B})$ placed above the leading ones of $\mathbf{A}_1 = \mathbf{B}_1$ must be zeros. This shows that $\alpha_j = 0$ for $j \geq 2$, implying that $\mathrm{row}_1(\mathbf{A}) = \alpha_1 \mathrm{row}_1(\mathbf{B})$. Since the leading ones of the first rows of $\mathbf{A}$ and $\mathbf{B}$ are placed in one column, we conclude that $\alpha_1 = 1$. $\qquad \square$

Given an $m \times n$ matrix $\mathbf{A}$ we denote by

$$\mathrm{rref}(\mathbf{A}) \qquad (1.69)$$

the unique row equivalent matrix for $\mathbf{A}$, which is in reduced row echelon form.

Theorem 1.22 *Two $m \times n$ matrices $\mathbf{A}$ and $\mathbf{B}$ are row equivalent if and only if their row spaces* $\mathbf{row}(\mathbf{A})$ *and* $\mathbf{row}(\mathbf{B})$ *coincide.*

Proof. If $\mathbf{A} \sim \mathbf{B}$, then $\mathbf{row}(\mathbf{A}) = \mathbf{row}(\mathbf{B})$ by Theorem 1.7. If $\mathbf{row}(\mathbf{A}) = \mathbf{row}(\mathbf{B})$, then by Theorem 1.7

$$\mathbf{row}(\mathrm{rref}(\mathbf{A})) = \mathbf{row}(\mathbf{A}) = \mathbf{row}(\mathbf{B}) = \mathbf{row}(\mathrm{rref}(\mathbf{B})).$$

By Theorem 1.21 we obtain that $\mathrm{rref}(\mathbf{A}) = \mathrm{rref}(\mathbf{B})$, which implies that

$$\mathbf{A} \sim \mathrm{rref}(\mathbf{A}) = \mathrm{rref}(\mathbf{B}) \sim \mathbf{B} \qquad \square$$

The reduced row echelon form has a number of properties, which are useful in practical questions. We illustrate them by the following examples.

$$\mathbf{A} = \begin{pmatrix} 9 & 18 & 2 & 11 & 2 \\ 3 & 6 & 1 & 4 & 0 \\ 5 & 10 & 1 & 6 & 1 \\ 4 & 8 & 1 & 5 & 1 \end{pmatrix} \Rightarrow \mathrm{rref}(\mathbf{A}) = \begin{pmatrix} 1 & 2 & 0 & 1 & 0 \\ 0 & 0 & 1 & 1 & 0 \\ 0 & 0 & 0 & 0 & 1 \\ 0 & 0 & 0 & 0 & 0 \end{pmatrix}. \qquad (1.70)$$

$$\mathbf{A}_3 = \begin{pmatrix} 9 & 18 & 2 \\ 3 & 6 & 1 \\ 5 & 10 & 1 \\ 4 & 8 & 1 \end{pmatrix} \Rightarrow \mathrm{rref}(\mathbf{A}_3) = \begin{pmatrix} 1 & 2 & 0 \\ 0 & 0 & 1 \\ 0 & 0 & 0 \\ 0 & 0 & 0 \end{pmatrix}.$$

$$\mathbf{A}_4 = \begin{pmatrix} 9 & 18 & 2 & 11 \\ 3 & 6 & 1 & 4 \\ 5 & 10 & 1 & 6 \\ 4 & 8 & 1 & 5 \end{pmatrix} \Rightarrow \mathrm{rref}(\mathbf{A}_4) = \begin{pmatrix} 1 & 2 & 0 & 1 \\ 0 & 0 & 1 & 1 \\ 0 & 0 & 0 & 0 \\ 0 & 0 & 0 & 0 \end{pmatrix}.$$

The leading list for $\mathbf{A}$ is $\{1, 3, 5\}$, for $\mathbf{A}_3$ is $\{1, 3\}$, and for $\mathbf{A}_4$ is also $\{1, 3\}$.

Theorem 1.23 (The First Column Rule) *Let $\mathbf{A}$ be an $m \times n$ matrix with the reduced row echelon form $\mathrm{rref}(\mathbf{A})$, k be a positive integer, $k < n$, and $\mathbf{A}_k$ be the $m \times k$ matrix made of the first k columns of $\mathbf{A}$. Then $\mathrm{rref}(\mathbf{A}_k)$ is the $m \times k$ matrix made of the first k columns of $\mathrm{rref}(\mathbf{A})$.*

Proof. The standard algorithm of the reduced row echelon form yields equal results for $\mathbf{A}$ and $\mathbf{A}_k$ in the first k columns.   □

The First Column Rule is especially important when one evaluates the reduced row echelon form of the augmented matrix $(\mathbf{A} \mid \mathbf{b})$. If we apply it with $k = n$, then we obtain that

$$\text{rref}\,((\mathbf{A} \mid \mathbf{b})) = (\text{rref}(\mathbf{A}) \mid \mathbf{d}) \tag{1.71}$$

for some vector $\mathbf{d}$ in $\mathbb{R}^m$. Hence the matrix $\text{rref}(\mathbf{A})$ can be recovered automatically.

By Theorem 1.1 the solutions for $\text{rref}(\mathbf{A})\mathbf{x} = \mathbf{d}$ are solutions for the linear system $\mathbf{A}\mathbf{x} = \mathbf{b}$ and vice-versa. By Corollary 1.1 we have $\text{rref}(\mathbf{A}|\mathbf{0}) = (\text{rref}(\mathbf{A})|\mathbf{0})$. Therefore, to save time one usually drops the last zero column of $(\mathbf{A}|\mathbf{0})$ in row operations.

We denote by J the complete list of all indices j_s

$$1 \le j_1 < j_2 < \cdots < j_r \le n \tag{1.72}$$

such that the column $\text{col}_{j_s}\,(\text{rref}(\mathbf{A})) = \mathbf{e}_s$, contains a leading one of $\text{rref}(\mathbf{A})$. The list (1.72) is called the **leading list** for $\mathbf{A}$. The **complementary list** J' is the increasing sequence of positive integers not exceeding n, which are not in J. The complementary list may be empty. The leading list is empty only for zero matrices. Since the leading list is determined by $\text{rref}(\mathbf{A})$, it is the same for any row equivalent matrices.

Theorem 1.24 (The Second Column Rule) *Let $\mathbf{A}$ be an $m \times n$ matrix. Let $\mathbf{C}$ be an $m \times (n - k)$ matrix obtained from $\text{rref}(\mathbf{A})$ by deleting k columns with indices in the complementary list. Let $\mathbf{B}$ be the $m \times (n - k)$ matrix obtained from $\mathbf{A}$ by deleting k columns with the same indices. Then*

$$\mathbf{B} \sim \mathbf{C} = \text{rref}(\mathbf{B}).$$

Proof. The standard algorithm of reduced row echelon form applied to $\mathbf{A}$ gives the same entries in $\mathbf{B}$, since the leading ones of $\text{rref}(\mathbf{A})$ are the same as the leading ones of $\mathbf{C}$.   □

If we delete the second column of $\mathbf{A}$ in (1.70), then we obtain that

$$\text{rref}\begin{pmatrix} \overset{1\ 3\ 4\ 5}{\underset{\downarrow\downarrow\downarrow\downarrow}{}} \\ \begin{pmatrix} 9 & 2 & 11 & 2 \\ 3 & 1 & 4 & 0 \\ 5 & 1 & 6 & 1 \\ 4 & 1 & 5 & 1 \end{pmatrix} \end{pmatrix} = \begin{matrix} \overset{1\ 3\ 4\ 5}{\underset{\downarrow\downarrow\downarrow\downarrow}{}} \\ \begin{pmatrix} 1 & 0 & 1 & 0 \\ 0 & 1 & 1 & 0 \\ 0 & 0 & 0 & 1 \\ 0 & 0 & 0 & 0 \end{pmatrix} \end{matrix}. \tag{1.73}$$

If we delete the second column in $\mathbf{A}_4$ above, then we conclude that

$$\mathbf{B} = \begin{pmatrix} 9 & 2 & 11 \\ 3 & 1 & 4 \\ 5 & 1 & 6 \\ 4 & 1 & 5 \end{pmatrix} \Rightarrow \mathrm{rref}(\mathbf{B}) = \begin{pmatrix} 1 & 0 & 1 \\ 0 & 1 & 1 \\ 0 & 0 & 0 \\ 0 & 0 & 0 \end{pmatrix}.$$

Let $\mathbf{A}$ be a matrix in reduced row echelon form. If we draw a line between any two rows then the two matrices obtained are both in reduced row echelon form:

$$\mathbf{A} = \left(\begin{array}{c} 1\ 0\ 1\ 0 \\ 0\ 1\ 1\ 0 \\ \hline 0\ 0\ 0\ 1 \\ 0\ 0\ 0\ 0 \end{array} \right) \tag{1.74}$$

Theorem 1.25 (The Row Rule) *Let* $\mathbf{A}$ *be an* $m \times n$ *such that* $\mathbf{A} = \mathrm{rref}(\mathbf{A})$. *Let* $\mathbf{B}$ *be an* $(m - k) \times n$ *matrix obtained by deleting any* k *rows from* $\mathbf{A}$. *Then*

$$\mathbf{B} = \mathrm{rref}(\mathbf{B}).$$

Proof. Dropping any row in a reduced row echelon form either do not affect the leading ones (if we drop the zero row) or decreases their number by one. The remaining leading ones are still moved right for the rows with greater indices. □

For example, if we drop the first and the forth rows in the matrix

$$\begin{pmatrix} 1 & 2 & 0 & 1 & 0 \\ 0 & 0 & 1 & 1 & 0 \\ 0 & 0 & 0 & 0 & 1 \\ 0 & 0 & 0 & 0 & 0 \end{pmatrix},$$

then we obtain the matrix

$$\begin{pmatrix} 0 & 0 & 1 & 1 & 0 \\ 0 & 0 & 0 & 0 & 1 \end{pmatrix},$$

which is in reduced row echelon form.

Problems

Prob. 58 — Matrices

$$A = \begin{pmatrix} * & * & 2 & * & 4 \\ * & * & 3 & * & 1 \\ * & * & * & * & 5 \end{pmatrix} \text{ and } B = \begin{pmatrix} * & * & 2 & * & 1 \\ * & * & 1 & * & 1 \\ * & * & * & * & 1 \end{pmatrix}$$

are in reduced row echelon form. Find **A** and **B**.

Prob. 59 — Matrices

$$A = \begin{pmatrix} * & * & 4 \\ * & * & 2 \end{pmatrix} \text{ and } B = \begin{pmatrix} 1 & 2 & * & * \\ 0 & 0 & * & * \\ * & * & * & 1 \end{pmatrix}$$

are in reduced row echelon form. Find **A** and **B**.

Prob. 60 — Matrices

$$A = \begin{pmatrix} * & * & * & 6 \\ * & * & 0 & 4 \\ 0 & 0 & 1 & 2 \\ * & * & * & * \end{pmatrix} \text{ and } B = \begin{pmatrix} 1 & * & 2 & * & 0 \\ * & 1 & 2 & * & 5 \\ * & * & 0 & 1 & 3 \end{pmatrix}$$

are in reduced row echelon form. Find **A** and **B**.

Prob. 61 — Find the reduced row echelon form for the matrix

$$\begin{pmatrix} 5 & 2 & -1 \\ 2 & 2 & 2 \\ -1 & 2 & 5 \end{pmatrix}.$$

Prob. 62 — Find the reduced row echelon form for the matrix

$$\begin{pmatrix} 3 & 2 & -1 & 2 & 0 \\ 4 & 1 & 0 & -3 & 0 \\ 2 & -1 & -2 & 1 & 1 \\ 3 & 1 & 3 & -9 & -1 \\ 3 & -1 & -5 & 7 & 2 \end{pmatrix}.$$

Prob. 63 — Describe all possible types of reduced row echelon forms for 2×2 matrices.

Prob. 64 — Evaluate the reduced row echelon form of the matrix

$$\begin{pmatrix} 2 & 1 & -1 & -1 & 1 & 1 \\ 1 & -1 & 1 & 1 & -2 & 0 \\ 3 & 3 & -3 & 3 & 4 & 2 \\ 4 & 5 & -5 & -5 & 7 & 3 \end{pmatrix}$$

1.9 The Leading and Complementary Lists

Problem 1.26 Find the leading and complementary lists for the matrix

$$\mathbf{A} = \begin{pmatrix} 1 & 1 & 5 & -4 & 2 \\ 8 & 4 & 28 & -19 & 14 \\ 3 & 1 & 9 & -9 & 4 \\ 3 & 2 & 12 & -9 & 5 \end{pmatrix}$$

Solutions: Elementary row operations show that

$$\mathrm{rref}(\mathbf{A}) = \begin{pmatrix} 1 & 0 & 2 & 0 & 0 \\ 0 & 1 & 3 & 0 & 0 \\ 0 & 0 & 0 & 1 & 0 \\ 0 & 0 & 0 & 0 & 1 \end{pmatrix}.$$

It follows that $j_1 = 1, j_2 = 2, j_3 = 4, j_4 = 5$. Hence $J = \{1,2,4,5\}$ and $J' = \{3\}$. □

Problem 1.27 The matrix

$$\mathbf{A} = \begin{pmatrix} * & * & * & 1 & * & 2 \\ * & * & * & 2 & * & 1 \\ * & * & * & 3 & * & 2 \\ * & * & * & * & 1 & 1 \end{pmatrix}$$

is in a reduced row echelon form. Find $\mathbf{A}$ and its leading list.

Solutions: There are no zero rows in the 4×6 matrix $\mathbf{A}$. It follows that there are four leading ones. The first three rows may have the leading ones only in the first three columns. It follows that

$$a_{12} = a_{13} = a_{21} = a_{23} = a_{31} = a_{32} = 0 \quad \text{and} \quad a_{11} = a_{22} = a_{33} = 1.$$

The column $\mathrm{col}_4(\mathbf{A})$ cannot contain the leading one of $\mathbf{A}$, which implies that $a_{44} = 0$ and that $a_{45} = 1$ is the leading one of $\mathrm{row}_4(\mathbf{A})$. Hence $a_{15} = a_{25} = a_{35} = 0$. It follows that $J = \{1,2,3,5\}$ and that

$$\mathbf{A} = \begin{pmatrix} 1 & 0 & 0 & 1 & 0 & 2 \\ 0 & 1 & 0 & 2 & 0 & 1 \\ 0 & 0 & 1 & 3 & 0 & 2 \\ 0 & 0 & 0 & 0 & 1 & 1 \end{pmatrix} \quad □$$

Definition 1.21 Given a linear system $\mathbf{Ax} = \mathbf{b}$ of m equations in n unknowns, the ith coordinate x_i of $\mathbf{x}$ is called a **leading variable** if i is in the leading list J. Then the ith column of $\mathbf{A}$ is called the **leading column** of $\mathbf{A}$.

The ith coordinate x_i of $\mathbf{x}$ is called a **non-leading variable** if i is in the complementary list J'. Then the ith column of $\mathbf{A}$ is called the **non-leading column** of $\mathbf{A}$. Non-leading variables are often called **free variables**.

Remark Notice that the **leading columns** are columns of the initial matrix $\mathbf{A}$. We determine whether a given column $\text{col}_j(\mathbf{A})$ is leading by inspecting the column $\text{col}_j(\text{rref}(\mathbf{A}))$. So, $\text{col}_j(\mathbf{A})$ is leading if and only if $\text{col}_j(\text{rref}(\mathbf{A}))$ contains a leading one.

We can determine whether a variable x_i is leading or not only if we find the reduced row echelon form for the coefficient matrix of a given system. In the reduced echelon form given below

$$\begin{pmatrix} \mathbf{x_1} & \mathbf{x_2} & \mathbf{x_3} & \mathbf{x_4} & \mathbf{x_5} & \mathbf{x_6} & \mathbf{x_7} & \mathbf{x_8} & \mathbf{x_9} & \mathbf{x_{10}} \\ \downarrow & \downarrow & \downarrow & \downarrow & \downarrow & \downarrow & \downarrow & \downarrow & \downarrow & \downarrow \\ 0 & 1 & 2 & 0 & 1 & 3 & 0 & 1 & 2 & 0 \\ 0 & 0 & 0 & 1 & 1 & 2 & 0 & 5 & 4 & 0 \\ 0 & 0 & 0 & 0 & 0 & 0 & 1 & 2 & 3 & 0 \\ 0 & 0 & 0 & 0 & 0 & 0 & 0 & 0 & 0 & 1 \end{pmatrix} \tag{1.75}$$

x_2, x_4, x_7, x_{10} are the **leading variables**. The total number of leading variables for any matrix equals the number of non-zero rows of the reduced row echelon form.

Definition 1.22 The number $r = \text{rank}(\mathbf{A})$ of non-zero rows in $\text{rref}(\mathbf{A})$ is called the **rank** of $\mathbf{A}$.

For example, the rank of the matrix in (1.75) is four. We denote the leading variables by $x_{j_1}, x_{j_2}, \ldots, x_{j_r}$, where $r = \text{rank}(\mathbf{A})$ and $1 \le j_1 < j_2 < \cdots < j_r \le n$ is the leading list.

The example (1.75) shows that the cases $j_1 > 1$ and $j_r = n$ are possible. Indeed, in (1.75) we have

$$j_1 = 2, \quad j_2 = 4, \quad j_3 = 7, \quad j_4 = 10.$$

The complementary list for (1.75) is

$$\{1, 3, 5, 6, 8, 9\}.$$

If $j_1 > 1$ then x_1 enters all equations with coefficient 0 implying that x_1 is an arbitrary number. It is convenient to allow such an option, especially in theoretical arguments. For instance, we faced it in the proof of Theorem 1.21. If we remove the first row in the matrix (1.75), then the new matrix will have three zero columns at the beginning.

Problems

Prob. 65 — Find the leading and complementary lists for the matrix

$$A = \begin{pmatrix} 1 & -1 & -1 & -1 & 1 & 2 \\ -1 & 1 & 1 & 1 & 1 & 0 \\ 1 & -1 & 1 & 3 & 1 & 4 \end{pmatrix}.$$

Prob. 66 — The matrix below is in reduced row echelon form and its rank is three. Using this information recover the matrix:

$$\begin{pmatrix} * & 2 & * & 1 & * \\ * & * & * & 3 & * \\ * & * & * & * & * \end{pmatrix}.$$

1.10 The Rank of a Matrix

Since the first left non-zero entry of any non-zero row of $\mathrm{rref}(A)$ is the leading one, the rank equals the total number of leading ones in $\mathrm{rref}(A)$. The rank of a matrix equals the number of indices in the leading list. This section includes a number of important facts about the rank of a matrix.

Theorem 1.26 *If* A *is* $m \times n$ *matrix, then*

$$0 \leq \mathrm{rank}(A) \leq \min(m, n).$$

Proof. By definition, $\mathrm{rank}(A)$ is the number of non-zero rows of $\mathrm{rref}(A)$, implying that $\mathrm{rank}(A) \leq m$. Each such a row begins with the leading one. The leading ones are places in separate columns of $\mathrm{rref}(A)$. It follows that $\mathrm{rank}(A) \leq n$.   □

Theorem 1.27 *Let* A *be an* $m \times n$ *matrix with* m *linear independent rows. Then* $\mathrm{rank}(A) = m$ *and* $m \leq n$.

Proof. By Theorem 1.8 the rows of rref$(\mathbf{A})$ are linearly independent. Hence all rows of rref$(\mathbf{A})$ are non-zero, which implies that rank$(\mathbf{A}) = m$. Theorem 1.26 shows that

$$m = \text{rank}(\mathbf{A}) \le \min(m,n) \le n. \qquad \square.$$

Corollary 1.5 *Let $S = \{\mathbf{v}_1,\ldots,\mathbf{v}_m\}$ be any set of rows in $^n\mathbb{R}$ and let $m > n$. Then the set S is linearly dependent.*

Proof. Suppose that the set S is linearly independent and consider the $m \times n$ matrix $\mathbf{A}$ with row$_i(\mathbf{A}) = v_i$, $i = 1,\ldots,m$. By Theorem 1.27, $m \le n$, which is a contradiction. $\qquad \square$

If $\mathbf{A}$ is an $m \times n$ matrix of rank r then the $r \times n$ matrix

$$\text{rref}_p(\mathbf{A})$$

made of the first non-zero r rows of rref$(\mathbf{A})$ is called the **principal part** of rref$(\mathbf{A})$. It is clear that

$$\mathbf{A} \sim \text{rref}(\mathbf{A}) = \begin{pmatrix} \text{rref}_p(\mathbf{A}) \\ \mathbf{0}_{(m-r)\times n} \end{pmatrix}. \qquad (1.76)$$

By the Row Rule the matrix rref$_p(\mathbf{A})$ is in reduced row echelon form.

Theorem 1.28 *Let $\mathbf{A}$ and $\mathbf{B}$ be matrices with equal number of columns. Then* rref$_p(\mathbf{A}) = $ rref$_p(\mathbf{B})$ *if and only if* $\mathbf{row}(\mathbf{A}) = \mathbf{row}(\mathbf{B})$.

Proof. Let $\mathbf{A}$ be an $m \times n$ matrix and $\mathbf{B}$ be a $k \times n$ matrix. By Theorem 1.26 the rank rank$(\mathbf{A})$ satisfies inequalities rank$(\mathbf{A}) \le \min(m,n)$. Similarly, rank$(\mathbf{B}) \le \min(k,n)$ By Theorem 1.22 and by (1.76) we have

$$\mathbf{row}(\mathbf{A}) = \mathbf{row}\left(\text{rref}_p(\mathbf{A})\right) \quad \text{and} \quad \mathbf{row}(\mathbf{B}) = \mathbf{row}\left(\text{rref}_p(\mathbf{B})\right). \qquad (1.77)$$

It follows that $\mathbf{row}(\mathbf{A}) = \mathbf{row}(\mathbf{B})$ if rref$_p(\mathbf{A}) = $ rref$_p(\mathbf{B})$.

Suppose that $\mathbf{row}(\mathbf{A}) = \mathbf{row}(\mathbf{B})$. Let

$$\mathbf{C} = \begin{pmatrix} \text{rref}_p(\mathbf{A}) \\ \mathbf{0}_{s\times n} \end{pmatrix} \quad \text{and} \quad \mathbf{D} = \begin{pmatrix} \text{rref}_p(\mathbf{A}) \\ \mathbf{0}_{t\times n} \end{pmatrix},$$

where the nonnegative integers s and t satisfy

$$\text{rank}(\mathbf{A}) + s = \text{rank}(\mathbf{B}) + t,$$

implying that the sizes of the matrices $\mathbf{C}$ and $\mathbf{D}$ are the same. Both matrices $\mathbf{C}$ and $\mathbf{D}$ are in reduced row echelon form. By (1.77) they have equal row spaces. By Theorem 1.21 we conclude that $\mathbf{C} = \mathbf{D}$. It follows that their principal parts must be equal too, as stated. $\square$

Theorem 1.29 (Kronecker-Capelli) *A system* $\mathbf{Ax} = \mathbf{b}$ *is consistent if and only if the rank of the coefficient matrix* $\mathbf{A}$ *of the system equals the rank of the augmented matrix* $(\mathbf{A}|\mathbf{b})$:

$$\text{rank}(\mathbf{A}) = \text{rank}((\mathbf{A}|\mathbf{b})). \tag{1.78}$$

Proof. Let $r = \text{rank}(\mathbf{A})$. Since $(\mathbf{A}|\mathbf{b})$ is obtained from $\mathbf{A}$ by adding only one column, its size is $m \times (n+1)$. By the First Column Rule, see Theorem 1.23 and (1.71), the elementary row operations which transform $(\mathbf{A}|\mathbf{b})$ to rref$((\mathbf{A}|\mathbf{b}))$ also transform $\mathbf{A}$ to rref$(\mathbf{A})$. The last column of rref$(\mathbf{A}|\mathbf{b})$ can either have zeros below the entry $(r, n + 1)$ or not. In the first case we obtain (1.78), whereas in the second the $(r + 1)$th row of rref$(\mathbf{A}|\mathbf{b})$ must be

$$\begin{pmatrix} 0 & 0 & \cdots & 0 & 1 \end{pmatrix}. \tag{1.79}$$

The corresponding equation

$$0 \cdot x_1 + 0 \cdot x_2 + \cdots + 0 \cdot x_n = 1$$

has no solutions implying the inconsistency of $\mathbf{Ax} = \mathbf{b}$. It follows that if $\mathbf{Ax} = \mathbf{b}$ is consistent, then (1.78) holds.

If (1.78) holds true then rref$(\mathbf{A}|\mathbf{b}) = (\text{rref}(\mathbf{A})|\mathbf{d})$ has r non-zero rows only. We put all non-leading variables to be equal zero. Then the system corresponding to the augmented matrix $(\text{rref}(\mathbf{A})|\mathbf{d})$ has a solution:

$$x_{j_1} = d_1, \ x_{j_2} = d_2, \ \ldots, \ x_{j_r} = d_r; \ x_i = 0 \ \text{for the rest indices } i, \tag{1.80}$$

implying the consistency of $\mathbf{Ax} = \mathbf{b}$. $\square$

In practice, the Kronecker-Capelli Theorem is used to determine the inconsistency of systems of linear equations. If, after a series of row operations applied to the augmented matrix of a system, one obtains a row like in (1.79), the system is definitely inconsistent. Furthermore, the Kronecker-Capelli Theorem states that it is both a necessary and sufficient condition for the inconsistency of a system. See Problem 1.4 as an example.

There is another important $n \times n$ matrix $\text{eref}(\mathbf{A})$ which is associated with $\text{rref}_p(\mathbf{A})$. The matrix $\text{rref}_p(\mathbf{A})$ has $r = \text{rank}(\mathbf{A})$ rows. By Theorem 1.26 the rank r of $\mathbf{A}$ cannot exceed n. There are exactly r numbers in the leading list J. Recall that $\mathbf{f}_i = \mathbf{e}_i^T$ is the row of length n with all entries zeros except for entry i, which is 1. Then the matrix $\text{eref}(\mathbf{A})$ is uniquely determined by its rows:

$$\text{row}_i\,(\text{eref}(\mathbf{A})) = \begin{cases} \mathbf{f}_i, \text{ if } i \text{ is in } J', \\ \text{row}_s\,(\text{rref}(\mathbf{A})) \text{ if } i = j_s \text{ is in } J. \end{cases} \tag{1.81}$$

The matrix $\text{eref}(\mathbf{A})$ is obtained from $\text{rref}_p(\mathbf{A})$ by inserting $n - r$ rows of the standard row basis in addition to the rows of $\text{rref}_p(\mathbf{A})$ so that all leading ones become placed on the main diagonal.

> **Definition 1.23** The $n \times n$ matrix $\text{eref}(\mathbf{A})$ is called the **extended row echelon form** for the matrix $\mathbf{A}$.

For matrices $\mathbf{A}$ and $\mathbf{B}$ in (1.66), and for $\mathbf{C}$ in (1.67) we have

$$\mathbf{A} = \begin{pmatrix} 0\ 1\ 0\ 4 \\ 0\ 0\ 1\ 2 \\ 0\ 0\ 0\ 0 \end{pmatrix} \begin{matrix} \leftarrow 2 \\ \leftarrow 3 \end{matrix} \Rightarrow \text{eref}(\mathbf{A}) = \begin{pmatrix} 1\ 0\ 0\ 0 \\ 0\ 1\ 0\ 4 \\ 0\ 0\ 1\ 2 \\ 0\ 0\ 0\ 1 \end{pmatrix} \begin{matrix} \\ \leftarrow 2 \\ \leftarrow 3 \\ \ \end{matrix}$$

$$\mathbf{B} = \begin{pmatrix} 1\ 9\ 0\ 0\ 8 \\ 0\ 0\ 1\ 0\ 4 \\ 0\ 0\ 0\ 1\ 0 \end{pmatrix} \begin{matrix} \leftarrow 1 \\ \leftarrow 3 \\ \leftarrow 4 \end{matrix} \Rightarrow \text{eref}(\mathbf{B}) = \begin{pmatrix} 1\ 9\ 0\ 0\ 8 \\ 0\ 1\ 0\ 0\ 0 \\ 0\ 0\ 1\ 0\ 4 \\ 0\ 0\ 0\ 1\ 0 \\ 0\ 0\ 0\ 0\ 1 \end{pmatrix} \begin{matrix} \leftarrow 1 \\ \\ \leftarrow 3 \\ \leftarrow 4 \\ \ \end{matrix} \tag{1.82}$$

$$\mathbf{C} = \begin{pmatrix} 1\ 1\ 0\ 1\ 0 \\ 0\ 0\ 1\ 0\ 0 \\ 0\ 0\ 0\ 0\ 1 \\ 0\ 0\ 0\ 0\ 0 \\ 0\ 0\ 0\ 0\ 0 \end{pmatrix} \begin{matrix} \leftarrow 1 \\ \leftarrow 3 \\ \leftarrow 5 \end{matrix} \Rightarrow \text{eref}(\mathbf{C}) = \begin{pmatrix} 1\ 1\ 0\ 1\ 0 \\ 0\ 1\ 0\ 0\ 0 \\ 0\ 0\ 1\ 0\ 0 \\ 0\ 0\ 0\ 1\ 0 \\ 0\ 0\ 0\ 0\ 1 \end{pmatrix} \begin{matrix} \leftarrow 1 \\ \\ \leftarrow 3 \\ \\ \leftarrow 5 \end{matrix} \tag{1.83}$$

If $\mathbf{A} = \mathbf{0}_{m \times n}$, then the complementary list is $\{1, 2, \dots, n\}$, which implies by (1.81) that $\text{eref}(\mathbf{0}_{m \times n}) = \mathbf{I}_n$. Notice that $\text{rref}_p(\mathbf{0}_{m \times n})$ is the matrix with no rows in it.

The matrix $\text{eref}(\mathbf{A})$ is uniquely determined by $\text{rref}(\mathbf{A})$. However, in general, it is not possible to recover $\text{rref}(\mathbf{A})$ by the matrix $\text{eref}(\mathbf{A})$. Consider, for example, the linear system

$$0 \cdot x_1 + 0 \cdot x_2 + 1 \cdot x_1 = 0.$$

Its coefficient matrix is

$$\mathbf{A} = \begin{pmatrix} 0\ 0\ 1 \end{pmatrix} = \text{rref}(\mathbf{A}).$$

It follows that the leading list is $J = \{3\}$, the complementary list is $J' = \{1,2\}$. By (1.81)

$$\begin{aligned}\mathrm{row}_1(\mathrm{eref}(\mathbf{A})) &= \mathbf{f}_1 \\ \mathrm{row}_2(\mathrm{eref}(\mathbf{A})) &= \mathbf{f}_2 \\ \mathrm{row}_3(\mathrm{eref}(\mathbf{A})) &= \mathrm{row}_1(\mathrm{rref}(\mathbf{A}))\end{aligned} \quad \Rightarrow \mathrm{eref}(\mathbf{A}) = \begin{pmatrix} 1 & 0 & 0 \\ 0 & 1 & 0 \\ 0 & 0 & 1 \end{pmatrix}.$$

Similarly, if $\mathbf{A} = \begin{pmatrix} 0 & 1 & 0 \end{pmatrix}$ then $\mathrm{eref}(\mathbf{A})$ is also the identity matrix $\mathbf{I}_3$.

Let us analyse the example of matrix $\mathrm{eref}(\mathbf{B})$ in (1.82). Its first row includes three non-zero entries, implying that this row is the first row of $\mathbf{B}$. The second row of $\mathrm{eref}(\mathbf{B})$ is not the row of $\mathbf{B}$ since its leading one is placed below 9. Its third row is the second row of $\mathbf{B}$ since it includes two non-zero entries. As to the forth row, the example

$$\mathbf{B}_1 = \begin{pmatrix} 1 & 9 & 0 & 0 & 8 \\ 0 & 0 & 1 & 0 & 4 \\ 0 & 0 & 0 & 0 & 0 \end{pmatrix} \begin{matrix} \leftarrow 1 \\ \leftarrow 3 \\ \ \end{matrix} \Rightarrow \mathrm{eref}(\mathbf{B}_1) = \begin{pmatrix} 1 & 9 & 0 & 0 & 8 \\ 0 & 1 & 0 & 0 & 0 \\ 0 & 0 & 1 & 0 & 4 \\ 0 & 0 & 0 & 1 & 0 \\ 0 & 0 & 0 & 0 & 1 \end{pmatrix} \begin{matrix} \leftarrow 1 \\ \ \\ \leftarrow 3 \\ \ \\ \ \end{matrix}$$

shows that $\mathrm{eref}(\mathbf{B}_1) = \mathrm{eref}(\mathbf{B})$.

However, if the leading list is known, then one can recover $\mathrm{rref}(\mathbf{A})$ uniquely. For this it is sufficient to cross out in the matrix $\mathrm{eref}(\mathbf{A})$ the rows with indexes in the complementary list.

The matrix $\mathrm{hef}(\mathbf{A})$ is uniquely determined by its rows:

$$\mathrm{row}_i(\mathrm{hef}(\mathbf{A})) = \begin{cases} \mathbf{0}, & \text{if } i \text{ is in } J', \\ \mathrm{row}_s(\mathrm{rref}(\mathbf{A})) & \text{if } i = j_s \text{ is in } J. \end{cases} \tag{1.84}$$

Definition 1.24 The $n \times n$ matrix $\mathrm{hef}(\mathbf{A})$ is called the **Hermite echelon form** for the matrix $\mathbf{A}$.

$$\mathbf{C} = \begin{pmatrix} 1 & 1 & 0 & 1 & 0 \\ 0 & 0 & 1 & 0 & 0 \\ 0 & 0 & 0 & 0 & 1 \\ 0 & 0 & 0 & 0 & 0 \\ 0 & 0 & 0 & 0 & 0 \end{pmatrix} \begin{matrix} \leftarrow 1 \\ \leftarrow 3 \\ \leftarrow 5 \\ \ \\ \ \end{matrix} \Rightarrow \mathrm{hef}(\mathbf{C}) = \begin{pmatrix} 1 & 1 & 0 & 1 & 0 \\ 0 & 0 & 0 & 0 & 0 \\ 0 & 0 & 1 & 0 & 0 \\ 0 & 0 & 0 & 0 & 0 \\ 0 & 0 & 0 & 0 & 1 \end{pmatrix} \begin{matrix} \leftarrow 1 \\ \ \\ \leftarrow 3 \\ \ \\ \leftarrow 5 \end{matrix} \tag{1.85}$$

Problems

Prob. 67 — Suppose that the reduced echelon form of an $m \times n$ matrix $\mathbf{A}$ equals

$$\begin{pmatrix} \mathbf{B} \\ \mathbf{C} \end{pmatrix},$$

where $\mathbf{B}$ is a $k \times n$ matrix and $\mathbf{C}$ is an $(m - k) \times n$ matrix. Is it true that both $\mathbf{B}$ and $\mathbf{C}$ are in reduced row echelon form? Justify your answer.

Prob. 68 — Suppose that a $k \times n$ matrix $\mathbf{B}$ and an $(m - k) \times n$ matrix $\mathbf{C}$ are in reduced row echelon form. Is it true that the matrix

$$\begin{pmatrix} \mathbf{B} \\ \mathbf{C} \end{pmatrix}$$

is in its reduced echelon form? Justify your answer.

Prob. 69 — Let $\mathbf{A}$ and $\mathbf{B}$ be $m \times n$ matrices. Is it true that the $m \times 2n$ matrix

$$\left(\mathrm{rref}(\mathbf{A})\ \mathrm{rref}(\mathbf{B}) \right)$$

is in reduced row echelon form? Justify your answer.

Prob. 70 — Evaluate the reduced row echelon forms of the matrices below:

(a) $\begin{pmatrix} 1 & 1 & 2 & 3 & 1 \\ 3 & -1 & -1 & -2 & -4 \\ 2 & 3 & -1 & -1 & -6 \\ 1 & 2 & 3 & -1 & -4 \end{pmatrix}$
(b) $\begin{pmatrix} 3 & 3 & 2 & 12 & 1 \\ 1 & 1 & 1 & 5 & 1 \\ 2 & 2 & 0 & 4 & -2 \\ 4 & 4 & 0 & 8 & -4 \end{pmatrix}$

(c) $\begin{pmatrix} 1 & 1 & 2 & 3 & 1 \\ 3 & -1 & -1 & -2 & -4 \end{pmatrix}$
(d) $\begin{pmatrix} 2 & 2 & 0 & 4 & -2 \\ 4 & 4 & 0 & 8 & -4 \end{pmatrix}$

Prob. 71 — Solve the system of equations

$$\begin{cases} 2x_1 + 2x_2 + 8x_3 - x_4 + 4x_5 & = 9 \\ 4x_1 - x_2 + x_3 + 2x_4 + 11x_5 & = 10 \\ x_1 - x_2 - 2x_3 + x_4 + 3x_5 & = 1 \\ 5x_1 - 2x_2 - x_3 + 3x_4 + 14x_5 & = 11 \end{cases}$$

Prob. 72 — Solve the system of equations

$$\begin{cases} 4x_1 + 8x_2 + 10x_3 + 14x_4 & = 6 \\ 5x_1 + 10x_2 + 10x_3 + 15x_4 & = 8 \\ 9x_1 + 18x_2 + 22x_3 + 31x_4 & = 16 \\ 3x_1 + 6x_2 + 8x_3 + 11x_4 & = 5 \end{cases}$$

Prob. 73 — Find all values for the parameters a, b, c, d such the system of equations

$$\begin{cases} 4x_1 + 8x_2 + 10x_3 + 14x_4 & = a \\ 5x_1 + 10x_2 + 10x_3 + 15x_4 & = b \\ 9x_1 + 18x_2 + 22x_3 + 31x_4 & = c \\ 3x_1 + 6x_2 + 8x_3 + 11x_4 & = d \end{cases}$$

has a solution.

Prob. 74 — Evaluate the reduced row echelon form of the matrix

$$\mathbf{A} = \begin{pmatrix} 1 & 3 & -1 & 2 & 1 \\ -1 & -3 & 1 & -2 & -1 \\ 1 & 3 & 0 & 1 & 2 \\ 1 & 2 & 2 & -1 & 4 \end{pmatrix}. \tag{1.86}$$

Evaluate $\text{rank}(\mathbf{A})$ and $\text{rref}_p(\mathbf{A})$.

Prob. 75 — Run the proof of Theorem 1.21 to show that the reduced row echelon form of the matrix (1.86) is unique.

Prob. 76 — The reduced row echelon form of the matrix $\mathbf{C}$ below is $\mathbf{C}$. Unknown entries of $\mathbf{C}$ are indicated by $*$.

$$\mathbf{C} = \begin{pmatrix} * & * & * & 1 & 2 \\ * & 1 & * & 1 & 1 \\ * & * & * & 2 & 1 \end{pmatrix}.$$

Assuming that $\mathbf{C}$ is the reduced row echelon form of the augmented matrix of a system of linear equations, $\mathbf{Ax} = \mathbf{b}$ write down the solution of the system. Determine all vectors $\mathbf{b}$ in $\mathbb{R}^3$ such that the system $\mathbf{Cx} = \mathbf{b}$ has a solution.

Prob. 77 — Answer the following questions in True or False. Justify your answers.

(1) Every matrix has a unique row echelon form.
(2) Elementary row operations applied to a matrix in row echelon form result in a matrix in row echelon form.
(3) Any matrix in reduced row echelon form is a matrix in row echelon form.
(4) For any matrix $\mathbf{A}$
$$\text{rref}(\mathbf{A}^T) = \text{rref}(\mathbf{A})^T.$$

(5) Let $\mathbf{A}$ be an $n \times n$ matrix such that $\text{rref}(\mathbf{A})$ contains exactly n leading ones. Then the homogeneous system $\mathbf{Ax} = \mathbf{0}$ has only one solution $\mathbf{x} = \mathbf{0}$.

(6) Leading ones in a matrix in row echelon form are placed in different columns.

(7) Let $\mathbf{A}$ be an $m \times n$ matrix in reduced row echelon form such that each column of $\mathbf{A}$ contains a leading one. Then all entries of $\mathbf{A}$ that are not equal to leading ones are zeros.

(8) If a linear system has more unknowns than equations, then it has infinitely many solutions.

(9) Let $\mathbf{C}$ be the augmented matrix of the system $\mathbf{Ax} = \mathbf{b}$. If $\mathbf{C}$ has a row of zeros, then the system $\mathbf{Ax} = \mathbf{b}$ has infinitely many solutions.

(10) If $\mathbf{A}$ is an $n \times n$ matrix in row echelon form then $\mathbf{A}^T$ is also in row echelon form.

Prob. 78 — Let $\mathbf{A}$ be an $m \times n$ matrix of rank r. Is it true that the reduced row echelon form of $\mathrm{rref}(\mathbf{A})^T$ is the matrix

$$\begin{pmatrix} \mathbf{I}_r & \mathbf{0}_{r \times (n-r)} \\ \mathbf{0}_{(m-r) \times r} & \mathbf{0}_{(m-r) \times (n-r)} \end{pmatrix} ?$$

Prob. 79 — Let

$$\mathbf{A} = \begin{pmatrix} 1 & 2 & 1 & 5 & 3 \\ 2 & 3 & 2 & 0 & 1 \\ 4 & 3 & 1 & 3 & 2 \end{pmatrix}, \quad \mathbf{B} = \begin{pmatrix} 3 & 5 & 3 & 5 & 4 \\ 2 & 2 & 1 & 1 & 1 \\ 7 & 8 & 4 & 8 & 5 \\ 5 & 6 & 3 & 7 & 4 \\ 1 & 1 & 1 & -5 & -2 \end{pmatrix}.$$

Determine whether the row spaces of these matrices coincide.

Prob. 80 — Find all matrices $\mathbf{X}$ in reduced row echelon form such that $\mathrm{eref}(\mathbf{X}) = \mathrm{eref}(\mathbf{C})$, where $\mathbf{C}$ is a matrix in (1.83).

1.11 Portfolio Theory

A portfolio refers to a collection of financial assets, such as stocks, bonds, cash or other financial instruments, which are all evaluated in cash. It is typically expressed as a row of numbers

$$\begin{pmatrix} 1000 & 2000 & 2000 \end{pmatrix},$$

which, in this particular example, shows that somebody invested $\$1,000$ in stocks, $\$2,000$ in governmental bonds, and $\$2,000$ in a bank deposit. It is clear that the total sum of the investments equals $\$5,000$. Since the invested sum may vary, it is reasonable to consider not specific investments amounts, but rather the proportions of the total amount invested in each of the assets in question. In the example above they make the row

$$\begin{pmatrix} 0.2 & 0.4 & 0.4 \end{pmatrix}$$

of non-negative numbers with the total sum equal 1.

Definition 1.25 Given an $1 \times m$ row $\mathbf{Y}$ is called an **investment portfolio** if all its components y_i are non-negative and

$$y_1 + y_2 + \cdots + y_m = 1.$$

Each of the assets has an expected return. It can be represented as a column

$$\mathbf{R} = \begin{pmatrix} r_1 \\ r_2 \\ \vdots \\ r_m \end{pmatrix},$$

where $r_1, r_2, \ldots, r_m$ are the expected returns of the individual assets. The return $r_1 = 1.05$ means that the investment in the first asset is returned with the interest rate of 5%.

Problem 1.28 An investor plans to invest 40% of the total sum amounting $100,000 to stocks, 30% to governmental bonds, and 30% to a bank deposit. Evaluate the expected return if the interest rates are 5%, 7% and 6% for investments in stocks, bonds, and deposits correspondingly.

Solution The expected return equals:

$$\$100,000 \times \begin{pmatrix} 0.4 & 0.3 & 0.3 \end{pmatrix} \begin{pmatrix} 1.05 \\ 1.07 \\ 1.06 \end{pmatrix} = \$105900,$$

implying that the interest rate of such an investment is 5.9%. $\qquad\qquad\square$

More complicated problems of the Portfolio theory are related to returns matrices with more than one column. Since all investments are directed to the future and since the future may develop in a number of directions, there are usually more than one possible returns columns.

Problem 1.29 An investor plans to invest the total sum amounting $60,000 to stocks, governmental bonds, and a bank deposit. Depending on the results of national elections, the corresponding interest rates will be 4%, 7%, and 6% if the conservative party wins, and 6%, 5%, and 7% if the labour party wins. The investor wants to determine a risk-less portfolio which gives the same and the highest possible return independent of election outcomes. Find this portfolio.

Solution: Let x, y, z be the required proportions of the investment to stocks, bonds and a bank deposit and a be the expected returns for each of two possibilities mentioned. Then we obtain the system of equations

$$\begin{cases} 1.04x + 1.07y + 1.06z & = a \\ 1.06x + 1.05y + 1.07z & = a \,, \\ x + y + z & = 1 \end{cases} \qquad (1.87)$$

where x, y, z are non-negative numbers and a is a positive unknown. Now, we should create the augmented matrix for this system and find its reduced row echelon form. To avoid manual calculation, I recommend the method shown in the solution to Problem 1.3. I extend the coefficient matrix of the system (1.87) by adding the 3×3 identity matrix. I evaluate the reduced row echelon form of the obtained 3×6 matrix by using either a scientific calculator, or Microsoft Math Solver, or Wolfram's Mathematica:

$$\begin{pmatrix} 1.04 & 1.07 & 1.06 & | & 1 & 0 & 0 \\ 1.06 & 1.05 & 1.07 & | & 0 & 1 & 0 \\ 1 & 1 & 1 & | & 0 & 0 & 1 \end{pmatrix} \sim \begin{pmatrix} 1 & 0 & 0 & | & -40 & -20 & 63.8 \\ 0 & 1 & 0 & | & 20 & -40 & 21.6 \\ 0 & 0 & 1 & | & 20 & 60 & -84.4 \end{pmatrix}. \qquad (1.88)$$

The matrix on the right in (1.88) is the reduced row echelon form of the matrix on the left, which implies that

$$\begin{cases} x & = -40a - 20a + 63.8 = 63.8 - 60a \\ y & = 20 - 40a + 21.6 = 21.6 - 20a \\ z & = 20a + 60a - 84.4 = 80a - 84.4 \end{cases} \qquad (1.89)$$

Since the unknowns must be non-negative, we obtain the following restrictions on the parameter a:

$$\begin{cases} a \le \frac{63.8}{60} = 1.06333\ldots \\ a \le \frac{21.6}{20} = 1.08 \\ a \ge \frac{84.4}{80} = 1.055 \end{cases} \Rightarrow 1.055 \le a \le \frac{63.8}{60} = 1.06333\ldots$$

It follows that the maximal possible value of the parameter a is $63.8/60$ implying that $x = 0$, $y = 1/3$, and $z = 2/3$. Therefore, to get the optimal risk-less portfolio, the investor must invest \$20,00 to bonds and \$40,000 to a bank deposit. With such a portfolio the investor will get interest rate of $6.333\ldots\%$ independent of the election results.   □

Definition 1.26 Given an $m \times n$ returns matrix $\mathbf{R}$, a portfolio

$$\mathbf{Y} = \begin{pmatrix} y_1 & y_2 & \cdots & y_m \end{pmatrix}$$

is called a **risk-less portfolio** if all assets y_i are non-negative

$$y_1 + y_2 + \cdots + y_m = 1,$$

and the row $\mathbf{YR}$ has positive equal entries.

For example, in Problem 1.29 we obtained that the risk-less portfolio $\mathbf{Y}$ and the returns matrix $\mathbf{R}$ satisfy

$$\mathbf{YR} = \begin{pmatrix} 0 & 1/2 & 1/3 \end{pmatrix} \begin{pmatrix} 1.04 & 1.06 \\ 1.07 & 1.05 \\ 1.06 & 1.07 \end{pmatrix} = \begin{pmatrix} 1.06333 & 1.06333 \end{pmatrix}.$$

Theorem 1.30 *Let* $\mathbf{R}$ *be an* $m \times n$ *returns matrix and let* $\mathbf{1}_n$ *be the* $n \times 1$ *column with all entries equal* 1. *Then a risk-less portfolio exists if and only if the linear system*

$$\mathbf{R}^T \mathbf{X} = \mathbf{1}_n \tag{1.90}$$

has a solution $\mathbf{X}$ *with non-negative components.*

Proof. If the equation (1.90) has a solution

$$\mathbf{X} = \begin{pmatrix} x_1 & x_2 & \cdots & x_m \end{pmatrix}^T$$

with non-negative components x_i, then they all cannot be zeros, implying that $S = x_1 + x_2 + \cdots + x_m > 0$. Then

$$\mathbf{Y} = \begin{pmatrix} \frac{x_1}{S} & \frac{x_2}{S} & \cdots & \frac{x_m}{S} \end{pmatrix}$$

is a risk-less investment portfolio with return $1/S$ at every component: $\mathbf{YR} = \mathbf{1}^T / S$. The proof in the opposite direction is similar.   □

Let us solve Problem 1.29 iusing Theorem 1.30. The matrix $\mathbf{R}^T$ is a 2×3 matrix:

$$\mathbf{R}^T = \begin{pmatrix} 1.04 & 1.07 & 1.06 \\ 1.06 & 1.05 & 1.07 \end{pmatrix}.$$

Then the augmented matrix of the system $\mathbf{R}^T \mathbf{X} = \mathbf{1}$ is

$$\begin{pmatrix} 1.04 & 1.07 & 1.06 & | & 1 \\ 1.06 & 1.05 & 1.07 & | & 1 \end{pmatrix} \sim \begin{pmatrix} 1 & 0 & \frac{319}{422} & | & \frac{100}{211} \\ 0 & 1 & \frac{54}{211} & | & \frac{100}{211} \end{pmatrix}.$$

It follows that

$$\begin{cases} x_1 & = \frac{200 - 319s}{422} \\ x_2 & = \frac{200 - 108s}{422} \\ x_3 & = s \end{cases},$$

and we obtain non-negative solutions x_1, x_2, x_3 if the parameter s satisfies $0 \le s \le 200/319 \approx 0.626959\ldots$. This implies that

$$S = x_1+x_2+x_3 = \frac{400 - 5s}{422} \Rightarrow \max_{0 \le s \le 200/319} \frac{422}{400 - 5s} = \frac{422}{400 - 5\dfrac{200}{319}} = 1.06333\ldots \quad \square$$

Problem 1.30 Suppose that an investor has assets in $y_1 = $ stocks, $y_2 = $ bonds, and $y_3 = $ deposits, and the returns matrix is

$$\mathbf{R} = \begin{pmatrix} 1.03 & 0.97 & 1 \\ 1.03 & 1.03 & 1.03 \\ 1.1 & 1.16 & 1.13 \end{pmatrix} \tag{1.91}$$

Find the risk-lass portfolio, which gives the maximal value of return.

Solution: We apply Theorem 1.30 and find all solutions to the system $\mathbf{R}^T\mathbf{X} = \mathbf{1}$. The augmented matrix of this system is

$$\begin{pmatrix} 1.03 & 1.03 & 1 & | & 1 \\ 0.97 & 1.03 & 1.16 & | & 1 \\ 1 & 1.03 & 1.13 & | & 1 \end{pmatrix} \sim \begin{pmatrix} 1 & 0 & -1 & | & 0 \\ 0 & 1 & \frac{2.13}{1.03} & | & \frac{1}{1.03} \\ 0 & 0 & 0 & | & 0 \end{pmatrix} \Rightarrow \begin{cases} x_1 & = s \\ x_2 & = \frac{1-2.13s}{1.03} \\ x_3 & = s \end{cases}, 0 \le s \le \frac{1}{2.13}.$$

Therefore,

$$S = x_1 + x_2 + x_3 = \frac{1 - 0.07s}{1.03}, 0 \le s \le \frac{1}{2.13} \Rightarrow$$

$$\max_{0 \le s \le 1/2.13} \frac{1.03}{1 - 0.07s} = \frac{1.03}{1 - \dfrac{0.07}{2.13}} = 1.065.$$

By Theorem 1.30 all risk-less investment portfolio are described by the formula

$$\mathbf{Y} = \left(\frac{x_1}{S} \ \frac{x_2}{S} \ \frac{x_3}{S} \right) = \left(\frac{1.03s}{1-0.07s} \ \frac{1-2.13s}{1-0.07s} \ \frac{1.03s}{1-0.07s} \right), 0 \le s \le \frac{1}{2.13}.$$

Since the return takes the maximal value at $s = 1/2.13$, the best possible risk-less investment portfolio for thai returns matrix is

$$\mathbf{Y} = \begin{pmatrix} 0.5 & 0 & 0.5 \end{pmatrix}. \quad \square$$

Another interesting topic of the Portfolio theory is the arbitrage problem. Suppose that only two assets are available: stocks and banks deposits with the returns 10% in stocks and 5% in banks deposits. Therefore, one can borrow $\$1,000$ in a bank and invest these money in the asset of stocks, which leads to the portfolio $\begin{pmatrix} 1000 & -1000 \end{pmatrix}$. Then the return of this portfolio is

$$\begin{pmatrix} 1000 & -1000 \end{pmatrix} \begin{pmatrix} 1.1 & 1.05 \end{pmatrix}^T = 1100 - 1050 = 50,$$

which leads to the income of \$50 obtained from nothing. To avoid such business, banks increase the rate of borrowing money to 10%. In this case, the above calculations show that the income obtained this way reduces to zero. One can observe a similar system for rates of exchange for different currencies. Dollars are sold for one rate and are purchased for lower rates.

Definition 1.27 An **arbitrage portfolio** $\mathbf{Y} = \begin{pmatrix} y_1 & y_2 & \cdots & y_m \end{pmatrix}$ for an $m \times n$ returns matrix $\mathbf{R}$ is one, which costs nothing

$$y_1 + y_2 + \cdots + y_m = 0,$$

cannot lose, i.e. $\mathbf{YR}$ is a row with non-negative components, and makes profit at least in one of the assets.

Problem 1.31 Suppose that an investor has assets in $y_1 = $ stocks, $y_2 = $ bonds, and $y_3 = $ deposits, and the returns matrix is

$$\mathbf{R} = \begin{pmatrix} 1.03 & 0.97 & 1 \\ 1.03 & 1.03 & 1.03 \\ 1.1 & 1.16 & 1.13 \end{pmatrix} \tag{1.92}$$

Find the arbitrage portfolio, which gives the maximal values of returns at each of three assets.

Solution: The linear system which determines any arbitrage portfolio is given by

$$\begin{cases} y_1 + y_2 + y_3 & = 0 \\ 1.03y_1 + 1.03y_2 + 1.1y_3 & = c_1 \\ 0.97y_1 + 1.03y_2 + 1.16y_3 & = c_2 \\ y_1 + 1.03y_2 + 1.13y_3 & = c_3 \end{cases}, \tag{1.93}$$

where c_1, c_2, c_3 are non-negative numbers at least one of which positive. System (1.93) can be written in the matrix form as follows:

$$\begin{pmatrix} 1 & 1 & 1 \\ 103 & 103 & 110 \\ 97 & 103 & 116 \\ 100 & 103 & 113 \end{pmatrix} \begin{pmatrix} y_1 \\ y_2 \\ y_3 \end{pmatrix} = \begin{pmatrix} 1 & 0 & 0 & 0 \\ 0 & 1 & 0 & 0 \\ 0 & 0 & 1 & 0 \\ 0 & 0 & 0 & 1 \end{pmatrix} \begin{pmatrix} 0 \\ 100c_1 \\ 100c_2 \\ 100c_3 \end{pmatrix}. \tag{1.94}$$

Notice that I multiplied by hundred the last three equations in (1.93) to be able to apply Wolfram's Mathematica or a scientific calculator to get the exact formulas for $\mathbf{Y}$. An application of Wolfram's Mathematica shows that

$$\text{rref}\begin{pmatrix} 1 & 1 & 1 & 1\,0\,0\,0 \\ 103 & 103 & 110 & 0\,1\,0\,0 \\ 97 & 103 & 116 & 0\,0\,1\,0 \\ 100 & 103 & 113 & 0\,0\,0\,1 \end{pmatrix} = \begin{pmatrix} 1\,0\,0 & -\frac{103}{7} & 0 & -\frac{10}{21} & \frac{13}{21} \\ 0\,1\,0 & \frac{213}{7} & 0 & \frac{13}{21} & -\frac{19}{21} \\ 0\,0\,1 & -\frac{103}{7} & 0 & -\frac{1}{7} & \frac{2}{7} \\ 0\,0\,0 & 0 & 1 & 1 & -2 \end{pmatrix}, \qquad (1.95)$$

which implies that

$$\begin{pmatrix} 1\,0\,0 \\ 0\,1\,0 \\ 0\,0\,1 \\ 0\,0\,0 \end{pmatrix} \begin{pmatrix} y_1 \\ y_2 \\ y_3 \end{pmatrix} = \begin{pmatrix} -\frac{103}{7} & 0 & -\frac{10}{21} & \frac{13}{21} \\ \frac{213}{7} & 0 & \frac{13}{21} & -\frac{19}{21} \\ -\frac{103}{7} & 0 & -\frac{1}{7} & \frac{2}{7} \\ 0 & 1 & 1 & -2 \end{pmatrix} \begin{pmatrix} 0 \\ 100c_1 \\ 100c_2 \\ 100c_3 \end{pmatrix}. \qquad (1.96)$$

It follows that

$$\begin{cases} y_1 & = \frac{-1000c_2+1300c_3}{21} \\ y_2 & = \frac{1300c_2-1900c_3}{21} \\ y_3 & = \frac{-100c_2+200c_3}{7} \\ 0 & = 100c_1 + 100c_2 - 200c_3 \end{cases} \qquad (1.97)$$

By Kronecker-Capelli Theorem, see Theorem 1.29, the system (1.93) is consistent if and only if $2c_3 = c_1 + c_2$. Substituting this formula in (1.97), we get

$$\begin{cases} y_1 & = \frac{50}{21}(13c_1 - 7c_2) \\ y_2 & = \frac{50}{21}(7c_2 - 19c_1) \, . \\ y_3 & = \frac{50}{21}(6c_1) \end{cases} \qquad (1.98)$$

Formulas (1.98) show that investment y_3 must be non-negative. Since

$$\mathbf{Y} = \begin{pmatrix} y_1 & y_2 & y_3 \end{pmatrix}$$

is an arbitrage portfolio, the borrowing can be performed from the assets y_1 and y_2. In practice, the size of borrowed sums is usually restricted by some fixed amount $50a/21$, where $a > 0$. In terms of the notation adopted, the returns c_1 and c_2 must satisfy the system of inequalities:

$$\begin{cases} 13c_1 - 7c_2 & \geq -a \\ 7c_2 - 19c_1 & \geq -a \end{cases} \Leftrightarrow \begin{cases} 7c_2 - 13c_1 & \leq a \\ 19c_1 - 7c_2 & \leq a \end{cases}. \qquad (1.99)$$

The set $\mathbf{C}$ of all possible pairs, satisfying (1.99) is bounded by two straight lines and the coordinate axis and is shown in Figure 1.9. It is clear that the largest possible returns for a given a are given by the formulas:

$$c_1 = \frac{a}{3}, c_2 = \frac{16a}{21}, c_3 = \frac{23a}{42}.$$

If, for example, the investor can borrow not more than \$5,000, then $a = 2100$ and the maximal possible returns of the arbitrage portfolio are $c_1 = \$700$, $c_2 =$

$1,600$, $c_3 = \$1,150$. Using formulas (1.98), we conclude that to get such returns, the investor must borrow $\$5,000$ in the assets y_1 and y_2 and invest $\$10,000$ to the asset y_3. $\square$

Problem 1.31 shows that the question of describing all possible arbitrage portfolios is not a simple task. However, there is a relatively simple test to check whether an arbitrage portfolio for a given returns matrix $\mathbf{R}$ exists or not.

Definition 1.28 Let $\mathbf{R}$ be an $m \times n$ matrix, and let

$$\mathbf{1}_m = \left(1\ 1\ \cdots\ 1\right)^T \in \mathbb{R}^m.$$

If the system $\mathbf{RX} = \mathbf{1}_m$ has a solution

$$\mathbf{p} = \left(p_1\ p_2\ \cdots\ p_m\right)^T$$

with all components $p_i > 0$, then these components are called **state prices** and the investor is said to be taken part in a "fair game."

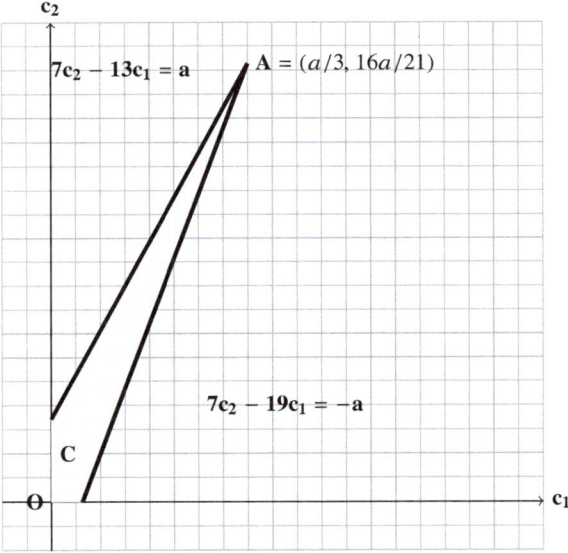

Fig. 1.9: Solutions to (1.99)

Theorem 1.31 (The Non-Arbitrage Theorem) *A necessary and sufficient condition for the **non-existence** of arbitrage portfolios is the **existence** of state prices.*

Proof: State Prices exist $\Rightarrow$ No-Arbitrage Portfolio. This part is a direct consequence of Theorem 1.4 and the following formula. Given an $m \times n$ matrix $\mathbf{R}$, for any row $\mathbf{y} \in {}^m\mathbb{R}$ and any column $\mathbf{p} \in \mathbb{R}^n$ we have

$$\sum_{j=1}^{n}(\mathbf{yR})_j p_j = (\mathbf{yR})\mathbf{p} = \mathbf{y}(\mathbf{Rp}) = \mathbf{y}\mathbf{1}_m = \sum_{i=1}^{m} y_i. \tag{1.100}$$

For any arbitrage vector $\mathbf{y}$ the right-hand side of (1.100) is zero, whereas the left-hand side is positive if a state price vector $\mathbf{p}$ exists. $\square$

The matrix $\mathbf{R}$ in Problem 1.31 has many arbitrage portfolios. Therefore, by already proved part of Theorem 1.31 it cannot have state prices. It is easy to see that

$$\text{rref}\begin{pmatrix} 103 & 97 & 100 & 100 \\ 103 & 103 & 103 & 100 \\ 110 & 116 & 113 & 100 \end{pmatrix} = \begin{pmatrix} 1 & 0 & 1/2 & 0 \\ 0 & 1 & 1/2 & 0 \\ 0 & 0 & 0 & 1 \end{pmatrix}.$$

By Kronecker-Cappelli Theorem, state prices do not exist.

Proof: No-Arbitrage Portfolio $\Rightarrow$ State Prices exist. To get some idea how to prove the opposite direction, let us consider a simple returns matrix:

$$\mathbf{R} = \begin{pmatrix} 1.05 & 1.01 \\ 1.5 & 1.1 \end{pmatrix} \Rightarrow \mathbf{Rp} = p_1 \begin{pmatrix} 1.05 \\ 1.5 \end{pmatrix} + p_2 \begin{pmatrix} 1.01 \\ 1.1 \end{pmatrix} = p_1\mathbf{v}_1 + p_2\mathbf{v}_2. \tag{1.101}$$

Let us observe that the right-hand side of (1.101) spans the cone colored in gray in Figure 1.10. The diagonal of the first quadrant shown in this figure is spanned by the vector $(1,1)^T$. The vector $\mathbf{Y}$ being perpendicular to the diagonal, is a vector of an arbitrage portfolio, since it makes positive angles with the vectors $\mathbf{v}_1$ and $\mathbf{v}_2$. The vector $\mathbf{Y}$ is not an arbitrage portfolio, if the diagonal lies between $\mathbf{v}_1$ and $\mathbf{v}_2$.

The formal proof runs as follows. Let us consider the closed cone

$$\mathbb{R}^{n+1}_+ = \{\mathbf{x} \in \mathbb{R}^{n+1} \mid x_j \geq 0, \, j = 0, 1, \ldots, n\}$$

and a linear subspace L in $\mathbb{R}^{n+1}$ defined by:

$$L = \left\{ \left(-\mathbf{y} \cdot \mathbf{1}_m \, (\mathbf{y} \cdot \mathbf{R})_1 \, \cdots \, (\mathbf{y}\mathbf{R})_n \right)^T \mid \mathbf{y} \in \mathbb{R}^m \right\}.$$

Assuming that arbitrage portfolios do not exist, we have $L \cap \mathbb{R}^{n+1}_+ = \mathbf{0}_{n+1}$. Indeed, if there exists a vector $\mathbf{v} \in L$ with non-negative coordinates and at least one coordinate positive, then $v_j = (\mathbf{y} \cdot \mathbf{R})_j$ are non-negative for every $j = 1, \cdots n$. If $-\mathbf{y} \cdot \mathbf{1}_m = 0$,

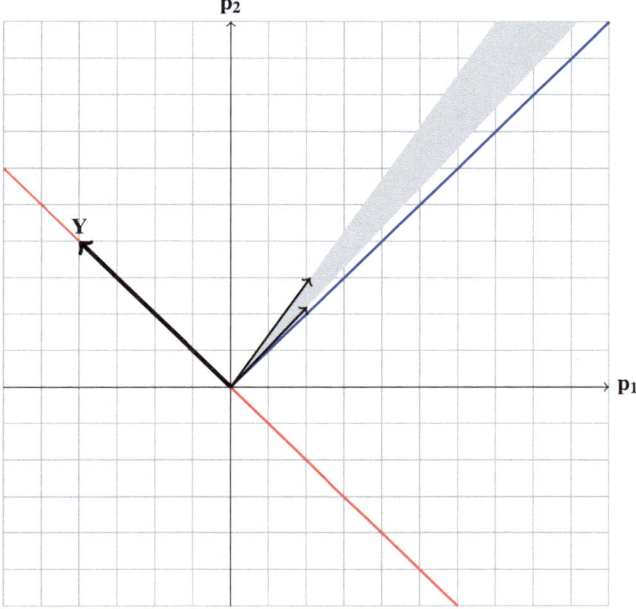

Fig. 1.10: Non-Arbitrage Theorem for $m = 2$

then the portfolio $\mathbf{y}$ is of total zero investment and $(\mathbf{y} \cdot \mathbf{R})_j > 0$ at least for one j, which means that $\mathbf{y}$ is an arbitrage portfolio contradicting our assumption.

If $-\mathbf{y} \cdot \mathbf{1}_m = k > 0$, then we consider the vector $\mathbf{x} = \mathbf{y} + \frac{k}{m} \mathbf{1}_m^T$. Then

$$\mathbf{x} \cdot \mathbf{1}_m = \mathbf{y} \cdot \mathbf{1}_m + \frac{k}{m} \mathbf{1}_m^T \mathbf{1}_m = -k + k = 0$$

and

$$\mathbf{x} \cdot \mathbf{R}_j = \mathbf{y} \cdot \mathbf{R}_j + \frac{k}{m} \mathbf{1}_m^T \cdot \mathbf{R}_j > 0, \ j = 1, \ldots, n,$$

which implies that $\mathbf{x}$ is an arbitrage portfolio, contradicting our assumption.

Theorem 1.32 (Separating Theorem) *For every linear subspace L of $\mathbb{R}^{n+1}$ such that $L \cap \mathbb{R}_+^{n+1} = \mathbf{0}_{n+1}$ there exists a hyperplane H such that $L \subset H$ and $H \cap \mathbb{R}_+^{n+1} = \mathbf{0}_{n+1}$.*

See Narici L. (Beckenstein E.), Theorem 7.7.4, for the proof of this fact of Convex Analysis. A proof is also available in Rockafellar (1990), see Theorem 11.2, Corollary 11.4.2 and Theorem 11.7.

The general formula for a hyperplane in $\mathbb{R}^{n+1}$ defined by a normal vector $\lambda \in \mathbb{R}^{n+1}$ is given by

$$H_\lambda = \left\{ \mathbf{x} \mid \lambda_0 x_0 + \lambda_1 x_1 + \cdots + \lambda_n x_n = 0 \right\}.$$

The condition $H_\lambda \cap \mathbb{R}^{n+1}_+ = \mathbf{0}_{n+1}$ implies that all coefiicients λ_j are non-zero and of the same sign. Since $L \subset H_\lambda$, we obtain that

$$-\lambda_0 \mathbf{y} \cdot \mathbf{1}_m + \sum_{j=1}^{n} \lambda_j \mathbf{y} \cdot \mathbf{R}_j = 0.$$

Since $\mathbf{y}$ is an arbitrary vector in $^m\mathbb{R}$, we get the identity

$$\mathbf{1}_m = \sum_{j=1}^{n} \frac{\lambda_j}{\lambda_0} \mathbf{R}_j,$$

where $p_j = \lambda_j / \lambda_0$ are state prices. $\qquad\square$

See Theorem 2.25 (Farkas' Lemma) for another proof.

Problems

Prob. 81 — Given the returns matrix

$$\mathbf{R} = \begin{pmatrix} 1.05 & 0.95 & 1 \\ 1.05 & 1.05 & 1.05 \\ 1.2 & 1.26 & 1.23 \end{pmatrix},$$

find all risk-less portfolios and arbitrage portfolios if they exist.

Prob. 82 — The director of a fund plans to invest $100 000 before the elections, considering three options: bank deposits, governmental bonds, and land. There are two equally likely outcomes for the elections: either laborists or conservators win. The expected yields from each investment option vary based on the elections outcomes. Bank deposits are 4% under laborists and 5% under conservators. Governmental bonds yield correspondingly 10% and 9%. The land yields 10% and 11%.

(1) Do there exist state prices?
(2) Describe all arbitrage portfolios if exist.
(3) If the director is going to loan $100 000 only from one of the three options specified above and invest this sum into to others then what may be director's best possible investment and the profit?
(4) What is the best possible way for the fund to invest $100 000?

Prob. 83 — Suppose that there are three assets and three states, and the returns matrix is

$$\mathbf{R} = \begin{pmatrix} 1.00 & 1.00 & 1.20 \\ 1.05 & 1.05 & 1.05 \\ 0.95 & 1.10 & 0.95 \end{pmatrix}.$$

Determine if there exists an arbitrage portfolio for the returns matrix $\mathbf{R}$.

Prob. 84 — An investor invests her money in three different assets - land, bonds and stocks. Three possible states can occur. The return matrix for investments is

$$\mathbf{R} = \begin{pmatrix} 1.05 & 1.20 & 1.10 \\ 1.05 & 1.05 & 1.05 \\ 0.90 & 1.05 & 0.95 \end{pmatrix}.$$

(a) Find a state price vector if it exists.
(b) Determine all risk-less portfolios if they exist.
(c) Determine all arbitrage portfolios and find the best portfolio among them, provided that only shares for an amount not exceeding $1000 can be sold.

Chapter 2
Gauss Bases

2.1 Introduction

In Chapter 2 we delve into the geometrical aspects of Gauss-Jordan elimination. The focus is on bases in the row and column spaces of matrices. We discuss how any subspace of $\mathbb{R}^n$ is the null-space of a uniquely determined principle reduced row echelon form $\mathrm{rref}_p(\mathbf{A})$ of some matrix $\mathbf{A}$, resulting in the existence of a unique column Gauss basis for such a subspace. This basis can be constructed using simple operations on $\mathrm{rref}_p(\mathbf{A})$. Similarly, the rows of $\mathrm{rref}_p(\mathbf{A})$ form the basis for the row space of $\mathbf{A}$. Considering that the elements of the null-space $\mathrm{N}(\mathbf{A})$ are orthogonal to the transposes of the rows of $\mathbf{A}$, we arrive at the following geometrical formula

$$\mathbb{R}^n = \mathrm{row}(\mathbf{A})^T \oplus \mathrm{N}(\mathbf{A}),$$

which holds for any $n \times n$ matrix $\mathbf{A}$.

Section 2.3 applies these geometrical concepts to develop a novel method for solving systems of linear equations, employing the so-called extended Gaussian matrix $\mathbf{G}$ associated with the system.

Section 2.4 reveals the ancient problem of the Luoshu Magic Squares, analyzing it through the lens of Linear Algebra.

The geometry of column and row spaces is the focus of Sections 2.5 and 2.6 respectively.

The Rank-Nullity Theorem is presented in Section 2.7.

In Section 2.8, we explore affine subspaces as solutions sets of linear systems.

We examine symmetric matrices from the perspective of Gauss-Jordan elimination in Section 2.9 and establish the Fredholm alternative in Section 2.10.

Section 2.11 solves numerous concrete problems on different versions of Gauss-Jordan elimination.

The chapter concludes with Section 2.12, where we introduce Linear Programming, an important application of Linear Algebra to Mathematical Economics. This field gained prominence with the work of Leonid Kantorovich and Tjalling C. Koopmans, who were jointly awarded the Nobel Prize in Economics in 1975 for their

S. Khrushchev, *Linear Algebra with Applications to Economics*, Classroom Companion: Economics, https://doi.org/10.1007/978-3-031-68682-5_2

contributions to the development of Linear Programming. This section ties in the geometrical ideas developed throughout Chapter 2.

2.2 Subspaces and Bases

Every subspace of **row**(**A**) contains the zero row. Indeed, by Definition 1.10 any subspace is non-empty. Hence there is a row **u** in this subspace. By Definition 1.10 the linear combination $1 \cdot \mathbf{u} - 1 \cdot \mathbf{u} = \mathbf{0}$ is in the subspace as stated.

Definition 2.1 A finite set $\{\mathbf{v}_1, \mathbf{v}_2, \ldots \mathbf{v}_k\}$ of rows (columns) in a subspace V of $\mathbf{row}(\mathbf{I}_n) = {}^n\mathbb{R}$ $(\mathbf{col}(\mathbf{I}_m) = \mathbb{R}^m)$ is called a **basis** for V if every row (column) **v** in V can be **uniquely** represented as the sum

$$\mathbf{v} = \alpha_1 \mathbf{v}_1 + \alpha_2 \mathbf{v}_2 + \cdots + \alpha_k \mathbf{v}_k,$$

where $\alpha_1, \alpha_2, \ldots, \alpha_k$ are real numbers.

For example, the standard basis $\{\mathbf{e}_1, \ldots, \mathbf{e}_m\}$ of columns of $\mathbf{I}_m$ is a basis for $\mathbb{R}^m$. The standard basis $\{\mathbf{f}_1, \ldots, \mathbf{f}_n\}$ of rows of $\mathbf{I}_n$ is a basis for ${}^n\mathbb{R}$. Notice that any linearly independent set $\{\mathbf{v}_1, \mathbf{v}_2, \ldots \mathbf{v}_k\}$ of vectors is a basis for the subspace

$$\text{Lin}\,(\mathbf{v}_1, \mathbf{v}_2, \ldots, \mathbf{v}_k)\,.$$

Any basis is linearly independent. Indeed, the zero row **0** can be represented with $\alpha_i = 0$ for every i. By Definition 2.1 this representation is unique.

Problem 2.1 Let

$$\mathbf{A} = \begin{pmatrix} 5 & 2 & 9 \\ 10 & 4 & 18 \\ 2 & 1 & 3 \\ 7 & 3 & 12 \\ 12 & 5 & 21 \end{pmatrix}.$$

Find rows of **A** which make a basis for **row**(**A**).

Solution: By (1.33)

$$\mathbf{0} = \alpha_1 \text{row}_1(\mathbf{A}) + \alpha_2 \text{row}_2(\mathbf{A}) + \alpha_3 \text{row}_3(\mathbf{A}) + \alpha_4 \text{row}_4(\mathbf{A}) + \alpha_5 \text{row}_5(\mathbf{A}) \quad (2.1)$$

if and only if $\alpha\mathbf{A} = \mathbf{0}$. In other words, (2.1) holds if and only if

$$\begin{cases} 5\alpha_1 + 10\alpha_2 + 2\alpha_3 + 7\alpha_4 + 12\alpha_5 & = 0 \\ 2\alpha_1 + 4\alpha_2 + \alpha_3 + 3\alpha_4 + 5\alpha_5 & = 0 \,. \\ 9\alpha_1 + 18\alpha_2 + 3\alpha_3 + 12\alpha_4 + 21\alpha_5 & = 0 \end{cases} \quad (2.2)$$

Simple calculations show that

$$
\mathrm{rref}\begin{pmatrix} 5 & 10 & 2 & 7 & 12 \\ 2 & 4 & 1 & 3 & 5 \\ 9 & 18 & 3 & 12 & 21 \end{pmatrix} = \begin{matrix} \alpha_1 \ \alpha_2 \ \alpha_3 \ \alpha_4 \ \alpha_5 \\ \downarrow \ \downarrow \ \downarrow \ \downarrow \ \downarrow \\ \begin{pmatrix} 1 & 2 & 0 & 1 & 2 \\ 0 & 0 & 1 & 1 & 1 \\ 0 & 0 & 0 & 0 & 0 \end{pmatrix} \end{matrix}
$$

Let $\alpha_2 = s$, $\alpha_4 = t$, $\alpha_5 = u$. Then $\alpha_1 = -2s - t - 2u$, $\alpha_3 = -t - u$. It follows that

$$
\begin{pmatrix} \alpha_1 \\ \alpha_2 \\ \alpha_3 \\ \alpha_4 \\ \alpha_5 \end{pmatrix} = \begin{pmatrix} -2s - t - 2u \\ s \\ -t - u \\ t \\ u \end{pmatrix} = s\begin{pmatrix} -2 \\ 1 \\ 0 \\ 0 \\ 0 \end{pmatrix} + t\begin{pmatrix} -1 \\ 0 \\ -1 \\ 1 \\ 0 \end{pmatrix} + u\begin{pmatrix} -2 \\ 0 \\ -1 \\ 0 \\ 1 \end{pmatrix} = s\mathbf{v}_1 + t\mathbf{v}_2 + u\mathbf{v}_3
$$

describes all possible solutions to the system (2.2). The particular solution to (2.2) determined by vector $\mathbf{v}_1$ is responsible for the relation

$$
\mathrm{row}_2(\mathbf{A}) = 2\mathrm{row}_1(\mathbf{A}).
$$

The particular solution to (2.2) determined by vector $\mathbf{v}_2$ is responsible for the relation

$$
\mathrm{row}_4(\mathbf{A}) = \mathrm{row}_1(\mathbf{A}) + \mathrm{row}_3(\mathbf{A}).
$$

The particular solution to (2.2) determined by vector $\mathbf{v}_3$ is responsible for the relation

$$
\mathrm{row}_5(\mathbf{A}) = 2\mathrm{row}_1(\mathbf{A}) + \mathrm{row}_3(\mathbf{A}).
$$

The formulas above show that $\mathrm{row}_i(\mathbf{A})$, $i = 2, 4, 5$, can be excluded. It follows that

$$
\mathbf{row}(\mathbf{A}) = \mathrm{Lin}\,(\mathrm{row}_1(\mathbf{A}), \mathrm{row}_3(\mathbf{A})).
$$

Let $\alpha_2 = \alpha_4 = \alpha_5 = 0$. Then $s = t = u = 0$, which implies that $\alpha_1 = -2s - t - 2u = 0$ and $\alpha_3 = -t - u = 0$. Hence the representation of the zero in (2.1) is possible if and only if $\alpha_1 = \alpha_3 = 0$. This shows that $\mathrm{row}_1(\mathbf{A})$ and $\mathrm{row}_3(\mathbf{A})$ make a basis for $\mathbf{row}(\mathbf{A})$. $\qquad\square$

Problem 2.2 Let

$$
\mathbf{A} = \begin{pmatrix} 5 & 2 & 9 \\ 10 & 4 & 18 \\ 2 & 1 & 3 \\ 7 & 3 & 12 \\ 12 & 5 & 21 \end{pmatrix}.
$$

Find columns of $\mathbf{A}$ which make a basis for $\mathbf{col}(\mathbf{A})$.

Solution: By Theorem 1.10 the columns of $\mathbf{A}$ are linearly independent if and only if the equation

$$0 = A\alpha = \alpha_1 \text{col}_1(A) + \alpha_2 \text{col}_2(A) + \alpha_3 \text{col}_3(A)$$

has only zero solution. Elementary calculations show that

$$\text{rref}(A) = \begin{pmatrix} 1 & 0 & 3 \\ 0 & 1 & -3 \\ 0 & 0 & 0 \\ 0 & 0 & 0 \\ 0 & 0 & 0 \end{pmatrix}.$$

Let $\alpha_3 = s$. Then $\alpha_1 = -3s$, $\alpha_2 = 3s$:

$$\begin{pmatrix} \alpha_1 \\ \alpha_2 \\ \alpha_3 \end{pmatrix} = \begin{pmatrix} -3s \\ 3s \\ s \end{pmatrix} = s \begin{pmatrix} -3 \\ 3 \\ 1 \end{pmatrix} = s\mathbf{v}_1.$$

The solution to the equation $A\alpha = 0$ defined by $\mathbf{v}_1$ shows that

$$3\text{col}_1(A) - 3\text{col}_2(A) = \text{col}_3(A),$$

implying that

$$\mathbf{col}(A) = \text{Lin}(\text{col}_1(A), \text{col}_2(A)).$$

Let $\alpha_3 = s = 0$. Then $\alpha_1 = \alpha_2 = 0$. It follows that $\text{col}_1(A)$ and $\text{col}_2(A)$ make a basis for $\mathbf{col}(A)$. □

Lemma 2.1 *Two non-zero columns (rows) $\mathbf{v}_1$ and $\mathbf{v}_2$ are linearly independent if and only if they are not proportional.*

Proof. We consider the case of columns. The case of rows is obtained by transposition of rows into columns $\mathbf{v} \rightarrow \mathbf{v}^T$. Suppose that $\mathbf{v}_1$ and $\mathbf{v}_2$ are linearly independent. Then both $\mathbf{v}_1$ and $\mathbf{v}_2$ are non-zero. If, for instance, $\mathbf{v}_2 = c\mathbf{v}_1$ then $c \neq 0$ and we obtain a non-trivial representation of the zero column: $0 = c\mathbf{v}_1 - \mathbf{v}_2$, which is impossible. Hence $\mathbf{v}_1$ and $\mathbf{v}_2$ cannot be proportional.

Let $\mathbf{v}_1$ and $\mathbf{v}_2$ be non-zero and not proportional. If there is a non-trivial representation of the zero column

$$0 = \alpha_1\mathbf{v}_1 + \alpha_2\mathbf{v}_2,$$

then $\alpha_i \neq 0$ at least for one i, $i = 1, 2$. If, for instance, $\alpha_1 \neq 0$, then $\mathbf{v}_1 = -(\alpha_2/\alpha_1)\mathbf{v}_2$, implying that the rows are proportional. Since this is impossible, the proof is completed. □

Lemma 2.1 can be extended to an arbitrary number of rows (columns).

Lemma 2.2 *A finite set $S = \{\mathbf{v}_1, \mathbf{v}_2, \dots, \mathbf{v}_k\}$ of non-zero columns (rows) in $\mathbb{R}^m$ ($^n\mathbb{R}$) is linearly independent if and only if for every integer i, $i \leq k$, the column (row) $\mathbf{v}_i$ does not belong to $\text{Lin}(\mathbf{v}_1, \mathbf{v}_2, \dots, \mathbf{v}_{i-1})$.*

Proof. Notice that a column $\mathbf{v}_i$ is in $\text{Lin}(\mathbf{v}_1, \mathbf{v}_2, \dots, \mathbf{v}_{i-1})$ if and only if

$$\mathbf{v}_i = \alpha_1 \mathbf{v}_1 + \cdots + \alpha_{i-1} \mathbf{v}_{i-1} \Leftrightarrow \mathbf{v}_i - \alpha_1 \mathbf{v}_1 - \cdots - \alpha_{i-1} \mathbf{v}_{i-1} = \mathbf{0}.$$

Since the last formula gives a non-trivial representation for $\mathbf{0}$, this cannot happen for a linear independent family $\{\mathbf{v}_1, \mathbf{v}_2, \ldots \mathbf{v}_k\}$, which completes the proof in one direction.

Suppose now that $\mathbf{v}_i$ does not belong to $\mathrm{Lin}\,(\mathbf{v}_1, \mathbf{v}_2, \ldots, \mathbf{v}_{i-1})$ for every i, $i = 2, \ldots, k$. If the set $\{\mathbf{v}_1, \mathbf{v}_2, \ldots \mathbf{v}_k\}$ is linearly dependent, then there is a non-trivial representation of the zero:

$$\mathbf{0} = \alpha_1 \mathbf{v}_1 + \cdots + \alpha_k \mathbf{v}_k.$$

Let i be the greatest index such that $\alpha_i \neq 0$. Then

$$\mathbf{v}_i = -\frac{\alpha_1}{\alpha_i} \mathbf{v}_1 - \cdots - \frac{\alpha_{i-1}}{\alpha_i} \mathbf{v}_{i-1} \Rightarrow \mathbf{v}_i \text{ is in } \mathrm{Lin}\,(\mathbf{v}_1, \mathbf{v}_2, \ldots, \mathbf{v}_{i-1}),$$

which is a contradiction. $\square$

Theorem 2.1 *Every non-zero subspace V of $^n\mathbb{R}$ or $\mathbb{R}^m$ has a basis*

$$S = \{\mathbf{v}_1, \ldots, \mathbf{v}_k\}$$

with $k \leq n$ ($k \leq m$).

Proof. Since the case of columns is reduced to the case of rows by transposition, we consider the case of rows only. Since $V \neq \{\mathbf{0}_m\}$, there is a non-zero row $\mathbf{v}_1$ in V. If $V = \mathrm{Lin}(\mathbf{v}_1)$ then the proof is finished. Otherwise, there is a non-zero row $\mathbf{v}_2$ in V, which is not in $\mathrm{Lin}(\mathbf{v}_1)$. By Lemma 2.2 the system $\{\mathbf{v}_1, \mathbf{v}_2\}$ is linearly independent and hence is a basis for $\mathrm{Lin}(\mathbf{v}_1, \mathbf{v}_2)$. If $V = \mathrm{Lin}(\mathbf{v}_1, \mathbf{v}_2)$, then the proof is finished. Otherwise, we continue by induction. Suppose that we constructed i linearly independent vectors $\{\mathbf{v}_1, \ldots, \mathbf{v}_i\}$ in V. Since any system of linearly independent vectors makes a basis for its linear span, the proof is finished if $V = \mathrm{Lin}(\mathbf{v}_1, \ldots, \mathbf{v}_i)$. Otherwise, there is a non-zero vector $\mathbf{v}_{i+1}$ in V, which is not in $\mathrm{Lin}(\mathbf{v}_1, \ldots, \mathbf{v}_i)$. By Lemma 2.2 the system $\{\mathbf{v}_1, \ldots \mathbf{v}_{i+1}\}$ is linearly independent and hence is a basis for $\mathrm{Lin}(\mathbf{v}_1, \ldots, \mathbf{v}_{i+1})$. By Theorem 1.27 we obtain that $i + 1 \leq n$. Therefore, not more than in k steps, $k \leq n$, this construction is finished, which completes the proof of the Theorem. $\square$

Corollary 2.1 *For any non-zero subspace V of $^m\mathbb{R}$ or $\mathbb{R}^n$ there is a matrix $\mathbf{A}$ with linearly independent rows (columns) satisfying*

$$V = \mathbf{row}(\mathbf{A}) \ (V = \mathbf{col}(\mathbf{A})).$$

Proof. By Theorem 2.1 there is a finite basis $\{\mathbf{v}_1, \ldots, \mathbf{v}_k\}$ of V. Let $\mathbf{A}$ be a $k \times n$ matrix with rows $\mathbf{row}(\mathbf{A}) = \mathbf{v}_i$, $i = 1, \ldots, k$. Then $V = \mathbf{row}(\mathbf{A})$. The case of columns is obtained by the transposition. $\square$

Theorem 2.2 *For any non-zero subspace V of $^n\mathbb{R}$ or $\mathbb{R}^m$ the number of elements in any two bases is the same.*

Proof. By Corollary 2.1 any basis with k elements determines a $k \times n$ matrix $\mathbf{A}$ with linearly independent rows equal to the rows of the basis. By Theorem 1.7 the row space of $\mathrm{rref}(\mathbf{A})$ is V. By Theorem 1.8 the rows of $\mathrm{rref}(\mathbf{A})$ are linearly independent. In particular, all rows of $\mathrm{rref}(\mathbf{A})$ are non-zero and, therefore, contain leading ones. Hence $k = \mathrm{rank}(\mathbf{A})$. By Theorem 1.21 there is only one matrix in reduced row echelon form with the row space V. It follows that the number of elements in any basis for V equals the rank of this matrix. $\square$

Definition 2.2 Let V be a subspace of $^n\mathbb{R}$ or $\mathbb{R}^m$. If $V = \{\mathbf{0}\}$, then we say that the dimension of V is zero. If $V \neq \{\mathbf{0}\}$, then the number of elements $\dim(V)$ in any basis for V is called the **dimension** of V. Notice that the subspace $\{\mathbf{0}\}$ does not have a basis. Therefore, we put $\dim(\{\mathbf{0}\}) = 0$.

If V is a subspace of $^n\mathbb{R}$, then $\dim(V) = \mathrm{rank}(\mathbf{A})$ for any matrix $\mathbf{A}$ with the row space V.

Theorem 2.3 *Let V be a subspace of $^n\mathbb{R}$ or $\mathbb{R}^m$. Then every basis for V has $\dim(V)$ elements.*

Proof. Since $\mathbf{v} \rightarrow \mathbf{v}^T$ is a one-to-one linear operation mapping rows to columns, the result follows by Theorem 2.2. $\square$

Corollary 2.2 $\dim(^n\mathbb{R}) = n$ *and* $\dim(\mathbb{R}^m) = m$.

Proof. The number of elements of the standard basis for $^n\mathbb{R} = \mathbf{row}(\mathbf{I}_n)$ equals the number of rows of $\mathbf{I}_n$ which is n. The number of elements of the standard basis for $\mathbb{R}^m = \mathbf{col}(\mathbf{I}_m)$ equals the number of columns of $\mathbf{I}_m$ which is m. $\square$

Corollary 2.3 *Let* $S = \{v_1, \ldots, v_n\}$ *be a set of linearly independent vectors in* $^n\mathbb{R}$ *or in* $\mathbb{R}^n$. *Then* S *is a basis.*

Proof. In case of $^n\mathbb{R}$ we consider the $n \times n$ matrix $\mathbf{A}$ with $\text{row}_i(\mathbf{A}) = v_i$ for $i = 1, \ldots, n$. By Theorem 1.8, n rows of $\text{rref}(\mathbf{A})$ are linearly independent, implying that $\text{rref}(\mathbf{A}) = \mathbf{I}_n$. Hence the row space of $\mathbf{A}$ is $\text{row}(\mathbf{I}_n) = {}^n\mathbb{R}$. Being linearly independent the vectors of S make a basis for $\text{row}(\mathbf{A}) = {}^n\mathbb{R}$. The case of $\mathbb{R}^n$ is reduced to this one by transposition. □

Theorem 2.4 *Given an* $m \times n$ *non-zero matrix* $\mathbf{A}$ *there exists a finite set* $\{v_1, v_2, \ldots, v_k\}$ *of linearly independent rows of* $\mathbf{A}$, *which makes a basis for* $\text{row}(\mathbf{A})$.

Proof. Let v_1 be the first (from the top) non-zero row of $\mathbf{A}$. If $\text{Lin}(v_1) = \text{row}(\mathbf{A})$, then the proof is finished with $k = 1$. If it is not the case, then there must be the first (following v_1 in the bottom direction) non-zero row such that v_2 is not in $\text{Lin}(v_1)$. By Lemma 2.1 the set $\{v_1, v_2\}$ is linearly independent. If $\text{Lin}(v_1, v_2) = \text{row}(\mathbf{A})$, then the proof is finished with $k = 2$. If it is not the case we continue this construction by induction. Suppose that a set of linearly independent rows $\{v_1, v_2, \ldots, v_i\}$ of $\mathbf{A}$ is constructed. Suppose also that they are numbered in the direction from the top to the bottom so that any row of $\mathbf{A}$ placed above row v_i is in

$$W = \text{Lin}(v_1, v_2, \ldots, v_i). \tag{2.3}$$

If $W = \text{row}(\mathbf{A})$ then the prof is finished with $k = i$. Otherwise, there is the first row v_{i+1} of $\mathbf{A}$ placed below v_i, which is not in W. By Lemma 2.2 the set

$$\{v_1, v_2, \ldots, v_{i+1}\}$$

is linearly independent. Since the total number of rows of $\mathbf{A}$ is m, we conclude that $i + 1 \leq m$. It follows that for some $i \leq m$ we must have $W = \text{row}(\mathbf{A})$ (see (2.3)), which completes the proof. □

Corollary 2.4 *Let* $\mathbf{A}$ *be an* $m \times n$ *matrix. Then the rank* $\text{rank}(\mathbf{A})$ *of* $\mathbf{A}$ *is a maximal number of linearly independent rows of* $\mathbf{A}$.

Problems

Prob. 85 — Apply Lemma 2.2 to find a basis for the **row space** of the matrix

$$\mathbf{B} = \begin{pmatrix} 1 & 1 & 0 & 1 & 1 \\ 0 & 1 & 1 & 2 & 3 \\ 1 & 1 & 0 & 1 & 1 \end{pmatrix}^T .$$

Prob. 86 — Find a basis for $\mathbf{col}(\mathbf{B}^T)$.

Prob. 87 — Given that $\{\mathbf{e}_1, \mathbf{e}_2, \mathbf{e}_3\}$ is the standard basis for $\mathbf{col}(\mathbf{I}_3)$ prove that the set

$$\{\mathbf{e}_1 + \mathbf{e}_2, \mathbf{e}_2 + \mathbf{e}_3, \mathbf{e}_3 + \mathbf{e}_1\}$$

is also a basis for $\mathbf{col}(\mathbf{I}_3)$.

Prob. 88 — Given that $\{\mathbf{e}_1, \mathbf{e}_2, \mathbf{e}_3\}$ is the standard basis for $\mathbf{col}(\mathbf{I}_3)$ find necessary and sufficient conditions on real numbers a, b, c in order that

$$\{\mathbf{e}_1 + a\mathbf{e}_2, \mathbf{e}_2 + b\mathbf{e}_3, \mathbf{e}_3 + c\mathbf{e}_1\}$$

be a basis for $\mathbf{col}(\mathbf{I}_3)$. *Hint:* Apply Theorem 1.10.

Prob. 89 — Let $\{\mathbf{v}_1, \mathbf{v}_2, \ldots, \mathbf{v}_k\}$ be a basis for a subspace V of the row space $\mathbf{row}(\mathbf{I}_n)$. Let $\beta_1, \ldots, \beta_k$ be real numbers such that $\beta_j \neq 0$ for some j, $1 \leq j \leq k$. Let

$$\mathbf{u}_i = \begin{cases} \beta_1 \mathbf{v}_1 + \cdots + \beta_j \mathbf{v}_j + \cdots + \beta_k \mathbf{v}_k, & \text{if } i = j \\ \mathbf{v}_i \text{ if } i \neq j \end{cases}$$

Then $\{\mathbf{u}_1, \mathbf{u}_2, \ldots, \mathbf{u}_k\}$ is a basis for V.

Prob. 90 — Let

$$\{\mathbf{v}_1, \mathbf{v}_2, \ldots, \mathbf{v}_k\}$$

be a finite set of rows of an $m \times n$ matrix $\mathbf{A}$ which is a basis for $\mathbf{row}(\mathbf{A})$. Let $\mathbf{E}$ be an elementary $m \times m$. Determine k rows of $\mathbf{EA}$ which make a basis for $\mathbf{row}(\mathbf{EA})$.

2.3 Gauss-Jordan Elimination by the Gauss Matrix

We begin with an example.

Problem 2.3 Find all solutions to the system:

$$\begin{cases} x_1 + x_2 + 5x_3 - x_4 + 2x_5 & = 4 \\ 2x_1 + 4x_3 - x_4 + 4x_5 & = 7 \\ x_1 - x_2 - x_3 - x_4 & = 0 \\ x_1 + 2x_3 - x_4 + x_5 & = 2 \end{cases} \tag{2.4}$$

Solution: We evaluate the reduced row echelon form of the augmented matrix $(\mathbf{A}|\mathbf{b})$ for the system (2.4):

$$
\begin{pmatrix}
1 & 1 & 5 & -1 & 2 & | & 4 \\
2 & 0 & 4 & -1 & 4 & | & 7 \\
1 & -1 & -1 & -1 & 0 & | & 0 \\
1 & 0 & 2 & -1 & 1 & | & 2
\end{pmatrix}
\begin{matrix}
r_1 := r_1 - r_4 \\
r_2 := r_2 - 2r_4 \\
r_3 := r_3 - r_4 \\
\sim
\end{matrix}
\begin{pmatrix}
0 & 1 & 3 & 0 & 1 & | & 2 \\
0 & 0 & 0 & 1 & 2 & | & 3 \\
0 & -1 & -3 & 0 & -1 & | & -2 \\
1 & 0 & 2 & -1 & 1 & | & 2
\end{pmatrix}
\begin{matrix} r_3:=r_3+r_1 \\ \sim \end{matrix}
$$

$$
\begin{pmatrix}
0 & 1 & 3 & 0 & 1 & | & 2 \\
0 & 0 & 0 & 1 & 2 & | & 3 \\
0 & 0 & 0 & 0 & 0 & | & 0 \\
1 & 0 & 2 & -1 & 1 & | & 2
\end{pmatrix}
\begin{matrix} r_4=r_4+r_2 \\ \sim \end{matrix}
\begin{pmatrix}
0 & 1 & 3 & 0 & 1 & | & 2 \\
0 & 0 & 0 & 1 & 2 & | & 3 \\
0 & 0 & 0 & 0 & 0 & | & 0 \\
1 & 0 & 2 & 0 & 3 & | & 5
\end{pmatrix}
\begin{matrix} \mathbf{RO2} \\ \sim \end{matrix}
\begin{pmatrix}
1 & 0 & 2 & 0 & 3 & | & 5 \\
0 & 1 & 3 & 0 & 1 & | & 2 \\
0 & 0 & 0 & 1 & 2 & | & 3 \\
0 & 0 & 0 & 0 & 0 & | & 0
\end{pmatrix}
= \text{rref}\,((\mathbf{A}|\mathbf{b})).
$$

To find solutions for (2.4), we color in red the columns of rref $((\mathbf{A}|\mathbf{b}))$ with the leading ones, in green the columns which do not contain leading ones, and in yellow the data column:

$$
\begin{pmatrix}
1 & 0 & 2 & 0 & 3 & 5 \\
0 & 1 & 3 & 0 & 1 & 2 \\
0 & 0 & 0 & 1 & 2 & 3 \\
0 & 0 & 0 & 0 & 0 & 0
\end{pmatrix}
\tag{2.5}
$$

In (2.5), the variables x_1, x_2 and x_4 are **leading variables**, x_3 and x_5 are **non-leading variables**. We assign x_3, x_5 **arbitrary** real values s, t, and then find the values of leading variables by (2.5):

$$
\boxed{x_1 = 5 - 2s - 3t, \quad x_2 = 2 - 3s - t, \quad x_3 = s, \quad x_4 = 3 - 2t, \quad x_5 = t}.
$$

The expression of this solution in a vector form looks as follows:

$$
\mathbf{x} = \begin{pmatrix} x_1 \\ x_2 \\ x_3 \\ x_4 \\ x_5 \end{pmatrix} = \begin{pmatrix} 5 - 2s - 3t \\ 2 - 3s - t \\ s \\ 3 - 2t \\ t \end{pmatrix} = \begin{pmatrix} 5 \\ 2 \\ 0 \\ 3 \\ 0 \end{pmatrix} + s \begin{pmatrix} -2 \\ -3 \\ 1 \\ 0 \\ 0 \end{pmatrix} + t \begin{pmatrix} -3 \\ -1 \\ 0 \\ -2 \\ 1 \end{pmatrix} = \mathbf{p} + s\mathbf{v}_1 + t\mathbf{v}_2. \tag{2.6}
$$

Notice that the coordinates of the particular solution $\mathbf{p}$ obtained in (2.6) vanish at the non-leading list indexes. □

We extend the approach of Problem 2.3 to systems $\mathbf{Ax} = \mathbf{b}$, where $\mathbf{A}$ is an arbitrary $m \times n$ matrix. By Theorem 1.19 the augmented matrix $(\mathbf{A}|\mathbf{b})$ has a unique reduced row echelon form $(\text{rref}(\mathbf{A})|\mathbf{d})$. Here $\mathbf{d}$ is a column of length m.

Definition 2.3 The column $\mathbf{x} = \mathbf{p}$ in (2.6) is called the **Gauss particular solution** to the system of equations $\mathbf{Ax} = \mathbf{b}$.

Every vector $\mathbf{e}_{j_s}$, $s = 1, 2, \ldots, r$, is a Gauss particular solution to the system

$$\mathbf{Ax} = \mathrm{col}_{j_s}(\mathbf{A}). \tag{2.7}$$

Every Gauss particular solution to the system $\mathbf{Ax}$ is a linear combination

$$\mathbf{p} = d_1 \mathbf{e}_{j_1} + d_2 \mathbf{e}_{j_2} + \cdots + d_r \mathbf{e}_{j_r}. \tag{2.8}$$

Since the vectors of the standard basis in $\mathbb{R}^n$ are linearly independent, the system $\{\mathbf{e}_{j_1}, \ldots, \mathbf{e}_{j_r}\}$ is a basis in its linear span.

Definition 2.4 The basis $\{\mathbf{e}_{j_1}, \ldots, \mathbf{e}_{j_r}\}$ for $\mathrm{Lin}\left(\mathbf{e}_{j_1}, \ldots, \mathbf{e}_{j_r}\right)$ is called the **Gauss basis of particular solutions**, whereas the subspace

$$P(\mathbf{A}) = \mathrm{Lin}\left(\mathbf{e}_{j_1}, \ldots, \mathbf{e}_{j_r}\right)$$

is called the **Gauss subspace of particular solutions**.

Problem 2.4 Find the Gauss subspace $P(\mathbf{A})$ of particular solutions for the matrix

$$\mathbf{A} = \begin{pmatrix} 1 & 1 & 5 & -4 & 2 \\ 8 & 4 & 28 & -19 & 14 \\ 3 & 1 & 9 & -9 & 4 \\ 3 & 2 & 12 & -9 & 5 \end{pmatrix}$$

Solutions: Elementary row operations show that

$$\mathrm{rref}(\mathbf{A}) = \begin{pmatrix} 1 & 0 & 2 & 0 & 0 \\ 0 & 1 & 3 & 0 & 0 \\ 0 & 0 & 0 & 1 & 0 \\ 0 & 0 & 0 & 0 & 1 \end{pmatrix}.$$

It follows that $j_1 = 1, j_2 = 2, j_3 = 4, j_4 = 5$. Hence $P(\mathbf{A}) = \mathrm{Lin}(\mathbf{e}_1, \mathbf{e}_2, \mathbf{e}_4, \mathbf{e}_5)$. □

Definition 2.5 Given an $m \times n$ matrix $\mathbf{A}$ let $N(\mathbf{A})$ be the set of all $n \times 1$ matrices (columns of length n) $\mathbf{x}$ such that $\mathbf{Ax} = \mathbf{0}$. Then $N(\mathbf{A})$ is called the **null space** of $\mathbf{A}$.

Lemma 2.3 $N(\mathbf{A})$ *is a subspace of* $\mathbb{R}^n = \mathbf{col}(\mathbf{I}_n)$.

Proof. If $\mathbf{x}$ and $\mathbf{y}$ are columns in $N(\mathbf{A})$, and α and β are real numbers, then $\alpha\mathbf{x} + \beta\mathbf{y}$ is a column in $N(\mathbf{A})$: $\mathbf{A}(\alpha\mathbf{x} + \beta\mathbf{y}) = \alpha\mathbf{A}\mathbf{x} + \beta\mathbf{A}\mathbf{y} = \mathbf{0}$. $\qquad\square$

Using the reduced row echelon form $\mathbf{B} = \mathrm{rref}(\mathbf{A})$ we can easily find $n - r$ solutions to $\mathbf{A}\mathbf{x} = \mathbf{0}$. Each of these solutions corresponds to a non-leading variable x_j, with j in J'. It is clear that non-leading variables exist if and only if $r < n$. The matrix $\mathbf{B} = (b_{ij})$ is an $m \times n$. Since $\mathbf{B} \sim \mathbf{A}$, we obtain that $N(\mathbf{A}) = N(\mathbf{B})$. Let a **non-leading variable** x_j take the value $x_j = 1$, whereas all other non-leading variables take the value zero. Then the system $\mathbf{B}\mathbf{x} = \mathbf{0}$ is equivalent to the system of equations

$$\begin{cases} x_{j_1} + b_{1j} = 0 \\ x_{j_2} + b_{2j} = 0 \\ \vdots \quad \vdots \quad \vdots \quad \vdots \\ x_{j_r} + b_{rj} = 0 \end{cases} \Rightarrow \begin{cases} x_{j_1} = -b_{1j} \\ x_{j_2} = -b_{2j} \\ \vdots \quad \vdots \\ x_{j_r} = -b_{rj} \end{cases} \qquad x_i = \begin{cases} 0 & i \text{ is in } J', i \neq j \\ 1 & i = j \end{cases}. \qquad (2.9)$$

The variables x_i in (2.9) are all non-leading. The entries b_{kj} of the jth column of $\mathbf{B}$ in (2.9) are placed above $b_{jj} = 0$. It follows that $b_{kj} = 0$ for $k \geq j$. Then the vector

$$\mathbf{g}_j = \left(\cdots - b_{1j}\ 0\ \cdots - b_{j-1,j}\ 0 \cdots \underset{j}{1}\ 0 \cdots 0 \right)^T \qquad (2.10)$$

satisfies $\mathbf{B}\mathbf{g}_j = \mathbf{0}$. It follows that $\mathbf{g}_j$ is in the null space $N(\mathbf{B})$ for every j in J'. If $\mathbf{B}\mathbf{x} = \mathbf{0}$, then

$$\mathbf{x} = \mathbf{p} + \sum_{j \in J'} x_j \mathbf{g}_j = \sum_{j \in J} x_j \mathbf{e}_j + \sum_{j \in J'} x_j \mathbf{g}_j. \qquad (2.11)$$

Since $\mathbf{B}\mathbf{g}_j = \mathbf{0}$ for every j in J', we obtain that $\mathbf{p}$ is a particular solution for $\mathbf{B}\mathbf{p} = \mathbf{0}$. Hence $\mathbf{p} = \mathbf{0}$. It follows that every vector in $N(\mathbf{A}) = N(\mathbf{B})$ is a linear combination of $n - r$ vectors $\mathbf{g}_j$ with j in J'. It is also clear that any solution to $\mathbf{B}\mathbf{x} = \mathbf{d}$ is given by the formula (2.11). This method is called the **method of null-space**.

The last formula in (2.9) shows that vectors $\mathbf{g}_j$ are linearly independent. Since they span $N(\mathbf{B})$, they make a basis for $N(\mathbf{A})$. By (2.10) this basis is uniquely determined by $\mathrm{rref}(\mathbf{A})$.

Definition 2.6 The finite ordered set of $n - r$ vectors $\mathbf{g}_j$, where j is in the complementary list J', is called the **Gauss basis for the null-space** $N(\mathbf{A})$.

The columns $\mathbf{g}_j$, $j \in J'$, and $\mathbf{e}_j$, $j \in J$, can be arranged into a matrix $\mathbf{G}$. By (2.9) this matrix is obtained from $\mathrm{rref}_p(\mathbf{A})$ by changing the signs of all entries to opposite except for the leading ones, moving the rows obtained to the rows of $\mathbf{G}$ with indexes $j_1 < \cdots < j_n$, and inserting the rows $\mathbf{f}_i$ for $i \in J'$.

For example, for the matrix (1.75)

$$\begin{pmatrix} x_1 & x_2 & x_3 & x_4 & x_5 & x_6 & x_7 & x_8 & x_9 & x_{10} \\ \downarrow & \downarrow & \downarrow & \downarrow & \downarrow & \downarrow & \downarrow & \downarrow & \downarrow & \downarrow \\ 0 & 1 & 2 & 0 & 1 & 3 & 0 & 1 & 2 & 0 \\ 0 & 0 & 0 & 1 & 1 & 2 & 0 & 5 & 4 & 0 \\ 0 & 0 & 0 & 0 & 0 & 0 & 1 & 2 & 3 & 0 \\ 0 & 0 & 0 & 0 & 0 & 0 & 0 & 0 & 0 & 1 \end{pmatrix}$$

with the leading list $J = \{2, 4, 7, 10\}$ and the complementary list $J' = \{1, 3, 5, 6, 8, 9\}$ the matrix **G** looks as follows:

$$
\begin{array}{rccccccccccc}
 & g_1 & g_2 & g_3 & g_4 & g_5 & g_6 & g_7 & g_8 & g_9 & g_{10} \\
1 \rightarrow & 1 & 0 & 0 & 0 & 0 & 0 & 0 & 0 & 0 & 0 \\
2 & 0 & 1 & -2 & 0 & -1 & -3 & 0 & -1 & -2 & 0 \\
3 \rightarrow & 0 & 0 & 1 & 0 & 0 & 0 & 0 & 0 & 0 & 0 \\
4 & 0 & 0 & 0 & 1 & -1 & -2 & 0 & -5 & -4 & 0 \\
5 \rightarrow & 0 & 0 & 0 & 0 & 1 & 0 & 0 & 0 & 0 & 0 \\
6 \rightarrow & 0 & 0 & 0 & 0 & 0 & 1 & 0 & 0 & 0 & 0 \\
7 & 0 & 0 & 0 & 0 & 0 & 0 & 1 & -2 & -3 & 0 \\
8 \rightarrow & 0 & 0 & 0 & 0 & 0 & 0 & 0 & 1 & 0 & 0 \\
9 \rightarrow & 0 & 0 & 0 & 0 & 0 & 0 & 0 & 0 & 1 & 0 \\
10 & 0 & 0 & 0 & 0 & 0 & 0 & 0 & 0 & 0 & 1
\end{array}
\tag{2.12}
$$

Here the columns with indexes in J interlace the columns g_j with j in J'. The columns with the indexes in J are colored in red, and those with indexes in J' are colored in black. It is clear that the 10×10 matrix

$$\mathbf{G} = \begin{pmatrix} g_1 & g_2 & g_3 & g_4 & g_5 & g_6 & g_7 & g_8 & g_9 & g_{10} \end{pmatrix}$$

is **upper triangular** with 1's on the main diagonal.

Definition 2.7 Let $\mathbf{A}$ be an $m \times n$ matrix and $\mathbf{f}_i$ be vectors of the standard basis for the row space ${}^n\mathbb{R}$. The **Gauss matrix** of solutions for $\mathbf{A}$ is the $n \times n$ matrix $\mathbf{G} = \mathbf{G_A}$, which is obtained from the principal part $\mathrm{rref}_p(\mathbf{A})$ of the reduced row echelon form of $\mathbf{A}$ following the two steps indicated below.

Step 1. We put $\mathrm{row}_i(\mathbf{G}) = \mathbf{f}_i$ for every i in the complementary list J'.

Step 2. We put $\mathrm{row}_i(\mathbf{G}) = 2\mathbf{f}_i - \mathrm{row}_s(\mathrm{rref}(\mathbf{A}))$ for every $i = j_s$ in the leading list J.

In real practice one just inserts the rows $\mathbf{f}_i$ between the rows of $\mathrm{rref}_p(\mathbf{A})$ to obtain 1's on the main diagonal of $\mathbf{G_A}$, and changes signs to the opposite at the non-zero entries of $\mathrm{rref}(\mathbf{A})$ to the right of the leading ones. Here is an example.

Problem 2.5 Find the Gauss basis for the null-space of the matrix in (2.5):

$$\text{rref}(\mathbf{A}) = \begin{pmatrix} 1 & 0 & 2 & 0 & 3 & 5 \\ 0 & 1 & 3 & 0 & 1 & 2 \\ 0 & 0 & 0 & 1 & 2 & 3 \\ 0 & 0 & 0 & 0 & 0 & 0 \end{pmatrix}.$$

Solution: The leading list for $\text{rref}(\mathbf{A})$ is $J = \{1,2,4\}$, implying that $j_1 = 1$, $j_2 = 2$, $j_3 = 4$. Hence $J' = \{3,5,6\}$. By **Step 1** and **Step 2**:

$$\text{rref}(\mathbf{A}) \Rightarrow \begin{matrix} \\ \\ 3 \rightarrow \\ \\ 5 \rightarrow \\ 6 \rightarrow \end{matrix} \begin{pmatrix} 1 & 0 & 2 & 0 & 3 & 5 \\ 0 & 1 & 3 & 0 & 1 & 2 \\ 0 & 0 & 1 & 0 & 0 & 0 \\ 0 & 0 & 0 & 1 & 2 & 3 \\ 0 & 0 & 0 & 0 & 1 & 0 \\ 0 & 0 & 0 & 0 & 0 & 1 \end{pmatrix} \Rightarrow \begin{matrix} 1 \\ 2 \\ \\ 4 \\ \\ \end{matrix} \begin{pmatrix} 1 & 0 & -2 & 0 & -3 & -5 \\ 0 & 1 & -3 & 0 & -1 & -2 \\ 0 & 0 & 1 & 0 & 0 & 0 \\ 0 & 0 & 0 & 1 & -2 & -3 \\ 0 & 0 & 0 & 0 & 1 & 0 \\ 0 & 0 & 0 & 0 & 0 & 1 \end{pmatrix}.$$

Then

$$\mathbf{g}_3 = \begin{pmatrix} -2 \\ -3 \\ 1 \\ 0 \\ 0 \\ 0 \end{pmatrix}, \ \mathbf{g}_5 = \begin{pmatrix} -3 \\ -1 \\ 0 \\ -2 \\ 1 \\ 0 \end{pmatrix}, \ \mathbf{g}_6 = \begin{pmatrix} -5 \\ -2 \\ 0 \\ -3 \\ 0 \\ 1 \end{pmatrix}, \tag{2.13}$$

is the Gauss basis for the null-space of $\text{rref}(\mathbf{A})$, and $\mathbf{e}_1, \mathbf{e}_2, \mathbf{e}_4$ is the Gauss basis for particular solutions.

Notice that the last 1 from the top of the first column in (2.13) is placed in the third row, which is reflected in the notation as $\mathbf{g}_3$. Similarly in the second column $\mathbf{g}_5$ this 1 is in the fifth row. Therefore, if we are given some vectors like in (2.13), the complementary leading list is recovered by the positions of the last 1's in the columns shown in (2.13). The leading list is uniquely recovered by the complementary one. Then we can construct the matrix $\mathbf{G}$ shown in (2.13) by inserting the columns of the standard basis at the positions of the leading list. The next step is to remove all rows with the indexes in the complementary list. Finally, we change all signs at entries to the right of leading ones. The obtained matrix is in reduced row echelon form with the null-space spanned by vectors in (2.13). □

Lemma 2.4 *Let* $\mathbf{V}$ *be an upper triangular* $n \times n$ *matrix with non-zero diagonal entries:* $v_{ii} \neq 0$, $i = 1, \ldots, n$. *Then the columns of* $\mathbf{V}$ *make a basis in* $\mathbb{R}^n$.

Proof. Applying **RO1** we see that $\mathbf{V}$ is row equivalent to a matrix in row echelon form with n leading ones. It follows that $\text{rref}(\mathbf{V}) = \mathbf{I}_n$. Hence $N(\mathbf{V}) = \{\mathbf{0}\}$. This implies that the columns of $\mathbf{V}$ are linearly independent. Since there are n columns, they make a basis for $\mathbb{R}^n$. □

Corollary 2.5 *Columns of any Gauss matrix* $\mathbf{G}$ *make a basis for* $\mathbb{R}^n$.

Proof. Apply Lemma 2.4. □

Theorem 2.5 *Every vector* $\mathbf{x}$ *in* $\mathbb{R}^n$ *can be uniquely represented as the sum*

$$\mathbf{x} = \mathbf{p} + \mathbf{u},$$

where $\mathbf{p}$ *is a Gauss particular solution in* $P(\mathbf{A})$ *and* $\mathbf{u}$ *is a vector in the null space* $N(\mathbf{A})$ *of the matrix* $\mathbf{A}$.

Proof. By Corollary 2.5 the vector $\mathbf{x}$ can be uniquely represented as the sum

$$\mathbf{x} = \alpha_1 \mathbf{g}_1 + \cdots + \alpha_n \mathbf{g}_n = \mathbf{p} + \mathbf{u},$$

where $\mathbf{p}$ is the sum of all $\alpha_j \mathbf{g}_j$ such that j is in the leading list, and $\mathbf{u}$ is the sum of all $\alpha_j \mathbf{g}_j$ such that j is in the complementary list.   □

Theorem 2.6 *Let* $\mathbf{A}$ *be an* $m \times n$ *matrix,* $J = \{j_1, \ldots, j_r\}$ *be the leading list for* $\mathbf{A}$. *Then columns* $\mathbf{g}_j$, *where* j *is in the complementary list* J', *make a basis for the null-space* $N(\mathbf{A})$

Proof. By (2.10) every vector $\mathbf{g}_j$ is in $N(\mathbf{A})$ for any j in J'. By Corollary 2.5 vectors $\mathbf{g}_j$ are linearly independent. If $\mathbf{Ax} = \mathbf{0}$, then by Theorem 2.5, $\mathbf{x} = \mathbf{p} + \mathbf{u}$, where $\mathbf{p} = \mathbf{0}$ and $\mathbf{u}$ is a linear combinations of vectors $\mathbf{g}_j$ with j is in J'.   □

If $\mathrm{rref}(\mathbf{A})$ is considered as the augmented matrix of a system, as it is the case with (2.5), then we can easily obtain the solution to the system

$$s\mathbf{v}_3 + t\mathbf{v}_5 + \mathbf{p} = s \begin{pmatrix} -2 \\ -3 \\ 1 \\ 0 \\ 0 \end{pmatrix} + t \begin{pmatrix} -3 \\ -1 \\ 0 \\ -2 \\ 1 \end{pmatrix} + \begin{pmatrix} 5 \\ 2 \\ 0 \\ 3 \\ 0 \end{pmatrix}.$$

Notice that the number of variables in this case is decreased by one, which results in elimination of the last entry in each of the columns $\mathbf{v}_3$, $\mathbf{v}_5$, $\mathbf{v}_6$. Finally, since the last column $\mathbf{v}_6$ in $\mathrm{rref}(\mathbf{A})$ is the column of data (in case if $\mathbf{A}$ is an augmented matrix), it must be multiplied by -1. Therefore, this new method is efficient even for solutions of general linear systems. We call this method as the **method of Gauss matrix**.

Definition 2.8 The union of the Gauss basis for the null-space with the Gauss basis of particular solutions is called the **Gauss basis for solutions**. By Corollary 2.5, Gauss basis for solutions is a basis for $\mathbb{R}^n$.

The Gauss matrix of any matrix $\mathbf{A}$ has an important property.

Theorem 2.7 *If* $\mathbf{G}$ *is the Gauss matrix of a matrix* $\mathbf{A}$, *then* $\mathbf{G} = \mathbf{I} + \mathbf{P}$, *where* $\mathbf{P}$ *is an upper triangular matrix satisfying* $\mathbf{P}^2 = \mathbf{0}$.

Proof. Let $\mathbf{P} = \mathbf{G} - \mathbf{I} = (p_{ij})$. Since $\mathbf{G}$ is an upper triangular and has zero entries on the main diagonal, we see that $p_{ij} = 0$ for $i \geq j$. We also have $p_{ij} = g_{ij}$ for $j > i$. If $i > j$ then

$$p_{ij} = g_{ij} = \begin{cases} 0 & \text{if } j \text{ is in } J, \\ 0 & \text{if } i \text{ is in } J'. \end{cases} \tag{2.14}$$

Since the square of any upper triangular matrix is upper triangular (see Problem 27), it is sufficient to show that the (i, j) entry c_{ij} of $\mathbf{P}^2$ is zero for any pair with $i \leq j$. If $i < j$, then $p_{ik} = 0$ for $k \leq i$ and $p_{kj} = 0$ for $k \geq j$, implying that

$$c_{ij} = \sum_{k=1}^{n} p_{ik} p_{kj} = \sum_{i < k < j} p_{ik} p_{kj}.$$

Since by (2.14) the entry p_{ik} is zero for k in J, and $p_{kj} = 0$ if k is in J', we conclude that $c_{ij} = 0$. If $i = j$, then $c_{ii} = p_{ii}^2 = 0$.   □

It is useful to observe the relationship of the Gauss matrix $\mathbf{G} = \mathbf{I}_n + \mathbf{P}$ with the extended row echelon form $\text{eref}(\mathbf{A}) = \mathbf{I}_n - \mathbf{P}$. Then by Theorem 2.7

$$\mathbf{G} \cdot \text{eref}(\mathbf{A}) = \text{eref}(\mathbf{A}) \cdot \mathbf{G} = \mathbf{I}_n - \mathbf{P}^2 = \mathbf{I}_n. \tag{2.15}$$

Problem 2.6 Find the Gauss basis for the null-space of the matrix:

$$\text{rref}(\mathbf{A}) = \begin{pmatrix} 1 & 0 & 0 & 3 \\ 0 & 1 & 0 & -2 \\ 0 & 0 & 1 & -1 \\ 0 & 0 & 0 & 0 \\ 0 & 0 & 0 & 0 \end{pmatrix}.$$

Solution: The leading list for $\text{rref}(\mathbf{A})$ is $J = \{1,2,3\}$. The complementary list is $J' = \{4\}$. It follows that

$$\mathbf{G} = \begin{pmatrix} 1 & 0 & 0 & -3 \\ 0 & 1 & 0 & 2 \\ 0 & 0 & 1 & 1 \\ 0 & 0 & 0 & 1 \end{pmatrix},$$

implying that the Gauss basis for $N(\mathbf{A})$ is generated by the last column $\mathbf{g}_4$ of $\mathbf{G}$. □

Problem 2.7 Using the method of Gauss matrix solve the system

$$\begin{cases} x_1 + 3x_2 + x_3 + 4x_4 & = 2 \\ x_1 + 3x_2 + 2x_4 & = 1 \\ x_1 + 3x_2 - x_3 & = 0 \\ x_1 + 3x_2 + 2x_4 & = 1 \end{cases} \tag{2.16}$$

Solution: First, we find the reduced row echelon form of the augmented matrix of the system (2.16):

$$\text{rref} \begin{pmatrix} 1 & 3 & 1 & 4 & | & 2 \\ 1 & 3 & 0 & 2 & | & 1 \\ 1 & 3 & -1 & 0 & | & 0 \\ 1 & 3 & 0 & 2 & | & 1 \end{pmatrix} = \begin{pmatrix} 1 & 3 & 0 & 2 & 1 \\ 0 & 0 & 1 & 2 & 1 \\ 0 & 0 & 0 & 0 & 0 \\ 0 & 0 & 0 & 0 & 0 \end{pmatrix}.$$

The leading list for the reduced row echelon form considered as the coefficient matrix of a homogeneous system is $J = \{1,3\}$. The complementary list is $J' = \{2,4,5\}$. By **Step 1** and **Step 2**:

$$\begin{matrix} \\ 2 \rightarrow \\ \\ 4 \rightarrow \\ 5 \rightarrow \end{matrix} \begin{pmatrix} 1 & 3 & 0 & 2 & 1 \\ 0 & 1 & 0 & 0 & 0 \\ 0 & 0 & 1 & 2 & 1 \\ 0 & 0 & 0 & 1 & 0 \\ 0 & 0 & 0 & 0 & 1 \end{pmatrix} \longmapsto \begin{matrix} 1 \\ \\ 3 \\ \\ \\ \end{matrix} \begin{pmatrix} 1 & -3 & 0 & -2 & -1 \\ 0 & 1 & 0 & 0 & 0 \\ 0 & 0 & 1 & -2 & -1 \\ 0 & 0 & 0 & 1 & 0 \\ 0 & 0 & 0 & 0 & 1 \end{pmatrix}.$$

It follows that

$$\mathbf{x} = s\mathbf{v}_2 + t\mathbf{v}_4 + \mathbf{p} = s \begin{pmatrix} -3 \\ 1 \\ 0 \\ 0 \end{pmatrix} + t \begin{pmatrix} -2 \\ 0 \\ -2 \\ 1 \end{pmatrix} + \begin{pmatrix} 1 \\ 0 \\ 1 \\ 0 \end{pmatrix}$$

is the solution to the system (2.16).   □

Problem 2.8 Using the method of Gauss matrix solve the system

$$\begin{cases} 2x_1 + 2x_2 + 8x_3 - x_4 + 4x_5 & = 9 \\ 4x_1 - x_2 + x_3 + 2x_4 + 11x_5 & = 10 \\ x_1 - x_2 - 2x_3 + x_4 + 3x_5 & = 1 \\ 5x_1 - 2x_2 - x_3 + 3x_4 + 14x_5 & = 11 \end{cases}$$

Solution: The reduced row echelon form of the augmented matrix of the system is transformed into the **extended Gauss matrix**:

$$
\begin{pmatrix}
1 & 0 & 1 & 0 & 2 & \big| & 2 \\
0 & 1 & 3 & 0 & 1 & \big| & 4 \\
0 & 0 & 0 & 1 & 2 & \big| & 3 \\
0 & 0 & 0 & 0 & 0 & \big| & 0
\end{pmatrix}
\Rightarrow \mathbf{G} =
\begin{pmatrix}
1 & 0 & -1 & 0 & -2 & \big| & 2 \\
0 & 1 & -3 & 0 & -1 & \big| & 4 \\
0 & 0 & 1 & 0 & 0 & \big| & 0 \\
0 & 0 & 0 & 1 & -2 & \big| & 3 \\
0 & 0 & 0 & 0 & 1 & \big| & 0
\end{pmatrix} .
$$

Hence we obtain the solution:

$$
\mathbf{x} = \underbrace{\begin{pmatrix} 2 \\ 4 \\ 0 \\ 3 \\ 0 \end{pmatrix}}_{\mathbf{p}} + s \underbrace{\begin{pmatrix} -1 \\ -3 \\ 1 \\ 0 \\ 0 \end{pmatrix}}_{\mathbf{u}} + t \begin{pmatrix} -2 \\ -1 \\ 0 \\ -2 \\ 1 \end{pmatrix} . \qquad \square
$$

Problems

Prob. 91 — Find the Gauss bases for each of the following subspaces of the corresponding Euclidean spaces $\mathbb{R}^n$. Construct the corresponding Gauss matrices $\mathbf{G}$.

$$
V_1 = \left\{ (x\ y)^T \in \mathbb{R}^2 \mid x + y = 0 \right\} \qquad
V_2 = \left\{ (x\ y\ z)^T \in \mathbb{R}^3 \mid x + y + z = 0 \right\}
$$

$$
V_3 = \mathrm{Lin} \left\{ \begin{pmatrix} 1 \\ 2 \\ 1 \end{pmatrix}, \begin{pmatrix} 1 \\ 1 \\ 0 \end{pmatrix} \right\}
$$

$$
V_4 = \mathrm{Lin} \left\{ \begin{pmatrix} 2 \\ 2 \\ 1 \end{pmatrix}, \begin{pmatrix} 1 \\ 1 \\ 0 \end{pmatrix}, \begin{pmatrix} 1 \\ 1 \\ 0 \end{pmatrix} \right\}
$$

2.4 The Luoshu Magic Square

The Luoshu **magic square** is the 3×3 matrix with nine entries filled by the integers 1 through 9 such that the sum of the entries in each row, each column, and each diagonal is one and the same:

4	9	2
3	5	7
8	1	6

See Swetz (2008) for the history of this question. A natural question is how one could find all such magic squares. We apply Linear Algebra to do this. Let us consider a magic square with unknown entries

x_1	x_2	x_3
x_4	x_5	x_6
x_7	x_8	x_9

and assume that $x_1, \ldots, x_9$ are positive numbers 1 through 9. Then

$$(x_1 + x_2 + x_3) + (x_4 + x_5 + x_6) + (x_7 + x_8 + x_9) = 1 + \cdots + 9 = 45,$$

implying that the magic sum of each row (and hence column and diagonal) is 15.

Using the rules of magic squares, we form the system of linear equations:

$$\begin{cases} x_1 + x_2 + x_3 & = 15 \\ x_4 + x_5 + x_6 & = 15 \\ x_7 + x_8 + x_9 & = 15 \\ x_1 + x_4 + x_7 & = 15 \\ x_2 + x_5 + x_8 & = 15 \\ x_3 + x_6 + x_9 & = 15 \\ x_1 + x_5 + x_9 & = 15 \\ x_3 + x_5 + x_7 & = 15 \end{cases} \tag{2.17}$$

corresponding to the augmented matrix

$$(\mathbf{A}|\mathbf{b}) = \begin{pmatrix} 1 & 1 & 1 & 0 & 0 & 0 & 0 & 0 & 0 & 15 \\ 0 & 0 & 0 & 1 & 1 & 1 & 0 & 0 & 0 & 15 \\ 0 & 0 & 0 & 0 & 0 & 0 & 1 & 1 & 1 & 15 \\ 1 & 0 & 0 & 1 & 0 & 0 & 1 & 0 & 0 & 15 \\ 0 & 1 & 0 & 0 & 1 & 0 & 0 & 1 & 0 & 15 \\ 0 & 0 & 1 & 0 & 0 & 1 & 0 & 0 & 1 & 15 \\ 1 & 0 & 0 & 0 & 1 & 0 & 0 & 0 & 1 & 15 \\ 0 & 0 & 1 & 0 & 1 & 0 & 1 & 0 & 0 & 15 \end{pmatrix}$$

Its reduced row echelon form is given by

$$\begin{pmatrix} 1 & 0 & 0 & 0 & 0 & 0 & 0 & 0 & 1 & 10 \\ 0 & 1 & 0 & 0 & 0 & 0 & 0 & 1 & 0 & 10 \\ 0 & 0 & 1 & 0 & 0 & 0 & 0 & -1 & -1 & -5 \\ 0 & 0 & 0 & 1 & 0 & 0 & 0 & -1 & -2 & -10 \\ 0 & 0 & 0 & 0 & 1 & 0 & 0 & 0 & 0 & 5 \\ 0 & 0 & 0 & 0 & 0 & 1 & 0 & 1 & 2 & 20 \\ 0 & 0 & 0 & 0 & 0 & 0 & 1 & 1 & 1 & 15 \\ 0 & 0 & 0 & 0 & 0 & 0 & 0 & 0 & 0 & 0 \end{pmatrix}$$

It follows that $x_8 = s$ and $x_9 = t$ are non-leading variables. Hence $x_5 = 5$ and

$$x_1 = 10-t, \ x_2 = 10-s, \ x_3 = s+t-5, \ x_4 = s+2t-10, \ x_6 = 20-s-2t, \ x_7 = 15-s-t.$$

Any choice of integer s and t leads to a magic square with the magic sum equal 15. However, we are interested only in those s and t, which make all variables different and equal to the numbers 1 through 9.

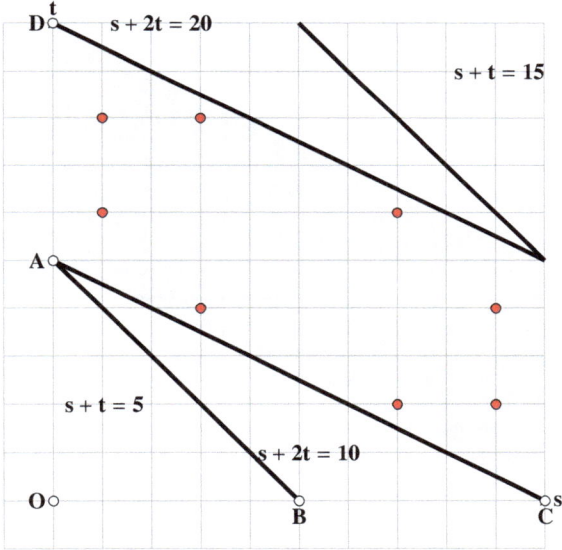

Fig. 2.1: Solutions leading to magic squares

The open gray domain G in Fig. 2.1 is described by the inequalities

$$10 < s + 2t < 20.$$

Every point (s,t) in G with integer coordinates defines a certain magic square. Only points marked in red define the required magic squares made of all integers 1 through 9. Below is the complete list of such squares. They all are obtained by the symmetries of the Loushu Magic Square ($s = 1, t = 6$). The symmetries of any square make the so-called dihedral group D_4. This explains why there are only eight magic squares of the required type:

4	9	2
3	5	7
8	1	6

Fig. 2.2:
$s = 1, t = 6$

2	9	4
7	5	3
6	1	8

Fig. 2.3:
$s = 1, t = 8$

6	7	2
1	5	9
8	3	4

Fig. 2.4:
$s = 3, t = 4$

2	7	6
9	5	1
4	3	8

Fig. 2.5:
$s = 3, t = 8$

8	3	4
1	5	9
6	7	2

Fig. 2.6:
$s = 7, t = 2$

4	3	8
9	5	1
2	7	6

Fig. 2.7:
$s = 7, t = 6$

8	1	6
3	5	7
4	9	2

Fig. 2.8:
$s = 9, t = 2$

6	1	8
7	5	3
2	9	4

Fig. 2.9:
$s = 9, t = 4$

Problems

Prob. 92 — Find the extended Gauss matrix for the Luoshu equation (2.17).

2.5 The Column Space of a Matrix

Definition 2.9 Let A be an $m \times n$ matrix. The set of all columns b in $\mathbb{R}^m$ such that the system $Ax = b$ is consistent is called the **range** of the matrix A. The range of A is denoted by $R(A)$.

Theorem 2.8 Let A be an $m \times n$ matrix and $J = \{j_1, \ldots, j_r\}$ be its leading list. Then

(a)
$$R(A) = \text{Lin}\left(\text{col}_{j_1}(A), \text{col}_{j_2}(A), \ldots, \text{col}_{j_r}(A)\right);$$

(b) the set of vectors $\{\text{col}_{j_1}(A), \text{col}_{j_2}(A), \ldots, \text{col}_{j_r}(A)\}$ is linearly independent and, therefore, makes a basis for $R(A)$.

Proof. By Theorem 2.5 and (2.8)

$$\mathbf{A}\mathbf{x} = \mathbf{A}\mathbf{p} + \mathbf{A}\mathbf{u} = \mathbf{A}\mathbf{p} = d_1\mathbf{A}\mathbf{e}_{j_1} + d_2\mathbf{A}\mathbf{e}_{j_2} + \cdots + d_r\mathbf{A}\mathbf{e}_{j_r} =$$
$$d_1\mathrm{col}_{j_1}(\mathbf{A}) + d_2\mathrm{col}_{j_2}(\mathbf{A}) + \cdots + d_r\mathrm{col}_{j_r}(\mathbf{A}),$$

which proves (**a**). If $\mathbf{b} = \mathbf{0}$ then $\mathrm{rref}((\mathbf{A}|\mathbf{0})) = (\mathrm{rref}(\mathbf{A})|\mathbf{0})$, implying that $\mathbf{d} = \mathbf{0}$, which proves (**b**). $\qquad\square$

Definition 2.10 Given an $m \times n$ matrix $\mathbf{A}$ the basis

$$\left\{\mathrm{col}_{j_1}(\mathbf{A}), \mathrm{col}_{j_2}(\mathbf{A}), \ldots, \mathrm{col}_{j_r}(\mathbf{A})\right\}$$

is called the **Gauss basis for the range** of matrix $\mathbf{A}$.

Contrary to the Gauss basis for the null-space, this basis depends on the matrix considered.

Theorem 2.9 *Let $\mathbf{A}$ be an $m \times n$ matrix and $1 \le j_1 < j_2 < \cdots < j_r \le n$ be the leading list for $\mathrm{rref}(\mathbf{A})$. Then every non-leading column $\mathrm{col}_j(\mathbf{A})$ is a linear combination of $\mathrm{col}_{j_s}(\mathbf{A})$ with $j_s < j$.*

Proof. Since j is in J', the vector $\mathbf{g}_j$ is in $\mathrm{N}(\mathbf{A})$. We obtain by (2.9)

$$\mathbf{0} = \mathbf{A}\mathbf{g}_j = \mathrm{col}_j(\mathbf{A}) - b_{1j}\mathrm{col}_{j_1}(\mathbf{A}) - \cdots - b_{rj}\mathrm{col}_{j_r}(\mathbf{A}),$$

which proves the Theorem. $\qquad\square$

Corollary 2.6 *If $\mathbf{A}$ is an $m \times n$ matrix then*

$$\dim(\mathbf{col}(\mathbf{A})) = \dim(\mathrm{R}(\mathbf{A})) = \mathrm{rank}(\mathbf{A}) = r.$$

Proof. Apply Theorems 2.8. $\qquad\square$

Corollary 2.7 *Let* **A** *be an* $m \times n$ *matrix. Then*

$$\text{rank}(\mathbf{A}^T) = \text{rank}(\mathbf{A}).$$

Proof. Since $\text{col}_j(\mathbf{A}) = \text{row}_j(\mathbf{A}^T)$, the result follows by Corollary 2.6. $\square$

Theorems 2.5 and 2.8 show that columns **b** in R(**A**) are in one-to-one linear correspondence with Gauss particular solutions **p**:

$$\mathbf{b} = \mathbf{Ap}.$$

By Theorem 2.5 every vector **x** is uniquely decomposed as a sum of a particular solution **p** and a homogeneous solution **u**. In this case **p** is not necessarily perpendicular to **u**. We write

$$P(\mathbf{A}) + N(\mathbf{A}) = \mathbb{R}^n. \tag{2.18}$$

The dimension $\dim(N(\mathbf{A}))$ is called the **nullity** of **A**:

$$\text{nullity}(\mathbf{A}) = \dim(N(\mathbf{A})) = n - r. \tag{2.19}$$

The following theorem is a typical example of the existence theorem in Linear Algebra. It does not give any formula for the null space of the transpose matrix if a formula for the null space of **A** is available. A typical example of an application of this theorem is central for the study Markov chains, see Corollary 5.4.

Theorem 2.10 *Let* **A** *be an* $n \times n$ *matrix. Then*

$$\text{nullity}(\mathbf{A}^T) = \text{nullity}(\mathbf{A}).$$

Proof. By corollary 2.7 both $n \times n$ matrices **A** and $\mathbf{A}^T$ have the ranks equal r. The result follows by formula (2.19). $\square$

Theorem 2.11 *Let* **A** *be an* $m \times p$ *matrix and* **B** *be a* $p \times n$ *matrix. Then*

$$\text{rank}(\mathbf{AB}) \leq \min(\text{rank}(\mathbf{A}), \text{rank}(\mathbf{B})). \tag{2.20}$$

Proof. By (1.30) the column space of $\mathbf{AB}$ is

$$\text{Lin}\left(\text{Acol}_1(\mathbf{B}), \cdots, \text{Acol}_p(\mathbf{B})\right).$$

Since every column $\text{Acol}_j(\mathbf{B})$ is a linear combination of columns of $\mathbf{A}$, the column space of $\mathbf{AB}$ is a subspace of the column space of $\mathbf{A}$. By Corollary 2.6 the dimension of the column space of any matrix equals its rank. It follows that

$$\text{rank}(\mathbf{AB}) \le \text{rank}(\mathbf{A}).$$

By Corollary 2.7 the ranks of a matrix and its transpose are equal. Applying Theorem 1.5 we obtain

$$\text{rank}(\mathbf{AB}) = \text{rank}\left((\mathbf{AB})^T\right) = \text{rank}(\mathbf{B}^T \mathbf{A}^T) \le \text{rank}(\mathbf{B}^T) = \text{rank}(\mathbf{B}). \quad \square$$

Problem 2.9 Find the range and the Gauss basis for the null-space for the matrix

$$\mathbf{A} = \begin{pmatrix} 1 & 2 & 0 & 3 & 0 & 4 \\ 0 & 0 & 1 & 5 & 0 & 6 \\ 0 & 0 & 0 & 0 & 1 & 1 \\ 0 & 0 & 0 & 0 & 0 & 0 \end{pmatrix}.$$

Solution: The leading list is $J = \{1, 3, 5\}$. This means that

$$R(\mathbf{A}) = \text{Lin}(\mathbf{e}_1, \mathbf{e}_3, \mathbf{e}_5) = P(\mathbf{A}).$$

To obtain the Gauss matrix $\mathbf{G}$ from the given matrix in reduced row echelon form one can just insert in this matrix the rows $\mathbf{f}_i$ with i not in the leading list and change signs to opposite for the entries of the reduced row echelon form to the right of leading ones:

$$
\begin{matrix}
 & \\
2 \rightarrow & \\
 & \\
4 \rightarrow & \\
 & \\
6 \rightarrow &
\end{matrix}
\begin{pmatrix}
1 & 2 & 0 & 3 & 0 & 4 \\
0 & 1 & 0 & 0 & 0 & 0 \\
0 & 0 & 1 & 5 & 0 & 6 \\
0 & 0 & 0 & 1 & 0 & 0 \\
0 & 0 & 0 & 0 & 1 & 1 \\
0 & 0 & 0 & 0 & 0 & 1
\end{pmatrix}
\longmapsto
\begin{matrix}
1 \\ \\ 3 \\ \\ 5 \\
\end{matrix}
\begin{pmatrix}
1 & -2 & 0 & -3 & 0 & -4 \\
0 & 1 & 0 & 0 & 0 & 0 \\
0 & 0 & 1 & -5 & 0 & -6 \\
0 & 0 & 0 & 1 & 0 & 0 \\
0 & 0 & 0 & 0 & 1 & -1 \\
0 & 0 & 0 & 0 & 0 & 1
\end{pmatrix}
= \mathbf{G}. \quad \square
$$

It is easy to see that we obtain the same formulas for $\mathbf{v}_2, \mathbf{v}_4, \mathbf{v}_6$ if we put $x_2 = s$, $x_4 = t$, $x_6 = u$. The advantage of the presented solution is that it clearly reveals the relationship of the entries of $\mathbf{v}_2, \mathbf{v}_4, \mathbf{v}_6$ to the entries of the reduced row echelon form. The algorithm shown in Problem 2.9 clearly indicates how to recover $\mathbf{G}$ by $\text{rref}(\mathbf{A})$ and vice-versa.

Problem 2.10 Find the representation of the vector

$$\mathbf{b} = \begin{pmatrix} 1\ 2\ 1\ 0\ 1\ 3 \end{pmatrix}^T$$

in the Gauss bases of $P(\mathbf{A})$ and $N(\mathbf{A})$ defined by the matrix

$$\mathbf{A} = \begin{pmatrix} 2 & -1 & 0 & 2 & 2 & 6 \\ 2 & 1 & 4 & 3 & 3 & 8 \\ 1 & 3 & 7 & 1 & 1 & 3 \\ 1 & 2 & 5 & 1 & 1 & 3 \end{pmatrix}.$$

Represent $\mathbf{b}$ as the sum of $\mathbf{p}$ in $P(\mathbf{A})$ and $\mathbf{u}$ in $N(\mathbf{A})$.

Solution: We have

$$\text{rref}(\mathbf{A}) = \begin{pmatrix} 1 & 0 & 1 & 0 & 0 & 1 \\ 0 & 1 & 2 & 0 & 0 & 0 \\ 0 & 0 & 0 & 1 & 1 & 2 \\ 0 & 0 & 0 & 0 & 0 & 0 \end{pmatrix}.$$

The leading list for the reduced row echelon form is $J = \{1,2,4\}$, the complementary list is $J' = \{3,5,6\}$. The Gauss matrix of $\mathbf{A}$ is

$$\mathbf{G} = \begin{pmatrix} 1 & 0 & -1 & 0 & 0 & -1 \\ 0 & 1 & -2 & 0 & 0 & 0 \\ 0 & 0 & 1 & 0 & 0 & 0 \\ 0 & 0 & 0 & 1 & -1 & -2 \\ 0 & 0 & 0 & 0 & 1 & 0 \\ 0 & 0 & 0 & 0 & 0 & 1 \end{pmatrix}. \tag{2.21}$$

It follows that $\mathbf{G}\alpha = \mathbf{b}$ if and only if α is the solution of the system in row echelon form:

$$\begin{cases} \alpha_1 - \alpha_3 - \alpha_6 & = 1 \\ \alpha_2 - 2\alpha_3 & = 2 \\ \alpha_3 & = 1 \\ \alpha_4 - \alpha_5 - 2\alpha_6 & = 0 \\ \alpha_5 & = 1 \\ \alpha_6 & = 3 \end{cases} \Rightarrow \begin{cases} \alpha_1 & = 5 \\ \alpha_2 & = 4 \\ \alpha_3 & = 1 \\ \alpha_4 & = 7 \\ \alpha_5 & = 1 \\ \alpha_6 & = 3 \end{cases}.$$

Hence

$$\mathbf{b} = 5\mathbf{g}_1 + 4\mathbf{g}_2 + \mathbf{g}_3 + 7\mathbf{g}_4 + \mathbf{g}_5 + 3\mathbf{g}_6.$$

Since $J = \{1,2,4\}$, we obtain that

$$\mathbf{p} = 5\mathbf{g}_1 + 4\mathbf{g}_2 + 7\mathbf{g}_4 = \begin{pmatrix} 5\ 4\ 0\ 7\ 0\ 0 \end{pmatrix}^T,$$
$$\mathbf{u} = \mathbf{g}_3 + \mathbf{g}_5 + 3\mathbf{g}_6 = \begin{pmatrix} -4\ -2\ 1\ -7\ 1\ 3 \end{pmatrix}^T.$$

Problem 2.11 Solve the system

$$\begin{cases} x_1 + x_2 + 3x_3 + 2x_4 + 7x_5 & = 9 \\ 2x_1 + 2x_2 + 6x_3 + 6x_5 & = 14 \\ x_1 + 2x_3 + x_4 + 4x_5 & = 5 \\ x_1 + x_2 + 3x_3 + 2x_4 + 7x_5 & = 9 \end{cases}$$

by the method of Gauss matrix. Find $R(A)$ for the coefficient matrix of this system. Find the Gauss basis for the null-space of A.

Solution: The reduced row echelon form of the augmented matrix of the given system is

$$\text{rref}\begin{pmatrix} 1 & 1 & 3 & 2 & 7 & | & 9 \\ 2 & 2 & 6 & 0 & 6 & | & 14 \\ 1 & 0 & 2 & 1 & 4 & | & 5 \\ 1 & 1 & 3 & 2 & 7 & | & 9 \end{pmatrix} = \begin{pmatrix} 1 & 0 & 2 & 0 & 2 & | & 4 \\ 0 & 1 & 1 & 0 & 1 & | & 3 \\ 0 & 0 & 0 & 1 & 2 & | & 1 \\ 0 & 0 & 0 & 0 & 0 & | & 0 \end{pmatrix}.$$

The leading list for the reduced row echelon form of A is $J = \{1,2,4\}$, the complementary list is $J' = \{3,5\}$. The extended Gauss matrix equals

$$G = \begin{pmatrix} 1 & 0 & -2 & 0 & -2 & | & 4 \\ 0 & 1 & -1 & 0 & -1 & | & 3 \\ 0 & 0 & 1 & 0 & 0 & | & 0 \\ 0 & 0 & 0 & 1 & -2 & | & 1 \\ 0 & 0 & 0 & 0 & 1 & | & 0 \end{pmatrix}.$$

It follows that

$$g_3 = \begin{pmatrix} -2 \\ -1 \\ 1 \\ 0 \\ 0 \end{pmatrix}, \quad g_5 = \begin{pmatrix} -2 \\ -1 \\ 0 \\ -2 \\ 1 \end{pmatrix},$$

and that

$$p + sg_3 + tg_5 = \begin{pmatrix} 4 \\ 3 \\ 0 \\ 1 \\ 0 \end{pmatrix} + s\begin{pmatrix} -2 \\ -1 \\ 1 \\ 0 \\ 0 \end{pmatrix} + t\begin{pmatrix} -2 \\ -1 \\ 0 \\ -2 \\ 1 \end{pmatrix}.$$

is the solution to the system. We see that $N(A) = \text{Lin}(g_3, g_5)$. Since $J = \{1,2,4\}$, the range of A equals $\text{Lin}(\text{col}_1(A), \text{col}_2(A), \text{col}_4(A))$. $\square$

Problems

Prob. 93 — Solve the system

$$\begin{cases} 2x_1 - x_2 - 3x_3 + 3x_4 + 2x_5 & = 9 \\ 2x_1 + x_2 - x_3 + x_4 + 3x_5 & = 13 \\ x_1 + 3x_2 + 2x_3 - 2x_4 + x_5 & = 8 \\ x_1 + 2x_2 + x_3 - x_4 + x_5 & = 7 \end{cases}$$

by the methods of free variables and the Gauss matrix. Find the range of the coefficient matrix. Find its null-space.

Prob. 94 — Solve the system

$$\begin{cases} 2x_1 - x_2 + x_3 + 4x_4 + 2x_5 & = 3 \\ 2x_1 + x_2 + 7x_3 + 8x_4 + 3x_5 & = 11 \\ x_1 + 3x_2 + 11x_3 + 9x_4 + x_5 & = 12 \\ x_1 + 2x_2 + 8x_3 + 7x_4 + x_5 & = 9 \end{cases}$$

by the methods of free variables and the Gauss matrix. Find the range of the augmented matrix of the system. Find its null-space.

Prob. 95 — Solve the system

$$\begin{cases} 4x_1 + \quad x_2 + \quad 2x_3 - \ 17x_4 - \ 5x_5 + \ 8x_6 & = 21 \\ 8x_1 + \ 9x_2 + \ 23x_3 + 15x_4 + 11x_5 + 39x_6 & = 593 \\ 24x_1 + 41x_2 + \ 9x_3 + \quad 3x_4 \quad + \quad x_6 & = 317 \\ 6x_1 + \ 5x_2 \quad - \quad x_4 + \ 3x_5 - \ 7x_6 & = 35 \\ 9x_1 + \ 11x_2 + 39x_3 + 23x_4 + \ 15x_5 & = 678 \\ 28x_1 + 49x_2 + \ 4x_3 + \ 5x_4 + \ 9x_5 + \ 7x_6 & = 593 \end{cases}$$

by the methods of free variables and the Gauss matrix.

Prob. 96 — Using Gauss elimination, solve the system of linear equations

$$\begin{cases} \quad\quad x_2 + x_3 + \ \dots \ + x_n & = 0 \\ x_1 \quad\quad + x_3 + \ \dots \ + x_n & = 1 \\ \dots\dots\dots\dots\dots\dots\dots\dots\dots \\ x_1 + x_2 + \dots + x_{n-1} \quad\quad & = n - 1 \end{cases}$$

2.6 The Geometry of Row Spaces

In this section we consider Gauss bases for row subspaces in $^n\mathbb{R}$ and in Section 2.7 for column subspaces in $\mathbb{R}^m$. The main result is that any such a non-zero subspace has a unique Gauss basis. We begin with bases for row subspaces.

Theorem 2.12 *Every row in* **row**(**A**) *can be uniquely represented as the sum*

$$\alpha_1 \text{row}_1 \left(\text{rref}(\mathbf{A}) \right) + \alpha_2 \text{row}_2 \left(\text{rref}(\mathbf{A}) \right) + \cdots + \alpha_r \text{row}_r \left(\text{rref}(\mathbf{A}) \right), \quad (2.22)$$

where $r = \text{rank}(\mathbf{A})$ *and* $\alpha_1, \alpha_2, \ldots, \alpha_r$ *are real numbers.*

Proof. By Theorem 1.7, **row**(**A**) = **row**(rref(**A**)). It follows that every row has a representation (2.22). To show that it is unique, we denote by $J = \{j_1, j_2, \ldots, j_r\}$ the leading list of rref(**A**). Suppose that $\mathbf{x} = \left(x_1 \cdots x_n \right)$ is a row in **row**(rref**A**). Since each column with index j_s has zero entries except the entry (s, j_s), which is 1, we conclude that $x_{j_s} = \alpha_s$ for $s = 1, \ldots, r$, which implies the uniqueness. □

Definition 2.11 Given an $m \times n$ matrix **A** the finite set

$$\left\{ \text{row}_1 \left(\text{rref}(\mathbf{A}) \right), \text{row}_2 \left(\text{rref}(\mathbf{A}) \right), \ldots, \text{row}_r \left(\text{rref}(\mathbf{A}) \right) \right\}, \quad (2.23)$$

where $r = \text{rank}(\mathbf{A})$ is called the **Gauss basis** for the row space **row**(**A**) of **A**.

By Theorem 1.21 the reduced row echelon form rref(**A**) of a matrix **A** of size $m \times n$ is **uniquely determined** by **row**(**A**). By Corollary 2.1 every non-zero subspace V of $^n\mathbb{R}$ is **row**(**A**) for some non-zero matrix **A** of size $m \times n$. It follows that every subspace V in $^n\mathbb{R}$ has a **unique Gauss basis of rows**.

Definition 2.12 Let V be a non-zero subspace V of $^n\mathbb{R}$. A finite set $\{\mathbf{r}_1, \ldots, \mathbf{r}_k\}$ of rows is called the Gauss basis for V if it is the Gauss basis for **row**(**A**) = V.

Problem 2.12 Find the Gauss basis for $^n\mathbb{R}$.

Solution: By definition, $^n\mathbb{R} = \text{row}(\mathbf{I}_n)$. Since $\mathbf{I}_n = \text{rref}(\mathbf{I}_n)$, we conclude that the standard basis

$$\{\mathbf{f}_1, \mathbf{f}_2, \ldots, \mathbf{f}_n\}$$

is the Gauss basis for $^n\mathbb{R}$. □

Problem 2.13 Find the Gauss basis in $^3\mathbb{R}$ for the plane $x + y + z = 0$.

Solution: The coefficient matrix of the system $x + y + z = 0$ is the 1×3 matrix $\mathbf{A} = \left(1\ 1\ 1 \right)$. Its Gauss matrix is

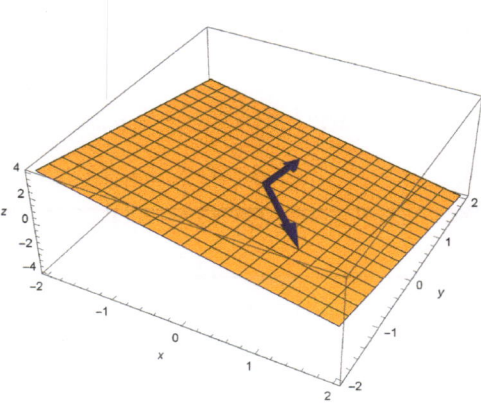

Fig. 2.10: The Gauss basis for $x + y + z = 0$ in $^3\mathbb{R}$.

$$G_A = \begin{pmatrix} 1 & -1 & -1 \\ 0 & 1 & 0 \\ 0 & 0 & 1 \end{pmatrix}.$$

Then $N(A) = \text{Lin}\{g_2, g_3\}$. It follows that the subspace of rows $\begin{pmatrix} x & y & z \end{pmatrix}$ satisfying the equation $x + y + z = 0$ is nothing but the row space of the matrix

$$\begin{pmatrix} -1 & 1 & 0 \\ -1 & 0 & 1 \end{pmatrix} \sim \begin{pmatrix} 1 & -1 & 0 \\ 0 & 1 & -1 \end{pmatrix}.$$

Since the last matrix is in reduced row echelon form, the Gauss basis in $^3\mathbb{R}$ for this plane is made of two rows listed in this very order:

$$\begin{pmatrix} 1 & -1 & 0 \end{pmatrix}, \quad \begin{pmatrix} 0 & 1 & -1 \end{pmatrix}.$$

The first vector of this basis looks longer on Fig. 2.10. By (1.51), these vectors make the angle of $120°$ degrees.   □

Problem 2.14 Find the Gauss basis in $^3\mathbb{R}$ for $\text{row}(A)$, where

$$A = \begin{pmatrix} 13 & 3 & 29 & 19 & 6 \\ 4 & 1 & 9 & 6 & 2 \\ 6 & 1 & 13 & 8 & 3 \\ 11 & 2 & 24 & 15 & 4 \end{pmatrix}.$$

Solution: Elementary row operations show that

$$\text{rref}(A) = \begin{pmatrix} 1 & 0 & 2 & 1 & 0 \\ 0 & 1 & 1 & 2 & 0 \\ 0 & 0 & 0 & 0 & 1 \\ 0 & 0 & 0 & 0 & 0 \end{pmatrix}.$$

By Definition 2.11 the Gauss basis for the row space is given by:

$$\left\{ \begin{pmatrix} 1\ 0\ 2\ 1\ 0 \end{pmatrix}, \begin{pmatrix} 0\ 1\ 1\ 2\ 0 \end{pmatrix}, \begin{pmatrix} 0\ 0\ 0\ 0\ 1 \end{pmatrix} \right\}. \quad \square$$

Since both the null space and the row space of any matrix are invariant under elementary row operations, it is interesting to study the relationship between $\mathbf{row(A)}$ and $N(\mathbf{A})$. To do this we send any row $\mathbf{x}$ in $\mathbf{row(A)}$ to $\mathbb{R}^n$ by the transposition $\mathbf{x} \to \mathbf{x}^T$. Then we obtain a subspace $\mathbf{row(A)}^T$ of $\mathbb{R}^n$ which we call the row space of $\mathbf{A}$ in $\mathbb{R}^n$. If $\mathbf{x}$ is any vector in $\mathbf{row(A)}^T$ then

$$\mathbf{x} = \alpha_1 \text{row}_1(\mathbf{A})^T + \alpha_2 \text{row}_2(\mathbf{A})^T + \cdots + \alpha_m \text{row}_m(\mathbf{A})^T, \tag{2.24}$$

where m denotes the number of rows of $\mathbf{A}$.

Lemma 2.5 *Any vector* $\mathbf{y}$ *in* $N(\mathbf{A})$ *is orthogonal to any* $\mathbf{x}$ *in* $\mathbf{row(A)}^T$.

Proof. Since $\mathbf{y}$ is in the null space of $\mathbf{A}$, it is perpendicular to any row of $\mathbf{A}$:

$$\langle \mathbf{x}, \mathbf{y} \rangle = \mathbf{x}^T \mathbf{y} = \alpha_1 \text{row}_1(\mathbf{A}) \mathbf{y} + \cdots + \alpha_n \text{row}_n(\mathbf{A}) \mathbf{y} = 0. \qquad \square$$

Definition 2.13 We say that two vector subspaces U and V in $\mathbb{R}^n$ are orthogonal if every vector $\mathbf{u}$ in U is orthogonal to any vector $\mathbf{v}$ in $\mathbf{V}$.

Orthogonal subspaces have an important property.

Lemma 2.6 *Suppose that U and V are orthogonal subspaces of $\mathbb{R}^n$. Let $\{\mathbf{u}_1, \ldots, \mathbf{u}_r\}$ be a linearly independent set of vectors in U and $\{\mathbf{v}_1, \ldots, \mathbf{v}_s\}$ in V. Then $\{\mathbf{u}_1, \ldots, \mathbf{u}_r, \mathbf{v}_1, \ldots, \mathbf{v}_s\}$ is linearly independent.*

Proof. Let $\alpha_1, \ldots, \alpha_r, \beta_1, \ldots, \beta_s$ be real numbers such that

$$\mathbf{0}_n = \underbrace{\alpha_1 \mathbf{u}_1 + \cdots + \alpha_r \mathbf{u}_r}_{\mathbf{u}} + \underbrace{\beta_1 \mathbf{v}_1 + \cdots + \beta_r \mathbf{v}_r}_{\mathbf{v}} = \mathbf{u} + \mathbf{v} \Rightarrow \mathbf{u} = -\mathbf{v}.$$

Then

$$|\mathbf{u}|^2 = \langle \mathbf{u}, \mathbf{u} \rangle = -\langle \mathbf{u}, \mathbf{v} \rangle = 0,$$

since $\mathbf{u} \perp \mathbf{v}$. It follows that $\mathbf{u} = \mathbf{v} = \mathbf{0}_n$. Observing that the vectors $\{\mathbf{u}_1, \ldots, \mathbf{u}_r\}$ are linearly independent we obtain that $\alpha_1 = \cdots = \alpha_r = 0$. Observing that the vectors $\{\mathbf{v}_1, \ldots, \mathbf{v}_s\}$ are linearly independent we obtain that $\beta_1 = \cdots = \beta_s = 0$. $\qquad \square$

In practice, it is important to know if the sums of all vectors in U and V make the whole space $\mathbb{R}^n$. The following Theorem gives an answer to this question.

Theorem 2.13 *Let U and V be orthogonal subspaces in $\mathbb{R}^n$. If*

$$\dim(U) + \dim(V) = n \tag{2.25}$$

then every vector **x** *in $\mathbb{R}^n$ can be uniquely decomposed into the sum*

$$\mathbf{x} = \mathbf{u} + \mathbf{v}$$

of two orthogonal vectors **u** *in U and* **v** *in V.*

Proof. If $U = \{\mathbf{0}_n\}$ then $\mathbf{u} = 0$ and there is nothing to prove. The same arguments work for the case $V = \{\mathbf{0}_n\}$. Suppose that $U \neq \{\mathbf{0}_n\}$ and $V \neq \{\mathbf{0}_n\}$. By Theorem 2.1 the subspace U has a basis $\{\mathbf{u}_1, \ldots, \mathbf{u}_r\}$ and V has a basis $\{\mathbf{v}_1, \ldots, \mathbf{v}_s\}$. By Lemma 2.6 the set of vectors

$$\{\mathbf{u}_1, \ldots, \mathbf{u}_r, \mathbf{v}_1, \ldots, \mathbf{v}_s\} \tag{2.26}$$

is linearly independent. By (2.25) we have $r + s = n$. By Corollary 2.3 the set (2.26) is a basis for $\mathbb{R}^n$. Then every **x** in $\mathbb{R}^n$ has a unique decomposition

$$\mathbf{x} = \underbrace{\alpha_1 \mathbf{u}_1 + \cdots + \alpha_r \mathbf{u}_r}_{\mathbf{u}} + \underbrace{\beta_1 \mathbf{v}_1 + \cdots + \beta_r \mathbf{v}_r}_{\mathbf{v}} = \mathbf{u} + \mathbf{v},$$

where **u** is in U and **v** is in V. $\qquad\square$

If U and V are orthogonal vector subspaces of $\mathbb{R}^n$ satisfying (2.25) then we use the notations for the **orthogonal sum** of U and V:

$$\mathbb{R}^n = U \oplus V = V \oplus U.$$

Since the vector addition is commutative, the order of subspaces in the above formula is not important. The following Lemma is useful in applications.

Lemma 2.7 *Let U, V_1, V_2 be subspaces in $\mathbb{R}^n$ such that*

$$U \oplus V_1 = U \oplus V_2. \tag{2.27}$$

Then $V_1 = V_2$.

Proof. Let **x** be any vector in V_1. By (2.27) it is the sum $\mathbf{u} + \mathbf{y}$, were **u** is a vector in U and **y** is a vector in V_2. It follows that $\mathbf{x} - \mathbf{y}$ is in V_2. Since both vectors **x** and **y** are perpendicular to U, their difference **u** in U is perpendicular to U too. It follows that $\mathbf{u} = \mathbf{0}$. This shows that V_1 is a subspace of V_2. Similarly, V_2 is a subspace of V_1. Hence $V_1 = V_2$ as stated. $\qquad\square$

Theorem 2.13 and the Rank-Nullity Theorem claim that for any $m \times n$ matrix $\mathbf{A}$

$$\mathbf{row}(\mathbf{A})^T \oplus N(\mathbf{A}) = \mathbb{R}^n. \tag{2.28}$$

We prove this theorem in the next section and after that consider Gauss bases for subspaces of $\mathbb{R}^n$.

2.7 The Rank-Nullity Theorem

Theorem 2.14 (The Rank-Nullity Theorem) *For any $m \times n$ matrix $\mathbf{A}$*

$$\mathrm{rank}(\mathbf{A}) + \mathrm{nullity}(\mathbf{A}) = n.$$

Proof. The nullity $\mathrm{nullity}(\mathbf{A})$ of $\mathbf{A}$ is the number of non-leading variables. The rank $\mathrm{rank}(\mathbf{A})$ of $\mathbf{A}$ is the number of leading variables. Since the total number of variables is n, the result follows.   □

Problem 2.15 Show that any $m \times n$ matrix $\mathbf{A}$ with linear independent rows and columns is a square matrix: $m = n$.   □

Solution: Since the rows of $\mathbf{A}$ are linearly independent, $\mathrm{rank}(\mathbf{A}) = m$. Since the columns of $\mathbf{A}$ are linearly independent, $\mathrm{nullity}(\mathbf{A}) = 0$. By the Rank-Nullity Theorem $n = m + 0 = m$ as stated.   □

Theorem 2.15 *For every $m \times n$ matrix $\mathbf{A}$*

$$\mathbf{row}(\mathbf{A})^T \oplus N(\mathbf{A}) = \mathbb{R}^n.$$

Proof. By Lemma 2.5 the subspaces $\mathbf{row}(\mathbf{A})^T$ and $N(\mathbf{A})$ are orthogonal in $\mathbb{R}^n$. By Theorem 2.14

$$\dim\left(\mathbf{row}(\mathbf{A})^T\right) + \dim\left(N(\mathbf{A})\right) = \mathrm{rank}(\mathbf{A}) + \mathrm{nullity}(\mathbf{A}) = n.$$

The result follows by Theorem 2.13.   □

Theorem 2.16 *Let V be a subspace of $\mathbb{R}^n$, $\dim(V) = n - r$. Then there is a $r \times n$ matrix $\mathbf{A}$ in reduced row echelon form such that $N(\mathbf{A}) = V$.*

Proof. By Corollary 2.1 the subspace V equals $\mathbf{row}(\mathbf{B})^T$ for some $(n - r) \times n$ matrix $\mathbf{B}$ with linearly independent rows. Theorem 2.15 shows that

$$\mathbb{R}^n = V \oplus N(\mathbf{B}) = N(\mathbf{B}) \oplus V. \tag{2.29}$$

By Theorem 2.14, $\dim(N(\mathbf{B})) = n - (n - r) = r$. By Corollary 2.1 there is some $r \times n$ matrix $\mathbf{A}$ such that $N(\mathbf{B}) = \mathbf{row}(\mathbf{A})^T$. By (2.29), Lemma 2.7, and by Theorem 2.15,

$$\mathbb{R}^n = \mathbf{row}(\mathbf{A})^T \oplus N(\mathbf{A}) = \left\{ \begin{array}{l} N(\mathbf{B}) \oplus N(\mathbf{A}) \\ N(\mathbf{B}) \oplus V \end{array} \right. \Rightarrow V = N(\mathbf{A}).$$

Then $\mathrm{rref}(\mathbf{A})$ is the matrix with the same row space and null-space as the matrix $\mathbf{A}$.   □

Theorem 2.17 *Two $m \times n$ matrices $\mathbf{A}$ and $\mathbf{B}$ are row equivalent if and only if $N(\mathbf{A}) = N(\mathbf{B})$.*

Proof. If $\mathbf{A} \sim \mathbf{B}$, then $N(\mathbf{A}) = N(\mathbf{B})$ by Corollary 1.1. If $N(\mathbf{A}) = N(\mathbf{B})$, then $\mathbf{row}(\mathbf{A}) = \mathbf{row}(\mathbf{B})$ by Theorem 2.15, which implies that $\mathbf{A} \sim \mathbf{B}$ by Theorem 1.28.   □

Theorem 2.18 *Let $\mathbf{A}$ and $\mathbf{B}$ be matrices with equal numbers of columns. Then*

$$\mathrm{rref}_p(\mathbf{A}) = \mathrm{rref}_p(\mathbf{B})$$

if and only if $N(\mathbf{A}) = N(\mathbf{B})$.

> *Proof.* By Theorem 2.15 the null-spaces $N(\mathbf{A})$ and $N(\mathbf{B})$ are equal if and only if $\mathbf{row}(\mathbf{A}) = \mathbf{row}(\mathbf{B})$. Theorem 1.28 completes the proof. □

By Theorem 2.16 every subspace V of $\mathbb{R}^n$ is the null space of an $r \times n$ matrix $\mathbf{A}$ in reduced row echelon form, where $r = \dim(V)$. By Theorem 2.18 this matrix is unique.

> **Definition 2.14** Let V be a subspace of $\mathbb{R}^n$ and let $r = \dim(V)$. Then the $r \times n$ matrix $\mathrm{rref}(V)$ is called the **reduced row echelon form** of V if it is in reduced row echelon form and its null-space is V. The leading list J of $\mathrm{rref}(V)$ is called the **leading list of a subspace**.

The plane $x + 2y + z = 0$ is a subspace V in $\mathbb{R}^3$. Its reduced row echelon form is $\begin{pmatrix} 1 & 2 & 1 \end{pmatrix}$. The leading list of V is $J = \{1\}$.

> **Definition 2.15** Let V be a subspace of $\mathbb{R}^n$, $\dim(V) = n - r$. Then the Gauss basis for the null-space of $\mathrm{rref}(V)$ is called the **Gauss basis for the subspace** V.

The Gauss basis for the subspace $x + 2y + z = 0$ is

$$\begin{pmatrix} -2 & 1 & 0 \end{pmatrix}^T, \quad \begin{pmatrix} -1 & 0 & 1 \end{pmatrix}^T.$$

The reduced row echelon form for the subspace $z = 0$ is the matrix $\begin{pmatrix} 0 & 0 & 1 \end{pmatrix}$. The leading list is $J = \{1,2\}$, implying that the Gauss matrix for the $\mathrm{rref}(V)$ is the identity matrix $\mathbf{I}_3$. The Gauss basis for V is, therefore, $\{\mathbf{e}_1, \mathbf{e}_2\}$.

If $\mathbf{A} = \mathbf{0}_{m \times n}$ then its Gauss matrix $\mathbf{G_A}$ is the identity matrix $\mathbf{I}_n$. Therefore, the Gauss basis for $\mathbb{R}^n$ is the standard basis

$$\{\mathbf{e}_1, \mathbf{e}_2, \ldots, \mathbf{e}_n\}.$$

Problem 2.16 Find the Gauss basis in $\mathbb{R}^3$ for the plane $x + y + z = 0$.

Solution: The coefficient matrix of the system $x + y + z = 0$ is the 1×3 matrix $\mathbf{A} = \begin{pmatrix} 1 & 1 & 1 \end{pmatrix}$. Its Gauss matrix is

$$\mathbf{G_A} = \begin{pmatrix} 1 & -1 & -1 \\ 0 & 1 & 0 \\ 0 & 0 & 1 \end{pmatrix}.$$

Since $N(\mathbf{A}) = \mathrm{Lin}\{\mathbf{g}_2, \mathbf{g}_3\}$ the Gauss bases in $\mathbb{R}^3$ for this plane is given by

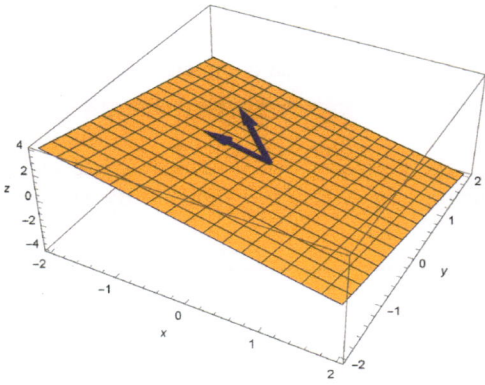

Fig. 2.11: The Gauss basis for $x + y + z = 0$ in $\mathbb{R}^3$.

$$\left(-1\ 1\ 0\right)^T,\ \left(-1\ 0\ 1\right)^T.$$

The first vector of this basis on Fig. 2.11 is shown on the left. By (1.51), these vectors make the angle of 60° degrees.   □

In general, given an $m \times n$ matrix $\mathbf{A}$ to find the Gauss basis for $\mathbf{row}(\mathbf{A})$ we construct the extended row echelon form $\mathrm{eref}(\mathbf{A})$. Then the required Gauss basis is made by the rows of $\mathrm{eref}(\mathbf{A})$ from the leading list. To find the Gauss basis for the null space $N(\mathbf{A})$ we construct the Gauss matrix $\mathbf{G_A}$. Then the Gauss basis for $N(\mathbf{A})$ is made by the columns of $\mathbf{G_A}$ from the complementary list.

2.8 Affine Subspaces

Definition 2.16 A non-empty subset W of $\mathbb{R}^n$ is called an **affine** subspace of $\mathbb{R}^n$ if there exist a subspace V of $\mathbb{R}^n$ and a vector $\mathbf{v}$ in $\mathbb{R}^n$ such that W equals the sum $\mathbf{v} + V$. In other words, affine subspaces are additive shifts of subspaces.

By Theorem 2.5 the set of solutions to a consistent system $\mathbf{Ax} = \mathbf{b}$ is described as $\mathbf{p} + N(\mathbf{A})$, where $\mathbf{p}$ is the Gauss particular solution. It follows that the set of all solutions for a consistent system $\mathbf{Ax} = \mathbf{b}$ is an **affine** subspace.

Theorem 2.19 *A non-empty subset W of $\mathbb{R}^n$ is an affine subspace of $\mathbb{R}^n$ if and only if there are a matrix $\mathbf{A}$ and a vector $\mathbf{p}$ in $P(\mathbf{A})$ such that $W = \mathbf{p} + N(\mathbf{A})$. The vector $\mathbf{p}$ is uniquely determined by W.*

Proof. By definition $W = \mathbf{v} + V$, where V is a subspace of $\mathbb{R}^n$. If $\dim(V) = n - r$, then by Theorem 2.16 there is a $r \times n$ matrix $\mathbf{A}$ such that $N(\mathbf{A}) = V$. Since the system $\mathbf{Ax} = \mathbf{Av}$ is consistent, it has the unique Gauss particular solution $\mathbf{p}$. Then $\mathbf{A}(\mathbf{p} - \mathbf{v}) = \mathbf{Av} - \mathbf{Av} = \mathbf{0}$. It follows that

$$W = \mathbf{v} + N(\mathbf{A}) = \mathbf{p} + \mathbf{v} - \mathbf{p} + N(\mathbf{A}) = \mathbf{p} + N(\mathbf{A}).$$

The set $\mathbf{W} = \mathbf{p} + N(\mathbf{A})$ is an affine subspace by definition. $\square$

Theorem 2.20 *If two consistent systems of m equations in n unknowns* $\mathbf{Ax} = \mathbf{b}$ *and* $\mathbf{Bx} = \mathbf{c}$ *have equal solutions sets then the augmented matrices* $(\mathbf{A}|\mathbf{b})$ *and* $(\mathbf{B}|\mathbf{c})$ *are row equivalent.*

Proof. The set of solutions of any consistent system $\mathbf{Ax} = \mathbf{b}$ is the affine subspace $\mathbf{p} + N(\mathbf{A})$. The set of solutions of $\mathbf{Bx} = \mathbf{c}$ is $\mathbf{q} + N(\mathbf{B})$. Since these affine sets coincide,

$$\mathbf{p} + N(\mathbf{A}) = \mathbf{q} + N(\mathbf{B}).$$

In particular, $\mathbf{q} + \mathbf{0}$ is in $\mathbf{p} + N(\mathbf{A})$, implying that $\mathbf{q} - \mathbf{p}$ is in $N(\mathbf{A})$. Similarly, $\mathbf{p} - \mathbf{q}$ is in $N(\mathbf{B})$. Then

$$\mathbf{q} + N(\mathbf{A}) = \mathbf{p} + (\mathbf{q} - \mathbf{p}) + N(\mathbf{A}) = \mathbf{p} + N(\mathbf{A}) = \mathbf{q} + N(\mathbf{B}).$$

It follows that $N(\mathbf{A}) = N(\mathbf{B})$ and $\mathbf{p} = \mathbf{q}$. Then $\mathrm{rref}(\mathbf{A}) = \mathrm{rref}(\mathbf{B})$ and, therefore,

$$(\mathbf{A}|\mathbf{b}) \sim (\mathrm{rref}(\mathbf{A})|\mathrm{rref}(\mathbf{A})\mathbf{p}) = (\mathrm{rref}(\mathbf{B})|\mathrm{rref}(\mathbf{B})\mathbf{p}) \sim (\mathbf{B}|\mathbf{c}).   \square$$

2.9 Symmetric Matrices

Definition 2.17 A matrix $\mathbf{A}$ is called **symmetric** if $\mathbf{A}^T = \mathbf{A}$.

If $\mathbf{A}$ is an $m \times n$ matrix then $\mathbf{A}^T$ is an $n \times m$ matrix. Therefore, $\mathbf{A}^T = \mathbf{A}$ implies that $m = n$. It follows that any symmetric matrix is a square matrix, i.e. it is an $n \times n$ matrix.

Theorem 2.21 *Let* $\mathbf{A}$ *be a symmetric matrix of size* $n \times n$. *Then*

$$R(\mathbf{A}) \oplus N(\mathbf{A}) = \mathbb{R}^n.$$

Proof. Since $\mathbf{A}^T = \mathbf{A}$, we have $\mathbf{row}(\mathbf{A})^T = \mathbf{col}(\mathbf{A})$. The proof is completed by Theorem 2.15. $\qquad\square$

Corollary 2.8 *Let* $\mathbf{A}$ *be a symmetric matrix. Then*

$$N(\mathbf{A}^k) = N(\mathbf{A}), \ k = 2,3,\ldots.$$

Proof. If $k \geq 2$ and $\mathbf{x}$ is in $N(\mathbf{A}^k)$, then

$$\mathbf{0} = \mathbf{A}^k \mathbf{x} = \mathbf{A}(\mathbf{A}^{k-1}\mathbf{x}),$$

implying that $\mathbf{A}^{k-1}\mathbf{x}$ is a vector in $N(\mathbf{A})$ and at the same time in $R(\mathbf{A})$. By Theorem 2.21 we conclude that $\mathbf{A}^{k-1}\mathbf{x} = \mathbf{0}$. If $k - 1 = 1$, then $\mathbf{x}$ is in $N(\mathbf{A})$. Otherwise, we can repeat the above arguments and in $k - 1$ steps conclude that $\mathbf{x}$ is in $N(\mathbf{A})$. $\quad\square$

Theorem 2.22 *Let* $\mathbf{A}$ *be an* $m \times n$ *matrix. Then*

$$\mathrm{rref}_p(\mathbf{A}^T \mathbf{A}) = \mathrm{rref}_p(\mathbf{A}).$$

Proof. The formula $\mathbf{x}^T \mathbf{A}^T \mathbf{A} \mathbf{x} = \langle \mathbf{A}\mathbf{x}, \mathbf{A}\mathbf{x} \rangle = |\mathbf{A}\mathbf{x}|^2$ shows that $N(\mathbf{A}) = N(\mathbf{A}^T \mathbf{A})$. By Theorem 2.15 we obtain that $\mathbf{row}(\mathbf{A}^T \mathbf{A}) = \mathbf{row}(\mathbf{A})$. The result follows by Theorem 1.28. $\qquad\square$

Corollary 2.9 *Let* $\mathbf{A}$ *be an* $m \times n$ *matrix. Then*

$$\mathrm{rank}(\mathbf{A}^T \mathbf{A}) = \mathrm{rank}(\mathbf{A}).$$

Corollary 2.10 *Let* $\mathbf{A}$ *be an* $n \times n$ *matrix. Then*

$$\mathrm{rref}((\mathbf{A}^T \mathbf{A})^k) = \mathrm{rref}(\mathbf{A}), \ k = 1,2,\ldots.$$

Proof. Apply Corollary 2.8. $\qquad\square$

2.10 Fredholm Alternative and Farkas' Lemma

Theorem 2.23 (Fredholm Alternative (1903)) *Let $\mathbf{A}$ be an $m \times n$ matrix and $\mathbf{b}$ be a vector in $\mathbb{R}^m$. Then exactly one of the following two statements is true.*

(a) The equation $\mathbf{A}\mathbf{x} = \mathbf{b}$ has a solution $\mathbf{x}$ in $\mathbb{R}^n$.
(b) There is a vector $\mathbf{y}$ in $\mathbb{R}^m$ such that

$$\mathbf{y}^T \mathbf{A} = \mathbf{0}_{1 \times n} \ \ and \ \ \mathbf{y}^T \mathbf{b} < 0.$$

Proof. If the system $\mathbf{A}\mathbf{x} = \mathbf{b}$ is inconsistent then $\mathbf{b}$ is not in

$$R(\mathbf{A}) = \mathbf{col}(\mathbf{A}) = \mathbf{row}(\mathbf{A}^T)^T.$$

By Theorem 2.15 applied to $\mathbf{A}^T$

$$\mathbb{R}^m = \mathbf{row}(\mathbf{A}^T)^T \oplus N(\mathbf{A}^T) = \mathbf{col}(\mathbf{A}) \oplus N(\mathbf{A}^T).$$

The set of all vectors $\mathbf{y}$ in $\mathbb{R}^m$ such that $\mathbf{y}^T \mathbf{A} = \mathbf{0}_{1 \times n}$ coincides with the set of all vectors in $N(\mathbf{A}^T)$:

$$\mathbf{0}_{n \times 1} = \mathbf{0}_{1 \times n}^T = (\mathbf{y}^T \mathbf{A})^T = \mathbf{A}^T \mathbf{y}.$$

Since $\mathbf{b}$ is not in $\mathbf{col}(\mathbf{A})$, we conclude that $\mathbf{b}$ cannot be orthogonal to the null-space of $\mathbf{A}^T$. It follows that $\mathbf{b} = \mathbf{u} \oplus \mathbf{v}$, where $\mathbf{u}$ is a vector in $\mathbf{col}(\mathbf{A})$ and $\mathbf{v}$ is a non-zero vector in $\mathbf{A}^T$. Then the vector $\mathbf{y} = -\mathbf{v}$ satisfies conditions stated in (**b**). Indeed, $\mathbf{v}$ is in $N(\mathbf{A}^T)$ and

$$\mathbf{v}^T \mathbf{b} = \mathbf{v}^T \mathbf{u} + \mathbf{v}^T \mathbf{v} = \mathbf{v}^T \mathbf{v} > 0. \qquad \square$$

Problem 2.17 Confirm the Fredholm Alternative for

$$\mathbf{A} = \begin{pmatrix} 2 & 1 & 4 & 2 & 9 \\ 2 & -1 & 0 & 3 & 5 \\ 1 & 3 & 7 & 1 & 12 \\ 1 & 2 & 5 & -1 & 5 \end{pmatrix} \text{ and } \mathbf{b}_t = \begin{pmatrix} 1 \\ 0 \\ 1 \\ t \end{pmatrix},$$

where t is a real number.

Solution: The augmented matrix $(\mathbf{A} \mid \mathbf{b}_t)$ is row equivalent to the following row echelon matrix

$$\begin{pmatrix} 1\,0\,1\,0\,1 & (2t-1) \\ 0\,1\,2\,0\,3 & (5-t)/15 \\ 0\,0\,0\,1\,2 & (5-7t)/15 \\ 0\,0\,0\,0\,0 & (2-t)/3 \end{pmatrix}.$$

By the Kronecker-Cappelli Theorem the system $\mathbf{Ax} = \mathbf{b}_t$ is consistent if and only if $t = 2$. The reduced row echelon form of $\mathbf{A}^T$ is

$$\begin{pmatrix} 1 & 0 & 0 & 3 \\ 0 & 1 & 0 & -2 \\ 0 & 0 & 1 & -1 \\ 0 & 0 & 0 & 0 \\ 0 & 0 & 0 & 0 \end{pmatrix} \Rightarrow \mathbf{g}_4 = \begin{pmatrix} -3 \\ 2 \\ 1 \\ 1 \end{pmatrix} \Rightarrow \mathrm{N}(\mathbf{A}^T) = \mathrm{Lin}(\mathbf{g}_4).$$

It follows that

$$\mathbf{g}_4^T \mathbf{b}_t = -3 + 1 + t \neq 0 \Leftrightarrow t \neq 2,$$

as it is stated by the Fredholm Alternative.   □

Remark. The options (**a**) and (**b**) in Theorem 2.23 cannot hold simultaneously. This is the reason for the name 'Fredholm Alternative'.

Theorem 2.24 *Let* $\mathbf{B}$ *be a* $k \times m$ *matrix and* $\mathbf{A}$ *be an* $m \times n$ *matrix. Then*

$$\mathrm{nullity}(\mathbf{BA}) \leq \mathrm{nullity}(\mathbf{B}) + \mathrm{nullity}(\mathbf{A}). \qquad (2.30)$$

Proof. The Gauss matrix of $\mathbf{A}$ determines the direct sum $P(\mathbf{A}) + N(\mathbf{A}) = \mathbb{R}^n$. Since the sum is direct, the matrix $\mathbf{A}$ maps $P(\mathbf{A})$ as a one-to-one mapping. Indeed, If $\mathbf{Ap}_1 = \mathbf{Ap}_2$ then $\mathbf{A}(\mathbf{p}_1 - \mathbf{p}_2) = \mathbf{0}$ implying that $\mathbf{p}_1 - \mathbf{p}_2$ belongs to both $P(\mathbf{A})$ and $N(\mathbf{A})$. It follows that $\mathbf{p}_1 = \mathbf{p}_2$ as stated.

Let $P_0(\mathbf{A})$ be the set off all vectors $\mathbf{p}$ in $P(\mathbf{A})$ such that $\mathbf{Ap}$ is a vector in $N(\mathbf{B})$. Then

$$N(\mathbf{BA}) = P_0(\mathbf{A}) + N(\mathbf{A}).$$

Indeed, if $\mathbf{BAx} = \mathbf{0}$ then $\mathbf{x} = \mathbf{p} + \mathbf{u}$, where $\mathbf{p}$ is in $P(\mathbf{A})$ and $\mathbf{u}$ is in $N(\mathbf{A})$. Since $\mathbf{BAu} = \mathbf{B0} = \mathbf{0}$, we conclude that $\mathbf{BAp} = \mathbf{0}$ implying that $\mathbf{p}$ is in $P_0(\mathbf{A})$ and $\mathbf{u}$ is in $N(\mathbf{A})$. If $\mathbf{x}$ is in $P_0(\mathbf{A}) + N(\mathbf{A})$ then obviously $\mathbf{x}$ is in $N(\mathbf{BA})$. It follows that

$$\dim (N(\mathbf{BA})) = \dim (P_0(\mathbf{A})) + \dim (N(\mathbf{A})).$$

Since $\mathbf{A}$ is a one-to-one linear mapping of $P_0(\mathbf{A})$ onto a subspace of $N(\mathbf{B})$, we conclude that

$$\dim (P_0(\mathbf{A})) \leq \dim (N(\mathbf{B})),$$

which completes the proof of the Theorem.   □

Corollary 2.11 (Sylvester's Inequality) *Let* $\mathbf{B}$ *be a* $k \times m$ *matrix and* $\mathbf{A}$ *be an* $m \times n$ *matrix. Then*

$$\text{rank}(\mathbf{B}) + \text{rank}(\mathbf{A}) \leq \text{rank}(\mathbf{BA}) + m.$$

Proof. By the Rank-Nullity Theorem

$$\text{rank}(\mathbf{B}) = m - \text{nullity}(\mathbf{B}), \ \text{rank}(\mathbf{A}) = n - \text{nullity}(\mathbf{A}), \ \text{rank}(\mathbf{BA}) = n - \text{nullity}(\mathbf{BA}).$$

It follows that

$$\text{rank}(\mathbf{BA}) + m - (\text{rank}(\mathbf{B}) + \text{rank}(\mathbf{A})) =$$
$$n - \text{nullity}(\mathbf{BA}) + m - (m - \text{nullity}(\mathbf{B}) + n - \text{nullity}(\mathbf{A})) =$$
$$\text{nullity}(\mathbf{B}) + \text{nullity}(\mathbf{A}) - \text{nullity}(\mathbf{BA}) \geq 0. \quad \square$$

By Theorem 2.11, $\text{rank}(\mathbf{BA}) \leq \text{rank}(\mathbf{A})$. If $\mathbf{B}$ is an invertible $m \times m$ matrix, then $\text{rank}(B) = m$ and Corollary 2.11 shows that $\text{rank}(\mathbf{A}) \leq \text{rank}(\mathbf{BA})$. It follows that $\text{rank}(\mathbf{BA}) = \text{rank}(\mathbf{A})$ for any invertible matrix $\mathbf{B}$. This is in agreement with Theorem 1.8 and Theorem 2.4.

Farkas' lemma (1902) is an analogue of the Fredholm Alternative for non-negative solutions to matrix equations.

Theorem 2.25 (Farkas' Lemma) *Let* A *be an* $m \times n$ *matrix and* **b** *be a vector in* $\mathbb{R}^m$. *Then exactly one of the following two statements is true.*

(a)*The equation* $\mathbf{Ax} = \mathbf{b}$ *has a **non-negative** solution* **x** *in* $\mathbb{R}^n$.
(b)*There is a vector* **y** *in* $\mathbb{R}^m$ *such that*

$$\mathbf{y}^T \mathbf{A} \geq \mathbf{0}_{1 \times n} \ \text{and} \ \mathbf{y}^T \mathbf{b} < 0.$$

See Rockafellar (1990), Corollary 22.3.1 for the detailed proof of this theorem. The idea of the proof is based on separation theorems in Convex Analysis. The set

$$C(\mathbf{A}) = \{\mathbf{Ax} \mid \mathbf{x} \geq \mathbf{0}\} = \left\{ \sum_{i=1}^{n} x_i \text{Col}_i(\mathbf{A}) \,\middle|\, x_i \geq 0 \right\}$$

is a closed cone in $\mathbb{R}^m$ generated by the columns of the matrix $\mathbf{A}$. If $\mathbf{b} \in C(\mathbf{A})$, then condition **(a)** holds. Otherwise, we have $\mathbf{b} \notin C(\mathbf{A})$. By Theorem 7.7.4 of Narici L. (Beckenstein E.), there is a hyperplane $H_{\mathbf{y}} = \{\mathbf{z} \in \mathbb{R}^m \mid \mathbf{z}^T \mathbf{y} = 0\}$ separating $C(A)$ and **b**, which is equivalent to **(b)**.

Farkas' Lemma can be used to prove the difficult part of the Non-Arbitrage Theorem (see Theorem 1.31). Let $\mathbf{A} = \mathbf{R}$ be a returns matrix and $\mathbf{b} = \mathbf{1}_m$. The absence of state prices for $\mathbf{R}$ means that $\mathbf{1}_m \notin C(\mathbf{R})$. By Farkas' Lemma, there is a portfolio **y** in $\mathbb{R}^m$ such that

$$\mathbf{y}^T \mathbf{R} \geq \mathbf{0}_{1 \times n} \text{ and } -c = \mathbf{y}^T \mathbf{1}_m = \sum_{i=1}^{m} y_i < 0.$$

Then $\mathbf{y} + c\mathbf{I}_m$ is an arbitrage portfolio.

2.11 Gauss Elimination: Problems

Problem 2.18 Find the Gauss bases for the row, range, null spaces, and particular solutions of the matrix

$$\mathbf{A} = \begin{pmatrix} 2 & 4 & 2 & 6 & -1 \\ 6 & 12 & -1 & 4 & 3 \\ 3 & 6 & 0 & 3 & 1 \end{pmatrix}. \tag{2.31}$$

Solution: Elementary calculations show that

$$\mathrm{rref}(\mathbf{A}) = \begin{pmatrix} 1 & 2 & 0 & 1 & 0 \\ 0 & 0 & 1 & 2 & 0 \\ 0 & 0 & 0 & 0 & 1 \end{pmatrix}.$$

The non-leading variables are $x_2 = s$, $x_4 = t$. It follows that $\mathbf{x}$ is in $N(\mathbf{A})$ if and only if

$$\mathbf{x} = \begin{pmatrix} -2s - t \\ s \\ -2t \\ t \\ 0 \end{pmatrix} = s \begin{pmatrix} -2 \\ 1 \\ 0 \\ 0 \\ 0 \end{pmatrix} + t \begin{pmatrix} -1 \\ 0 \\ -2 \\ 1 \\ 0 \end{pmatrix} = s\mathbf{v}_2 + t\mathbf{v}_4 \Rightarrow N(\mathbf{A}) = \mathrm{Lin}\{\mathbf{v}_2, \mathbf{v}_4\}.$$

The Gauss basis for $N(\mathbf{A})$ is $\{\mathbf{v}_2, \mathbf{v}_4\}$. The Gauss basis for the row space is made by the three rows of $\mathrm{rref}(\mathbf{A})$. The Gauss basis for the range of $\mathbf{A}$ is made by $\{\mathrm{col}_1(\mathbf{A}), \mathrm{col}_3(\mathbf{A}), \mathrm{col}_5(\mathbf{A})\}$. The Gauss basis for particular solutions is $\{\mathbf{e}_1, \mathbf{e}_3, \mathbf{e}_5\}$. Notice that

$$\mathrm{col}_2(\mathbf{A}) = 2\mathrm{col}_1(\mathbf{A}), \quad \mathrm{col}_4(\mathbf{A}) = \mathrm{col}_1(\mathbf{A}) + 2\mathrm{col}_3(\mathbf{A}),$$

which is in agreement with Theorem 2.9. $\square$

Problem 2.19 The subspace V is spanned by the following vectors

$$\mathbf{u}_1 = \begin{pmatrix} 1 & 2 & 1 & -1 & 2 & -2 \end{pmatrix}^T$$
$$\mathbf{u}_2 = \begin{pmatrix} 2 & 1 & -1 & -1 & 0 & 3 \end{pmatrix}^T$$
$$\mathbf{u}_3 = \begin{pmatrix} 1 & -1 & -2 & 0 & -2 & 5 \end{pmatrix}^T$$

Find the reduced row echelon form of a 5×6 matrix $\mathbf{A}$ with $N(\mathbf{A}) = V$. Determine the Gauss basis for V.

Solution: Any row $\begin{pmatrix} x_1 & x_2 & x_3 & x_4 & x_5 & x_6 \end{pmatrix}$ of $\mathbf{A}$ satisfies the system

$$\begin{cases} x_1 + 2x_2 + x_3 - x_4 + 2x_5 - 2x_6 & = 0 \\ 2x_1 + x_2 - x_3 - x_4 + 3x_6 & = 0 \\ x_1 - x_2 - 2x_3 - 2x_5 + 5x_6 & = 0 \end{cases} \qquad (2.32)$$

We find the reduced row echelon form of the coefficient matrix of the system (2.32) and determine its rank equal 2:

$$\mathrm{rref} \begin{pmatrix} 1 & 2 & 1 & -1 & 2 & -2 \\ 2 & 1 & -1 & -1 & 0 & 3 \\ 1 & -1 & -2 & 0 & -2 & 5 \end{pmatrix} = \begin{pmatrix} 1 & 0 & -1 & -1/3 & -2/3 & 8/3 \\ 0 & 1 & 1 & -1/3 & 4/3 & -7/3 \\ 0 & 0 & 0 & 0 & 0 & 0 \end{pmatrix}.$$

Then by the Rank-Nullity Theorem $\dim(V) = 6-2 = 4$. We see that $x_3 = s$, $x_4 = 3t$, $x_5 = 3u$, $x_6 = 3v$ are non-leading variables for (2.32). Hence

$$\mathbf{x} = \begin{pmatrix} s + t + 2u - 8v \\ -s + t - 4u + 7v \\ s \\ 3t \\ 3u \\ 3v \end{pmatrix} = s \begin{pmatrix} 1 \\ -1 \\ 1 \\ 0 \\ 0 \\ 0 \end{pmatrix} + t \begin{pmatrix} 1 \\ 1 \\ 0 \\ 3 \\ 0 \\ 0 \end{pmatrix} + u \begin{pmatrix} 2 \\ -4 \\ 0 \\ 0 \\ 3 \\ 0 \end{pmatrix} + v \begin{pmatrix} -8 \\ 7 \\ 0 \\ 0 \\ 0 \\ 3 \end{pmatrix}. \qquad (2.33)$$

Let

$$\mathbf{A} = \begin{pmatrix} 1 & -1 & 1 & 0 & 0 & 0 \\ 1 & 1 & 0 & 3 & 0 & 0 \\ 2 & -4 & 0 & 0 & 3 & 0 \\ -8 & 7 & 0 & 0 & 0 & 3 \\ 0 & 0 & 0 & 0 & 0 & 0 \end{pmatrix}$$

be the 5×6 matrix whose first four rows are the transposes of the basis found in (2.33). Since the first four rows of $\mathbf{A}$ are linearly independent, we conclude that $\mathrm{rank}(\mathbf{A}) = 4$. Elementary calculations show that

$$\mathrm{rref}(\mathbf{A}) = \begin{pmatrix} 1 & 0 & 0 & 0 & -7/6 & -2/3 \\ 0 & 1 & 0 & 0 & -4/3 & -1/3 \\ 0 & 0 & 1 & 0 & -1/6 & 1/3 \\ 0 & 0 & 0 & 1 & 5/6 & 1/3 \\ 0 & 0 & 0 & 0 & 0 & 0 \end{pmatrix}.$$

The leading list of $\mathrm{rref}(\mathbf{A})$ is $J = \{1,2,3,4\}$. So, to find the Gauss basis for $N(\mathbf{A})$ we put $x_5 = s$, $x_6 = t$. Then $\mathbf{x}$ is in $N(\mathbf{A})$ if and only if

$$\mathbf{x} = \begin{pmatrix} x_1 \\ x_2 \\ x_3 \\ x_4 \\ x_5 \\ x_6 \end{pmatrix} = \begin{pmatrix} 7s/6 + 2t/3 \\ 4s/3 + t/3 \\ s/6 - t/3 \\ -5s/6 - t/3 \\ s \\ t \end{pmatrix} = s \begin{pmatrix} 7/6 \\ 4/3 \\ 1/6 \\ -5/6 \\ 1 \\ 0 \end{pmatrix} + t \begin{pmatrix} 2/3 \\ 1/3 \\ -1/3 \\ -1/3 \\ 0 \\ 1 \end{pmatrix} = s\mathbf{g}_5 + t\mathbf{g}_6.$$

Hence $\dim(N(\mathbf{A})) = 2$ Observing that $x_5 = s$ and $x_6 = t$ in the linear combination $s\mathbf{g}_5 + t\mathbf{g}_6$, we find that

$$\mathbf{u}_1 = 2\mathbf{g}_5 - 2\mathbf{g}_6, \quad \mathbf{u}_2 = 3\mathbf{g}_6, \quad \mathbf{u}_3 = -2\mathbf{g}_5 + 5\mathbf{g}_6.$$

Hence V is a subspace of $N(\mathbf{A})$. Since their dimensions coincide, these subspaces must be equal. Notice that $\mathbf{u}_3 = \mathbf{u}_2 - \mathbf{u}_1$ implying that $\dim(V) = 2$ by a direct calculation. The Gauss basis for V equals $\{\mathbf{g}_5, \mathbf{g}_6\}$.   □

Problem 2.20 A 4×4 matrix $\mathbf{A}$ has two known and two unknown columns:

$$\mathbf{A} = \begin{pmatrix} 3 & * & 17 & * \\ 2 & * & 12 & * \\ 2 & * & 10 & * \\ 1 & * & 9 & * \end{pmatrix}$$

All solutions to the system $\mathbf{Ax} = \mathbf{b}$ are described by the formula

$$\mathbf{x} = \begin{pmatrix} 3 \\ 1 \\ -1 \\ 2 \end{pmatrix} + s \begin{pmatrix} 4 \\ 1 \\ 2 \\ 5 \end{pmatrix} + t \begin{pmatrix} 3 \\ 1 \\ 1 \\ 3 \end{pmatrix},$$

where s and t are arbitrary real numbers. Find $\mathbf{A}$, $\mathbf{b}$, and the Gauss particular solution for this $\mathbf{b}$.

Solution: Since the columns at s and t span the null space of $\mathbf{A}$, the nullity of $\mathbf{A}$ is two. By the Rank-Nullity Theorem the rank of $\mathbf{A}$ is also two. Hence $\mathrm{rref}(\mathbf{A})$ has only two nonzero rows. Every row $\begin{pmatrix} x_1 & x_2 & x_3 & x_4 \end{pmatrix}$ is a solution to the system

$$\begin{cases} 4x_1 + x_2 + 2x_3 + 5x_4 & = 0 \\ 3x_1 + x_2 + x_3 + 3x_4 & = 0 \end{cases}. \tag{2.34}$$

The reduced row echelon form of the coefficient matrix of this system is given by

$$\mathrm{rref} \begin{pmatrix} 4 & 1 & 2 & 5 \\ 3 & 1 & 1 & 3 \end{pmatrix} = \begin{pmatrix} 1 & 0 & 1 & 2 \\ 0 & 1 & -2 & -3 \end{pmatrix}.$$

It follows that x_1 and x_2 in (2.34) are leading variables and $x_3 = s$, $x_4 = t$ are non-leading variables. We find a formula for the solutions to (2.34):

$$\begin{pmatrix} x_1 \\ x_2 \\ x_3 \\ x_4 \end{pmatrix} = \begin{pmatrix} -s - 2t \\ 2s + 3t \\ s \\ t \end{pmatrix} = s \begin{pmatrix} -1 \\ 2 \\ 1 \\ 0 \end{pmatrix} + t \begin{pmatrix} -2 \\ 3 \\ 0 \\ 1 \end{pmatrix}.$$

Since rank(A) = 2, the two linear independent rows, which are transposes of the columns at s and t in the above formula, make a basis of the row space for A. It follows that

$$\begin{pmatrix} -1 & 2 & 1 & 0 \\ -2 & 3 & 0 & 1 \\ 0 & 0 & 0 & 0 \\ 0 & 0 & 0 & 0 \end{pmatrix} \sim \begin{pmatrix} 1 & 0 & 3 & -2 \\ 0 & 1 & 2 & -1 \\ 0 & 0 & 0 & 0 \\ 0 & 0 & 0 & 0 \end{pmatrix} = \text{rref}(A).$$

We find the Gauss basis for the null space of rref(A). The are two non-leading variables $x_3 = s$ and $x_4 = t$. Then x is in $N(A)$ if and only if

$$\begin{pmatrix} x_1 \\ x_2 \\ x_3 \\ x_4 \end{pmatrix} = \begin{pmatrix} -3s + 2t \\ -2s + t \\ s \\ t \end{pmatrix} = s \begin{pmatrix} -3 \\ -2 \\ 1 \\ 0 \end{pmatrix} + t \begin{pmatrix} 2 \\ 1 \\ 0 \\ 1 \end{pmatrix} = s v_3 + t v_4.$$

It follows that $\text{col}_3(A)$ is a linear combination of the first two columns:

$$0 = A v_3 = -3\text{col}_1(A) - 2\text{col}_2(A) + \text{col}_3(A) \Rightarrow \text{col}_2(A) = \begin{pmatrix} 4 \\ 3 \\ 2 \\ 3 \end{pmatrix}.$$

Similarly,

$$0 = A v_4 = 2\text{col}_1(A) + \text{col}_2(A) + \text{col}_4(A) \Rightarrow \text{col}_4(A) = \begin{pmatrix} -10 \\ -7 \\ -6 \\ -5 \end{pmatrix}.$$

Hence

$$A = \begin{pmatrix} 3 & 4 & 17 & -10 \\ 2 & 3 & 12 & -7 \\ 2 & 2 & 10 & -6 \\ 1 & 3 & 9 & -5 \end{pmatrix}, \quad b = \begin{pmatrix} 3 & 4 & 17 & -10 \\ 2 & 3 & 12 & -7 \\ 2 & 2 & 10 & -6 \\ 1 & 3 & 9 & -5 \end{pmatrix} \begin{pmatrix} 3 \\ 1 \\ -1 \\ 2 \end{pmatrix} = \begin{pmatrix} -24 \\ -17 \\ -14 \\ -13 \end{pmatrix}.$$

To find the Gauss particular solution p we observe that $j_1 = 1, j_2 = 2$. Then

$$p = \begin{pmatrix} 3 \\ 1 \\ -1 \\ 2 \end{pmatrix} + v_3 - v_4 = \begin{pmatrix} 3 - 3 - 4 \\ 1 - 2 - 2 \\ -1 + 1 \\ 2 - 2 \end{pmatrix} = \begin{pmatrix} -4 \\ -3 \\ 0 \\ 0 \end{pmatrix}. \quad \square$$

Problem 2.21 Find the Gauss bases for the row space, null space, and column space of the matrix

$$A = \begin{pmatrix} 1 & -3 & 2 \\ 2 & -5 & 1 \\ 5 & -14 & 7 \\ 5 & -11 & -2 \end{pmatrix}.$$

Solution: Evaluating rref(**A**), we find that the leading list J is $\{1,2\}$, the complementary list J' is $\{3\}$ and

$$\mathrm{rref}_p(\mathbf{A}) = \begin{pmatrix} 1 & 0 & -7 \\ 0 & 1 & -3 \end{pmatrix} \Rightarrow \mathrm{eref}(\mathbf{A}) = \begin{pmatrix} 1 & 0 & -7 \\ 0 & 1 & -3 \\ 0 & 0 & 1 \end{pmatrix}, \ \mathbf{G_A} = \begin{pmatrix} 1 & 0 & 7 \\ 0 & 1 & 3 \\ 0 & 0 & 1 \end{pmatrix}.$$

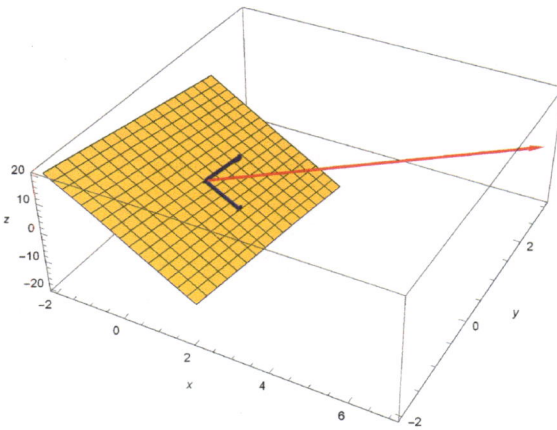

Fig. 2.12: The Gauss basis for $7x + 3y + z = 0$ in $\mathbb{R}^3$.

Then

$$\left\{ \begin{pmatrix} 1 & 0 & -7 \end{pmatrix}, \begin{pmatrix} 0 & 1 & -3 \end{pmatrix} \right\}$$

is the Gauss basis for **row(A)** (see blue vectors on Fig. 2.12) and $\begin{pmatrix} 7 & 3 & 1 \end{pmatrix}$ is the Gauss basis for N(**A**) (see the red vector on Fig. 2.12). Notice, that in this particular case we consider the blue rows as columns in $\mathbb{R}^3$. Otherwise, we couldn't plot them together with the red column.

Finally, we find the Gauss basis for the subspace **col(A)**. Since $J = \{1,2\}$, we conclude that the first two columns of **A** make a basis for **col(A)** = **row**$(\mathbf{A}^T)^T$. First, we find the Gauss basis for N($\mathbf{A}^T$). Since

$$\text{rref}(\mathbf{A}^T) = \begin{pmatrix} 1 & 0 & 3 & -3 \\ 0 & 1 & 1 & 4 \\ 0 & 0 & 0 & 0 \\ 0 & 0 & 0 & 0 \end{pmatrix} \Rightarrow \mathbf{G}_{\mathbf{A}^T} = \begin{pmatrix} 1 & 0 & -3 & 3 \\ 0 & 1 & -1 & -4 \\ 0 & 0 & 1 & 0 \\ 0 & 0 & 0 & 1 \end{pmatrix},$$

the Gauss basis for $N(\mathbf{A}^T)$ is the following:

$$\left\{ \left(-3 \ -1 \ 1 \ 0 \right)^T, \left(3 \ -4 \ 0 \ 1 \right)^T \right\}.$$

By Theorem 2.15, $\mathbb{R}^4 = \text{row}(\mathbf{A}^T)^T \oplus N(\mathbf{A}^T)$. Since $\text{col}(\mathbf{A}) = \text{row}(\mathbf{A}^T)^T$, the Gauss basis for the subspace $\text{col}(\mathbf{A})$ is the Gauss basis for the null-space of the matrix

$$\mathbf{B} = \begin{pmatrix} -3 & -1 & 1 & 0 \\ 3 & -4 & 0 & 1 \end{pmatrix} \sim \begin{pmatrix} 1 & 0 & -4/15 & 1/15 \\ 0 & 1 & -1/5 & -1/5 \end{pmatrix}.$$

Since

$$\mathbf{G_B} = \begin{pmatrix} 1 & 0 & 4/15 & -1/15 \\ 0 & 1 & 1/5 & 1/5 \\ 0 & 0 & 1 & 0 \\ 0 & 0 & 0 & 1 \end{pmatrix},$$

we obtain that

$$\left\{ \left(4/15 \ 1/5 \ 1 \ 0 \right)^T, \left(-1/15 \ 1/5 \ 0 \ 1 \right)^T \right\}$$

is the Gauss basis for $\text{col}(\mathbf{A})$.   □

Problems

Prob. 97 — Find all Gauss bases corresponding to the matrix

$$\begin{pmatrix} 0 & 7 & 14 & 0 & 7 & 21 & 0 & 7 & 14 & 3 \\ 0 & 4 & 8 & 1 & 5 & 14 & 2 & 13 & 18 & -1 \\ 0 & 3 & 6 & 0 & 3 & 9 & 1 & 5 & 9 & 0 \\ 0 & 2 & 4 & 0 & 2 & 6 & 0 & 2 & 4 & 1 \end{pmatrix}$$

Prob. 98 — Find the matrix

$$\mathbf{A} = \begin{pmatrix} 1 & 2 & * & * \\ -1 & 0 & * & * \\ * & * & * & * \\ * & * & * & * \end{pmatrix}$$

if its null space $N(\mathbf{A})$ satisfies the conditions

$$N(\mathbf{A}) = N(\mathbf{A}^T) = \text{Lin}\left\{ \begin{pmatrix} 1 \\ 0 \\ 1 \\ -1 \end{pmatrix}, \begin{pmatrix} 0 \\ 1 \\ 0 \\ 1 \end{pmatrix} \right\}.$$

2.12 Linear Programming

In 1939, Leonid Kantorovich, a Soviet mathematician and economist developed a method for optimization of linear economic functions under linear constraints. Simultaneously, Tjalling C. Koopmans, a Dutch-American economist, found a similar approach to solve classical economics problems. In 1975, Kantorovich and Koopmans shared the Nobel Prize in Economics for their work on resource allocation. The Vanderbei (2014) textbook is an excellent introduction to this topic. We consider the Linear Programming algorithms on an example of the following concrete **Resource Allocation Problem**.

Problem 2.22 A small firm makes two types of white boards made of aluminum, plastic, and wood. Market costs of one kilo of aluminum, plastic, and wood are $20, $10, and $5 correspondingly.

The first type of the board requires 1 kilo of aluminium, 5 kilo of plastic and 2 kilo of wood. The second type of the board requires 5 kilo of aluminium, 3 kilo of plastic and 2 kilo of wood. The daily resources of aluminum, plastic, and wood are 35, 30, and 16 kilos. The market price of the first board is $100 per board. The market price of the second board is $150 per board. Determine the levels of productions of two boards to maximize the profit.

Solution: Let $\mathbf{p} = \begin{pmatrix} 100 & 150 \end{pmatrix}$ be the row of market prices of the boards, and $\mathbf{r} = \begin{pmatrix} 20 & 10 & 5 \end{pmatrix}$ be the row of one kilo costs of the resources required for the boards production. The column and the 3×2 matrix below

$$\mathbf{b} = \begin{pmatrix} 35 \\ 30 \\ 16 \end{pmatrix}, \ \mathbf{A} = \begin{pmatrix} 1 & 5 \\ 5 & 3 \\ 2 & 2 \end{pmatrix}$$

represent the resources limits and the amounts of resources in kilos required for the production of each of two boards. Then the 1×2 row $\mathbf{rA}$ gives the costs of production of each board, whereas $\mathbf{p} - \mathbf{rA}$ is the row of profits which this company makes on selling each of the two types of boards. If x units of the first board and y units of the second board are sold, then the profit Π of the firm equals

$$\Pi = (\mathbf{p} - \mathbf{rA}) \begin{pmatrix} x \\ y \end{pmatrix} = 20x + 10y.$$

Since the resources of the firm are restricted by the column $\mathbf{b}$, the problem is to maximize the linear function Π in two variables x and y under the constraints

$$\mathbf{A} \begin{pmatrix} x \\ y \end{pmatrix} = \begin{pmatrix} x + 5y \\ 5x + 3y \\ 2x + 2y \end{pmatrix} \leq \mathbf{b} = \begin{pmatrix} 35 \\ 30 \\ 16 \end{pmatrix}, \ x \geq 0, \ y \geq 0. \tag{2.35}$$

The matrix inequality in (2.35) means that every component of thge left colomn does not exceed the corresponding component of the right column.

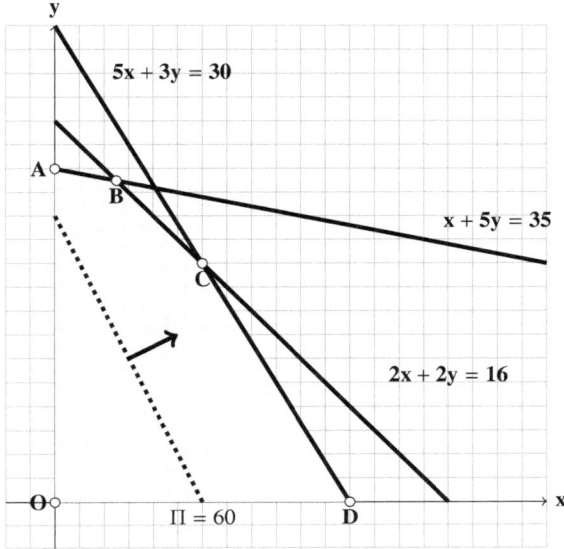

Fig. 2.13: The graph of the domain D.

The domain D on the plane described by (2.35) is not empty since it includes the origin ($x = y = 0$). The domain D is also convex, i.e. if $\left(x_1\ y_1\right)^T$, $\left(x_2\ y_2\right)^T$ are in D and $0 \le \alpha \le 1$, then

$$\begin{pmatrix} x \\ y \end{pmatrix} = \alpha \begin{pmatrix} x_1 \\ y_1 \end{pmatrix} + (1 - \alpha) \begin{pmatrix} x_2 \\ y_2 \end{pmatrix} \tag{2.36}$$

is also in D. Since $\Pi(\lambda x, \lambda y) = \lambda \Pi(x, y)$, the function Π attains its maximal value on the boundary of D. Since the function Π is linear, it attains its maximal value being considered on the segment defined by (2.36), at its ends ($\alpha = 0$ or $\alpha = 1$). Since the boundary of D is a polygon, the function Π attains its maximal value at the vertexes of this polygon, see Figure 2.13. Solving systems of two equations in two unknowns one can easily find the coordinates of all vertexes of polygon D:

$$\mathbf{A}(0,7),\ \mathbf{B}(1.25, 6.75),\ \mathbf{C}(3,5),\ \mathbf{D}(6,0).$$

Since the linear function $\Pi = 20x + 10y$ takes its maximal value at one of the vertexes:

$$20x + 10y\big|_{(0,7)} = 70,\ 20x + 10y\big|_{(1.25, 6.75)} = 92.5,$$
$$20x + 10y\big|_{(3,5)} = 110,\ 20x + 10y\big|_{(6,0)} = \mathbf{120},$$

we conclude that the firm should produce six boards for $100 each and no boards of the second type. We can arrive at the same conclusion investigating the level curves of the function Π, see the dotted line in Figure 2.13, which shows the set of points (x, y) in D such that $\Pi(x, y) = 60$. Moving this line in the direction of the gradient of Π, i.e. in the direction of the maximal rate of change of Π, we see that this line passes all vertexes of the polygon in the direction fro $\mathbf{A}$ to $\mathbf{D}$.   □

This method of the solution to Problem 2.22 is illustrative but is not useful if more than two variables are involved into considerations. Between 1946 and 1947, George B.Dantzig developed a general algorithm for solution problems of Linear Programming, which is called the **simplex method**.

Solution to Problem 2.22 with the simplex method: We rewrite the constrains with non-negative **slack variables** x_3, x_4, x_5:

$$\begin{cases} x_1 + 5x_2 + x_3 & = 35 \\ 5x_1 + 3x_2 + x_4 & = 30 \\ x_1 + x_2 + x_5 & = 8 \end{cases} \Rightarrow A = \begin{pmatrix} 1 & 5 & 1 & 0 & 0 \\ 5 & 3 & 0 & 1 & 0 \\ 1 & 1 & 0 & 0 & 1 \end{pmatrix},$$

where this time the matrix $\mathbf{A}$ is the extended matrix of the matrix in Problem 2.22.

A solution $x_1 = 0$, $x_2 = 0$, $x_3 = 35$, $x_4 = 30$, $x_5 = 8$ is called a **feasible solution**:

$$x_3 = 35 - x_1 - 5x_2$$
$$x_4 = 30 - 5x_1 - 3x_2$$
$$x_5 = 8 - x_1 - x_2$$

$$z = 2x_1 + x_2 \quad \leftarrow \text{ is the objective function}$$

The above formulas are called the **tableau**. The solution $x_3 = 35$, $x_4 = 30$, $x_5 = 8$ is called a **basic feasible solution**. Notice that $z = 0.1\Pi$.

Let $x_2 = 0$. Then the objective function increases in x_1. We find the maximal possible value for x_1 which keeps nonnegative all other variables:

$$0 \leq x_3 = 35 - x_1 - 5x_2 = 35 - x_1$$
$$0 \leq x_4 = 30 - 5x_1 - 3x_2 = 30 - 5x_1 \Leftrightarrow 0 \leq x_1 \leq 6 \Rightarrow \boxed{x_4}$$
$$0 \leq x_5 = 8 - x_1 - x_2 = 9 - x_1$$

A **feasible solution** is $x_1 = 6$, $x_2 = 0$, $x_3 = 29$, $x_4 = 0$, $x_5 = 2$.
   basic feasible solution is $x_1 = 6$, $x_3 = 29$, $x_5 = 2$.
   We move x_4 to the right and x_1 to the left-hand part:

$$x_1 = 6 - \frac{3}{5}x_2 - \frac{1}{5}x_4$$

$$x_3 = 35 - x_1 - 5x_2 = 35 - \left(6 - \frac{3}{5}x_2 - \frac{1}{5}x_4\right) - 5x_2 = 29 - \frac{22}{5}x_2 + \frac{1}{5}x_4$$

$$x_5 = 8 - x_1 - x_2 = 8 - \left(6 - \frac{3}{5}x_2 - \frac{1}{5}x_4\right) - x_2 = 2 - \frac{2}{5}x_2 + \frac{1}{5}x_4$$

$$z = 2x_1 + x_2 = 2\left(6 - \frac{3}{5}x_2 - \frac{1}{5}x_4\right) + x_2 = 12 - \frac{1}{5}x_2 - \frac{2}{5}x_4$$

Since $x_2 \geq 0$ and $x_4 \geq 0$, we conclude that max $z = 12$. The maximum is attained at $x_2 = x_4 = 0$, $x_1 = 6$, $x_3 = 29$, $x_5 = 2$. Hence, the point $x_1 = 6$, $x_2 = 0$ is the point of maximum.   □

Problem 2.23 Maximize $x_1 + 3x_2$ subject to

$$\begin{cases} -2x_1 + x_2 & \leq 1 \\ 2x_1 - x_2 & \leq 4 \\ x_1 + 2x_2 & \leq 7 \\ x_1, x_2, & \geq 0 \end{cases}$$

Solution: The convex domain G is bounded by the lines

$$x_1 = 0; \qquad x_1 = 0;$$
$$x_2 = 0; \qquad x_2 = 0;$$
$$-2x_1 + x_2 = 1; \Leftrightarrow x_2 = 2x_1 + 1;$$
$$2x_1 - x_2 = 4; \qquad x_2 = 2x_1 - 4;$$
$$x_1 + 2x_2 = 7. \qquad x_2 = 3.5 - 0.5x_1.$$

The plot of G is shown in Figure 2.14.

The coordinates of the vertexes are the following

$$\mathbf{A}(0,1), \ \mathbf{B}(1,3), \ \mathbf{C}(3,2), \ \mathbf{D}(2,0).$$

Then the grey domain is the convex hull of these vertexes. It follows that the linear function $x_1 + 3x_2$ takes its maximal value at one of the vertexes:

$$x_1 + 3x_2\big|_{(0,1)} = 3, \ x_1 + 3x_2\big|_{(1,3)} = \mathbf{10},$$

$$x_1 + 3x_2\big|_{(3,2)} = 9, \ x_1 + 3x_2\big|_{(2,0)} = 2.$$

It follows that the maximal value $\mathbf{10}$ of $x_1 + 3x_2$ taken in the gray domain is attained at $\mathbf{B}(1,3)$.

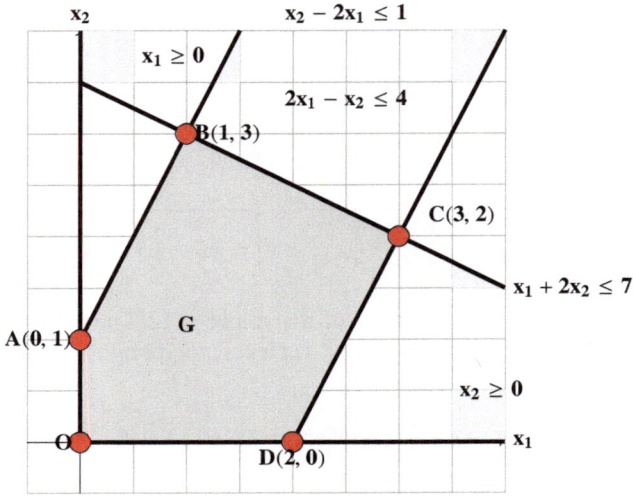

Fig. 2.14: The graph of the domain D.

To solve this problem by the simplex method, we introduce non-negative slack variables x_3, x_4, x_5 and construct the tableau:

$$x_3 = 1 + 2x_1 - x_2$$
$$x_4 = 4 - 2x_1 + x_2$$
$$x_5 = 7 - x_1 - 2x_2$$

$$z = x_1 + 3x_2 \quad \leftarrow \text{ \textbf{the objective function}}$$

If $x_1 = x_2 = x_3 = 0$, then $x_3 = 1$, $x_4 = 4$, $x_5 = 7$. It is a feasible solution which corresponds to the vertex $\mathbf{O(0,0)}$. If $x_2 = 0$, then z increases in x_1. Similarly, if $x_1 = 0$ then z increases in x_2. We cannot increase these variables indefinitely, since the slack variables must be non-negative.

Formulas for the slack variables give the following limits if $x_2 = 0$:

$$0 \le x_3 = 1 + 2x_1 \Rightarrow x_1 < +\infty$$
$$0 \le x_4 = 4 - 2x_1 \Rightarrow x_1 \le 2 \quad \Rightarrow x_1 \le 2 \Rightarrow z = x_1 \le 2.$$
$$0 \le x_5 = 7 - x_1 \Rightarrow x_1 \le 7$$

Formulas for the slack variables give the following limits if $x_1 = 0$:

$$0 \le x_3 = 1 - x_2 \Rightarrow x_2 \le 1$$
$$0 \le x_4 = 4 + x_2 \Rightarrow x_2 > \infty \quad \Rightarrow x_2 \le 1 \Rightarrow z = 3x_2 \le 3.$$
$$0 \le x_5 = 7 - 2x_2 \Rightarrow x_2 \le 3.5$$

In the last case the value of z is bigger. Therefore, we put $x_1 = 0$. In what follows we call the variables on the left-hand side as **basic variables** and two variables on the right-hand side as **non-basic variables**. In particular, variables x_3, x_4, x_5 are basic and the variables x_1, x_2 are non-basic.

The maximal possible value of x_2 under the constrain $x_1 = 0$ is determined from the equation for x_3. We put the symbol * in front of it and rewrite it to express x_2 as a function of x_1 and x_3. After that we replace x_2 by the formula (*) in the equation for the rest slack variables and z:

$$* \quad x_2 = 1 + 2x_1 - x_3$$
$$x_4 = 5 - x_3$$
$$x_5 = 5 - 5x_1 + 2x_3$$

$$z = 3 + 7x_1 - 3x_3 \quad \leftarrow \textbf{ the objective function}$$

This corresponds to the vertex $\mathbf{A(0,1)}$, where z takes the value 3. Indeed, $x_1 = x_3 = 0$ implies that $z = 3$ and $x_2 = 1$.

Since z increases only in x_1, we find the bounds for x_1 from the above equations. This will be $x_1 \leq 1$, which we obtain from the equation for x_5. We express x_1 as a function in x_3 and x_5:

$$x_2 = 3 - \frac{1}{5}x_3 - \frac{2}{5}x_5$$
$$x_4 = 5 - x_3$$
$$* \quad x_1 = 1 + \frac{2}{5}x_3 - \frac{1}{5}x_5$$

$$z = 10 - \frac{1}{5}x_3 - \frac{7}{5}x_5 \quad \leftarrow \textbf{ the objective function}$$

Since both coefficients at x_3 and x_5 in the formula for z are negative, the maximal value is $z = 10$, which is attained at $x_3 = x_5 = 0$. The formulas for x_2 and x_1 show that $x_1 = 1$, $x_2 = 3$. These are the coordinates of $\mathbf{B(1,3)}$.

It is clear that our calculations move us along the boundary of polygon through its vertexes:

$$\mathbf{O(0,0)} \longrightarrow \mathbf{A(0,1)} \longrightarrow \mathbf{B(1,3)}. \quad \square$$

Problem 2.24 Maximize $5x_1 + 4x_2 + 3x_3$ in the domain

$$\begin{cases} 2x_1 + 3x_2 + x_3 & \leq 5 \\ 4x_1 + x_2 + 2x_3 & \leq 11 \\ 3x_1 + 4x_2 + 2x_3 & \leq 8 \\ x_1,\, x_2,\, x_3 & \geq 0 \end{cases}$$

Solution: We introduce non-negative slack variables x_4, x_5, x_6 and construct the tableau:

$$x_4 = 5 - 2x_1 - 3x_2 - x_3$$
$$x_5 = 11 - 4x_1 - x_2 - 2x_3$$
$$x_6 = 8 - 3x_1 - 4x_2 - 2x_3$$

$$z = 5x_1 + 4x_2 + 3x_3 \quad \leftarrow \textbf{the objective function}$$

If $x_1 = x_2 = x_3 = 0$, then $x_4 = 5$, $x_5 = 11$, $x_6 = 8$ is a basic feasible solution. If $x_2 = x_3 = 0$, then the maximal positive value of x_1 is $5/2 = 2.5$. Then $x_4 = 0$ implying that $x_5 = 1$, $x_6 = \frac{1}{2}$. It follows that x_1, x_5, x_6 is a basic feasible solution, and x_2, x_3, x_4 is a non-basic solution. To have x_4 on the right and x_1 on the left we express x_1 in terms of x_2, x_3, x_4 and substitute to obtain the second tableau.

$$x_1 = \frac{5}{2} - \frac{3}{2}x_2 - \frac{1}{2}x_3 - \frac{1}{2}x_4$$
$$x_5 = 1 + 5x_2 + 2x_4$$
$$x_6 = \frac{1}{2} + \frac{1}{2}x_2 - \frac{1}{2}x_3 + \frac{3}{2}x_4$$

$$z = \frac{25}{2} - \frac{7}{2}x_2 + \frac{1}{2}x_3 - \frac{5}{2}x_4 \quad \leftarrow \textbf{the objective function}$$

Then $x_3 = x_4 = 0 \Rightarrow 0 \leq x_3 \leq 1$. The coefficient at x_3 at z is positive, implying that $z \uparrow$ if $x_3 \uparrow$. If $x_1 = 1$, then

$$x_1 = 1, \ x_2 = 0, \ x_3 = 1, \ x_4 = 0, \ x_5 = 1, \ x_6 = 1$$

is a feasible solution. So, we express x_3 as a function of x_2, x_4, x_6 and obtain the third tableau.

$$x_1 = 2 - 2x_2 - 2x_4 + x_6$$
$$x_3 = 1 + x_2 + 3x_4 - 2x_6$$
$$x_5 = 1 + 5x_2 + 2x_4$$

$$z = 13 - 3x_2 - x_4 - x_6 \quad \leftarrow \textbf{the objective function}$$

Hence max $z = 13$ and z takes this value at $x_2 = x_4 = x_6 = 0$, $x_3 = 1$, $x_1 = 1$, $x_5 = 1$. $\qquad\qquad\qquad\qquad\qquad\qquad\qquad\qquad\qquad\qquad\qquad\qquad\qquad\quad \square$

A more general problem is to maximize the sum

$$\sum_{i=1}^{n} c_i x_i$$

for given positive coefficients $c_1, \ldots, c_n$ subject to the restrictions

$$\sum_{j=1}^{n} a_{ij} x_j \leq b_i, \ i = 1, \ldots, m; \ x_j \geq 0, \ j = 1, \ldots, n.$$

In Problem 2.24 we have $n = 3$, $m = 3$. In general case, we introduce m non-negative slack variables $x_{n+1}, \ldots, x_{n+m}$ and construct the tableau for basic variables $x_{n+1}, \ldots, x_{n+m}$:

$$x_{n+1} = b_1 - \sum_{j=1}^{n} a_{1j} x_j$$

$$x_{n+2} = b_2 - \sum_{j=1}^{n} a_{2j} x_j$$

$$\vdots = \vdots$$

$$x_{n+m} = b_m - \sum_{j=1}^{n} a_{mj} x_j$$

$$z = \sum_{i=1}^{n} c_i x_i \quad \leftarrow \textbf{the objective function}$$

Since the objective function increases in x_1, we initially set $x_2 = \cdots = x_n = 0$ in the tableau and obtain m restrictions on the growth of x_1:

$$b_1 \geq a_{11} x_1, b_2 \geq a_{12} x_2, \ldots, b_m \geq a_{m1} x_1. \tag{2.37}$$

Then we find the index k such that the inequality $b_k \geq a_{k1} x_1$ implies other inequalities in (2.37). To place x_{n+k} on the right and x_1 on the left, we express x_1 from the equation

$$x_{n+k} = b_k - a_{k1} x_1 - a_{k2} x_2 - \cdots - a_{kn} x_n$$

and make the corresponding substitutions to obtain a new tableau, where x_1 is a basic variable and x_{n+k} is a non-basic variable. These substitution made in the objective function convert it into a sum of some constant C and a linear combination of new non-basic variables. If all coefficients in front of these variables are negative, then the maximum of the objective function is C. Otherwise, we pick a variable with a positive coefficient and construct a new tableau for it. In this case, the algorithm continues further, iterating this process until an optimal solution is found or no further improvement can be made, see Vanderbei (2014) for more details.

Problems

Prob. 99 — Maximize $3x_1 + 2x_2 + 4x_3$ subject to

$$\begin{cases} x_1 + x_2 + 2x_3 & \leq 4 \\ 2x_1 + 3x_3 & \leq 5 \\ 2x_1 + x_2 + 3x_3 & \leq 7 \\ x_1,\ x_2,\ x_3 & \geq 0 \end{cases}$$

Prob. 100 — Maximize $5x_1 + 6x_2 + 9x_3 + 8x_4$ subject to

$$\begin{cases} x_1 + 2x_2 + 3x_3 + x_4 & \leq 5 \\ x_1 + x_2 + 2x_3 + 3x_4 & \leq 3 \\ x_1,\ x_2,\ x_3\ x_4 & \geq 0 \end{cases}$$

Chapter 3
Inverse Matrices and Determinants

3.1 Introduction

Chapter 3 begins with some theory of invertible matrices, followed by Section 3.4, where the inverse problem of Gaussian elimination is addressed. The reduced row echelon form rref$(\mathbf{A})$ of any matrix $\mathbf{A}$ uniquely determines the null space $N(\mathbf{A})$, its row space $\mathbf{row}(\mathbf{A})$, and the leading list J. This means that row operations do not change these invariants of matrices with one reduced row echelon form. What the row operations do change are the columns at the positions of the leading list, which can be assigned to be an arbitrary linearly independent set of columns. In Section 3.4, one can find constructive formulas for such a recovery.

In Chapters 1 and 2, we developed the method of consecutive elimination of unknowns to solve systems of linear equations $\mathbf{Ax} = \mathbf{b}$. In case of square coefficient matrix $\mathbf{A}$, it is possible to exclude all variables except for a fixed one, x_j, and write a formula as seen in Section 3.6:

$$x_j = \frac{\lambda^T \mathbf{b}}{\lambda^T \mathrm{col}_j(\mathbf{A})}. \tag{3.1}$$

Using induction, we find explicit expressions for vectors λ in case $n = 2, 3$, and 4. These arguments for an arbitrary positive integer n lead to a formula which defines an important function in Linear Algebra called the determinant $\det(\mathbf{A})$ of an $n \times n$ matrix $\mathbf{A}$.

Since determinants appear in the formulas for solutions of type (3.1), they most likely must have simple behavior under the row operations of $\mathbf{A}$. The properties of determinants under row operations are studied in Section 3.9.

Formulas (3.1) are shaped into the so-called Cramer's Rules in Section 3.10 .

Beautiful applications of determinants to Analytic Geometry are considered in Section 3.11.

In Section 3.12 we discuss properties of characteristic polynomials of square matrices.

© The Author(s), under exclusive license to Springer Nature Switzerland AG 2024
S. Khrushchev, *Linear Algebra with Applications to Economics*, Classroom Companion:
Economics, https://doi.org/10.1007/978-3-031-68682-5_3

In Section 3.13 a detailed exposition of the Leontief' Input-Output model in Economics is given. This section depends mostly on the material presented in Chapter 3.

3.2 Inverse Matrices

Let $\mathbf{A}$ be an $n \times n$ matrix with zero null space $N(\mathbf{A}) = \{\mathbf{0}_n\}$. By the Rank-Nullity Theorem (see Theorem 2.14) $\text{rank}(\mathbf{A}) = n$. Since

$$\dim(R(\mathbf{A})) = \text{rank}(\mathbf{A}) = n,$$

we conclude that $R(\mathbf{A}) = \mathbb{R}^n$. Hence for every $\mathbf{b}$ in $\mathbb{R}^n$ the system $\mathbf{Ax} = \mathbf{b}$ has a unique solution $\mathbf{x}$. In particular, for every element $\mathbf{e}_i$ of the standard basis for $\mathbb{R}^n$ there is a unique vector $\mathbf{v}_i$ such that $\mathbf{Av}_i = \mathbf{e}_i$ for $i = 1, \ldots, n$. Let

$$\mathbf{B} = \begin{pmatrix} \mathbf{v}_1 \ \mathbf{v}_2 \ \cdots \ \mathbf{v}_n \end{pmatrix}. \tag{3.2}$$

Then by (1.30)

$$\mathbf{AB} = \begin{pmatrix} \mathbf{Av}_1 \ \mathbf{Av}_2 \ \cdots \ \mathbf{Av}_n \end{pmatrix} = \begin{pmatrix} \mathbf{e}_1 \ \mathbf{e}_2 \ \cdots \ \mathbf{e}_n \end{pmatrix} = \mathbf{I}_n.$$

Definition 3.1 Let $\mathbf{A}$ be an $n \times n$ matrix. An $n \times n$ matrix $\mathbf{B}$ is called an **inverse matrix** for $\mathbf{A}$ if
$$\mathbf{AB} = \mathbf{I}_n.$$

Lemma 3.1 *Let* $\mathbf{A}$ *be an* $n \times n$ *matrix. Then*

$$\mathbf{rref}(\mathbf{A}) = \mathbf{I}_n.$$

if and only if $N(\mathbf{A}) = \mathbf{0}$.

Proof. If $N(\mathbf{A}) = \mathbf{0}$ then all variables for $\mathbf{rref}(\mathbf{A})$ are leading, which means that all columns of $\mathbf{rref}(\mathbf{A})$ contain leading ones. Since $\mathbf{A}$ is a square matrix, this means that $\mathbf{rref}(\mathbf{A}) = \mathbf{I}_n$. If $\mathbf{rref}(\mathbf{A}) = \mathbf{I}_n$ then all variables are leading implying that $N(\mathbf{A}) = \mathbf{0}$.   □

Theorem 3.1 *Let* $\mathbf{A}$ *and* $\mathbf{B}$ *be two* $n \times n$ *matrices. Then*

$$\mathbf{BA} = \mathbf{I}_n \quad \Leftrightarrow \quad \mathbf{AB} = \mathbf{I}_n. \tag{3.3}$$

Proof. If $\mathbf{BA} = \mathbf{I}_n$, then $\mathbf{BAx} = \mathbf{x}$ for every $\mathbf{x}$ in $\mathbb{R}^n$. Hence $\mathbf{Ax} = \mathbf{0}$ implies that $\mathbf{x} = \mathbf{BAx} = \mathbf{B0} = \mathbf{0}$. We conclude that $N(\mathbf{A}) = \mathbf{0}$. By Lemma 3.1 we have $\mathbf{rref}(\mathbf{A}) = \mathbf{I}_n$. It follows that the column space $\mathbf{col}(\mathbf{A})$ of $\mathbf{A}$ is $\mathbb{R}^n$. Hence for every $\mathbf{b}$ in $\mathbb{R}^n$ the equation $\mathbf{Ax} = \mathbf{b}$ is consistent. Applying the left multiplication of $\mathbf{BA} = \mathbf{I}_n$ by $\mathbf{A}$, we obtain

$$\mathbf{ABA} = \mathbf{A} \Rightarrow \mathbf{ABAx} = \mathbf{Ax} \Rightarrow \mathbf{ABb} = \mathbf{b} \Rightarrow \mathbf{AB} = \mathbf{I}_n,$$

as stated. If $\mathbf{AB} = \mathbf{I}_n$ then we repeat the above arguments with substitution $\mathbf{A} \leftrightarrow \mathbf{B}$. ☐

Theorem 3.2 *The inverse matrix is unique.*

Proof. Suppose that $\mathbf{B}_1$ and $\mathbf{B}_2$ are inverse matrices for $\mathbf{A}$. Then

$$\mathbf{B}_1 = \mathbf{B}_1\mathbf{I}_n = \mathbf{B}_1\left(\mathbf{AB}_2\right) = \left(\mathbf{B}_1\mathbf{A}\right)\mathbf{B}_2 = \mathbf{I}_n\mathbf{B}_2 = \mathbf{B}_2.$$ ☐

Notice that if $\mathbf{B}$ is the inverse matrix for $\mathbf{A}$, then $\mathbf{A}$ is the inverse matrix for $\mathbf{B}$. We denote the inverse matrix for $\mathbf{A}$ by $\mathbf{A}^{-1}$. In this notation the statement above can be expressed by the following formula

$$\left(\mathbf{A}^{-1}\right)^{-1} = \mathbf{A}. \tag{3.4}$$

By Theorem 2.7 the Gauss matrix $\mathbf{G}$ is the sum of $\mathbf{I}_n$ and a matrix $\mathbf{P}$ satisfying $\mathbf{P}^2 = \mathbf{0}$.

Theorem 3.3 *For any Gauss matrix $\mathbf{G}$ the matrix $\mathbf{G}^{-1} = \mathbf{I} - \mathbf{P}$ is the inverse matrix:*

$$\mathbf{GG}^{-1} = \mathbf{I}.$$

Proof. We have

$$\mathbf{GG}^{-1} = \left(\mathbf{I} + \mathbf{P}\right)\left(\mathbf{I} - \mathbf{P}\right) = \mathbf{I} - \mathbf{P}^2 = \mathbf{I}.$$ ☐

Notice that $\mathbf{G}^{-1} = \mathrm{eref}(\mathbf{A})$.

Problem 3.1 Use Theorem 3.3 to solve Problem 2.10.

Solution: By (2.21)

$$\alpha = \mathbf{G}^{-1}\mathbf{b} = (\mathbf{I} + \mathbf{P})\,\mathbf{b} = \begin{pmatrix} 1 & 0 & 1 & 0 & 0 & 1 \\ 0 & 1 & 2 & 0 & 0 & 0 \\ 0 & 0 & 1 & 0 & 0 & 0 \\ 0 & 0 & 0 & 1 & 1 & 2 \\ 0 & 0 & 0 & 0 & 1 & 0 \\ 0 & 0 & 0 & 0 & 0 & 1 \end{pmatrix} \begin{pmatrix} 1 \\ 2 \\ 1 \\ 0 \\ 1 \\ 3 \end{pmatrix} = \begin{pmatrix} 5 \\ 4 \\ 1 \\ 7 \\ 1 \\ 3 \end{pmatrix}. \qquad \square$$

The following theorem is useful in applications.

Theorem 3.4 *Let* $\mathbf{A}$ *be an* $n \times n$ *matrix. Then the following conditions are equivalent:*

(**a**) $\mathrm{rref}(\mathbf{A}) = \mathbf{I}_n$;
(**b**) $\mathrm{nullity}(\mathbf{A}) = 0$;
(**c**) $\mathrm{rank}(\mathbf{A}) = n$;
(**d**) $\mathbf{A}$ *has the inverse matrix.*

Proof. (**a**) $\Rightarrow$ (**b**) If $\mathrm{rref}(\mathbf{A}) = \mathbf{I}_n$ then every variable is leading. Hence $\mathrm{nullity}(\mathbf{A}) = 0$.

(**b**) $\Rightarrow$ (**c**) If $\mathrm{nullity}(\mathbf{A}) = 0$ then every variable is leading implying that the number of leading variables is n. Hence $\mathrm{rank}(\mathbf{A}) = n$.

(**c**) $\Rightarrow$ (**d**) If $\mathrm{rank}(\mathbf{A}) = n$ then every variable is leading implying that $N(\mathbf{A}) = \mathbf{0}$. The matrix $\mathbf{B}$ defined in (3.2) is the inverse matrix for $\mathbf{A}$.

(**d**) $\Rightarrow$ (**a**) If $\mathbf{B}$ is the inverse matrix for $\mathbf{A}$ then $\mathbf{BA} = \mathbf{I}_n$ by Theorem 3.1. Therefore,

$$\mathbf{Ax} = \mathbf{0} \Rightarrow \mathbf{0} = \mathbf{BAx} = \mathbf{x},$$

implying that $N(\mathbf{A}) = \mathbf{0}$. It follows that every variable is leading. Hence $\mathrm{rref}(\mathbf{A}) = \mathbf{I}_n$. $\qquad \square$

Definition 3.2 A square matrix $\mathbf{A}$ is called **singular** if there is no inverse matrix for it. A square matrix $\mathbf{A}$ is called **non-singular** if there is the inverse matrix for it.

Corollary 3.1 *An* $n \times n$ *matrix* $\mathbf{A}$ *is singular if and only if*

$$\mathrm{rank}(\mathbf{A}) < n. \tag{3.5}$$

Proof. Apply (**c**) of Theorem 3.4. $\qquad \square$

Problem 3.2 Show that the inverse matrix for a 2×2 matrix

$$\mathbf{A} = \begin{pmatrix} a & b \\ c & d \end{pmatrix}$$

exists if and only if $ad - bc \neq 0$ and

$$\mathbf{A}^{-1} = \frac{1}{ad - bc} \begin{pmatrix} d & -b \\ -c & a \end{pmatrix}. \tag{3.6}$$

Solution: If $ad - bc = 0$ then

$$\begin{pmatrix} a & b \\ c & d \end{pmatrix} \begin{pmatrix} d \\ -c \end{pmatrix} = \begin{pmatrix} ad - bc \\ cd - dc \end{pmatrix} = \mathbf{0}_2.$$

If $\max(|c|, |d|) > 0$, then $N(\mathbf{A}) \neq \{\mathbf{0}\}$, implying that $\mathbf{A}$ is singular by Theorem 3.4. If $\max(|a|, |b|) > 0$, then $N(\mathbf{A}) \neq \{\mathbf{0}\}$, since

$$\begin{pmatrix} a & b \\ 0 & 0 \end{pmatrix} \begin{pmatrix} -b \\ a \end{pmatrix} = \begin{pmatrix} -ab + ba \\ 0 \end{pmatrix} = \mathbf{0}_2.$$

It follows that $\mathbf{A}$ is singular in this case as well. If $\mathbf{A} = \mathbf{0}_{2 \times 2}$, then $\text{nullity}(\mathbf{A}) = 2$, implying that $\mathbf{A}$ is singular.

If $ad - bc \neq 0$, then (3.6) is checked by a direct calculation. $\qquad \square$

Theorem 3.5 *Let $\mathbf{A}$ be an $n \times n$ matrix. Then $\mathbf{A}$ is non-singular if and only if $\mathbf{A}^T$ is non-singular. Moreover,*

$$\left(\mathbf{A}^T\right)^{-1} = \left(\mathbf{A}^{-1}\right)^T. \tag{3.7}$$

Proof. By Theorem 3.4, (**c**), a matrix $\mathbf{A}$ is non-singular if and only if $\text{rank}(\mathbf{A}) = n$. By Corollary 2.7 $\text{rank}(\mathbf{A}) = \text{rank}(\mathbf{A}^T)$, which proves the first part of the Theorem. To prove the second part we apply Theorem 1.5 and Theorem 3.1:

$$\mathbf{I}_n = \mathbf{A}\mathbf{A}^{-1} \Leftrightarrow \mathbf{I}_n = \left(\mathbf{A}\mathbf{A}^{-1}\right)^T \Leftrightarrow \mathbf{I}_n = \left(\mathbf{A}^{-1}\right)^T \mathbf{A}^T.$$

This implies (3.7). $\qquad \square$

Theorem 3.6 *Let* **A** *and* **B** *be two* $n \times n$ *matrices. Then* **AB** *is singular if and only if one of the matrices* **A** *and* **B** *is singular. If* **AB** *is non-singular, then*

$$(\mathbf{AB})^{-1} = \mathbf{B}^{-1}\mathbf{A}^{-1}. \tag{3.8}$$

Proof. By Corollary 3.1 one of the matrices **A** and **B** is singular if and only if

$$\min(\mathrm{rank}(\mathbf{A}), \mathrm{rank}(\mathbf{B})) < n.$$

If this condition holds then $\mathrm{rank}(\mathbf{AB}) < n$ by (2.20). Hence **AB** is singular by (**c**) of Theorem 3.4.

   If both **A** and **B** are non-singular then the inverse matrices $\mathbf{A}^{-1}$ and $\mathbf{B}^{-1}$ both exist. By Theorem 1.4

$$\left(\mathbf{B}^{-1}\mathbf{A}^{-1}\right)(\mathbf{AB}) = \mathbf{B}^{-1}\left(\mathbf{A}^{-1}\mathbf{A}\right)\mathbf{B} = \mathbf{B}^{-1}\left(\mathbf{I}_n\right)\mathbf{B} = \mathbf{B}^{-1}\left(\mathbf{I}_n\mathbf{B}\right) = \mathbf{B}^{-1}\mathbf{B} = \mathbf{I}_n,$$

implying that $\mathbf{B}^{-1}\mathbf{A}^{-1}$ is the inverse for **AB** as stated in (3.8).   □

Definition 3.3 Let **A** be an $m \times n$ matrix. Then the matrix $\left(\mathbf{A} \mid \mathbf{I}_m\right)$ is called the **extended** matrix for **A**.

   The following theorem gives a simple method to construct the inverse matrix for a given matrix **A**.

Theorem 3.7 *A square* $n \times n$ *matrix* **A** *is non-singular if and only if*

$$\mathbf{rref}\,(\mathbf{A}|\mathbf{I}_n) = (\mathbf{I}_n|\mathbf{B}). \tag{3.9}$$

If (3.9) *takes place, then* $\mathbf{A}^{-1} = \mathbf{B}$.

Proof. If an $n \times n$ matrix **A** is non-singular, then there exists the inverse matrix **B**, implying that $\mathbf{BA} = \mathbf{I}_n$. It follows that $N(\mathbf{A}) = \{\mathbf{0}\}$. By Lemma 3.1 we obtain that $\mathbf{rref}\,(\mathbf{A}) = \mathbf{I}_n$. The augmented matrix $(\mathbf{A}|\mathbf{I}_n)$ has size $n \times 2n$. Since its first n columns make matrix **A**, we obtain that (3.9) holds for some $n \times n$ matrix **B**.

Suppose now that (3.9) holds for some $n \times n$ matrix $\mathbf{B}$. Then by Theorem 1.2

$$\mathbf{A}\mathbf{x} = \mathbf{I}_n \mathbf{b} = \mathbf{b} \Leftrightarrow \mathbf{x} = \mathbf{B}\mathbf{b}.$$

Substituting here $\mathbf{b} = \mathbf{A}\mathbf{x}$, we obtain that $\mathbf{B}\mathbf{A}\mathbf{x} = \mathbf{x}$ for every vector $\mathbf{x}$ in $\mathbb{R}^n$. It follows that $\mathbf{B}\mathbf{A} = \mathbf{I}_n$. By Theorem 3.1 this implies that $\mathbf{B} = \mathbf{A}^{-1}$ as stated. □

Corollary 3.2 *A square matrix* $\mathbf{A}$ *is non-singular if and only if*

$$\mathbf{rref}\,(\mathbf{A}) = \mathbf{I}_n. \tag{3.10}$$

Proof. Since $\mathbf{A}$ is an $n \times n$ matrix, we see that

$$\mathbf{rref}\,(\mathbf{A}|\mathbf{I}_n) = (\mathbf{rref}\,(\mathbf{A})\,|\mathbf{C}) \tag{3.11}$$

for some $n \times n$ matrix $\mathbf{C}$. By Theorem 3.7 the matrix $\mathbf{A}$ is nonsingular if and only if (3.9) holds. Combining this with (3.11) we complete the proof of the Corollary. □

Problem 3.3 Show that the matrix

$$\mathbf{A} = \begin{pmatrix} 1 & 5 & 3 & 4 \\ 4 & 3 & 2 & 5 \\ 3 & 2 & 1 & 1 \\ -1 & 1 & -1 & 2 \end{pmatrix}$$

is non-singular and find $\mathbf{A}^{-1}$.

Solution: We evaluate the reduced row echelon form of the matrix $(\mathbf{A}|\mathbf{I}_4)$. Using elementary row operations we may annihilate all entries of the first column except for the first:

$$\begin{pmatrix} 1 & 5 & 3 & 4 & | & 1\,0\,0\,0 \\ 4 & 3 & 2 & 5 & | & 0\,1\,0\,0 \\ 3 & 2 & 1 & 1 & | & 0\,0\,1\,0 \\ -1 & 1 & -1 & 2 & | & 0\,0\,0\,1 \end{pmatrix} \sim \begin{pmatrix} 1 & 5 & 3 & 4 & | & 1\,0\,0\,0 \\ 0 & -17 & -10 & -11 & | & -4\,1\,0\,0 \\ 0 & -13 & -8 & -11 & | & -3\,0\,1\,0 \\ 0 & 6 & 2 & 6 & | & 1\,0\,0\,1 \end{pmatrix}.$$

We apply the elementary row operations $r_2 := r_2 + 3r_4$ and $r_3 := r_3 + 2r_4$ to diminish the entries of the matrix followed by the row operation $r_1 := r_1 - r_4$. We obtain a row equivalent matrix

$$\begin{pmatrix} 1 & -1 & 1 & -2 & | & 0\ \ 0\ \ 0\ -1 \\ 0 & 1 & -4 & 7 & | & -1\,1\,0\ \ 3 \\ 0 & -1 & -4 & 1 & | & -1\,0\,1\ \ 2 \\ 0 & 6 & 2 & 6 & | & 1\ \ 0\ 0\ \ 1 \end{pmatrix}.$$

Applying the row operations $r_1 := r_1 + r_2$, $r_3 := r_3 + r_2$, and $r_4 := r_4 - 6r_2$ we get

$$\begin{pmatrix} 1 & 0 & -3 & 5 & | & -1 & 1 & 0 & 2 \\ 0 & 1 & -4 & 7 & | & -1 & 1 & 0 & 3 \\ 0 & 0 & -8 & 8 & | & -2 & 1 & 1 & 5 \\ 0 & 0 & 26 & -36 & | & 7 & -6 & 0 & -17 \end{pmatrix}.$$

We apply now the row operation $r_4 := r_4 + 3r_3$:

$$\begin{pmatrix} 1 & 0 & -3 & 5 & | & -1 & 1 & 0 & 2 \\ 0 & 1 & -4 & 7 & | & -1 & 1 & 0 & 3 \\ 0 & 0 & -8 & 8 & | & -2 & 1 & 1 & 5 \\ 0 & 0 & 2 & -12 & | & 1 & -3 & 3 & -2 \end{pmatrix}$$

The row operations $r_1 := r_1 + r_4, r_2 := r_2 + 2r_4, r_3 := r_2 + 4r_4$ result in the following matrix

$$\begin{pmatrix} 1 & 0 & -1 & -7 & | & 0 & -2 & 3 & 0 \\ 0 & 1 & 0 & -17 & | & 1 & -5 & 6 & -1 \\ 0 & 0 & 0 & -40 & | & 2 & -11 & 13 & -3 \\ 0 & 0 & 2 & -12 & | & 1 & -3 & 3 & -2 \end{pmatrix},$$

which is row equivalent to the matrix

$$\begin{pmatrix} 1 & 0 & -1 & -7 & | & 0 & -2 & 3 & 0 \\ 0 & 1 & 0 & -17 & | & 1 & -5 & 6 & -1 \\ 0 & 0 & 1 & -6 & | & 1/2 & -3/2 & 3/2 & -1 \\ 0 & 0 & 0 & 1 & | & -2/40 & 11/40 & -13/40 & 3/40 \end{pmatrix} \underset{\sim}{r_1 := r_1 + r_3}$$

$$\begin{pmatrix} 1 & 0 & 0 & -13 & | & 1/2 & -7/2 & 9/2 & -1 \\ 0 & 1 & 0 & -17 & | & 1 & -5 & 6 & -1 \\ 0 & 0 & 1 & -6 & | & 1/2 & -3/2 & 3/2 & -1 \\ 0 & 0 & 0 & 1 & | & -2/40 & 11/40 & -13/40 & 3/40 \end{pmatrix} \begin{array}{l} r_1 := r_1 + 13r_4 \\ r_2 := r_2 + 17r_4 \\ r_3 := r_3 + 6r_4 \\ \underset{\sim}{} \end{array}$$

$$\begin{pmatrix} 1 & 0 & 0 & 0 & | & -6/40 & 3/40 & 11/40 & -1/40 \\ 0 & 1 & 0 & 0 & | & 6/40 & -13/40 & 19/40 & 11/40 \\ 0 & 0 & 1 & 0 & | & 8/40 & 6/40 & -18/40 & -22/40 \\ 0 & 0 & 0 & 1 & | & -2/40 & 11/40 & -13/40 & 3/40 \end{pmatrix} \Rightarrow$$

$$\mathbf{A}^{-1} = \frac{1}{40} \begin{pmatrix} -6 & 3 & 11 & -1 \\ 6 & -13 & 19 & 11 \\ 8 & 6 & -18 & -22 \\ -2 & 11 & -13 & 3 \end{pmatrix}.$$

Theorem 3.8 *For any $m \times n$ matrix $\mathbf{A}$ there is an invertible $m \times m$ matrix $\mathbf{B}$ such that*

$$\mathbf{A} = \mathbf{B} \cdot \mathrm{rref}(\mathbf{A}). \tag{3.12}$$

Proof. Since $\mathbf{A} \sim \mathrm{rref}(\mathbf{A})$, there is a finite chain of elementary matrices $\mathbf{E}_j$ such that

$$\mathbf{E}_r \cdots \mathbf{E}_2 \mathbf{E}_1 \mathbf{A} = \mathrm{rref}(\mathbf{A}) \Rightarrow \mathbf{A} = \mathbf{E}_1^{-1} \mathbf{E}_2^{-1} \cdots \mathbf{E}_r^{-1} \mathrm{rref}(\mathbf{A}).$$

Since any elementary matrix is invertible, a finite product $\mathbf{B}$ of elementary matrices is invertible by Theorem 3.6. □

Revise the solution to Problem 1.10. The following theorem describes all solutions to the equation (3.12) in $\mathbf{B}$.

Theorem 3.9 *Let $\mathbf{A}$ be an $m \times n$ matrix with the leading list $\{j_1, j_2, \ldots, j_r\}$. Then an $m \times m$ matrix $\mathbf{P}$ satisfies the equation*

$$\mathbf{A} = \mathbf{P} \, \mathrm{rref}(\mathbf{A}) \tag{3.13}$$

if and only if for $s = 1, \ldots, r$,

$$\mathrm{col}_s(\mathbf{P}) = \mathrm{col}_{j_s}(\mathbf{A}). \tag{3.14}$$

Proof. If (3.13) holds then for $s = 1, \ldots, r$,

$$\mathrm{col}_{j_s}(\mathbf{A}) = \mathbf{A} \mathbf{e}_{j_s} = \mathbf{P} \, \mathrm{rref}(\mathbf{A}) \mathbf{e}_{j_s} = \mathbf{P}_m \mathbf{e}_s = \mathrm{col}_s(\mathbf{P}), \tag{3.15}$$

where $_m \mathbf{e}_s$ is the sth coordinate vector in $\mathbb{R}^m$. This implies (3.14). If (3.14) holds then

$$\mathbf{A} \mathbf{x} = \mathbf{P} \, \mathrm{rref}(\mathbf{A}) \mathbf{x} \tag{3.16}$$

for every $\mathbf{x} = \mathbf{e}_{j_s}$, $s = 1, \ldots, r$. It follows that (3.16) is true for every $\mathbf{x}$ in $P(\mathbf{A})$. Since $N(\mathbf{A}) = N(\mathrm{rref}(\mathbf{A}))$, (3.16) is also true for every $\mathbf{x}$ in $N(\mathbf{A})$. By (2.18) it is true for every $\mathbf{x}$ in $\mathbb{R}^m$, which implies (3.13). □

In the example below the leading list is $\{1, 3\}$. The columns $\mathrm{col}_3(\mathbf{P})$ and $\mathrm{col}_4(\mathbf{P})$ may be arbitrary.

$$\mathbf{A} = \begin{pmatrix} 10 & 10 & 3 & 26 & 13 \\ 3 & 3 & 1 & 8 & 4 \\ 2 & 2 & 0 & 4 & 2 \\ 7 & 7 & 2 & 18 & 9 \end{pmatrix} = \begin{pmatrix} 10 & 3 & 5 & 3 \\ 3 & 1 & 1 & 1 \\ 2 & 0 & 3 & 5 \\ 7 & 2 & 5 & 1 \end{pmatrix} \begin{pmatrix} 1 & 1 & 0 & 2 & 1 \\ 0 & 0 & 1 & 2 & 1 \\ 0 & 0 & 0 & 0 & 0 \\ 0 & 0 & 0 & 0 & 0 \end{pmatrix}. \qquad (3.17)$$

$$\underbrace{}_{\mathbf{P}} \qquad \underbrace{}_{\text{rref}(\mathbf{A})}$$

The reason for this phenomenon is clear. When we take the dot product of the ith row of $\mathbf{P}$ with the jth column of rref$(\mathbf{A})$ all entries of this column below the second row in rref$(\mathbf{A})$ are zeros. Hence dot product does not depend on the entries of the kth columns of $\mathbf{P}$ for $k > 2$, implying that col$_3(\mathbf{P})$ and col$_4(\mathbf{P})$ do not influence the product $\mathbf{P}$rref$(\mathbf{A})$. In (3.17) the choice of these columns is made to make the columns of $\mathbf{P}$ be linearly independent. Then $\mathbf{P}$ is invertible, since its null-space is zero.

Theorem 3.10 *An $n \times n$ matrix $\mathbf{A}$ is invertible if and only if it is a finite product of elementary matrices.*

Proof. By Corollary 3.2 the matrix $\mathbf{A}$ is invertible if and only if it is row equivalent to $\mathbf{I}_n$. If $\mathbf{A} = \mathbf{I}_n$, then $\mathbf{A} = \mathbf{E}^2$, where $\mathbf{E}$ is an elementary matrix interchanging two different rows of $\mathbf{I}_n$. If $\mathbf{A} \neq \mathbf{I}_n$, then $\mathbf{A} \sim \mathbf{I}_n$. By Theorem 1.6 there is a finite number of elementary matrices $\mathbf{E}_1, \ldots, \mathbf{E}_r$ such that

$$\mathbf{E}_r \cdots \mathbf{E}_2 \mathbf{E}_1 \mathbf{A} = \mathbf{I}_n \Rightarrow \mathbf{A} = \mathbf{E}_1^{-1} \mathbf{E}_2^{-1} \cdots \mathbf{E}_r^{-1}.$$

If matrix $\mathbf{A}$ is the product of elementary matrices, then it is invertible by Theorem 3.6, since any elementary matrix invertible.   □

Problems

Prob. 101 — Find the inverse matrix for the matrix

$$\mathbf{A} = \begin{pmatrix} 3 & 1 & -1 \\ 1 & 1 & 0 \\ 2 & 1 & 2 \end{pmatrix}$$

by evaluating rref$(\mathbf{A}|\mathbf{I}_3)$.

Prob. 102 — Let

$$\mathbf{A} = \begin{pmatrix} 1 & 0 & 0 & -1 \\ 3 & 1 & 2 & 2 \\ 1 & 0 & -2 & 1 \\ 2 & 0 & 0 & 1 \end{pmatrix}.$$

Evaluate $\mathrm{rref}(\mathbf{A}|\mathbf{I}_4)$ and then find $\mathbf{A}^{-1}$.

Prob. 103 — Factor the matrix

$$\mathbf{A} = \begin{pmatrix} 1 & 1 & 1 & 1 & 1 & 3 \\ 2 & 1 & 1 & 1 & 2 & 4 \\ 1 & -1 & -1 & 1 & 1 & 5 \\ 1 & 0 & 0 & 1 & 1 & 4 \end{pmatrix}$$

as a product of an invertible matrix $\mathbf{P}$ and of $\mathrm{rref}(\mathbf{A})$.

3.3 Rank Factorization Theorem

Definition 3.4 Let $\mathbf{A}$ be an $m \times n$ matrix with $\mathrm{rank}(\mathbf{A}) = r$. A **rank factorization** of $\mathbf{A}$ is a product $\mathbf{BC} = \mathbf{A}$, where $\mathbf{B}$ is an $m \times r$ matrix and $\mathbf{C}$ is an $r \times n$ matrix.

Theorem 3.11 *Let $\mathbf{BC} = \mathbf{A}$ be a rank factorization of an $m \times n$ matrix $\mathbf{A}$. Then*

$$r = \mathrm{rank}(\mathbf{A}) = \mathrm{rank}(\mathbf{B}) = \mathrm{rank}(\mathbf{C}).$$

Proof. The range $R(\mathbf{A})$ of the matrix $\mathbf{A} = \mathbf{BC}$ is a subspace of $R(\mathbf{B})$. Since $\mathbf{B}$ is an $m \times r$ matrix, we conclude that $\mathrm{rank}(\mathbf{B}) \leq r$ by Theorem 1.26. It follows that

$$r = \mathrm{rank}(\mathbf{A}) = \dim(R(\mathbf{A})) \leq \dim(R(\mathbf{B})) = \mathrm{rank}(\mathbf{B}) \leq r.$$

In this chain of inequalities we have equalities implying that $\mathrm{rank}(\mathbf{B}) = r$.

By Corollary 2.7 the matrices $\mathbf{A}$ and $\mathbf{A}^T$ have equal ranks. By definition $\mathbf{C}^T$ is an $n \times r$ matrix and $\mathbf{B}^T$ is an $r \times m$ matrix. By Theorem 1.5

$$\mathbf{A}^T = (\mathbf{BC})^T = \mathbf{C}^T \mathbf{B}^T$$

is a rank factorization of $\mathbf{A}^T$, implying that $\mathrm{rank}(\mathbf{C})^T = \mathrm{rank}(\mathbf{C}) = r$ by the first part of the proof.   □

Lemma 3.2 *Let* $\mathbf{A}$ *be an* $m \times n$ *matrix,* $r = \mathrm{rank}(\mathbf{A})$, *and* $\mathbf{A} = \mathbf{BC}$ *be a rank factorization of* $\mathbf{A}$. *Then* $\mathrm{N}(\mathbf{B}) = \{\mathbf{0}\}$ *and* $\mathrm{N}(\mathbf{A}) = \mathrm{N}(\mathbf{C})$.

Proof. The first multiplier $\mathbf{B}$ in the rank factorization of $\mathbf{A}$ has r columns. By Theorem 3.11 we conclude that $\mathrm{rank}(\mathbf{B}) = r$. Then $\dim(\mathbf{col}(\mathbf{B})) = r$ by Corollary 2.6, implying that the columns of $\mathbf{B}$ make a basis of the column space. Hence $\mathrm{N}(\mathbf{B}) = \{\mathbf{0}\}$ by Theorem 1.10. It follows that $\mathbf{Ax} = \mathbf{0}$ if and only if $\mathbf{Cx} = \mathbf{0}$. Hence $\mathrm{N}(\mathbf{A}) = \mathrm{N}(\mathbf{C})$. $\qquad\square$

By Theorem 2.15 we obtain that $\mathbf{row}(\mathbf{A}) = \mathbf{row}(\mathbf{C})$. By Theorem 1.21 the matrix $\mathrm{rref}(\mathbf{C})$ equals the part of $\mathrm{rref}(\mathbf{A})$ made by the first r rows. These arguments give hints how one can find a rank factorization for a given matrix $\mathbf{A}$.

Theorem 3.12 *Let* $\mathbf{A}$ *be an* $m \times n$ *matrix,* $r = \mathrm{rank}(\mathbf{A})$ *be its rank, and* $\mathrm{rref}_p(\mathbf{A})$ *be the principal part of its reduced row echelon form. Let* $1 \le j_1 < j_2 < \cdots < j_r \le n$ *be the leading list for* $\mathbf{A}$, *and let* $\mathbf{B}_r$ *be the* $m \times r$ *matrix*

$$\mathbf{B}_r = \left(\mathrm{col}_{j_1}(\mathbf{A})\; \mathrm{col}_{j_2}(\mathbf{A}) \cdots \mathrm{col}_{j_r}(\mathbf{A})\right).$$

Then $\mathbf{A} = \mathbf{B}_r \mathrm{rref}_p(\mathbf{A})$ *is a rank factorization of* $\mathbf{A}$.

Proof. By Lemma 1.2, $\mathbf{A} = \mathbf{B}_r \mathrm{rref}_p(\mathbf{A})$ if and only if

$$\mathbf{Ax} = \mathbf{B}_r \mathrm{rref}_p(\mathbf{A})\mathbf{x} \tag{3.18}$$

for every $\mathbf{x}$ in $\mathbb{R}^n$. Since $\mathbb{R}^n = \mathrm{P}(\mathbf{A}) + \mathrm{N}(\mathbf{A})$, every $\mathbf{x}$ in $\mathbb{R}^n$ is the sum of a unique vector $\mathbf{p}$ in $\mathrm{P}(\mathbf{A})$ and $\mathbf{u}$ in $\mathrm{N}(\mathbf{A})$, see Theorem 2.5 and (2.18). If $\mathbf{x} = \mathbf{u}$ then both parts of the equality in (2.5) are zeros. Indeed, $\mathbf{Au} = \mathbf{0}$ by the definition of the null space. The row space of $\mathrm{rref}_p(\mathbf{A})$ equals the row space of $\mathrm{rref}(\mathbf{A})$. Since $\mathrm{N}(\mathbf{A}) \perp \mathrm{rref}(\mathbf{A})$, we conclude that $\mathrm{N}(\mathbf{A}) \perp \mathrm{rref}_p(\mathbf{A})$. Hence $\mathrm{N}(\mathrm{rref}_p(\mathbf{A})) = \mathrm{N}(\mathbf{A})$ implying that the right hand part of the equality in (3.18) is zero. Since

$$\mathrm{P}(\mathbf{A}) = \mathrm{Lin}\left(\mathbf{e}_{j_1}, \ldots, \mathbf{e}_{j_r}\right)$$

and both parts of (3.18) are linear in $\mathbf{x}$, it is sufficient to establish (3.18) only for $\mathbf{x} = \mathbf{e}_{j_s}$, $s = 1, \ldots, r$. Since $\mathrm{rref}_p(\mathbf{A}) = \mathrm{rref}(\mathrm{rref}_p(\mathbf{A}))$ and $\mathrm{rref}(\mathbf{A})$ have equal non-zero rows, we see that $\mathrm{rref}_p(\mathbf{A})\mathbf{e}_{j_s} = \mathbf{e}_s$. Then

$$\mathbf{B}_r \mathrm{rref}_p(\mathbf{A})\mathbf{e}_{j_s} = \mathbf{B}_r \mathbf{e}_s = \mathrm{col}_{j_s}(\mathbf{A}) = \mathbf{A}\mathbf{e}_{j_s}.$$

If we put now $\mathbf{x} = \mathbf{e}_k$ in (3.18) we obtain that the kth columns of matrices $\mathbf{A}$ and $\mathbf{B}_r \mathrm{rref}_p(\mathbf{A})$ are identical. Since $k = 1,\dots,n$, we conclude that both matrices have equal columns with equal indices and, therefore, are equal. The matrix $\mathrm{rref}_p(\mathbf{A})$ is in a reduced row echelon form and has r nonzero rows. Hence its rank is r. The columns of $\mathbf{B}_r$ are linearly independent, since they make the Gauss basis for $\mathrm{col}(\mathbf{A})$. Hence the rank of $\mathbf{B}_r$ is also r. □

Definition 3.5 The rank factorization $\mathbf{A} = \mathbf{B}_r \mathrm{rref}_p(\mathbf{A})$ is called the **standard rank factorization**.

Using the standard rank factorization of an $n \times n$ matrix $\mathbf{A} = \mathbf{B}_r \mathrm{rref}_p(\mathbf{A})$ we describe all rank factorizations of $\mathbf{A}$.

Theorem 3.13 *Let $\mathbf{A} = \mathbf{BC}$ be a rank factorization of a given $m \times n$ matrix and let $r = \mathrm{rref}(\mathbf{A})$. Then there exists an invertible $r \times r$ matrix $\mathbf{D}$ such that*

$$\mathbf{B} = \mathbf{B}_r \mathbf{D}^{-1}, \quad \mathbf{C} = \mathbf{D}\mathrm{rref}_p(\mathbf{A}).$$

Proof. By Theorem 3.12 we have $\mathbf{A} = \mathbf{B}_r \mathrm{rref}_p(\mathbf{A})$. By Lemma 3.2 $\mathrm{N}(\mathbf{A}) = \mathrm{N}(\mathbf{C}) = \mathrm{N}(\mathrm{rref}_p(\mathbf{A}))$. Since $\mathbf{C}$ and $\mathrm{rref}_p(\mathbf{A})$ are both of size $r \times n$, we see that $\mathrm{rref}(\mathbf{C}) = \mathrm{rref}_p(\mathbf{A})$. By Theorem 3.8 there exists an invertible a $r \times r$ matrix $\mathbf{D}$ such that $\mathbf{C} = \mathbf{D}\mathrm{rref}_p(\mathbf{A})$. By Theorem 3.12 we have $\mathbf{A} = \mathbf{B}_r \mathrm{rref}_p(\mathbf{A})$. It follows that

$$\mathbf{A} = \mathbf{B}_r \mathrm{rref}_p(\mathbf{A}) = \mathbf{B}_r \mathbf{D}^{-1}\mathbf{D}\mathrm{rref}_p(\mathbf{A}) = \mathbf{B}\mathbf{D}\mathrm{rref}_p(\mathbf{A}) \Rightarrow \mathbf{B}_r \mathbf{D}^{-1}\mathbf{y} = \mathbf{B}\mathbf{y}$$

for every $\mathbf{y}$ in $\mathrm{R}(\mathbf{D}\mathrm{rref}_p(\mathbf{A}))$. Since $\mathrm{rref}_p(\mathbf{A})$ has rank r, the dimension of $\mathrm{R}(\mathrm{rref}_p(\mathbf{A}))$ is r. But $\mathrm{R}(\mathrm{rref}_p(\mathbf{A}))$ is a subspace of $\mathbb{R}^r$. Therefore, $\mathrm{R}(\mathrm{rref}_p(\mathbf{A})) = \mathbb{R}^r$. Since $\mathbf{D}$ is an invertible $r \times r$ matrix, we conclude that $\mathrm{R}(\mathbf{D}\mathrm{rref}_p(\mathbf{A})) = \mathbb{R}^r$. Then $\mathbf{B} = \mathbf{B}_r \mathbf{D}^{-1}$ by Lemma 1.2. □

An Algorithm of a Rank Factorization
Step 1. Evaluate $\mathrm{rref}(\mathbf{A})$.
Step 2. Determine the leading list $J = \{j_1,\dots,j_r\}$.
Step 3. Define the matrices

$$\mathbf{B} = \left(\mathrm{Col}_{j_1}(\mathbf{A})\ \mathrm{Col}_{j_2}(\mathbf{A})\ \cdots\ \mathrm{Col}_{j_r}(\mathbf{A})\right), \quad \mathbf{C} = \mathrm{rref}_p(\mathbf{A}).$$

Then $\mathbf{A} = \mathbf{BC}$ is a rank factorization of $\mathbf{A}$.

Problem 3.4 Find a rank factorization of the matrix

$$A = \begin{pmatrix} 3 & 3 & 2 & 12 & -2 \\ 2 & 2 & 1 & 7 & 0 \\ 1 & 1 & -1 & -1 & 1 \\ 6 & 6 & -2 & 6 & 2 \end{pmatrix}.$$

Solution: Step 1. We find the matrix

$$\text{rref}(A) = \begin{pmatrix} 1 & 1 & 0 & 2 & 0 \\ 0 & 0 & 1 & 3 & 0 \\ 0 & 0 & 0 & 0 & 1 \\ 0 & 0 & 0 & 0 & 0 \end{pmatrix}.$$

Step 2. It follows that $\text{rank}(A) = 3$ and $J = \{1, 3, 5\}$.
    Step 3. We define

$$B = \begin{pmatrix} 3 & 2 & -2 \\ 2 & 1 & 0 \\ 1 & -1 & 1 \\ 6 & -2 & 2 \end{pmatrix}, \quad C = \begin{pmatrix} 1 & 1 & 0 & 2 & 0 \\ 0 & 0 & 1 & 3 & 0 \\ 0 & 0 & 0 & 0 & 1 \end{pmatrix}.$$

Hence $A = BC$ is a rank factorization for A.   □

Problems

Prob. 104 — Find a rank factorization for the matrix

$$A = \begin{pmatrix} 1 & -1 & 2 \\ 0 & 2 & -2 \\ 1 & 1 & 0 \end{pmatrix}.$$

Prob. 105 — Show that the formula below is a rank factorization of the matrix A:

$$A = \begin{pmatrix} 1 & 1 & 0 \\ 0 & 1 & 1 \\ -1 & 1 & 2 \\ 1 & 0 & -1 \end{pmatrix} = \begin{pmatrix} 1 & 1 \\ 0 & 1 \\ -1 & 1 \\ 1 & 0 \end{pmatrix} \begin{pmatrix} 1 & 0 & -1 \\ 0 & 1 & 1 \end{pmatrix} = BC.$$

Prob. 106 — Find a rank factorization for the matrix

$$A = \begin{pmatrix} 1 & 3 & -2 & 0 & 2 & 0 & 0 \\ 2 & 6 & -5 & -2 & 4 & -3 & -1 \\ 0 & 0 & 5 & 10 & 0 & 15 & 5 \\ 2 & 6 & 0 & 8 & 4 & 18 & 6 \end{pmatrix}.$$

3.4 The Inverse Problem of Gaussian Elimination

The Gauss-Jordan theory gives a complete description of solutions to an arbitrary system $\mathbf{A}\mathbf{x} = \mathbf{b}$ of m linear equations in n unknowns. In this section, we apply this theory in the inverse direction. To begin with we summarize what was already done. With any linear system of m equations in n unknowns we associate an $m \times n$ matrix $\mathbf{A}$ which is called the coefficient matrix of the system. Then the system can be written in matrix form as $\mathbf{A}\mathbf{x} = \mathbf{b}$. Using only elementary row operations one can always find a unique matrix $\mathrm{rref}(\mathbf{A})$ which is called the reduced row echelon form for $\mathbf{A}$. The reduced row echelon form determines uniquely six Gauss bases: the basis for the null space $N(\mathbf{A})$, for the row space $\mathbf{row}(\mathbf{A})$, for the range $R(\mathbf{A})$, for the column space $\mathbf{col}(\mathbf{A})$, for particular solutions $P(\mathbf{A})$, and the Gauss basis of solutions defined as the set of n columns of the Gauss $n \times n$ matrix $\mathbf{G}$. The Gauss bases for $\mathbf{row}(\mathbf{A})$ and $\mathbf{col}(\mathbf{A})$ have $r = \mathrm{rank}(\mathbf{A})$ vectors. The Gauss basis for $N(\mathbf{A})$ has $n - r$ elements. The Gauss basis for $P(\mathbf{A})$ is a subset $\{\mathbf{e}_{j_1}, \ldots, \mathbf{e}_{j_r}\}$ of the standard basis in $\mathbb{R}^n$, where $j_1 < \cdots < j_r$ is the leading list for $\mathbf{A}$. The Gauss basis of solutions is a basis for $\mathbb{R}^n$ and hence has n vectors. These constructions are summarized by the diagram:

$$
\mathrm{rref}\left((\mathbf{A}|\mathbf{b})\right) = (\mathrm{rref}(\mathbf{A})|\mathbf{d}) \Rightarrow
\begin{cases}
N(\mathbf{A}), \\
\mathbf{row}(\mathbf{A}) \oplus N(\mathbf{A}) = \mathbb{R}^n, \\
\mathbf{col}(\mathbf{A}) = \mathrm{Lin}\left(\mathrm{col}_{j_1}(\mathbf{A}), \ldots, \mathrm{col}_{j_r}(\mathbf{A})\right), \\
P(\mathbf{A}) = \mathrm{Lin}\left(\mathbf{e}_{j_1}, \ldots, \mathbf{e}_{j_r}\right), \\
P(\mathbf{A}) + N(\mathbf{A}) = \mathbb{R}^n.
\end{cases}
$$

The **Inverse Problem of Gaussian Elimination** is to recover a linear system $\mathbf{A}\mathbf{x} = \mathbf{b}$ by a given null space V and by given columns $\mathbf{c}_{j_1}, \mathbf{c}_{j_2}, \ldots, \mathbf{c}_{j_r}$ of $\mathbf{A}$. Notice that $\mathrm{rref}(\mathbf{A})$ being uniquely recovered by the null space $N(\mathbf{A})$ determines the leading list $1 \leq j_1 < \cdots < j_r \leq n$ for $\mathbf{A}$. We find the solution to this problem by rank factorizations of matrices. This method will also be used later in the theory of Strong Generalized Inverses.

Theorem 3.14 *Given a subspace V of $\mathbb{R}^n$, $\dim(V) = n - r$, a positive integer $m \geq r$, and an arbitrary system $\mathbf{b}_1, \mathbf{b}_2, \ldots, \mathbf{b}_r$ of linearly independent vectors in $\mathbb{R}^m$ there is a unique matrix $\mathbf{A}$ of size $m \times n$ such that*

$$\mathrm{col}_{j_1}(\mathbf{A}) = \mathbf{b}_1, \quad \mathrm{col}_{j_2}(\mathbf{A}) = \mathbf{b}_2, \quad \ldots, \quad \mathrm{col}_{j_r}(\mathbf{A}) = \mathbf{b}_r, \quad N(\mathbf{A}) = V,$$

where $1 \leq j_1 < j_2 < \cdots < j_r \leq n$ is the leading list for $\mathbf{A}$.

Proof. There is a unique $m \times n$ matrix $\mathbf{C}$ such that

$$\text{rref}(\mathbf{C}) = \mathbf{C}, \quad \text{rank}(\mathbf{C}) = r, \quad N(\mathbf{C}) = V.$$

Let

$$\mathbf{B} = \begin{pmatrix} \mathbf{b}_1 & \mathbf{b}_2 & \cdots & \mathbf{b}_r \end{pmatrix}.$$

Then $\mathbf{A} = \mathbf{BC}$. So, such a matrix exists. It is unique, since by Theorem 3.12 any such $\mathbf{A}$ has exactly this rank factorization.   □

The following corollary, in fact, says that among all matrices of fixed rank r with a given reduced row echelon form there always exist matrices with $\mathbf{R}(\mathbf{A})$ equal to an arbitrary given subspace of $\mathbb{R}^m$ of dimension r.

Corollary 3.3 *Let V be a subspace of $\mathbb{R}^n$, $\dim(V) = n - r$, where $0 \le r \le n$, and W be a subspace of $\mathbb{R}^m$, $\dim(W) = r$, where $r \le m$. Then there exists an $m \times n$ matrix $\mathbf{A}$ such that $N(\mathbf{A}) = V$ and $R(\mathbf{A}) = W$.*

Corollary 3.3 together with the Rank-Nullity Theorem (see Theorem 2.14) gives necessary and sufficient conditions, in order there would exist a matrix with given range and null-space. Moreover, this matrix can be constructed explicitly.

Problem 3.5 Find a matrix $\mathbf{A}$ with the range defined by the matrix $\mathbf{X}$ and the null-space defined by the matrix $\mathbf{Y}$, where

$$\mathbf{X} = \begin{pmatrix} 2 & 1 & 4 & 7 & 2 \\ 2 & -1 & 0 & 1 & 3 \\ 1 & 3 & 7 & 11 & 1 \\ 1 & 2 & 5 & 8 & -1 \end{pmatrix}, \quad \mathbf{Y} = \begin{pmatrix} 1 & 1 & 3 & 2 & 7 \\ 2 & -1 & 0 & 3 & 7 \\ 0 & 2 & 4 & 1 & 5 \\ 1 & 2 & 5 & -1 & 6 \end{pmatrix}.$$

Solution: We find the reduced row echelon forms for $\mathbf{X}$ and $\mathbf{Y}$:

$$\text{rref}(\mathbf{X}) = \begin{pmatrix} 1 & 0 & 1 & 2 & 0 \\ 0 & 1 & 2 & 3 & 0 \\ 0 & 0 & 0 & 0 & 1 \\ 0 & 0 & 0 & 0 & 0 \end{pmatrix}, \quad \text{rref}(\mathbf{Y}) = \begin{pmatrix} 1 & 0 & 1 & 0 & 3 \\ 0 & 1 & 2 & 0 & 2 \\ 0 & 0 & 0 & 1 & 1 \\ 0 & 0 & 0 & 0 & 0 \end{pmatrix}.$$

The leading list for $\mathbf{X}$ is $\{1,2,5\}$. Then

$$\begin{pmatrix} 2 & 1 & 2 \\ 2 & -1 & 3 \\ 1 & 3 & 1 \\ 1 & 2 & -1 \end{pmatrix} \begin{pmatrix} 1 & 0 & 1 & 0 & 3 \\ 0 & 1 & 2 & 0 & 2 \\ 0 & 0 & 0 & 1 & 1 \end{pmatrix} = \begin{pmatrix} 2 & 1 & 4 & 2 & 10 \\ 2 & -1 & 0 & 3 & 7 \\ 1 & 3 & 7 & 1 & 10 \\ 1 & 2 & 5 & -1 & 6 \end{pmatrix} = \mathbf{A}$$

is a rank factorization of $\mathbf{A}$. Hence $\text{rref}(\mathbf{A}) = \text{rref}(\mathbf{Y})$ implying that $N(\mathbf{A}) = N(\mathbf{Y})$. The leading list for $\mathbf{A}$ is the same as for $\mathbf{Y}$, i.e. it is $\{1,2,4\}$. The leading list of $\mathbf{X}$ is $\{1,2,5\}$. Since

$$\text{col}_1(\mathbf{A}) = \text{col}_1(\mathbf{X}), \ \text{col}_2(\mathbf{A}) = \text{col}_2(\mathbf{X}), \ \text{col}_4(\mathbf{A}) = \text{col}_5(\mathbf{X}),$$

we conclude that

$$R(\mathbf{A}) = \mathbf{col}(\mathbf{A}) = \mathbf{col}(\mathbf{X}) = R(\mathbf{X}). \quad \square$$

Theorem 3.14 allows one to recover the system of linear equations by the given set of solutions $\mathbf{p} + \mathbf{u}$, and by the given list of leading columns. We illustrate this by the following problem.

Problem 3.6 The systems

$$\begin{cases} 7x_1 + 2x_2 + 11x_3 + 2x_4 + 9x_5 &= 27 \\ 6x_1 + x_2 + 8x_3 + 2x_4 + 7x_5 &= 23 \\ 3x_1 + 3x_3 + x_4 + 3x_5 &= 11 \\ x_1 - 2x_2 - 3x_3 - x_5 &= 1 \end{cases} \qquad \begin{cases} x_1 + 2x_2 + *x_3 + 2x_4 + *x_5 &= * \\ -x_1 + x_2 + *x_3 + 2x_4 + *x_5 &= * \\ 2x_1 - x_2 + *x_3 + x_4 + *x_5 &= * \\ x_1 + *x_3 - x_4 + *x_5 &= * \end{cases}$$

are equivalent. Determine the second system.

Solution: The reduced row echelon form of the augmented matrix for the given system

$$\mathrm{rref} \begin{pmatrix} 7 & 2 & 11 & 2 & 9 & 27 \\ 6 & 1 & 8 & 2 & 7 & 23 \\ 3 & 0 & 3 & 1 & 3 & 11 \\ 1 & -2 & -3 & 0 & -1 & 1 \end{pmatrix} = \begin{pmatrix} 1 & 0 & 1 & 0 & 1 & 3 \\ 0 & 1 & 2 & 0 & 1 & 1 \\ 0 & 0 & 0 & 1 & 0 & 2 \\ 0 & 0 & 0 & 0 & 0 & 0 \end{pmatrix}$$

shows that its rank $r = 3$ and the leading list of the coefficient matrix is: $j_1 = 1$, $j_2 = 2$, $j_3 = 4$, whereas the particular solution is $\mathbf{p} = (3\ 1\ 0\ 2\ 0)^T$. By Theorem 3.12 the coefficient matrix $\mathbf{A}$ of the second system and $\mathbf{Ap}$ are given by

$$\mathbf{A} = \begin{pmatrix} 1 & 2 & 2 \\ -1 & 1 & 2 \\ 2 & -1 & 1 \\ 1 & 0 & -1 \end{pmatrix} \begin{pmatrix} 1 & 0 & 1 & 0 & 1 \\ 0 & 1 & 2 & 0 & 1 \\ 0 & 0 & 0 & 1 & 0 \end{pmatrix} = \begin{pmatrix} 1 & 2 & 5 & 2 & 3 \\ -1 & 1 & 1 & 2 & 0 \\ 2 & -1 & 0 & 1 & 1 \\ 1 & 0 & 1 & -1 & 1 \end{pmatrix},$$

$$\mathbf{Ap} = \begin{pmatrix} 1 & 2 & 5 & 2 & 3 \\ -1 & 1 & 1 & 2 & 0 \\ 2 & -1 & 0 & 1 & 1 \\ 1 & 0 & 1 & -1 & 1 \end{pmatrix} \begin{pmatrix} 3 \\ 1 \\ 0 \\ 2 \\ 0 \end{pmatrix} = \begin{pmatrix} 9 \\ 2 \\ 7 \\ 1 \end{pmatrix}.$$

It follows that the system equivalent to the given system is

$$\begin{cases} x_1 + 2x_2 + 5x_3 + 2x_4 + 3x_5 &= 9 \\ -x_1 + x_2 + x_3 + 2x_4 &= 2 \\ 2x_1 - x_2 + x_4 + x_5 &= 7 \\ x_1 + x_3 - x_4 + x_5 &= 1 \end{cases} \qquad \square$$

Problem 3.7 Let $\mathbf{A}$ be an $n \times n$ matrix and $r = \mathrm{rank}(\mathbf{A})$. Prove that a consistent system $\mathbf{Ax} = \mathbf{b}$ is row equivalent to a system $\mathbf{Bx} = \mathbf{d}$ with $\mathbf{B}^2 = \mathbf{0}$ and some $\mathbf{d}$ if and only if $2r \leq n$.

Solution: We have $\mathbf{B}^2 = \mathbf{0}$ if and only if $R(\mathbf{B}) = \mathbf{col}(\mathbf{B})$ is a subspace of $N(\mathbf{B})$. The augmented matrix $(\mathbf{A} \,|\, \mathbf{b})$ is row equivalent to $(\mathbf{B} \,|\, \mathbf{d})$ with some $\mathbf{d}$ if and only if $\mathbf{A} \sim \mathbf{B}$, equivalently, $N(\mathbf{A}) = N(\mathbf{B})$. If $\mathbf{A}\mathbf{x} = \mathbf{b}$ is consistent and $2r \leq n$ (equivalently, $r \leq n - r$) then r elements of the Gauss basis for $N(\mathbf{A})$ by Theorem 3.14 can be columns of some matrix $\mathbf{B}$ satisfying $N(\mathbf{B}) = N(\mathbf{A})$. Since $\mathbf{col}(\mathbf{B})$ is a subspace of $N(\mathbf{B})$, we conclude that $\mathbf{B}^2 = \mathbf{0}$. Since $\mathbf{B} \sim \mathbf{A}$, there is $\mathbf{d}$ such that the system $\mathbf{B}\mathbf{x} = \mathbf{d}$ is row equivalent to $\mathbf{A}\mathbf{x} = \mathbf{b}$.

If $(\mathbf{A} \,|\, \mathbf{b}) \sim (\mathbf{B} \,|\, \mathbf{d})$ with some d, then $\mathbf{A} \sim \mathbf{B}$, implying that $N(\mathbf{B}) = N(\mathbf{A})$. Since $\mathbf{B}^2 = \mathbf{0}$, we see that $\mathbf{col}(\mathbf{B})$ is a subspace of $N(\mathbf{A})$. It follows that $r \leq n - r$, equivalently, $2r \leq n$.   □

Problems

Prob. 107 — Find a matrix $\mathbf{A}$ with the range defined by the null-space of the matrix $\mathbf{X}$ and the null-space defined by the null-space of the matrix $\mathbf{Y}$, where

$$\mathbf{X} = \begin{pmatrix} 1 & 3 & -2 & 0 & 2 & 0 \\ 2 & 6 & -5 & -2 & 4 & -3 \\ 0 & 0 & 5 & 10 & 0 & 15 \\ 2 & 6 & 0 & 8 & 4 & 18 \end{pmatrix}, \quad \mathbf{Y} = \begin{pmatrix} 1 & 1 & 1 & 1 & 1 & 3 \\ 2 & 1 & 1 & 1 & 2 & 4 \\ 1 & -1 & -1 & 1 & 1 & 5 \\ 1 & 0 & 0 & 1 & 1 & 4 \end{pmatrix}.$$

3.5 Extended Matrices

The reduced row echelon form $\left(\mathrm{rref}(\mathbf{A}) \,|\, \mathbf{B}\right)$ of the extended matrix $\left(\mathbf{A} \,|\, \mathbf{I}_m\right)$ can be used to find all important subspaces associated with the matrix $\mathbf{A}$: $\mathbf{row}(\mathbf{A})$, $N(\mathbf{A})$, $\mathbf{col}(\mathbf{A}) = R(\mathbf{A})$, $P(\mathbf{A})$.

Since $\left(\mathbf{A} \,|\, \mathbf{I}_m\right) \sim \left(\mathrm{rref}(\mathbf{A}) \,|\, \mathbf{B}\right)$, we have $\mathbf{I}_m \sim \mathbf{B}$. By Corollary 3.2 the $m \times m$ matrix $\mathbf{B}$ is **non-singular** and, therefore, has the inverse matrix $\mathbf{B}^{-1}$.

If $r = \mathrm{rank}(\mathbf{A}) = m$ then there are no zero rows in $\mathrm{rref}(\mathbf{A})$, i.e. $\mathrm{rref}(\mathbf{A})$ equals to its principle part $\mathrm{rref}_p(\mathbf{A})$, implying that the system $\mathbf{A}\mathbf{x} = \mathbf{y}$ is consistent for every $\mathbf{y}$ in $\mathbb{R}^m$.

If $r = \mathrm{rank}(\mathbf{A} < m$ then the last $m - r$ rows of $\mathrm{rref}(\mathbf{A})$ are zero rows. In this case $\left(\mathrm{rref}(\mathbf{A}) \,|\, \mathbf{B}\right)$ has the following block structure:

$$\left(\mathrm{rref}(\mathbf{A}) \,|\, \mathbf{B}\right) = \left(\begin{array}{c|c} \mathrm{rref}_p(\mathbf{A}) & \mathbf{B}_p \\ \hline \mathbf{0}_{(m-r) \times n} & \mathbf{B}_c \end{array}\right). \tag{3.19}$$

Here $\mathbf{B}_p$ is the $r \times m$ matrix with the rows equal to the first r rows of $\mathbf{B}$ and $\mathbf{B}_c$ is the $(m - r) \times m$ matrix with the rows equal to the last $m - r$ rows of $\mathbf{B}$:

$$\mathrm{row}_i(\mathbf{B}) = \begin{cases} \mathrm{row}_i(\mathbf{B}_p), & \text{if } 1 \leq i \leq r, \\ \mathrm{row}_{i-r}(\mathbf{B}_c), & \text{if } r < i \leq m. \end{cases}$$

Theorem 3.15 *The system* $\mathbf{A}\mathbf{x} = \mathbf{y}$ *is consistent if and only if either* $\mathrm{rank}(\mathbf{A})$ *is* m *or* $\mathbf{y}$ *is a vector in* $\mathrm{N}(\mathbf{B}_c)$*. In particular,*

$$R(\mathbf{A}) = \mathbf{col}(\mathbf{A}) = N(\mathbf{B}_c).$$

Proof. The system $\mathbf{A}\mathbf{x} = \mathbf{y}$ is row equivalent to

$$\mathrm{rref}(\mathbf{A})\mathbf{x} = \mathbf{B}\mathbf{y}. \tag{3.20}$$

Since $\mathbf{B}$ is a non-singular matrix, its rows are linearly independent. It follows that $\mathrm{rank}(\mathbf{B}_c) = m - r$. By the Rank-Nullity Theorem $\mathrm{nullity}(\mathbf{B}_c) = m - (m - r) = r$. By the Kronecker-Cappelli Theorem (see Theorem 1.29) the vector $\mathbf{y}$ is in $\mathrm{R}(\mathbf{A}) = \mathbf{col}(\mathbf{A})$ if and only if $\mathbf{y}$ is in $\mathrm{N}(\mathbf{B}_c)$. $\qquad\square$

Theorem 3.16 *The matrix* $\mathbf{B}_c$ *in* (3.19) *is in reduced row echelon form.*

Proof. Since the leading one for any ith row of $\left(\mathrm{rref}(\mathbf{A})|\mathbf{B}\right)$ with $i > r$ is the entry of $\mathbf{B}$, we conclude that $\mathrm{rref}(\mathbf{B}_c) = \mathbf{B}_c$. $\qquad\square$

Let J be the leading list $1 \le j_1 < j_2 < \cdots < j_r \le n$ for $\mathbf{A}$. We denote by $\mathbf{E}_J$ the $n \times m$ matrix

$$\mathbf{E}_J = \left(\mathbf{e}_{j_1} \cdots \mathbf{e}_{j_r}\ \mathbf{0}_n \cdots \mathbf{0}_n\right). \tag{3.21}$$

Since $r \le \min(m,n)$, the matrix $\mathbf{E}_J$ can always be defined. The matrix $\mathbf{E}_J$ is called the **leading list matrix**. It is easy to see that $\mathbf{E}_J\mathbf{B}$ satisfies

$$\mathrm{row}_i(\mathbf{E}_J\mathbf{B}) = \begin{cases} \mathrm{row}_k(\mathbf{B}) \text{ if } i = j_k, \\ \mathbf{0} \text{ if } i \ne j_1, \cdots, j_r. \end{cases} \tag{3.22}$$

Theorem 3.17 *For every* $\mathbf{y}$ *in* $\mathrm{N}(\mathbf{B}_c)$ *the vector* $\mathbf{p} = \mathbf{E}_J\mathbf{B}\mathbf{y}$ *is the Gauss particular solution for the system* $\mathbf{A}\mathbf{x} = \mathbf{y}$*.*

Proof. By Theorem 3.15 the system $\mathbf{Ax} = \mathbf{y}$ is consistent for every $\mathbf{y}$ in $N(\mathbf{B}_c)$. By (3.20) $\mathbf{p} = \mathbf{E}_J \mathbf{By}$ is the Gauss particular solution for the system $\mathbf{Ax} = \mathbf{y}$. □

Problem 3.8 Given an 4×5 matrix

$$\mathbf{A} = \begin{pmatrix} 10 & 10 & 3 & 26 & 13 \\ 3 & 3 & 1 & 8 & 4 \\ 2 & 2 & 0 & 4 & 2 \\ 7 & 7 & 2 & 18 & 9 \end{pmatrix} \tag{3.23}$$

find Gauss bases for $\mathbf{row}(\mathbf{A})$, $N(\mathbf{A})$, $\mathbf{col}(\mathbf{A})$, $P(\mathbf{A})$.

Solution: The reduced row echelon form of $\left(\mathbf{A}|\mathbf{I}_4\right)$ is

$$\begin{pmatrix} 1 & 1 & 0 & 2 & 1 & 0 & 0 & 1/2 & 0 \\ 0 & 0 & 1 & 2 & 1 & 0 & 0 & -7/4 & 1/2 \\ 0 & 0 & 0 & 0 & 0 & 1 & 0 & 1/4 & -3/2 \\ 0 & 0 & 0 & 0 & 0 & 0 & 1 & 1/4 & -1/2 \end{pmatrix}. \tag{3.24}$$

It follows that

$$\mathbf{B} = \begin{pmatrix} 0 & 0 & 1/2 & 0 \\ 0 & 0 & -7/4 & 1/2 \\ 1 & 0 & 1/4 & -3/2 \\ 0 & 1 & 1/4 & -1/2 \end{pmatrix}, \quad \mathbf{B}_c = \begin{pmatrix} 1 & 0 & 1/4 & -3/2 \\ 0 & 1 & 1/4 & -1/2 \end{pmatrix}.$$

A vector $\mathbf{y}$ is in $R(\mathbf{A})$ if and only if $\mathbf{y}$ is in $N(\mathbf{B}_c)$:

$$\begin{cases} 4y_1 + y_3 - 6y_4 = 0 \\ 4y_2 + y_3 - 2y_4 = 0 \end{cases} \Leftrightarrow \begin{cases} y_1 = \frac{3}{2}y_4 - \frac{1}{4}y_3 \\ y_2 = \frac{1}{2}y_4 - \frac{1}{4}y_3 \end{cases}. \tag{3.25}$$

It is easy to check that the columns of $\mathbf{A}$ satisfy (3.25). The Gauss basis for the subspace $N(\mathbf{B}_c)$

$$\left\{ \begin{pmatrix} -1/4 \\ -1/4 \\ 1 \\ 0 \end{pmatrix}, \begin{pmatrix} 3/2 \\ 1/2 \\ 0 \\ 1 \end{pmatrix} \right\}$$

does not equal the Gauss basis $\{\mathbf{col}_1(\mathbf{A}), \mathbf{col}_3(\mathbf{A})\}$ for the range $\mathbf{col}(\mathbf{A}) = N(\mathbf{B}_c)$ of $\mathbf{A}$. The Gauss basis for the row space of $\mathbf{A}$ is

$$\left\{ \begin{pmatrix} 1 & 1 & 0 & 2 & 1 \end{pmatrix}, \begin{pmatrix} 0 & 0 & 1 & 2 & 1 \end{pmatrix} \right\}.$$

The Gauss basis for $N(\mathbf{A})$ equals

$$\left\{ \begin{pmatrix} -1 \\ 1 \\ 0 \\ 0 \\ 0 \end{pmatrix}, \begin{pmatrix} -2 \\ 0 \\ -2 \\ 1 \\ 0 \end{pmatrix}, \begin{pmatrix} -1 \\ 0 \\ 0 \\ -1 \\ 0 \\ 1 \end{pmatrix} \right\}$$

To describe $P(A)$ we observe that by (3.24) the leading list J is $\{1,3\}$, the leading list matrix is the 5×4 matrix

$$\mathbf{E}_J = \begin{pmatrix} \mathbf{e}_1 \ \mathbf{e}_3 \ \mathbf{0}_4 \ \mathbf{0}_4 \end{pmatrix}.$$

Then the Gauss particular solution for $\mathbf{Ax} = \mathbf{y}$ can be recovered by the formula $\mathbf{p} = \mathbf{E}_J \mathbf{By}$. For example, the particular solution $\mathbf{p}$ for $\mathbf{y} = \begin{pmatrix} 1 \ 0 \ 2 \ 1 \end{pmatrix}^T$ equals

$$\mathbf{p} = \begin{pmatrix} 1 & 0 & 0 & 0 \\ 0 & 0 & 0 & 0 \\ 0 & 1 & 0 & 0 \\ 0 & 0 & 0 & 0 \\ 0 & 0 & 0 & 0 \end{pmatrix} \begin{pmatrix} 0 & 0 & 1/2 & 0 \\ 0 & 0 & -7/4 & 1/2 \\ 1 & 0 & 1/4 & -3/2 \\ 0 & 1 & 1/4 & -1/2 \end{pmatrix} \begin{pmatrix} 1 \\ 0 \\ 2 \\ 1 \end{pmatrix} = \begin{pmatrix} 1 \\ 0 \\ -3 \\ 0 \\ 0 \end{pmatrix}. \quad \square$$

Problem 3.9 Describe the range of the matrix

$$\mathbf{A} = \begin{pmatrix} 3 & 6 & 1 & 8 \\ 2 & 4 & 1 & 6 \\ 7 & 14 & 2 & 18 \end{pmatrix}.$$

Solution: The reduced row echelon form of the extended matrix can be found by elementary row operations:

$$\begin{pmatrix} 3 & 6 & 1 & 8 & | & 1 & 0 & 0 \\ 2 & 4 & 1 & 6 & | & 0 & 1 & 0 \\ 7 & 14 & 2 & 18 & | & 0 & 0 & 1 \end{pmatrix} \sim \begin{pmatrix} 1 & 2 & 0 & 2 & | & 0 & -2/3 & 1/3 \\ 0 & 0 & 1 & 2 & | & 0 & 7/3 & -2/3 \\ 0 & 0 & 0 & 0 & | & 1 & -1/3 & -1/3 \end{pmatrix}.$$

It follows that $\mathbf{b} = \begin{pmatrix} b_1 \ b_2 \ b_3 \end{pmatrix}^T$ is in $N(\mathbf{B}_c)$ if and only if

$$b_1 - \frac{1}{3}b_2 - \frac{1}{3}b_3 = 0.$$

Since $b_2 = 3s$ and $b_3 = 3t$ are non-leading variables, the vector $\mathbf{b}$ is in $N(\mathbf{B}_c)$ if and only if

$$\begin{pmatrix} b_1 \\ b_2 \\ b_3 \end{pmatrix} = \begin{pmatrix} s+t \\ 3s \\ 3t \end{pmatrix} = s \begin{pmatrix} 1 \\ 3 \\ 0 \end{pmatrix} + t \begin{pmatrix} 1 \\ 0 \\ 3 \end{pmatrix} = s\mathbf{v}_2 + t\mathbf{v}_3.$$

The linear span $\mathrm{Lin}(\mathbf{v}_2, \mathbf{v}_3)$ of the Gauss basis for $\mathbf{B}_c$ is a basis for $R(\mathbf{A})$. Notice that the Gauss basis $\{\mathrm{col}_1(\mathbf{A}), \mathrm{col}_3(\mathbf{A})\}$ for the column space of $\mathbf{A}$ is another basis for $R(\mathbf{A})$. $\quad \square$

By Theorem 3.8 for every $m \times n$ matrix $\mathbf{A}$ there is a non-singular $m \times m$ matrix $\mathbf{P}$ such that

$$\mathbf{A} = \mathbf{P}\,\mathrm{rref}(\mathbf{A}).$$

In Theorem 3.8 a matrix $\mathbf{P}$ was obtained as a product of elementary matrices. This method is difficult to apply in practice. Here we show how one can find a matrix $\mathbf{P}$ by the method of extended matrices.

Theorem 3.18 *Let $\mathbf{A}$ be an $m \times n$ matrix and let*

$$\left(\mathbf{A}|\mathbf{I}_m\right) \sim \left(\mathrm{rref}(\mathbf{A}) \mid \mathbf{B}\right). \tag{3.26}$$

Then

$$\mathbf{A} = \mathbf{B}^{-1}\,\mathrm{rref}(\mathbf{A}). \tag{3.27}$$

Proof. By (3.20) for every x in $\mathbb{R}^n$ the equation $\mathbf{A}\mathbf{x} = \mathbf{y}$ is equivalent to $\mathrm{rref}(\mathbf{A})\mathbf{x} = \mathbf{B}\mathbf{y}$. Since $\mathbf{B} \sim \mathbf{I}_m$, the matrix $\mathbf{B}$ is not singular. It follows that

$$\mathbf{B}^{-1}\mathrm{rref}(\mathbf{A})\mathbf{x} = \mathbf{y} = \mathbf{A}\mathbf{x}.$$

Since $\mathbf{x}$ is any vector of $\mathbb{R}^n$, we obtain (3.27) by Lemma 1.2. □

For example, for the matrix in (3.8) we found that

$$\mathbf{B} = \begin{pmatrix} 0 & 0 & 1/2 & 0 \\ 0 & 0 & -7/4 & 1/2 \\ 1 & 0 & 1/4 & -3/2 \\ 0 & 1 & 1/4 & -1/2 \end{pmatrix}.$$

By the method of extended matrices

$$\mathbf{B}^{-1} = \begin{pmatrix} 10 & 3 & 1 & 0 \\ 3 & 1 & 0 & 1 \\ 2 & 0 & 0 & 0 \\ 7 & 2 & 0 & 0 \end{pmatrix}.$$

It is easy to check that

$$\mathbf{A} = \begin{pmatrix} 10 & 3 & 1 & 0 \\ 3 & 1 & 0 & 1 \\ 2 & 0 & 0 & 0 \\ 7 & 2 & 0 & 0 \end{pmatrix}\begin{pmatrix} 1 & 1 & 0 & 2 & 1 \\ 0 & 0 & 1 & 2 & 1 \\ 0 & 0 & 0 & 0 & 0 \\ 0 & 0 & 0 & 0 & 0 \end{pmatrix} = \mathbf{B}^{-1}\mathrm{rref}(\mathbf{A}). \tag{3.28}$$

We observe that $\{j_1 = 1, j_2 = 3\}$ is the leading list for $\mathbf{A}$ and

$$\mathrm{col}_1\left(\mathbf{B}^{-1}\right) = \mathrm{col}_{j_1}(\mathbf{A}), \; ; \mathrm{col}_2\left(\mathbf{B}^{-1}\right) = \mathrm{col}_{j_2}(\mathbf{A}).$$

Theorem 3.19 *Let* $\mathbf{A}$ *be an* $m \times n$ *matrix of rank* $r = \mathrm{rank}(\mathbf{A})$. *Then there are an invertible* $m \times m$ *matrix* $\mathbf{P}$ *and an invertible* $n \times n$ *matrix* $\mathbf{Q}$ *such that*

$$\mathbf{A} = \mathbf{P}\left(\begin{array}{c|c} \mathbf{I}_r & \mathbf{0}_{r \times (n-r)} \\ \hline \mathbf{0}_{(m-r) \times r} & \mathbf{0}_{(m-r) \times (n-r)} \end{array}\right) \mathbf{Q}. \tag{3.29}$$

Proof. By Theorem 3.8 there is an invertible $m \times m$ matrix $\mathbf{P}$ such that $\mathbf{A} = \mathbf{P}\,\mathrm{rref}(\mathbf{A})$. Applying elementary row operations to $\mathrm{rref}(\mathbf{A})^T$ we find that

$$\mathrm{rref}\left(\mathrm{rref}(\mathbf{A})^T\right) = \left(\begin{array}{c|c} \mathbf{I}_r & \mathbf{0}_{r \times (m-r)} \\ \hline \mathbf{0}_{(n-r) \times r} & \mathbf{0}_{(n-r) \times (m-r)} \end{array}\right).$$

By Theorem 3.8 there is an invertible $n \times n$ matrix $\mathbf{Q}^T$ such that

$$\mathrm{rref}(\mathbf{A})^T = \mathbf{Q}^T\left(\begin{array}{c|c} \mathbf{I}_r & \mathbf{0}_{r \times (m-r)} \\ \hline \mathbf{0}_{(n-r) \times r} & \mathbf{0}_{(n-r) \times (m-r)} \end{array}\right) \Rightarrow$$

$$\mathrm{rref}(\mathbf{A}) = \left(\begin{array}{c|c} \mathbf{I}_r & \mathbf{0}_{r \times (n-r)} \\ \hline \mathbf{0}_{(m-r) \times r} & \mathbf{0}_{(m-r) \times (n-r)} \end{array}\right) \mathbf{Q}. \quad \square$$

Problem 3.10 Find the factorization (3.29) for the matrix $\mathbf{A}$ defined in (3.23).

Solution: By (3.28) we have

$$\mathbf{P} = \begin{pmatrix} 10 & 3 & 1 & 0 \\ 3 & 1 & 0 & 1 \\ 2 & 0 & 0 & 0 \\ 7 & 2 & 0 & 0 \end{pmatrix}, \quad \mathrm{rref}(\mathbf{A})^T = \begin{pmatrix} 1 & 0 & 0 & 0 \\ 1 & 0 & 0 & 0 \\ 0 & 1 & 0 & 0 \\ 2 & 2 & 0 & 0 \\ 1 & 1 & 0 & 0 \end{pmatrix}.$$

Since

$$\left(\mathrm{rref}(\mathbf{A})^T \mid \mathbf{I}_5\right) = \left(\begin{array}{cccc|ccccc} 1 & 0 & 0 & 0 & 0 & 0 & -1 & 0 & 1 \\ 0 & 1 & 0 & 0 & 0 & 0 & 1 & 0 & 0 \\ 0 & 0 & 0 & 0 & 1 & 0 & 1 & 0 & -1 \\ 0 & 0 & 0 & 0 & 0 & 1 & 1 & 0 & -1 \\ 0 & 0 & 0 & 0 & 0 & 0 & 0 & 1 & -2 \end{array}\right),$$

we obtain that

$$\mathbf{Q}^T = \begin{pmatrix} 0 & 0 & -1 & 0 & 1 \\ 0 & 0 & 1 & 0 & 0 \\ 1 & 0 & 1 & 0 & -1 \\ 0 & 1 & 1 & 0 & -1 \\ 0 & 0 & 0 & 1 & -2 \end{pmatrix}^{-1} = \begin{pmatrix} 1 & 0 & 1 & 0 & 0 \\ 1 & 0 & 0 & 1 & 0 \\ 0 & 1 & 0 & 0 & 0 \\ 2 & 2 & 0 & 0 & 1 \\ 1 & 1 & 0 & 0 & 0 \end{pmatrix}.$$

It follows that

$$\mathbf{Q} = \begin{pmatrix} 1 & 1 & 0 & 2 & 1 \\ 0 & 0 & 1 & 2 & 1 \\ 1 & 0 & 0 & 0 & 0 \\ 0 & 1 & 0 & 0 & 0 \\ 0 & 0 & 0 & 1 & 0 \end{pmatrix}, \quad \mathrm{rref}\left(\mathrm{rref}(\mathbf{A})^T\right)^T = \begin{pmatrix} 1 & 0 & 0 & 0 & 0 \\ 0 & 1 & 0 & 0 & 0 \\ 0 & 0 & 0 & 0 & 0 \\ 0 & 0 & 0 & 0 & 0 \end{pmatrix}.$$

It is easy to check that

$$\mathbf{A} = \mathbf{P} \begin{pmatrix} 1 & 0 & 0 & 0 & 0 \\ 0 & 1 & 0 & 0 & 0 \\ 0 & 0 & 0 & 0 & 0 \\ 0 & 0 & 0 & 0 & 0 \end{pmatrix} \mathbf{Q}. \quad \square$$

3.6 Simultaneous Elimination

In contrast to Gauss-Jodan elimination, which is a method of consecutive elimination of unknowns, the method of determinants does this simultaneously. It is limited by square matrices only. In this section, we show the main ideas of this method.

Let $\mathbf{A}$ be an invertible $n \times n$ matrix. Then $\mathrm{rank}(\mathbf{A}) = n$. By the Rank-Nullity Theorem, the null space of $\mathbf{A}$ equals zero implying that n columns of $\mathbf{A}$ make a basis for $\mathbb{R}^n$. Given $\mathbf{b}$ in $\mathbb{R}^n$ we want to find, say, the first coordinate x_1 of the solution $\mathbf{x}$ for the system $\mathbf{Ax} = \mathbf{b}$. Let $\mathbf{A}_1$ be an $n \times (n-1)$ matrix defined by

$$\mathbf{A}_1 = \left(\mathrm{col}_2(\mathbf{A}) \cdots \mathrm{col}_n(\mathbf{A})\right).$$

The columns of $\mathbf{A}_1$ being the columns of $\mathbf{A}$ are linearly independent implying that $\mathrm{rank}(\mathbf{A}_1) \geq n-1$. Since the size of $\mathbf{A}_1$ equals $n \times (n-1)$, we conclude that $\mathrm{rank}(\mathbf{A}_1) = n-1$. Since $\mathbf{A}_1^T$ is an $(n-1) \times n$ matrix, we conclude by the Rank-Nullity Theorem that $\mathrm{nullity}(\mathbf{A}_1^T) = 1$. It follows that the Gauss basis for $N(\mathbf{A}_1^T)$ is made of only one non-zero vector λ. Since $\lambda \perp \mathrm{col}_j(\mathbf{A})$ for $j = 2, \ldots, n$, we see that $\lambda^T \mathrm{col}_1(\mathbf{A}) \neq 0$. Indeed, otherwise λ must be in $N(\mathbf{A}) = \{\mathbf{0}\}$. Multiplying the equation $\mathbf{Ax} = \mathbf{b}$ by λ^T on the left we obtain a formula for x_1:

$$\lambda^T \mathbf{b} = \lambda^T \mathbf{Ax} = \lambda^T \mathrm{col}_1(\mathbf{A}) x_1 \Rightarrow \boxed{x_1 = \frac{\lambda^T \mathbf{b}}{\lambda^T \mathrm{col}_1(\mathbf{A})}}. \tag{3.30}$$

To find a formula for x_j, we define an $n \times (n-1)$ matrix $\mathbf{A}_j$. It is obtained from $\mathbf{A}$ by dropping $\mathrm{col}_j(\mathbf{A})$. Then $N(\mathbf{A}_j^T)$ is spanned by only one non-zero vector λ. As above, we obtain that

$$x_j = \frac{\lambda^T \mathbf{b}}{\lambda^T \mathrm{col}_j(\mathbf{A})}. \tag{3.31}$$

Problem 3.11 Using simultaneous elimination solve the system:

$$\begin{cases} 5x_1 + 2x_2 + 13x_3 + 4x_4 &= b_1 \\ 2x_1 + x_2 + 6x_3 + 2x_4 &= b_2 \\ 4x_1 + 2x_2 + 13x_3 + 4x_4 &= b_3 \\ 6x_1 + 3x_2 + 21x_3 + 7x_4 &= b_4 \end{cases}$$

Solution: We have

$$\mathbf{A}_1 = \begin{pmatrix} 2 & 13 & 4 \\ 1 & 6 & 2 \\ 2 & 13 & 4 \\ 3 & 21 & 7 \end{pmatrix}, \quad \mathbf{A}_1^T = \begin{pmatrix} 2 & 1 & 2 & 3 \\ 13 & 6 & 13 & 2 \\ 4 & 2 & 4 & 7 \end{pmatrix}, \quad \mathrm{rref}(\mathbf{A}_1^T) = \begin{pmatrix} 1 & 0 & 1 & 0 \\ 0 & 1 & 0 & 0 \\ 0 & 0 & 0 & 1 \end{pmatrix}.$$

The Gauss basis for $N(\mathbf{A}_1^T)$ is $\left\{ \begin{pmatrix} -1 & 0 & 1 & 0 \end{pmatrix}^T \right\}$. Then

$$x_1 = \frac{\lambda^T \mathbf{b}}{\lambda^T \mathrm{col}_1(\mathbf{A})} = \frac{-b_1 + b_3}{-5 + 4} = b_1 - b_3.$$

Similarly,

$$\mathbf{A}_2 = \begin{pmatrix} 5 & 13 & 4 \\ 2 & 6 & 2 \\ 4 & 13 & 4 \\ 6 & 21 & 7 \end{pmatrix}, \quad \mathbf{A}_2^T = \begin{pmatrix} 5 & 2 & 4 & 6 \\ 13 & 6 & 13 & 21 \\ 4 & 2 & 4 & 7 \end{pmatrix}, \quad \mathrm{rref}(\mathbf{A}_2^T) = \begin{pmatrix} 1 & 0 & 0 & -1 \\ 0 & 1 & 0 & 7/2 \\ 0 & 0 & 1 & 1 \end{pmatrix}.$$

The Gauss basis for $N(\mathbf{A}_2^T)$ is $\left\{ \begin{pmatrix} 1 & -7/2 & -1 & 1 \end{pmatrix}^T \right\}$,

$$x_2 = \frac{\lambda^T \mathbf{b}}{\lambda^T \mathrm{col}_2(\mathbf{A})} = \frac{b_1 - 7b_2/2 - b_3 + b_4}{-1/2} = -2b_1 + 7b_2 + 2b_3 - 2b_4.$$

It follows that

$$x_3 = -2b_2 + b_3, \quad x_4 = 3b_2 - 3b_3 + b_4. \quad \square$$

In practice, the method of simultaneous elimination is not so useful, unless one needs to find only a few unknowns x_j. Indeed, the method of extended matrices shows that

$$\begin{pmatrix} 5 & 2 & 13 & 4 \\ 2 & 1 & 6 & 2 \\ 4 & 2 & 13 & 4 \\ 6 & 3 & 21 & 7 \end{pmatrix}^{-1} = \begin{pmatrix} 1 & 0 & -1 & 0 \\ -2 & 7 & 2 & -2 \\ 0 & -2 & 1 & 0 \\ 0 & 3 & -3 & 1 \end{pmatrix}.$$

Then

$$\begin{pmatrix} x_1 \\ x_2 \\ x_3 \\ x_4 \end{pmatrix} = \begin{pmatrix} 1 & 0 & -1 & 0 \\ -2 & 7 & 2 & -2 \\ 0 & -2 & 1 & 0 \\ 0 & 3 & -3 & 1 \end{pmatrix} \begin{pmatrix} b_1 \\ b_2 \\ b_3 \\ b_4 \end{pmatrix} = \begin{pmatrix} b_1 - b_3 \\ -2b_1 + 7b_2 + 2b_3 - 2b_4 \\ -2b_2 + b_3 \\ 3b_2 - 3b_3 + b_4 \end{pmatrix}.$$

However, the method of simultaneous elimination indicates that there exists an explicit formula for x_j in terms of the entries a_{ij} of the coefficient matrix $\mathbf{A}$. Moreover, it gives a preliminary form (3.31) of this formula. What remains is to find a formula for λ. Notice that λ is perpendicular to all columns of the matrix $\mathbf{A}$ except for the column corresponding to the unknown x_j which is to be found. Therefore, we also obtain a formula for such a vector.

3.7 Determinants: Motivation

We begin with the case $n = 2$:

$$\begin{cases} a_{11}x_1 + a_{12}x_2 = b_1 \\ a_{21}x_1 + a_{22}x_2 = b_2 \end{cases} \tag{3.32}$$

To eliminate x_2 we multiply the first equation by λ_1, and the second by λ_2:

$$\begin{matrix} \lambda_1 \times & | & a_{11}x_1 + a_{12}x_2 = b_1 \\ \lambda_2 \times & | & a_{21}x_1 + a_{22}x_2 = b_2 \end{matrix} \Rightarrow \begin{matrix} \lambda_1 a_{11}x_1 + \lambda_1 a_{12}x_2 = \lambda_1 b_1 \\ \lambda_2 a_{21}x_1 + \lambda_2 a_{22}x_2 = \lambda_2 b_2 \end{matrix}$$

Taking the sum of these two equations, we obtain

$$(\lambda_1 a_{11} + \lambda_2 a_{21})x_1 + (\lambda_1 a_{12} + \lambda_2 a_{22})x_2 = (\lambda_1 b_1 + \lambda_2 b_2).$$

If $\lambda_1 a_{12} + \lambda_2 a_{22} = 0$, for instance $\lambda_1 = a_{22}$, $\lambda_2 = -a_{12}$, then

$$(a_{11}a_{22} - a_{12}a_{21})x_1 = (b_1 a_{22} - a_{12}b_2). \tag{3.33}$$

If $\lambda_1 = -a_{21}$, $\lambda_2 = a_{11}$, then

$$(a_{11}a_{22} - a_{12}a_{21})x_2 = (a_{11}b_2 - b_1 a_{21}). \tag{3.34}$$

Let

$$\Delta \overset{\text{def}}{=} \begin{vmatrix} a_{11} & a_{12} \\ a_{21} & a_{22} \end{vmatrix} = a_{11}a_{22} - a_{12}a_{21} \tag{3.35}$$

be the **determinant** of the 2×2 matrix $\mathbf{A}$. By (3.33) and (3.34)

$$\begin{vmatrix} a_{11} & a_{12} \\ a_{21} & a_{22} \end{vmatrix} x_1 = \begin{vmatrix} b_1 & a_{12} \\ b_2 & a_{22} \end{vmatrix}, \quad \begin{vmatrix} a_{11} & a_{12} \\ a_{21} & a_{22} \end{vmatrix} x_2 = \begin{vmatrix} a_{11} & b_1 \\ a_{21} & b_2 \end{vmatrix}. \tag{3.36}$$

If $\Delta \neq 0$, then we obtain the explicit formulas for the solutions to the system (3.32):

$$x_1 = \frac{\begin{vmatrix} b_1 & a_{12} \\ b_2 & a_{22} \end{vmatrix}}{\Delta}, \quad x_2 = \frac{\begin{vmatrix} a_{11} & b_1 \\ a_{21} & b_2 \end{vmatrix}}{\Delta}.$$

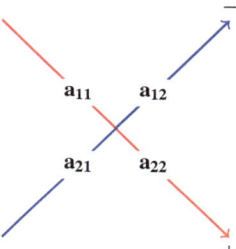

Fig. 3.1: 2×2 determinants

The products $\pm a_{1\sigma 1} a_{2\sigma 2}$ in (3.35) correspond to one-to-one mappings σ of the set $\{1, 2\}$ onto itself:

$$a_{11}a_{22} \longmapsto \begin{pmatrix} 1 & 2 \\ 1 & 2 \end{pmatrix}, \quad -a_{12}a_{21} \longmapsto \begin{pmatrix} 1 & 2 \\ 2 & 1 \end{pmatrix}.$$

The sign "-" at $a_{12}a_{21}$ corresponds to an **inversion** in the second row of the second matrix above. In other words, a bigger number 2 in the row precedes 1. There is no inversion in the second row of the first matrix.

Let us consider now a system of three linear equations in three unknowns

$$\begin{array}{l} \lambda_1 \times \\ \lambda_2 \times \\ \lambda_3 \times \end{array} \left\{ \begin{array}{l} a_{11}x_1 + a_{12}x_2 + a_{13}x_3 = b_1 \\ a_{21}x_1 + a_{22}x_2 + a_{23}x_3 = b_2 \\ a_{31}x_1 + a_{32}x_2 + a_{33}x_3 = b_3 \end{array} \right. \tag{3.37}$$

where $\lambda_1, \lambda_2, \lambda_3$ are yet unknown numbers, which must vanish coefficients at x_2 and x_3 in the sum of the equations of this system. Unknown multipliers $\lambda_1, \lambda_2, \lambda_3$ must satisfy the following system of two linear equations

$$\begin{aligned} \textbf{Elimination of } x_2 \rightarrow & \begin{cases} a_{12}\lambda_1 + a_{22}\lambda_2 + a_{32}\lambda_3 & = 0 \\ a_{13}\lambda_1 + a_{23}\lambda_2 + a_{33}\lambda_3 & = 0 \end{cases} \end{aligned} \tag{3.38}$$

We rewrite the system in the form

$$\begin{cases} a_{12}\lambda_1 + a_{22}\lambda_2 & = -a_{32}\lambda_3 \\ a_{13}\lambda_1 + a_{23}\lambda_2 & = -a_{33}\lambda_3 \end{cases}$$

and apply (3.36) to find that:

$$\begin{aligned} \begin{vmatrix} a_{12} & a_{22} \\ a_{13} & a_{23} \end{vmatrix} \lambda_1 &= \begin{vmatrix} -a_{32}\lambda_3 & a_{22} \\ -a_{33}\lambda_3 & a_{23} \end{vmatrix} = \begin{vmatrix} a_{22} & a_{32} \\ a_{23} & a_{33} \end{vmatrix} \lambda_3, \\[2mm] \begin{vmatrix} a_{12} & a_{22} \\ a_{13} & a_{23} \end{vmatrix} \lambda_2 &= \begin{vmatrix} a_{12} & -a_{32}\lambda_3 \\ a_{13} & -a_{33}\lambda_3 \end{vmatrix} = - \begin{vmatrix} a_{12} & a_{32} \\ a_{13} & a_{33} \end{vmatrix} \lambda_3. \end{aligned} \tag{3.39}$$

Since the choice of λ_3 is free, we may put

$$\lambda_1 = \begin{vmatrix} a_{22} & a_{32} \\ a_{23} & a_{33} \end{vmatrix}, \quad \lambda_2 = - \begin{vmatrix} a_{12} & a_{32} \\ a_{13} & a_{33} \end{vmatrix}, \quad \lambda_3 = \begin{vmatrix} a_{12} & a_{22} \\ a_{13} & a_{23} \end{vmatrix}. \tag{3.40}$$

Direct computations show that

$$a_{12}\lambda_1 + a_{22}\lambda_2 + a_{32}\lambda_3 = a_{12} \begin{vmatrix} a_{22} & a_{32} \\ a_{23} & a_{33} \end{vmatrix} - a_{22} \begin{vmatrix} a_{12} & a_{32} \\ a_{13} & a_{33} \end{vmatrix} + a_{32} \begin{vmatrix} a_{12} & a_{22} \\ a_{13} & a_{23} \end{vmatrix} =$$
$$a_{12}(a_{22}a_{33} - a_{23}a_{32}) - a_{22}(a_{12}a_{33} - a_{13}a_{32}) + a_{32}(a_{12}a_{23} - a_{13}a_{22}) = 0.$$

Similarly,

$$a_{13}\lambda_1 + a_{23}\lambda_2 + a_{33}\lambda_3 = a_{13} \begin{vmatrix} a_{22} & a_{32} \\ a_{23} & a_{33} \end{vmatrix} - a_{23} \begin{vmatrix} a_{12} & a_{32} \\ a_{13} & a_{33} \end{vmatrix} + a_{33} \begin{vmatrix} a_{12} & a_{22} \\ a_{13} & a_{23} \end{vmatrix} =$$
$$a_{13}(a_{22}a_{33} - a_{23}a_{32}) - a_{23}(a_{12}a_{33} - a_{13}a_{32}) + a_{33}(a_{12}a_{23} - a_{13}a_{22}) = 0.$$

Thus the choice indicated in (3.40) leads to (3.38) even if the coefficient at λ_1 and λ_2 in (3.39) equals zero. Then the coefficient Δ at x_1 is called the **determinant** of $\mathbf{A} = (a_{ij})$, see (3.37):

$$\Delta = a_{11}\lambda_1 + a_{21}\lambda_2 + a_{31}\lambda_3 =$$

$$a_{11} \begin{vmatrix} a_{22} & a_{23} \\ a_{32} & a_{33} \end{vmatrix} - a_{21} \begin{vmatrix} a_{12} & a_{13} \\ a_{32} & a_{33} \end{vmatrix} + a_{31} \begin{vmatrix} a_{12} & a_{13} \\ a_{22} & a_{23} \end{vmatrix} \tag{3.41}$$

By (3.35) and (3.41) we obtain the following formula for the determinant of a 3×3 matrix:

$$\Delta \overset{\text{def}}{=} \begin{vmatrix} a_{11} & a_{12} & a_{13} \\ a_{21} & a_{22} & a_{23} \\ a_{31} & a_{32} & a_{33} \end{vmatrix} = \begin{aligned} &+a_{11}a_{22}a_{33} + a_{12}a_{23}a_{31} + a_{13}a_{21}a_{32} \\ &-a_{13}a_{22}a_{31} - a_{12}a_{21}a_{33} - a_{11}a_{23}a_{32}. \end{aligned} \tag{3.42}$$

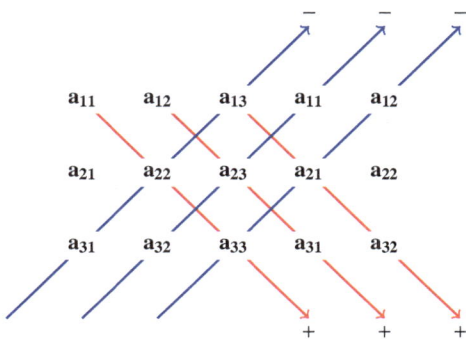

Fig. 3.2: **Sarrus Rule** for 3×3 determinants

Formula (3.41) shows that the products which contain a_{21} as a multiplier are associated with matrices

$$\begin{pmatrix} 1 & 2 & 3 \\ * & 1 & * \end{pmatrix}.$$

Since 1 is the smallest number, the number of inversions containing 1 is one. It is clear that this 1 generates $(-1)^1$ at a_{21} in (3.41). So that $n(\sigma)$ is the sum of 1 with the number of inversions in the second row in the pair $(*, *)$. It is 0 for $(2, 3)$ and 1 for $(3, 2)$. There are $6 = 3 \cdot 2$ products $\pm a_{1\sigma 1}a_{2\sigma 2}a_{3\sigma 3}$, three of which correspond to $+$:

products with $+$ $\rightarrow$ $\quad a_{11}a_{22}a_{33} \quad a_{12}a_{23}a_{31} \quad a_{13}a_{21}a_{32}$

$$\sigma \rightarrow \qquad \begin{pmatrix} 1 & 2 & 3 \\ 1 & 2 & 3 \end{pmatrix} \quad \begin{pmatrix} 1 & 2 & 3 \\ 2 & 3 & 1 \end{pmatrix} \quad \begin{pmatrix} 1 & 2 & 3 \\ 3 & 1 & 2 \end{pmatrix} \tag{3.43}$$

number of inversions $\rightarrow$ $\qquad 0 \qquad\qquad 2 \qquad\qquad 2$

and three correspond to $-$

products with $-$ $\rightarrow$ $\quad -a_{13}a_{22}a_{31} \;\; -a_{12}a_{21}a_{33} \;\; -a_{11}a_{23}a_{32}$

$$\sigma \rightarrow \qquad \begin{pmatrix} 1 & 2 & 3 \\ 3 & 2 & 1 \end{pmatrix} \quad \begin{pmatrix} 1 & 2 & 3 \\ 2 & 1 & 3 \end{pmatrix} \quad \begin{pmatrix} 1 & 2 & 3 \\ 1 & 3 & 2 \end{pmatrix} \tag{3.44}$$

number of inversions $\rightarrow$ $\quad 3 \qquad\qquad 1 \qquad\qquad 1$

Let $\mathfrak{S}_3$ be the set of all one-to-one mappings σ of $\{1, 2, 3\}$ onto itself and $n(\sigma)$ be the number of inversions in the second row of σ. Then (3.42) can be written in a compact form:

$$\Delta = \sum_{\sigma \in \mathfrak{S}_3} (-1)^{n(\sigma)} a_{1\sigma 1}a_{2\sigma 2}a_{3\sigma 3}.$$

In case of $n = 3$ the choice of $+$ and $-$ in each of six products can also be made by the **Sarrus Rule**. According to this rule one considers the following 3×5 matrix:

$$\text{col}_1(\mathbf{A}) \ \text{col}_2(\mathbf{A}) \ \text{col}_3(\mathbf{A}) \ \text{col}_1(\mathbf{A}) \ \text{col}_2(\mathbf{A}).$$

The products of the entries placed along the arrows $\searrow$ must be taken with sign $+$, and the entries placed along the lines $\nearrow$ with sign $-$, see Fig. 3.2.

If $\Delta \neq 0$, then

$$x_1 = \frac{\begin{vmatrix} b_1 & a_{12} & a_{13} \\ b_2 & a_{22} & a_{23} \\ b_3 & a_{32} & a_{33} \end{vmatrix}}{\Delta}, \quad x_2 = \frac{\begin{vmatrix} a_{11} & b_1 & a_{13} \\ a_{21} & b_2 & a_{23} \\ a_{31} & b_3 & a_{33} \end{vmatrix}}{\Delta}, \quad x_3 = \frac{\begin{vmatrix} a_{11} & a_{12} & b_1 \\ a_{21} & a_{22} & b_2 \\ a_{31} & a_{32} & b_3 \end{vmatrix}}{\Delta}.$$

Following (3.41) we define the determinant of an 4×4 matrix by the formula

$$\Delta = a_{11} \begin{vmatrix} a_{22} & a_{23} & a_{24} \\ a_{32} & a_{33} & a_{34} \\ a_{42} & a_{43} & a_{44} \end{vmatrix} - a_{21} \begin{vmatrix} a_{12} & a_{13} & a_{14} \\ a_{32} & a_{33} & a_{34} \\ a_{42} & a_{43} & a_{44} \end{vmatrix} +$$

$$a_{31} \begin{vmatrix} a_{12} & a_{13} & a_{14} \\ a_{22} & a_{23} & a_{24} \\ a_{42} & a_{43} & a_{44} \end{vmatrix} - a_{41} \begin{vmatrix} a_{12} & a_{13} & a_{14} \\ a_{22} & a_{23} & a_{24} \\ a_{32} & a_{33} & a_{34} \end{vmatrix}. \quad (3.45)$$

Then by Sarrus Rule

$$\Delta = a_{11}a_{22}a_{33}a_{44} + a_{11}a_{23}a_{34}a_{42} + a_{11}a_{24}a_{32}a_{43} - a_{11}a_{24}a_{33}a_{42} - a_{11}a_{23}a_{32}a_{44} -$$
$$a_{11}a_{22}a_{34}a_{42} - a_{12}a_{21}a_{33}a_{44} - a_{13}a_{21}a_{34}a_{42} - a_{14}a_{21}a_{32}a_{43} + a_{14}a_{21}a_{33}a_{42} +$$
$$a_{13}a_{21}a_{32}a_{44} + a_{12}a_{21}a_{34}a_{43} + a_{12}a_{23}a_{31}a_{44} + a_{13}a_{24}a_{31}a_{42} + a_{14}a_{22}a_{31}a_{43} -$$
$$a_{14}a_{23}a_{31}a_{42} - a_{13}a_{22}a_{31}a_{44} - a_{12}a_{24}a_{31}a_{43} - a_{12}a_{23}a_{34}a_{41} - - a_{13}a_{24}a_{32}a_{41} -$$
$$a_{14}a_{22}a_{33}a_{41} + a_{14}a_{23}a_{32}a_{41} + a_{13}a_{22}a_{34}a_{41} + a_{12}a_{24}a_{33}a_{41}. \quad (3.46)$$

Notice that Δ is the sum of 24 products $\pm a_{1\sigma 1}a_{2\sigma 2}a_{3\sigma 3}a_{4\sigma 4}$, where σ runs over all one-to one mappings of the set $\{1, 2, 3, 4\}$ onto itself. The products including a_{21} enter (3.45) with minus times the determinant of the 3×3 matrix shown on Fig. 3.3. Since the second column of σ is $2 \to 1$, the number of inversions associated with 1 equals one, which confirms the sign minus at a_{21} in (3.45). For instance,

$$-a_{14}a_{21}a_{32}a_{43} \longmapsto \sigma = \begin{pmatrix} 1 & 2 & 3 & 4 \\ 4 & 1 & 2 & 3 \end{pmatrix} \Rightarrow n(\sigma) = 3.$$

Thus we obtain a formula for the determinant of size 4×4:

$$\Delta = \sum_{\sigma \in \mathfrak{S}_4} (-1)^{n(\sigma)} a_{1\sigma 1} a_{2\sigma 2} a_{3\sigma 3} a_{4\sigma 4}.$$

For an arbitrary positive integer n we denote by $\mathbb{N}_n \overset{\text{def}}{=} \{1, 2, \ldots, n\}$ the set of all positive integers less or equal n, and by $\mathfrak{S}_n$ the set of all one-to-one mappings of $\mathbb{N}_n$ onto itself. Let

$$P(x_1, x_2, \cdots, x_n) = \prod_{i<j}(x_i - x_j)$$

be a formal polynomial in n unknowns $x_1, x_2, \ldots, x_n$. The symbol $\prod$ denotes the product. For every σ in $\mathfrak{S}_n$ we define its **signature** by the formula:

$$\varepsilon(\sigma) = \frac{P(x_1, \ldots, x_n)}{P(x_{\sigma 1}, \ldots, x_{\sigma n})} = \prod_{i<j} \frac{x_i - x_j}{x_{\sigma i} - x_{\sigma j}}. \tag{3.47}$$

It is clear that $\varepsilon(\sigma) = \pm 1$.

Definition 3.6 Let $\mathbf{A} = (a_{ij})$ be an $n \times n$ matrix. Then the determinant of $\mathbf{A}$ is defined by the formula

$$\det(\mathbf{A}) = \sum_{\sigma \in \mathfrak{S}_n} \varepsilon(\sigma) a_{1\sigma 1} a_{2\sigma 2} \cdots a_{n\sigma n} = \sum_{\sigma \in \mathfrak{S}_n} \varepsilon(\sigma) \prod_{k=1}^{n} a_{k\sigma k}. \tag{3.48}$$

IMPORTANT !! DO NOT APPLY SARRUS RULE TO THE DETERMINANTS OF HIGHER ORDERS. THIS DOES NOT WORK.

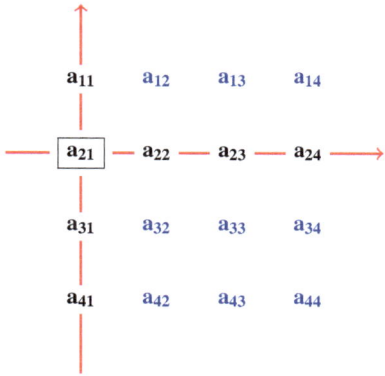

Fig. 3.3: Definition of 4×4 determinants

3.8 The group $\mathfrak{S}_n$

Any σ in $\mathfrak{S}_n$, being a mapping, can be written as a matrix:

$$\sigma = \begin{pmatrix} 1 & 2 & \cdots & n \\ \sigma 1 & \sigma 2 & \cdots & \sigma n \end{pmatrix}. \tag{3.49}$$

Here the entry at $(2, j)$ indicates the image of j under σ. Therefore, any permutation of the columns determines one and the same σ. For example, the matrices

$$\begin{pmatrix} 1 & 2 & 3 & 4 \\ 3 & 2 & 1 & 4 \end{pmatrix} \quad \text{and} \quad \begin{pmatrix} 2 & 1 & 4 & 3 \\ 2 & 3 & 4 & 1 \end{pmatrix}$$

represent one mapping σ. If one agrees to write σ in the form (3.49), then the second row of the matrix (3.49) uniquely determines the first. To save space, one can identify σ with the second row of the matrix (3.49) called a **permutation**

$$\begin{pmatrix} \sigma 1 & \sigma 2 & \cdots & \sigma n \end{pmatrix}$$

of the first row $\{1 < 2 < \cdots < n\}$. To summarise, any one-to-one mapping $\sigma :$ $\mathbb{N}_n \to \mathbb{N}_n$ can be represented as

$$\text{a matrix} \quad \begin{pmatrix} 1 & 2 & \cdots & n \\ \sigma 1 & \sigma 2 & \cdots & \sigma n \end{pmatrix} \quad \text{or a permutation} \quad \begin{pmatrix} \sigma 1 & \sigma 2 & \cdots & \sigma n \end{pmatrix}.$$

For any pair σ and τ in $\mathfrak{S}_n$ one can consider the superposition (**product**) of these mappings:

$$\sigma \tau k = \sigma (\tau(k)).$$

Problem 3.12 Find the product $\sigma \tau$, where

$$\sigma = \begin{pmatrix} 1 & 2 & 3 & 4 \\ 2 & 3 & 4 & 1 \end{pmatrix}, \quad \tau = \begin{pmatrix} 1 & 2 & 3 & 4 \\ 3 & 4 & 1 & 2 \end{pmatrix}.$$

Solution:

$$\begin{pmatrix} 1 & 2 & 3 & 4 \\ 2 & 3 & 4 & 1 \end{pmatrix} \begin{pmatrix} 1 & 2 & 3 & 4 \\ 3 & 4 & 1 & 2 \end{pmatrix} = \begin{pmatrix} 1 & 2 & 3 & 4 \\ 4 & 1 & 2 & 3 \end{pmatrix}. \quad \square \tag{3.50}$$

Remark. Notice that the product of two matrices in (3.50) is defined as the superposition of two mappings and has nothing in common with the matrix product.

Definition 3.7 A **group** is a non-empty set G satisfying the following properties.

(1) For every pair elements x and y in G there exists $z = x \cdot y$ in G.
(2) $(x \cdot y) \cdot z = x \cdot (y \cdot z)$.
(3) There exists an element e in G called the unit of G such that

$$e \cdot x = x \cdot x = x$$

for every x in G.

(4) For every x in G there exists an x^{-1} (called the **inverse** for x) such that

$$x \cdot x^{-1} = x^{-1} \cdot x = e.$$

Theorem 3.20 *The set $\mathfrak{S}_n$ with operation of composition of its elements is a group.*

Proof. If x, y, and z are mappings in $\mathfrak{S}_n$ then they satisfy **(2)**, since they are mappings of $\mathbf{N}_n$ onto itself. The unit e in $\mathfrak{S}_n$ in the matrix notation is

$$e = \begin{pmatrix} 1 & 2 & \cdots & n \\ 1 & 2 & \cdots & n \end{pmatrix}.$$

It is easy to see that the inverse element σ^{-1} for σ in the matrix notation is defined by the formula

$$\begin{pmatrix} 1 & 2 & \cdots & n \\ \sigma 1 & \sigma 2 & \cdots & \sigma n \end{pmatrix}^{-1} = \begin{pmatrix} \sigma 1 & \sigma 2 & \cdots & \sigma n \\ 1 & 2 & \cdots & n \end{pmatrix}. \qquad \square$$

By Cayley's Theorem every finite group is isomorphic to a subgroup of some $\mathfrak{S}_n$, which demonstrates the importance of $\mathfrak{S}_n$ used in the definition of determinants. See Meier (2008) for the proof.

The simplest mappings in $\mathfrak{S}_n$ are mappings swapping only two numbers in $\mathbf{N}_n$ and leaving others unchanged. Such mappings are called **transpositions**. If σ swaps i and j, $j \neq i$, then we write $\sigma = (i\ j)$. For example,

$$(2\ 4)\begin{pmatrix} 1 & 2 & 3 & 4 \\ 3 & \mathbf{2} & 1 & \mathbf{4} \end{pmatrix} = \begin{pmatrix} 1 & 2 & 3 & 4 \\ 3 & \mathbf{4} & 1 & \mathbf{2} \end{pmatrix} \Leftrightarrow (2\ 4)\begin{pmatrix} 3\ \mathbf{2}\ 1\ \mathbf{4} \end{pmatrix} = \begin{pmatrix} 3\ \mathbf{4}\ 1\ \mathbf{2} \end{pmatrix}$$

is a transposition $(2\ 4)$ of the permutation $\begin{pmatrix} 3\ 2\ 1\ 4 \end{pmatrix}$.

Lemma 3.3 *Every σ in $\mathfrak{S}_n$ is a product of a finite number of transpositions.*

Proof. Since any σ in $\mathfrak{S}_n$ is identified with the permutation

$$\begin{pmatrix} \sigma 1\ \sigma 2\ \cdots\ \sigma n \end{pmatrix},$$

we see that either $\sigma n = n$ or $\sigma n = k$, $k < n$. In the first case we do nothing and in the second case we apply to σ the transposition $(k\ n)$. In both cases the result is that the last element is n. After not more than n steps we obtain the unit permutation $(1\ 2\ \cdots\ n)$. Reversing the order of these operations we obtain a decomposition of σ into a product of not more than n transpositions.   □

For example,

$$(3\ 4\ 2\ 1) = (1\ 2)(1\ 3)(2\ 4) = (3\ 4)(1\ 4)(2\ 3).$$

Indeed,

$$(1\ 2\ 3\ 4) \overset{(2\ 4)}{\rightarrow} (1\ 4\ 3\ 2) \overset{(1\ 3)}{\rightarrow} (3\ 4\ 1\ 2) \overset{(1\ 2)}{=} (3\ 4\ 2\ 1),$$

$$(1\ 2\ 3\ 4) \overset{(2\ 3)}{\rightarrow} (1\ 3\ 2\ 4) \overset{(1\ 4)}{\rightarrow} (4\ 3\ 2\ 1) \overset{(3\ 4)}{=} (3\ 4\ 2\ 1).$$

The order of transpositions in their product is not important if they do not have common numbers. Otherwise, it is important:

$$(1\ 2)(1\ 4) = (4\ 1\ 3\ 2) \neq (2\ 4\ 3\ 1) = (1\ 4)(1\ 2).$$

Definition 3.8 A permutation $(\sigma 1\ \sigma 2\ \cdots\ \sigma n)$ is said to have an **inversion** at $r < s$ if $\sigma r > \sigma s$. We denote by $\mathrm{inv}(\sigma)$ the total number of inversions in the permutation σ.

Lemma 3.4 *For every σ in $\mathfrak{S}_n$*

$$\varepsilon(\sigma) = (-1)^{\mathrm{inv}(\sigma)}.$$

Proof. By (3.47) the numerator of the product, which defines $\varepsilon(\sigma)$, is made of the products $x_i - x_j$ with $i < j$. Any inversion $\sigma r > \sigma s$ at $r < s$ defines the multiplier $(x_{\sigma r} - x_{\sigma s})$ of the denominator, which uniquely corresponds to the multiplier $(x_{\sigma s} - x_{\sigma r})$ in the numerator. Then the quotient

$$\frac{x_{\sigma s} - x_{\sigma r}}{x_{\sigma r} - x_{\sigma s}} = -1$$

in the product (3.47) contributes (-1) in the value of $\varepsilon(\sigma)$. All other pairs $\sigma r < \sigma s$ contribute $(+1)$. It follows that $\varepsilon(\sigma)$ is the product of $\mathrm{inv}(\sigma)$ multipliers (-1) as stated.   □

Problem 3.13 Find the signature of the permutation

$$(3\ 2\ 1\ 4). \tag{3.51}$$

Solution: The first number 3 contributes 2 inversions, namely, $3 > 2$ and $3 > 1$. The second number 2 contributes one inversion $2 > 1$. The numbers 1 and 4 contribute none. Since the total number of inversions is $2 + 1 = 0 + 0 = 3$, the signature of this permutation is $(-1)^3 = -1$. Notice that second indices in the product $a_{13}a_{22}a_{31}a_{44}$ make the permutation (3.51) and its signature equals the sign at $a_{13}a_{22}a_{31}a_{44}$ in (3.46). □

Problem 3.14 Find the signature of the permutation

$$\begin{pmatrix} 1 \ 4 \ 2 \ 3 \end{pmatrix}. \tag{3.52}$$

Solution: Being the smallest number 1 makes no inversions, 4 makes 2, and 2 and 3 make none. It follows that $\varepsilon(\sigma) = +1$. The product $a_{11}a_{24}a_{32}a_{33}$ corresponding to the permutation (3.52) enters the formula (3.46) with sigh +. □
There is one more way to evaluate $\varepsilon(\sigma)$.

Lemma 3.5 *Let* $\varepsilon_1, \varepsilon_2, \ldots, \varepsilon_k$ *be a sequence of* ± 1. *Then the integer*

$$\xi_k = \sum_{j=1}^{k} \varepsilon_j$$

is of the same parity as k.

Proof. The parity of $\xi_1 = \varepsilon_1$ is odd, i.e coincides with the parity of $k = 1$. If the Lemma is proved for k, then

$$\text{parity}(\xi_{k+1}) = \text{parity}(\xi_k + \varepsilon_{k+1}) = \text{parity}(k + \varepsilon_{k+1}) = \text{parity}(k + 1). \qquad □$$

Lemma 3.6 *Any transposition of a permutation*

$$\sigma = \begin{pmatrix} j_1 \ j_2 \ \cdots \ j_n \end{pmatrix},$$

either increases $\text{inv}(\sigma)$ *or decreases it by an odd number. In other words, this operation changes the sign of* $\varepsilon(\sigma)$.

Proof. Swapping j_r and j_{r+1} we increase or decrease $\text{inv}(\sigma)$ exactly by 1, depending on whether $j_r < j_{r+1}$ or $j_r > j_{r+1}$. It takes k such simple steps to move j_r to the position of j_{r+k}. As result j_{r+k} is moved to the position preceding j_r. Therefore, it takes only $k - 1$ steps to move it back to the starting position of j_r. The increment in the number of inversions after all these operations is a sum of $2k - 1$ terms $+1$ or -1. By Lemma 3.5 the signature of the permutation obtained is of the opposite sign. □

Corollary 3.4 *Let* $t(\sigma)$ *be a number of transpositions required to transform a given permutation* $\sigma = \begin{pmatrix} j_1 \ j_2 \ \cdots \ j_n \end{pmatrix}$ *into the identity* $\begin{pmatrix} 1 \ 2 \ \cdots \ n \end{pmatrix}$. *Then*

$$\varepsilon(\sigma) = (-1)^{t(\sigma)}.$$

Proof. The signature of $\left(1\ 2\ \cdots\ n\right)$ is 1. By Lemma 3.6 any transposition changes the sign of a permutation to the opposite. Since after $t(\sigma)$ transpositions we obtain the identity permutation with signature 1, the signature of σ must be $(-1)^{t(\sigma)}$.   □

Problem 3.15 Find the signature of the permutation $\left(4\ 1\ 2\ 3\right)$.

Since
$$\left(4\ 1\ 2\ 3\right) \overset{1}{\to} \left(1\ 4\ 2\ 3\right) \overset{2}{\to} \left(1\ 2\ 4\ 3\right) \overset{3}{\to} \left(1\ 2\ 3\ 4\right),$$
by Corollary 3.4 we obtain that $\varepsilon(\sigma) = (-1)^3 = -1$. Notice that $a_{14}a_{21}a_{32}a_{43}$ enters the formula (3.46) with sign $-$.   □

Theorem 3.21 *Let σ and τ be two one-to-one mappings in $\mathfrak{S}_n$ and let $\sigma\tau$ be a composite one-to-one mapping of the set $\{1,2,\ldots,n\}$. Then*

$$\varepsilon(\sigma\tau) = \varepsilon(\sigma)\varepsilon(\tau).$$

Proof. We have

$$\varepsilon(\sigma\tau) = \frac{P(x_1,\ldots,x_n)}{P(x_{\sigma\tau 1},\ldots,x_{\sigma\tau n})} = \frac{P(x_1,\ldots,x_n)}{P(x_{\tau 1},\ldots,x_{\tau n})}\frac{P(x_{\tau 1},\ldots,x_{\tau n})}{P(x_{\sigma\tau 1},\ldots,x_{\sigma\tau n})} = \varepsilon(\tau)\varepsilon(\sigma).$$

Corollary 3.5 *For every σ in $\mathfrak{S}_n$*

$$\varepsilon(\sigma^{-1}) = \varepsilon(\sigma).$$

Proof. Since $\sigma\sigma^{-1}$ corresponds to the identity permutation, the result follows by Theorem 3.21.   □

3.9 Row Operations and Determinants

Lemma 3.7 *The number $\mathrm{Card}(\mathfrak{S}_n)$ of elements in $\mathfrak{S}_n$ is*

$$\mathrm{Card}(\mathfrak{S}_n) = 1\cdot 2\cdot\ldots\cdot n = n!$$

Proof. The formula is true for $n = 2$ and for $n = 3$ as Sarrus formulas show. Suppose now that it holds for n and prove it for $n + 1$. Any element σ of $\mathfrak{S}_{n+1}$ lies in one of $n + 1$ disjoint subsets of $\mathfrak{S}_{n+1}$:

$$\mathfrak{A}_1,\ \mathfrak{A}_2,\ \ldots,\mathfrak{A}_{n+1},$$

where $\mathfrak{A}_k$ denotes the set of all substitutions moving 1 to k. Then each subset $\mathfrak{A}_k$ has $n!$ elements. It follows that $\mathfrak{S}_{n+1}$ has $(n + 1) \cdot n! = (n + 1)!$ elements. $\quad\square$

The number of elements in $\mathfrak{S}_n$ increases very fast:

$$1! = 1$$
$$2! = 2 \cdot 1 = 2$$
$$3! = 3 \cdot 2 \cdot 1 = 6$$
$$4! = 4 \cdot \cdot 2 \cdot 1 = 24$$
$$5! = 5 \cdot 4 \cdot \cdot 2 \cdot 1 = 120$$
$$6! = 6 \cdot 5 \cdot 4 \cdot \cdot 2 \cdot 1 = 720$$
$$7! = 7 \cdot 6 \cdot 5 \cdot 4 \cdot \cdot 2 \cdot 1 = 5040$$
$$8! = 8 \cdot 7 \cdot 6 \cdot 5 \cdot 4 \cdot \cdot 2 \cdot 1 = 40\,320$$
$$9! = 9 \cdot 8 \cdot 7 \cdot 6 \cdot 5 \cdot 4 \cdot \cdot 2 \cdot 1 = 362\,880$$
$$10! = 10 \cdot 9 \cdot 8 \cdot 7 \cdot 6 \cdot 5 \cdot 4 \cdot \cdot 2 \cdot 1 = 3\,628\,800$$
$$11! = 11 \cdot 10 \cdot 9 \cdot 8 \cdot 7 \cdot 6 \cdot 5 \cdot 4 \cdot \cdot 2 \cdot 1 = \boxed{39\,916\,800}.$$

The factorial 11! exceeds the total number of citizens of New York (8 550 405 by October 29, 2018). It may take up to 5 minutes for every citizen of New York to evaluate $\varepsilon(\sigma)$ for a determinant of size 11×11, provided one succeeds in distribution of this task among citizens. Even if they all will use calculators, it will take another 5 minutes to complete the calculations of the products distributed among them. Notice that not all products will be found (since we simply do not have for this enough people) and it is absolutely unclear how to arrange the data obtained in one sum.

However, these calculations can be considerably simplified by Gauss-Jordan elimination. We are going to study the influence of elementary row operations on the values of determinants. Observing that the reduced row echelon form of any matrix is an upper triangular matrix, we arrive to the conclusion that this is a very promising method, provided we know answers to the two following questions:

- What is the value of the determinant of an upper triangular matrix?
- How do determinants change under elementary row operations?

Theorem 3.22 *If* $\mathbf{A} = (a_{ij})$ *is an upper triangular* $n \times n$ *matrix, then*

$$\det(\mathbf{A}) = \begin{vmatrix} a_{11} & a_{12} & a_{13} & \cdots & a_{1n} \\ 0 & a_{22} & a_{23} & \cdots & a_{2n} \\ 0 & 0 & a_{33} & \cdots & a_{3n} \\ \vdots & \vdots & \vdots & \ddots & \vdots \\ 0 & 0 & 0 & \cdots & a_{nn} \end{vmatrix} = a_{11}a_{22}a_{33} \cdots a_{nn}.$$

Proof. We show that only the product $a_{11}a_{22}a_{33}\cdots a_{nn}$ is a non-zero term in the sum (3.48) which defines $\det(\mathbf{A})$. Take any nonzero term $a_{1\sigma 1}a_{2\sigma 2}\cdots a_{n\sigma n}$ in the sum, which defines the determinant. Then $a_{n\sigma n}\neq 0$. The assumption $\sigma n < n$ implies that $a_{n\sigma n} = 0$, since $\mathbf{A}$ is upper triangular. It follows that $\sigma n = n$. Take now $a_{n-1\sigma(n-1)}\neq 0$. Then $\sigma(n-1)\neq \sigma n = n$. If $\sigma(n-1) < n-1$ then $a_{n-1\sigma(n-1)} = 0$. Therefore, $\sigma(n-1) = n-1$. Continuing by induction, we obtain that $\sigma k = k$ for $k = n, n-1, \ldots, 1$, as stated.   □

In practice, it is important to know what happens with a determinant after elementary column operations. The following theorem provides the required tool.

Theorem 3.23 *For any square matrix* $\mathbf{A}$

$$\det(\mathbf{A}^T) = \det(\mathbf{A}).$$

Proof. The correspondence $\sigma \to \sigma^{-1}$ is a one-to-one mapping of $\mathfrak{S}_n$ onto itself. If $\mathbf{A} = (a_{ij})$ then $\mathbf{A}^T = (a_{ji})$. It follows that

$$\det(\mathbf{A}^T) = \sum_{\sigma\in\mathfrak{S}_n}\varepsilon(\sigma)\prod_{k=1}^{n}a_{\sigma k\,k} = \sum_{\sigma^{-1}\in\mathfrak{S}_n}\varepsilon(\sigma)\prod_{k=1}^{n}a_{k\sigma^{-1}k}\overset{\sigma:=\sigma^{-1}}{=}$$

$$\sum_{\sigma\in\mathfrak{S}_n}\varepsilon(\sigma^{-1})\prod_{k=1}^{n}a_{k\sigma k} = \sum_{\sigma\in\mathfrak{S}_n}\varepsilon(\sigma)\prod_{k=1}^{n}a_{k\sigma k} = \det(\mathbf{A})$$

by Corollary 3.5.   □

Theorem 3.24 *If* $\mathbf{B}$ *is obtained from* $\mathbf{A}$ *by multiplying a row (column) by a real number c, then*

$$\det(\mathbf{B}) = c\cdot\det(\mathbf{A})$$

Proof. By Theorem 3.23 we may consider only the case of rows. Let

$$b_{ij} = \begin{cases} c\cdot a_{rj} & \text{if } i = r \\ a_{ij} & \text{if } i \neq r \end{cases}.$$

$$\det(\mathbf{B}) = \sum_{\sigma \in \mathfrak{S}_n} \varepsilon(\sigma) a_{1\sigma 1} \cdots (c \cdot a_{r\sigma r}) \cdots a_{n\sigma n} = c \cdot \det(\mathbf{A}). \qquad \square$$

Corollary 3.6 *If* $\mathrm{row}_r(\mathbf{A}) = \mathbf{0}$ *then* $\det(\mathbf{A}) = 0$.

Proof. Apply Theorem 3.24 to $\mathrm{row}_r(\mathbf{A}) = \mathbf{0}$ with $c = 0$. $\qquad \square$

Theorem 3.25 *If an* $n \times n$ *matrix* $\mathbf{B}$ *is obtained from a matrix* $\mathbf{A}$ *by swapping two rows (columns) of* $\mathbf{A}$, *then*

$$\det(\mathbf{B}) = -\det(\mathbf{A}).$$

Proof. Since $\det(\mathbf{A}^T) = \det(\mathbf{A})$, we may consider only the case or rows. Suppose that $\mathbf{B}$ is obtained from $\mathbf{A}$ by swapping the rows $\mathrm{row}_r(\mathbf{A})$ and $\mathrm{row}_s(\mathbf{A})$, where $r < s$. Then

$$b_{ij} = \begin{cases} a_{sj} & \text{if } i = r, \\ a_{rj} & \text{if } i = s, \end{cases}$$

and $b_{ij} = a_{ij}$ if $i \neq r, s$. Therefore,

$$\det(\mathbf{B}) = \sum_{\sigma \in \mathfrak{S}_n} \varepsilon(\sigma) b_{1\sigma 1} \cdots b_{r\sigma r} \cdots b_{s\sigma s} \cdots b_{n\sigma n} =$$
$$\sum_{\sigma \in \mathfrak{S}_n} \varepsilon(\sigma) a_{1\sigma 1} \cdots a_{s\sigma r} \cdots a_{r\sigma s} \cdots a_{n\sigma n} =$$
$$\sum_{\sigma \in \mathfrak{S}_n} \varepsilon(\sigma) a_{1\sigma 1} \cdots a_{r\sigma s} \cdots a_{s\sigma r} \cdots a_{n\sigma n}.$$

By Lemma 3.6 the permutations

$$\sigma = \big(\sigma 1 \cdots \sigma r \cdots \sigma s \cdots \sigma n\big)$$
$$\tau = \big(\sigma 1 \cdots \sigma s \cdots \sigma r \cdots \sigma n\big)$$

have opposite signatures. It is also clear that for a given pair of indices (r, s) the mapping $\sigma \mapsto \tau$ of $\mathfrak{S}_n$ is one-to-one. $\qquad \square$

Theorem 3.26 *If two different rows (columns) of* $\mathbf{A}$ *are equal, then*

$$\det(\mathbf{A}) = 0.$$

Proof. If $\text{row}_r(\mathbf{A}) = \text{row}_s(\mathbf{A})$ for $r < s$ then $\mathbf{A}$ does not change after swapping these rows and at the same time the determinant changes its sign by Theorem 3.25.   □

Theorem 3.27 *If* $\mathbf{B}$ *is obtained from* $\mathbf{A}$ *by the elementary row operation*

$$\text{row}_r(\mathbf{A}) \to \text{row}_r(\mathbf{A}) + c \cdot \text{row}_s(\mathbf{A}),$$

where $r \neq s$ *and* $c \in \mathbb{R}$, *then* $\det(\mathbf{A}) = \det(\mathbf{B})$.

Proof. We write the formula for $\det(\mathbf{A})$ and apply Theorem 3.26:

$$\det(\mathbf{B}) = \sum_{\sigma \in \mathfrak{S}_n} \varepsilon(\sigma) a_{1\sigma 1} \cdots (a_{r\sigma r}) + c \cdot a_{s\sigma s}) \cdots a_{n\sigma n} =$$

$$\det(\mathbf{A}) + c \cdot \sum_{\sigma \in \mathfrak{S}_n} \varepsilon(\sigma) a_{1\sigma 1} \cdots \mathbf{a}_{s\sigma s} \cdots \mathbf{a}_{s\sigma s} \cdots a_{n\sigma n} = \det(\mathbf{A}). \qquad □$$

Theorem 3.28 *An* $n \times n$ *matrix* $\mathbf{A}$ *is singular if and only if* $\det(\mathbf{A}) = 0$.

Proof. Since $\text{rref}(\mathbf{A}) = \mathbf{C}$ is obtained by a finite number of elementary row operations from the matrix $\mathbf{A}$,

$$\det(\mathbf{A}) = k \cdot \det(\mathbf{C}), \quad k \neq 0.$$

By Corollary 3.5 an $n \times n$ matrix $\mathbf{A}$ is singular if and only if $\text{rank}(\mathbf{A}) < n$, i.e. the nth row of $\text{rref}(\mathbf{A})$ consists only of zeros. Since $\mathbf{C} = \text{rref}(\mathbf{A})$ is upper triangular, Theorem 3.22 says that

$$\det(\mathbf{C}) = c_{11} c_{22} \cdots c_{nn}.$$

The last row of $\mathrm{rref}(\mathbf{A}) = \mathbf{C}$ is zero if and only if $c_{nn} = 0$. This proves the Theorem. □

Theorem 3.29 *For any $n \times n$ matrices $\mathbf{A}$ and $\mathbf{B}$*

$$\det(\mathbf{AB}) = \det(\mathbf{A}) \cdot \det(\mathbf{B}). \qquad (3.53)$$

Proof. If $\mathbf{AB}$ is singular then by Theorem 3.6 one of the matrices $\mathbf{A}$ and $\mathbf{B}$ is singular implying that both sides of (3.53) are zeros by Theorem 3.28.

If $\mathbf{AB}$ is non-singular, then by Theorem 3.6 both matrices $\mathbf{A}$ and $\mathbf{B}$ are non- singular. By Theorems 3.24, 3.25, and 3.27 the Theorem holds for any elementary matrix $\mathbf{A} = \mathbf{E}$. By Theorem 3.10 the non-singular matrix $\mathbf{A}$ is a product $\mathbf{E}_r \mathbf{E}_{r-1} \cdots \mathbf{E}_1$ of elementary matrices. Hence

$$\det(\mathbf{A}) = \det(\mathbf{E}_r \mathbf{E}_{r-1} \cdots \mathbf{E}_1) = \det(\mathbf{E}_r) \det(\mathbf{E}_{r-1} \cdots \mathbf{E}_1) =$$
$$\det(\mathbf{E}_r) \det(\mathbf{E}_{r-1}) \cdots \det(\mathbf{E}_1).$$
$$\det(\mathbf{E}_r \mathbf{E}_{r-1} \cdots \mathbf{E}_1 \mathbf{B}) = \det(\mathbf{E}_r) \det(\mathbf{E}_{r-1}) \cdots \det(\mathbf{E}_1) \det(\mathbf{E}_1) \det(\mathbf{B}) =$$
$$\det(\mathbf{A}) \cdot \det(\mathbf{B}). \qquad □$$

The following uniqueness theorem is a simple corollary of Theorems 3.24, 3.25, and 3.27.

Theorem 3.30 *Suppose $\mathbf{A} \mapsto D(\mathbf{A})$ is a mapping from the set of all $n \times n$ matrices to $\mathbb{R}$ satisfying the following conditions:*

(a) *If any row of $\mathbf{A}$ is multiplied by a constant c then $D(\mathbf{A})$ is also multiplied by c.*
(b) *If a multiple of any row is added to another row then $D(\mathbf{A})$ does not change.*
(c) *If two rows are interchanged, then $D(\mathbf{A})$ is multiplied by -1.*
(d) *$D(\mathbf{I}_n) = 1$.*

Then

$$D(\mathbf{A}) = \det(\mathbf{A}).$$

3.10 Cramer's Rule

Definition 3.9 Let $\mathbf{A} = (a_{ij})$ be an $n \times n$ matrix. Let $\mathbf{M}_{ij}$ be the $(n-1) \times (n-1)$ matrix obtained by crossing out the ith row and the jth column of A. The determinant $\det(\mathbf{M}_{ij})$ is called the **minor** of a_{ij}. Then the **cofactor** C_{ij} of a_{ij} is defined by

$$C_{ij} = (-1)^{i+j} \det(\mathbf{M}_{ij}) .$$

Problem 3.16 For

$$\mathbf{A} = \begin{pmatrix} 14 & 3 & 7 \\ 9 & 2 & 5 \\ 4 & 1 & 2 \end{pmatrix}$$

evaluate C_{11}, C_{21}, C_{31}.

Solution: Since

$$\det(\mathbf{M}_{11}) = \begin{vmatrix} 2 & 5 \\ 1 & 2 \end{vmatrix} = -1, \ \det(\mathbf{M}_{21}) = \begin{vmatrix} 3 & 7 \\ 1 & 2 \end{vmatrix} = -1, \ \det(\mathbf{M}_{31}) = \begin{vmatrix} 3 & 7 \\ 2 & 5 \end{vmatrix} = 1,$$

we conclude that $C_{11} = -1, C_{21} = 1, C_{31} = 1$. Notice that

$$a_{11}C_{11} + a_{21}C_{21} + a_{31}C_{31} = -14 + 9 + 4 = -1 = \det(\mathbf{A}),$$

which is agrees with formula (3.41).   □

Theorem 3.31 *Let $A = (a_{ij})$ be an $n \times n$ matrix. Then for every $j = 1, 2, \ldots, n$ we have a cofactor expansion by the jth column:*

$$\det(\mathbf{A}) = a_{1j}C_{1j} + a_{2j}C_{2j} + \ldots + a_{nj}C_{nj}. \qquad (3.54)$$

Proof. The main difficulty in the proof of this theorem is that i is not an index of a row for $\mathbf{M}_{ij}$ and j is not an index of a column. In other words, rows and columns of $\mathbf{M}_{ij}$ are enumerated with gaps at i and j correspondingly. However, the rows and the columns of $\mathbf{M}_{ij}$ are enumerated in the increasing order, true with gaps at i and j. Therefore, we may rewrite the standard formula for $\det(\mathbf{M}_{ij})$ in these notations with gaps. Let $\mathbb{N} \setminus \{k\}$ denote the ordered set $\mathbb{N}_n$ without number k. Then the rows of the $(n-1) \times (n-1)$ matrix $\mathbf{M}_{ij}$ are enumerated by $\mathbb{N} \setminus \{i\}$. Similarly, the columns of $\mathbf{M}_{ij}$ are enumerated by $\mathbb{N} \setminus \{j\}$. We denote by $\mathfrak{S}_n(ij)$ the set of all one-to-one mappings $\tau : \mathbb{N} \setminus \{i\} \to \mathbb{N} \setminus \{j\}$. It follows that

$$\det(\mathbf{M}_{ij}) = \sum_{\tau \in \mathfrak{S}_n(ij)} \varepsilon_{ij}(\tau) \prod_{k \neq i} a_{k\tau k}.$$

Then the sum in (3.54) is

$$\sum_{i=1}^{n} a_{ij}(-1)^{i+j} \det(\mathbf{M}_{ij}) = \sum_{i=1}^{n} a_{ij} \sum_{\tau \in \mathfrak{S}_n(ij)} (-1)^{i+j} \varepsilon_{ij}(\tau) \prod_{k \neq i} a_{k\tau k}. \qquad (3.55)$$

To compare the right-hand part of (3.55) with (3.48) we single out a_{ij} for $i = 1, \ldots, n$ in the products and rewrite (3.48) as follows:

$$\det(\mathbf{A}) = \sum_{i=1}^{n} a_{ij} \sum_{\sigma \in \mathfrak{S}_n, \, \sigma i = j} \varepsilon(\sigma) \prod_{k \neq i} a_{k\sigma k}.$$

Notice that every σ in $\mathfrak{S}_n$ such that $\sigma i = j$ is uniquely determined by its restriction τ to $\mathbb{N} \setminus \{i\}$. Now the result follows by the lemma.

Lemma 3.8 *Let σ in $\mathfrak{S}_n$ satisfy $\sigma i = j$ and let τ be the restriction of σ to $\mathbb{N} \setminus \{i\}$. Then*

$$\mathrm{inv}(\sigma) = i + j + \mathrm{inv}(\tau) + 2k, \qquad (3.56)$$

where k is an integer.

Proof. Every σ satisfying $\sigma i = j$ can be obtained by a finite number of transpositions applied consequently to $(i \; j)$ so that each such transposition do not move j. If $j > i$ then

$$(i \; j) = \begin{pmatrix} 1 & \cdots & i & \cdots & j & \cdots & n \\ 1 & \cdots & \mathbf{j} & \cdots & \mathbf{i} & \cdots & n \end{pmatrix},$$

and the complete list of inversions is given by the pairs

$$(j, i+1), \; \cdots, \; (j, j-1), \; (j, i) \Rightarrow \mathrm{inv}((i \; j)) = j - i = j + i - 2i.$$

If $j < i$ then, similarly, $\mathrm{inv}((j \; i)) = i - j = j + i - 2j$. It follows that (3.56) holds for $\sigma = (j \; i)$.

Suppose σ, $\sigma i = j$ satisfies (3.56) and τ is the restriction of σ on $\mathbb{N} \setminus \{i\}$. Let $(r \; s)$, $r < s$, be a transposition which does not move j. Then $r \neq j$ and $s \neq j$. If $s < j$ or $j < r$ then $(r \; s)$ contributes only to the inversions of τ. If $r < j < s$ then the application of $(r \; s)$ to σ results in two inversions $s > j$ and $j > r$. Hence the number of inversions at j increases by 2. So, if $\sigma' = (r \; s)\sigma$ then its restriction to $\mathbb{N} \setminus \{i\}$ is $\tau' = (r \; s)\sigma$ and, therefore,

$$\mathrm{inv}(\sigma') = i + j + \mathrm{inv}(\tau') + 2k',$$

which proves the Lemma. $\square$

□

Corollary 3.7 *Let* $\mathbf{A} = (a_{ij})$ *be an* $n \times n$ *matrix. Then*

$$\det(\mathbf{A}) = a_{11}C_{11} + a_{21}C_{21} + \cdots + a_{n1}C_{n1}. \qquad (3.57)$$

Proof. Put $j = 1$ in Theorem 3.31   □

Remark. Compare (3.57) with (3.41) and (3.45). Notice that by the Sarrus Rule the sum of the products in (3.46) which contain a_{31} equals

$$a_{12}a_{23}a_{31}a_{44} + a_{13}a_{24}a_{31}a_{42} + a_{14}a_{22}a_{31}a_{43}$$

$$- a_{14}a_{23}a_{31}a_{42} - a_{13}a_{22}a_{31}a_{44} - a_{12}a_{24}a_{31}a_{43} =$$

$$a_{31} \left(a_{12}a_{23}a_{44} + a_{13}a_{24}a_{42} + a_{14}a_{22}a_{43} - a_{14}a_{23}a_{42} - a_{13}a_{22}a_{44} - a_{12}a_{24}a_{43} \right) =$$

$$a_{31} \begin{vmatrix} a_{12} & a_{13} & a_{14} \\ a_{22} & a_{23} & a_{24} \\ a_{42} & a_{43} & a_{44} \end{vmatrix} = a_{31}C_{31}.$$

Corollary 3.8 *Let* $A = (a_{ij})$ *be an* $n \times n$ *matrix. Then for every* $i = 1, 2, \ldots, n$ *we have a cofactor expansion by the* i*th row:*

$$\det(\mathbf{A}) = a_{i1}C_{i1} + a_{i2}C_{i2} + \ldots + a_{in}C_{in}. \qquad (3.58)$$

Proof. Apply Theorem 3.23.   □

Corollary 3.9 *If the cofactors of one row are multiplied by the entries of a different row and added, then the result is* 0. *That is, if* $i \neq j$, *then*

$$a_{j1}C_{i1} + a_{j2}C_{i2} + \cdots + a_{jn}C_{in} = 0.$$

Proof. Let $\mathbf{B}$ be the matrix which differs from $\mathbf{A}$ only in the ith row:

$$\text{row}_i(\mathbf{B}) = \text{row}_j(\mathbf{A}).$$

Since $\text{row}_i(\mathbf{B}) = \text{row}_j(\mathbf{B})$, the determinant of $\mathbf{B}$ is zero by Theorem 3.26. Then

$$a_{j1}C_{i1} + a_{j2}C_{i2} + \cdots + a_{jn}C_{in} = \det(\mathbf{B}) = 0. \qquad \Box$$

Corollary 3.10 *Let $\mathbf{A}$ be an $n \times n$ matrix with linearly independent columns and j be the index of any column. Then the vector*

$$\lambda = \begin{pmatrix} C_{1j} & C_{2j} & \cdots & C_{nj} \end{pmatrix}^T$$

is perpendicular to any column $\text{col}_k(\mathbf{A})$ with $k \neq j$ and the dot product $\lambda^T \cdot \text{col}_j(\mathbf{A})$ equals the determinant of $\mathbf{A}$.

Notice that Corollary 3.10 gives the formula for the vector λ in (3.31).

Definition 3.10 Let $\mathbf{A}$ be an $n \times n$ matrix. The **cofactor matrix** $\mathbf{C} = \mathbf{C}(\mathbf{A})$ for $\mathbf{A}$ is the matrix with entry C_{ij} at (i, j). The **adjoint matrix** (also sometimes called the adjugate) $\text{adj}(\mathbf{A})$ of $\mathbf{A}$ is the transpose of the cofactor matrix:

$$\text{adj}(\mathbf{A}) = \mathbf{C}^T = \begin{pmatrix} C_{11} & C_{21} & \cdots & C_{n1} \\ C_{12} & C_{22} & \cdots & C_{n2} \\ \vdots & \vdots & \ddots & \vdots \\ C_{1n} & C_{2n} & \cdots & C_{nn} \end{pmatrix}.$$

Theorem 3.32 *Let $\mathbf{A}$ be an $n \times n$ matrix such that $\det(\mathbf{A}) \neq 0$. Then*

$$\boxed{\mathbf{A}^{-1} = \frac{1}{\det(\mathbf{A})}\text{adj}(\mathbf{A})}.$$

Proof. By Corollaries 3.8 and 3.10

$$\mathbf{A} \cdot \text{adj}(\mathbf{A}) = \begin{pmatrix} a_{11} & a_{12} & \cdots & a_{1n} \\ a_{21} & a_{22} & \cdots & a_{2n} \\ \vdots & \vdots & \ddots & \vdots \\ a_{n1} & a_{n2} & \cdots & a_{nn} \end{pmatrix}\begin{pmatrix} C_{11} & C_{21} & \cdots & C_{n1} \\ C_{12} & C_{22} & \cdots & C_{n2} \\ \vdots & \vdots & \ddots & \vdots \\ C_{1n} & C_{2n} & \cdots & C_{nn} \end{pmatrix} = \det(\mathbf{A})\mathbf{I}_n, \qquad (3.59)$$

which proves the Theorem.   □

Remark Dealing with concrete problems it is important to keep in mind that $\text{adj}(\mathbf{A}) = \mathbf{C}^T$.

Theorem 3.33 (Cramer's Rule) *Let $\mathbf{A}$ be an $n \times n$ matrix with $\det(\mathbf{A}) \neq 0$ and let $\mathbf{b} \in \mathbb{R}^n$. Let $\mathbf{A}_i$ be the matrix obtained from $\mathbf{A}$ by replacement of $\text{col}_i(\mathbf{A})$ with $\mathbf{b}$. Then the solution $\mathbf{x} = (x_1, x_2, \ldots, x_n)^T$ to the linear system $\mathbf{A}\mathbf{x} = \mathbf{b}$ is given by the formulas*

$$x_i = \frac{\det(\mathbf{A}_i)}{\det(\mathbf{A})}, \quad i = 1, 2, \ldots, n.$$

Proof. By Theorem 3.32

$$\mathbf{x} = \begin{pmatrix} x_1 \\ x_2 \\ \vdots \\ x_n \end{pmatrix} = \frac{1}{\det(\mathbf{A})} \begin{pmatrix} C_{11} & C_{21} & \cdots & C_{n1} \\ C_{12} & C_{22} & \cdots & C_{n2} \\ \vdots & \vdots & \ddots & \vdots \\ C_{1n} & C_{2n} & \cdots & C_{nn} \end{pmatrix} \begin{pmatrix} b_1 \\ b_2 \\ \vdots \\ b_n \end{pmatrix}.$$

It follows that

$$x_i = \frac{b_1 C_{1i} + b_2 C_{2i} + \cdots + b_n C_{ni}}{\det(\mathbf{A})} = \frac{\det(\mathbf{A}_i)}{\det(\mathbf{A})}. \qquad \square$$

Compare this with formula (3.31).

Theorem 3.34 (Vandermonde determinant)

$$\begin{vmatrix} 1 & x_1 & x_1^2 & \cdots & x_1^{n-1} \\ 1 & x_2 & x_2^2 & \cdots & x_2^{n-1} \\ 1 & x_3 & x_3^2 & \cdots & x_3^{n-1} \\ \vdots & \vdots & \vdots & \ddots & \vdots \\ 1 & x_n & x_n^2 & \cdots & x_n^{n-1} \end{vmatrix} = \prod_{1 \leq i < j \leq n} (x_i - x_j).$$

Proof. Let us subtract from each column starting from the last the previous one multiplied by x_n. Then

$$\det(\mathbf{V}_n) = \begin{vmatrix} 1 & x_1 & x_1^2 & \cdots & x_1^{n-1} \\ 1 & x_2 & x_2^2 & \cdots & x_2^{n-1} \\ 1 & x_3 & x_3^2 & \cdots & x_3^{n-1} \\ \vdots & \vdots & \vdots & \ddots & \vdots \\ 1 & x_n & x_n^2 & \cdots & x_n^{n-1} \end{vmatrix} =$$

$$\begin{vmatrix} 1 & x_1 - x_n & x_1(x_1 - x_n) & \cdots & x_1^{n-2}(x_1 - x_n) \\ 1 & x_2 - x_n & x_2(x_2 - x_n) & \cdots & x_2^{n-2}(x_2 - x_n) \\ 1 & x_3 - x_n & x_3(x_3 - x_n) & \cdots & x_3^{n-2}(x_3 - x_n) \\ \vdots & \vdots & \vdots & \ddots & \vdots \\ 1 & 0 & 0 & \cdots & 0 \end{vmatrix}.$$

Applying the cofactor expansion about the last row, we obtain that $\det(\mathbf{V}_n) =$

$$(-1)^{n+1} \begin{vmatrix} x_1 - x_n & x_1(x_1 - x_n) & \cdots & x_1^{n-2}(x_1 - x_n) \\ x_2 - x_n & x_2(x_2 - x_n) & \cdots & x_2^{n-2}(x_2 - x_n) \\ \vdots & \vdots & \ddots & \vdots \\ x_{n-1} - x_n & x_{n-1}(x_{n-1} - x_n) & \cdots & x_{n-1}^{n-2}(x_{n-1} - x_n) \end{vmatrix} =$$

$$(-1)^{n-1}(x_1 - x_n)(x_2 - x_n) \cdots (x_{n-1} - x_n) \times$$

$$\times \begin{vmatrix} 1 & x_1 & x_1^2 & \cdots & x_1^{n-2} \\ 1 & x_2 & x_2^2 & \cdots & x_2^{n-2} \\ \vdots & \vdots & \vdots & \ddots & \vdots \\ 1 & x_{n-1} & x_{n-1}^2 & \cdots & x_{n-1}^{n-2} \end{vmatrix} = \prod_{j=1}^{n-1}(x_n - x_j)\det(\mathbf{V}_{n-1}).$$

$\square$

Theorem 3.35 *Let $\{x_1, \ldots, x_n\}$ by n different points on the real line $\mathbb{R}$ and $\{b_1, \ldots, b_n\}$ be arbitrary real numbers. Then there exists a unique polynomial*

$$p(x) = a_0 + a_1 x + \cdots + a_{n-1} x^{n-1}$$

of degree $n - 1$ such that

$$p(x_k) = b_k, \quad k = 1, \ldots, n. \tag{3.60}$$

Proof. We have the linear system

$$\begin{cases} a_0 + a_1 x_1 + \cdots + a_{n-1} x_1^{n-1} & = b_1 \\ a_0 + a_1 x_2 + \cdots + a_{n-1} x_2^{n-1} & = b_2 \\ \quad\cdots\cdots\cdots\cdots\cdots\cdots\cdots \quad \cdots \\ a_0 + a_1 x_n + \cdots + a_{n-1} x_n^{n-1} & = b_n \end{cases} \tag{3.61}$$

in unknowns $a_0, a_1, \cdots, a_{n-1}$. The coefficient matrix of (3.61) is $\mathbf{V}_n$, implying that its determinant is nonzero. We find the unknowns by Cramer's Rule. To simplify calculations we add the row $\begin{pmatrix} 1 & x & x^2 & \cdots & x^{n-1} & p(x) \end{pmatrix}$ above the augmented matrix of the system (3.61). By Corollary 3.8 the determinant

$$\begin{vmatrix} 1 & x & x^2 & \cdots & x^{n-1} & p(x) \\ 1 & x_1 & x_1^2 & \cdots & x_1^{n-1} & b_1 \\ 1 & x_2 & x_2^2 & \cdots & x_2^{n-1} & b_2 \\ \vdots & \vdots & \vdots & \ddots & \vdots \\ 1 & x_n & x_n^2 & \cdots & x_n^{n-1} & b_n \end{vmatrix}$$

is a polynomial of degree $n - 1$. By (3.60) for every $x = x_k$ the determinant has two equal rows. Theorem 3.26 says that it is zero. Since a polynomial of degree $n - 1$ cannot have n different roots, the determinant is zero for every x:

$$\begin{vmatrix} p(x) & 1 & x & x^2 & \cdots & x^{n-1} \\ b_1 & 1 & x_1 & x_1^2 & \cdots & x_1^{n-1} \\ b_2 & 1 & x_2 & x_2^2 & \cdots & x_2^{n-1} \\ \vdots & \vdots & \vdots & & \ddots & \vdots \\ b_n & 1 & x_n & x_n^2 & \cdots & x_n^{n-1} \end{vmatrix} = 0 \Rightarrow$$

$$p(x)V(x_1,\ldots,x_n) - b_1 V(x,x_2,\ldots,x_n) + \ldots + (-1)^n V(x,x_1,\ldots,x_{n-1}) = 0 \Rightarrow$$

$$p(x) = \sum_{k=1}^{n} b_k \frac{V(x_1,\ldots x_{k-1},x,x_{k+1},\ldots,x_n)}{V(x_1,\ldots,x_n)} =$$

$$\sum_{k=1}^{n} b_k \frac{(x-x_1)\cdots(x-x_{k-1})(x-x_{k+1})\cdots(x-x_n)}{(x_k-x_1)\cdots(x_k-x_{k-1})(x_k-x_{k+1})\cdots(x_k-x_n)}. \qquad \square$$

Problems

Prob. 108 — Let

$$\mathbf{A} = \begin{pmatrix} 1 & 0 & 0 & -1 \\ 3 & 1 & 2 & 2 \\ 1 & 0 & -2 & 1 \\ 2 & 0 & 0 & 1 \end{pmatrix}.$$

Using Cramer's formulas find x_1 for the system $\mathbf{Ax} = \begin{pmatrix} 1 & 1 & 1 & 1 \end{pmatrix}^T$.

3.11 Analytic Geometry and Determinants

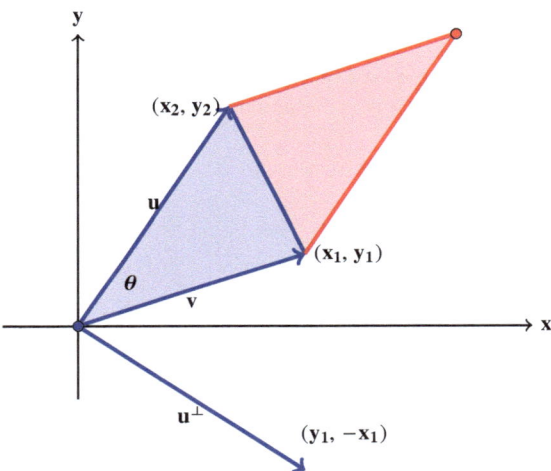

Fig. 3.4: 2×2 determinants and area

Let $\mathbf{u}$ and $\mathbf{v}$ be two vectors in $\mathbb{R}^2$, which make a basis for this plane. Then $\mathbf{v}$ can be obtained from $\mathbf{u}$ by a rotation $\mathbf{R}_\theta$, see Fig 4.3 where $-\pi < \theta < \pi$. If $0 < \theta < \pi$, then we say that $\mathbf{v}$ is obtained from $\mathbf{u}$ by an anticlockwise rotation. If $-\pi < \theta < 0$, then we say that it is obtained by a clockwise rotation. Notice that vectors $\mathbf{u}$ and $\mathbf{v}$ are linearly independent. This excludes the cases $\theta = \pm\pi$ and $\theta = 0$.

In analytic geometry the standard basis of $\mathbb{R}^2$ is traditionally denoted by $\mathbf{i} = \mathbf{e}_1$, $\mathbf{j} = \mathbf{e}_2$. If $\mathbf{u} = \begin{pmatrix} x_1 & y_1 \end{pmatrix}^T$, then the vector

$$\mathbf{u}^\perp = \begin{vmatrix} \mathbf{i} & \mathbf{j} \\ x_1 & y_1 \end{vmatrix} = y_1\mathbf{i} - x_1\mathbf{j}$$

is perpendicular to $\mathbf{u}$: $\langle \mathbf{u}, \mathbf{u}^\perp \rangle = x_1 y_1 - y_1 x_1 = 0$. In particular,

$$\mathbf{i}^\perp = -\mathbf{j}, \ \mathbf{j}^\perp = \mathbf{i} \Rightarrow (x\mathbf{i} + y\mathbf{j})^\perp = y\mathbf{i} - x\mathbf{j}.$$

Similarly,

$$\langle \mathbf{u}^\perp, \mathbf{u}^\perp \rangle = \begin{vmatrix} y_1 & -x_1 \\ x_1 & y_1 \end{vmatrix} = x_1^2 + y_1^2 = |\mathbf{u}|^2.$$

If **v** is obtained from **v** by a clockwise rotation followed by a compression, see (4.13), then

$$\mathbf{v} = \begin{pmatrix} x_2 \\ y_2 \end{pmatrix} = r\mathbf{R}_{-\theta}\mathbf{u} = r\begin{pmatrix} \cos\theta & \sin\theta \\ -\sin\theta & \cos\theta \end{pmatrix}\begin{pmatrix} x_1 \\ y_1 \end{pmatrix} = r\begin{pmatrix} x_1\cos\theta + y_1\sin\theta \\ -x_1\sin\theta + y_1\cos\theta \end{pmatrix}.$$

It follows that

$$\det\begin{pmatrix} \mathbf{u} & \mathbf{v} \end{pmatrix} = \det\begin{pmatrix} \mathbf{u} & \mathbf{v} \end{pmatrix}^T = \langle \mathbf{u}^\perp, \mathbf{v} \rangle = \begin{vmatrix} x_2 & y_2 \\ x_1 & y_1 \end{vmatrix} = r(x_1^2 + y_1^2)\sin\theta = |\mathbf{u}||\mathbf{v}|\sin\theta \quad (3.62)$$

is the area of the blue triangle on Fig. 3.4 made by vectors **u** and **v**. The vector $\mathbf{u}^\perp$ has the same length as **u** and is perpendicular to **u**:

$$\mathbf{u}^\perp \cdot \mathbf{u} = -y_1 x_1 + x_1 y_1 = 0.$$

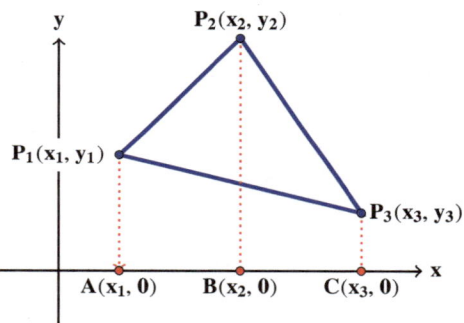

Fig. 3.5: 2×2 determinants

The area of the plane triangle is a combination of the areas of three trapezoids:

$$\text{Area}(AP_1P_2B) + \text{Area}(BP_2P_3C) - \text{Area}(AP_1P_3C) =$$

$$\frac{1}{2}\left\{(x_2 - x_1)(y_1 + y_2) + (x_3 - x_2)(y_2 + y_3) - (x_3 - x_1)(y_1 + y_3)\right\} =$$

$$\frac{1}{2}\left\{(x_3y_2 - x_2y_3) - (x_3y_1 - x_1y_3) + (x_2y_1 - x_1y_2)\right\} =$$

$$\frac{1}{2}\begin{vmatrix} y_2 & y_3 \\ x_2 & x_3 \end{vmatrix} - \frac{1}{2}\begin{vmatrix} y_1 & y_3 \\ x_1 & x_3 \end{vmatrix} + \frac{1}{2}\begin{vmatrix} y_1 & y_2 \\ x_1 & x_2 \end{vmatrix} = \frac{1}{2}\begin{vmatrix} 1 & 1 & 1 \\ y_1 & y_2 & y_3 \\ x_1 & x_2 & x_3 \end{vmatrix} = \frac{1}{2}\begin{vmatrix} 1 & 0 & 0 \\ y_1 & y_2 - y_1 & y_3 - y_1 \\ x_1 & x_2 - x_1 & x_3 - x_1 \end{vmatrix} \Rightarrow$$

$$\boxed{\text{Area}\,(\triangle\, P_1P_2P_3) = \frac{1}{2}\begin{vmatrix} 1 & 1 & 1 \\ y_1 & y_2 & y_3 \\ x_1 & x_2 & x_3 \end{vmatrix} = \frac{1}{2}\begin{vmatrix} y_2 - y_1 & y_3 - y_1 \\ x_2 - x_1 & x_3 - x_1 \end{vmatrix}.}$$

In analytic geometry the standard basis of $\mathbb{R}^3$ is traditionally denoted by $\mathbf{i} = \mathbf{e}_1$, $\mathbf{j} = \mathbf{e}_2$, $\mathbf{k} = \mathbf{e}_3$.

> **Definition 3.11** The **cross product** of two vectors $\mathbf{a} = (a_1, a_2, a_3)$, $\mathbf{b} = (b_1, b_2, b_3)$ in $\mathbb{R}^3$ is defined by the cofactor formula about the first row, see (3.58) with $i = 1$:
>
> $$\mathbf{a} \times \mathbf{b} = \begin{vmatrix} \mathbf{i} & \mathbf{j} & \mathbf{k} \\ a_1 & a_2 & a_3 \\ b_1 & b_2 & b_3 \end{vmatrix} = (a_2 b_3 - a_3 b_2)\mathbf{i} - (a_1 b_3 - a_3 b_1)\mathbf{j} + (a_1 b_2 - a_2 b_1)\mathbf{k}.$$

Notice that such a vector determinant makes sense for any vectors on places of, $\mathbf{i}$, $\mathbf{j}$, and $\mathbf{k}$. By Theorem 3.25

$$\mathbf{a} \times \mathbf{b} = -\mathbf{b} \times \mathbf{a}. \tag{3.63}$$

By Theorem 3.23

$$\begin{vmatrix} \mathbf{i} & \mathbf{j} & \mathbf{k} \\ a_1 & a_2 & a_3 \\ b_1 & b_2 & b_3 \end{vmatrix} = \begin{vmatrix} \mathbf{i} & a_1 & b_1 \\ \mathbf{j} & a_2 & b_2 \\ \mathbf{k} & a_3 & b_3 \end{vmatrix},$$

implying that $\mathbf{a} \times \mathbf{b}$ can also be evaluated by the cofactor formula about the first column. In particular, we obtain the formulas:

$$\mathbf{i} \times \mathbf{j} = \begin{vmatrix} \mathbf{i} & \mathbf{j} & \mathbf{k} \\ 1 & 0 & 0 \\ 0 & 1 & 0 \end{vmatrix} = \mathbf{k}, \quad \mathbf{k} \times \mathbf{i} = \begin{vmatrix} \mathbf{i} & \mathbf{j} & \mathbf{k} \\ 0 & 0 & 1 \\ 1 & 0 & 0 \end{vmatrix} = \mathbf{j}, \quad \mathbf{j} \times \mathbf{k} = \begin{vmatrix} \mathbf{i} & \mathbf{j} & \mathbf{k} \\ 0 & 1 & 0 \\ 0 & 0 & 1 \end{vmatrix} = \mathbf{i}.$$

Lemma 3.9 *For any vectors* $\mathbf{a}$, $\mathbf{b}$, $\mathbf{c}$ *in* $\mathbb{R}^3$

$$\langle \mathbf{a} \times \mathbf{b} , \mathbf{c} \rangle = \begin{vmatrix} c_1 & c_2 & c_3 \\ a_1 & a_2 & a_3 \\ b_1 & b_2 & b_3 \end{vmatrix}. \tag{3.64}$$

Proof. By Corollary 3.8

$$(a_2 b_3 - a_3 b_2)c_1 - (a_1 b_3 - a_3 b_1)c_2 + (a_1 b_2 - a_2 b_1) = \begin{vmatrix} c_1 & c_2 & c_3 \\ a_1 & a_2 & a_3 \\ b_1 & b_2 & b_3 \end{vmatrix}. \qquad \square$$

> **Corollary 3.11** *The vector* $\mathbf{a} \times \mathbf{b}$ *is perpendicular to both vectors* $\mathbf{a}$ *and* $\mathbf{b}$.

Proof. Put $c = a$, $c = b$ and observe that in both cases the determinant has two equal rows with different indices. Hence the determinant is zero.   □

Since the cross product $a \times b$ of two not proportional vectors is perpendicular to them both, it is also perpendicular to the unique plane through these vectors. It follows that $a \times b$ is the normal vector to this plane and

$$\begin{vmatrix} x & y & z \\ a_1 & a_2 & a_3 \\ b_1 & b_2 & b_3 \end{vmatrix} = 0$$

is a general equation of the plane.

Lemma 3.10 (Lagrange's identity)

$$(a_2 b_3 - b_2 a_3)^2 + (a_1 b_3 - b_1 a_3)^2 + (a_1 b_2 - b_1 a_2)^2 =$$
$$(a_1^2 + a_2^2 + a_3^2)(b_1^2 + b_2^2 + b_3^2) - (a_1 b_1 + a_2 b_2 + a_3 b_3)^2.$$

Proof. Evaluating both parts of Lagrange's identity, we obtain that they both are equal to

$$a_2^2 b_3^2 - 2a_2 a_2 b_2 b_3 + b_2^2 a_3^2 + a_1^2 b_3^2 - 2a_1 a_3 b_1 b_3 + b_1^2 a_3^2 + a_1^2 b_2^2 - 2a_1 a_2 b_1 b_2 + b_1^2 a_2^2.$$
□

Corollary 3.12 *If $c = a \times b$ then*

$$|a \times b|^2 = \begin{vmatrix} c_1 & c_2 & c_3 \\ a_1 & a_2 & a_3 \\ b_1 & b_2 & b_3 \end{vmatrix} = |a|^2 |b|^2 \sin^2 \varphi, \qquad (3.65)$$

where ϕ is the angle between a and b.

Proof. The first equality in (3.65) follows by (3.64). By the definition of the cross product and Lagrange's identity

$$\begin{vmatrix} c_1 & c_2 & c_3 \\ a_1 & a_2 & a_3 \\ b_1 & b_2 & b_3 \end{vmatrix} = (a_2b_3 - b_2a_3)^2 + (a_1b_3 - b_1a_3)^2 + (a_1b_2 - b_1a_2)^2 =$$

$$|\mathbf{a}|^2|\mathbf{b}|^2 - |\langle \mathbf{a}, \mathbf{b} \rangle|^2 = |\mathbf{u}|^2|\mathbf{v}|^2 - |\mathbf{u}|^2 \cdot |\mathbf{v}|^2 \cos^2 \phi = |\mathbf{u}|^2|\mathbf{v}|^2 \sin^2 \varphi. \quad \square$$

Corollary 3.13 *The Area of the Parallelogram spanned by two vectors* $\mathbf{a}$ *and* $\mathbf{b}$ *equals* $|\mathbf{a} \times \mathbf{b}|$.

Proof. By Corollary 3.12 the area is $|\mathbf{a}| \cdot |\mathbf{b}||\sin \theta| = |\mathbf{a} \times \mathbf{b}|$. $\square$

Corollary 3.14 *The cross product* $\mathbf{a} \times \mathbf{b}$ *is zero if and only if* $\mathbf{a}$ *and* $\mathbf{b}$ *are proportional vectors.*

Proof. It follows by (3.65). $\square$

By Corollary 3.12 the length of the cross product $\mathbf{a} \times \mathbf{b}$ is the area of the parallelogram determined by the vectors $\mathbf{a}$ and $\mathbf{b}$. By Corollary 3.11 the cross product is perpendicular to $\mathbf{a}$ and $\mathbf{b}$. The **right-hand rule** determines the orientation of the cross-product $\mathbf{a} \times \mathbf{b}$. It says that if one let the fingers of his or her right hand curl from $\mathbf{a}$ toward $\mathbf{b}$, then his or her thumb points in the direction of $\mathbf{a} \times \mathbf{b}$.

Theorem 3.36 (The Right-Hand Rule) *Let* $\mathbf{a}$ *and* $\mathbf{b}$ *be two not proportional vectors in* $\mathbb{R}^3$. *Let* $\mathfrak{P}$ *be the plane through* $\mathbf{a}$ *and* $\mathbf{b}$. *Then an observer at the head of the normal* $\mathbf{a} \times \mathbf{b}$ *to* $\mathfrak{P}$ *sees that the direction of the vector* $\mathbf{b}$ *is obtained by an anticlockwise rotation of the direction of* $\mathbf{a}$ *through an angle* φ, $0 < \varphi < \pi$.

Proof. Since $\mathbf{a}$ and $\mathbf{b}$ are not proportional, the cross-product $\mathbf{a} \times \mathbf{b}$ is not zero. Hence, at least one coordinate of $\mathbf{a} \times \mathbf{b}$ is nonzero. Without loss of generality we may assume that $\langle \mathbf{a} \times \mathbf{b}, \mathbf{k} \rangle \neq 0$. We consider first the case $\langle \mathbf{a} \times \mathbf{b}, \mathbf{k} \rangle > 0$. Then both $\mathbf{a} \times \mathbf{b}$ and $\mathbf{k}$ are above the planes $\mathfrak{P}$ and $\mathrm{Lin}(\mathbf{i}, \mathbf{j})$. Since $\langle \mathbf{a} \times \mathbf{b}, \mathbf{k} \rangle \neq 0$

there exist planes $\mathfrak{Q}_\mathbf{a}$ and $\mathfrak{Q}_\mathbf{b}$ through $\mathbf{a}, \mathbf{k}$ and through $\mathbf{b}, \mathbf{k}$. The intersection of $\mathfrak{Q}_\mathbf{a}$ with $\mathrm{Lin}(\mathbf{i}, \mathbf{j})$ is the line directed by the vector $a_1\mathbf{i} + a_2\mathbf{j}$. The intersection of $\mathfrak{Q}_\mathbf{b}$ with $\mathrm{Lin}(\mathbf{i}, \mathbf{j})$ is the line directed by the vector $b_1\mathbf{i} + b_2\mathbf{j}$. We have

$$0 < \langle \mathbf{a} \times \mathbf{b}, \mathbf{k} \rangle = \begin{vmatrix} a_1 & a_2 \\ b_1 & b_2 \end{vmatrix} = a_1 b_2 - b_1 a_2. \tag{3.66}$$

The plane vector (b_1, b_2) is obtained from (a_1, a_2) by a rotation and compression in $\mathrm{Lin}(\mathbf{i}, \mathbf{j})$:

$$\begin{pmatrix} b_1 \\ b_2 \end{pmatrix} = r\mathbf{R}_\phi \begin{pmatrix} a_1 \\ a_2 \end{pmatrix} = r \begin{pmatrix} \cos\phi & -\sin\phi \\ \sin\phi & \cos\phi \end{pmatrix} \begin{pmatrix} a_1 \\ a_2 \end{pmatrix} = r \begin{pmatrix} a_1\cos\phi - a_2\sin\phi \\ a_1\sin\phi + a_2\cos\phi \end{pmatrix}. \tag{3.67}$$

By (3.66) and (3.67) we obtain that

$$0 < \langle \mathbf{a} \times \mathbf{b}, \mathbf{k} \rangle = r(a_1^2 + a_2^2)\sin\phi,$$

implying that $0 < \phi < \pi$. This means that an observer at the head of $\mathbf{k}$ sees the rotation of the plane $\mathfrak{Q}_\mathbf{a}$ about $\mathbf{k}$ to the plane $\mathfrak{Q}_\mathbf{b}$ in the anticlockwise direction. In particular, the vector $\mathbf{a}$ moves anticlockwise to the vector $\mathbf{b}$. Since $\mathbf{a} \times \mathbf{b}$ is on the same side of $\mathfrak{P}$ as $\mathbf{k}$ the same is true for an observer at the head of $\mathbf{a} \times \mathbf{b}$. The case of $\langle \mathbf{a} \times \mathbf{b}, \mathbf{k} \rangle < 0$ is reduced to the considered one by the formula (3.63). $\qquad\square$

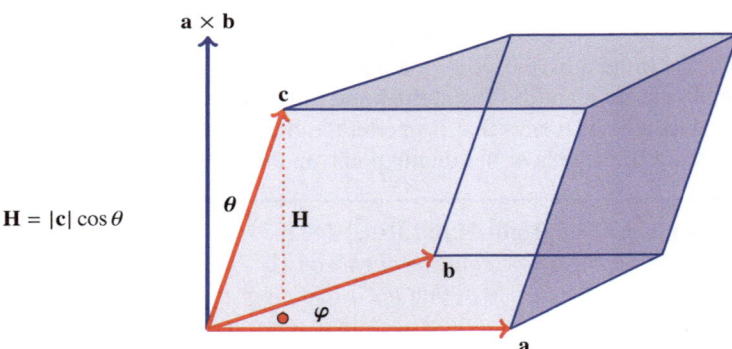

$$\mathbf{H} = |\mathbf{c}| \cos\theta$$

Fig. 3.6: 3×3 determinants and volumes

In Analytic Geometry any orthonormal basis in $\mathbb{R}^3$ is called a **coordinate system**.

Definition 3.12 A coordinate system $\{\mathbf{u}_1, \mathbf{u}_2, \mathbf{u}_3\}$ in $\mathbb{R}^3$ is said to be oriented positive if

$$\mathbf{u}_1 \times \mathbf{u}_2 = \mathbf{u}_3.$$

Problem 3.17 A zero vector can be represented as a linear combination of three vectors $\mathbf{a}$, $\mathbf{b}$, $\mathbf{c}$ in $\mathbb{R}^3$ with not all vectors equal $\mathbf{0}$ if and only if

$$\begin{vmatrix} c_1 & c_2 & c_3 \\ a_1 & a_2 & a_3 \\ b_1 & b_2 & b_3 \end{vmatrix} = 0.$$

Solution: Since the change of the order of $\mathbf{a}$, $\mathbf{b}$, $\mathbf{c}$ may only change the sign of the determinant, we may assume that $\mathbf{c} = \alpha\mathbf{a} + \beta\mathbf{b}$. Then

$$\begin{vmatrix} c_1 & c_2 & c_3 \\ a_1 & a_2 & a_3 \\ b_1 & b_2 & b_3 \end{vmatrix} = \alpha \begin{vmatrix} a_1 & a_2 & a_3 \\ a_1 & a_2 & a_3 \\ b_1 & b_2 & b_3 \end{vmatrix} + \beta \begin{vmatrix} b_1 & b_2 & b_3 \\ a_1 & a_2 & a_3 \\ b_1 & b_2 & b_3 \end{vmatrix} = 0.$$

Let now

$$\mathbf{A} = \begin{pmatrix} c_1 & c_2 & c_3 \\ a_1 & a_2 & a_3 \\ b_1 & b_2 & b_3 \end{pmatrix} \quad \text{and} \quad \det(\mathbf{A}) = 0.$$

Then $\det(\mathbf{A}^T) = 0$ implying that $\mathbf{rref}(\mathbf{A}^T) \neq \mathbf{I}$. Therefore, there is a nonzero solution $\mathbf{x} = (x_1, x_2, x_3)$ to the equation $\mathbf{A}^T\mathbf{x} = 0$. In other words $x_1\mathbf{c} + x_2\mathbf{a} + x_3\mathbf{b} = \mathbf{0}$, where not all real numbers x_1, x_2, x_3 are zeros. □

Formula (3.64) shows that 3×3 determinants are related through the formula for the volume of a parallelepiped. Indeed, Fig. 3.6 shows that the volume of a parallelepiped equals

$$\text{Vol} = H \cdot |\mathbf{a}| \cdot |\mathbf{b}| \sin \varphi = |\mathbf{c}| \cdot \cos \theta \cdot \mathbf{a} \cdot |\mathbf{b}| \sin \varphi = \pm \begin{vmatrix} c_1 & c_2 & c_3 \\ a_1 & a_2 & a_3 \\ b_1 & b_2 & b_3 \end{vmatrix}, \qquad (3.68)$$

where the sign is chosen to make the determinant positive. Formula (3.68) leads to the notion of a **signed volume**

$$\text{Vol}(\mathbf{A}^T) = \det(\mathbf{A}^T).$$

Here $\mathbf{A}$ is the matrix with the columns $\mathbf{a}$, $\mathbf{b}$, $\mathbf{c}$.

3.12 Characteristic Polynomials

Let λ be a real or complex variable and $\mathbf{A}$ be an $n \times n$ matrix. Then by (3.48) the function

$$P_{\mathbf{A}}(\lambda) = \det(\mathbf{A} - \lambda \mathbf{I}_n) \tag{3.69}$$

is a polynomial in λ of degree n. Its leading coefficient is $(-1)^n$. Its free coefficient is $\det(\mathbf{A})$ (put $\lambda = 0$ in (3.69)).

> **Definition 3.13** The polynomial $P_{\mathbf{A}}(\lambda)$ is called the **characteristic polynomial** of $\mathbf{A}$ and the equation $P_{\mathbf{A}}(\lambda) = 0$ is called the **characteristic equation** of $\mathbf{A}$.

> **Definition 3.14** Given an $n \times n$ matrix the sum of all diagonal entries of $\mathbf{A}$
>
> $$\mathrm{Tr}(\mathbf{A}) = a_{11} + a_{12} + \cdots + a_{nn}$$
>
> is called the **trace** of $\mathbf{A}$.

The characteristic polynomial of an 1×1 matrix $\mathbf{A}$ is the monomial $(a_{11} - \lambda)$. The characteristic polynomial of a 2×2 matrix $\mathbf{A}$ is the quadratic polynomial

$$P_{\mathbf{A}}(\lambda) = \det \begin{pmatrix} a_{11} - \lambda & a_{12} \\ a_{21} & a_{22} - \lambda \end{pmatrix} = \lambda^2 - \mathrm{Tr}(\mathbf{A})\lambda + \det(\mathbf{A}). \tag{3.70}$$

> **Theorem 3.37** *Let $\mathbf{A}$ be a 3×3 matrix. Then*
>
> $$P_{\mathbf{A}}(\lambda) = -\lambda^3 + \mathrm{Tr}(\mathbf{A})\lambda^2 - \mathrm{Tr}\,(\mathrm{adj}(\mathbf{A}))\,\lambda + \det(\mathbf{A}). \tag{3.71}$$

> *Proof.* To prove (3.71) we apply Corollary 3.8 with $i = 1$:
>
> $$P_{\mathbf{A}}(\lambda) = (a_{11} - \lambda) \begin{vmatrix} a_{22} - \lambda & a_{23} \\ a_{32} & a_{33} - \lambda \end{vmatrix} - a_{12} \begin{vmatrix} a_{21} & a_{23} \\ a_{31} & a_{33} - \lambda \end{vmatrix} + a_{13} \begin{vmatrix} a_{21} & a_{22} - \lambda \\ a_{31} & a_{32} \end{vmatrix}.$$
>
> Evaluating determinants and grouping terms we arrive at (3.71). □

Lemma 3.11 *If* **A** *is* 3×3 *matrix, then*

$$\operatorname{Tr}\left(\operatorname{adj}(\mathbf{A})\right) = C_{11} + C_{22} + C_{33} = \frac{1}{2}\left(\operatorname{Tr}(\mathbf{A}^2) - (\operatorname{Tr}(\mathbf{A}))^2\right).$$

Proof. The result follows by direct computations:

$$\operatorname{Tr}(\mathbf{A}^2) = a_{11}^2 + a_{22}^2 + a_{33}^2 + 2a_{12}a_{21} + 2a_{13}a_{31} + 2a_{23}a_{32},$$
$$(\operatorname{Tr}(\mathbf{A}))^2 = a_{11}^2 + a_{22}^2 + a_{33}^2 + 2a_{11}a_{22} + 2a_{11}a_{33} + 2a_{22}a_{33},$$
$$C_{11} + C_{22} + C_{33} = (a_{22}a_{33} - a_{23}a_{32}) + (a_{11}a_{33} - a_{13}a_{31}) + (a_{11}a_{22} - a_{12}a_{21}). \square$$

Theorem 3.38 *Let* **A** *be an* $n \times n$ *matrix. Then*

$$p_{\mathbf{A}}(\lambda) = (-1)^n \lambda^n + (-1)^{n-1}\operatorname{Tr}(\mathbf{A})\lambda^{n-1} + \cdots + \det(\mathbf{A}).$$

Proof. We observe that the diagonal of $\mathbf{A} - \lambda\mathbf{I}$ is occupied by the entries of the form:

$$a_{11} - \lambda, \; a_{22} - \lambda, \; \ldots, a_{nn} - \lambda. \tag{3.72}$$

By the definition of determinants, see (3.48), the power λ^{n-1} may appear only if $n - 1$ diagonal entries of $\mathbf{A} - \lambda\mathbf{I}$ listed in (3.72) are multiplied. The corresponding permutation σ being identity on $n - 1$ elements of the set $\{1, 2, \ldots, n\}$ must be the identity permutation, implying that the diagonal entry a_{ii} is multiplied by $(-1)^{n-1}\lambda^{n-1}$. This means that the coefficient at λ^{n-1} in the characteristic polynomial of $\mathbf{A}$ is the sum of the diagonal entries of $\mathbf{A}$ multiplied by $(-1)^{n-1}$. Sinse $p_{\mathbf{A}}(0) = \det(\mathbf{A} - 0\mathbf{I}) = \det(\mathbf{A})$, this completes the proof of this theorem. $\square$

Problem 3.18 Evaluate the characteristic polynomial of the matrix

$$\mathbf{A} = \begin{pmatrix} 1 & 2 & 3 \\ 2 & 3 & 1 \\ 3 & 1 & 2 \end{pmatrix}.$$

Solution: We have $\operatorname{Tr}(\mathbf{A}) = 1 + 3 + 2 = 6$,

$$C_{11} + C_{22} + C_{33} = (6 - 1) + (2 - 9) + (3 - 4) = -3,$$
$$\det(\mathbf{A}) = (6 - 1) - 2(4 - 3) + 3(2 - 9) = -18.$$

By (3.71) we obtain that

$$P_{\mathbf{A}}(\lambda) = -\lambda^3 + 6\lambda^2 + 3\lambda - 18. \quad \Box$$

If $\lambda = \mathbf{A}$ in (3.69), then we obtain that $p_{\mathbf{A}}(\mathbf{A}) = \mathbf{0}$. This, however, is not a correct operation. The right-hand part in (3.69) is a usual zero number. In the case of matrices this zero must be replaced by a zero $n \times n$ matrix. Therefore, other arguments must be applied. To begin with we observe that $n \times n$ matrices can be identified with $\mathbb{R}^{n^2}$. One can just enumerate entries in a linear way. First, we list the entries of the first row, then the entries of the second row, etc. All powers $\mathbf{A}^k$ of an $n \times n$ matrix $\mathbf{A}$ are then in $\mathbb{R}^{n^2}$. Since the dimension of $\mathbb{R}^{n^2}$ is n^2, there is a linear combination of powers $\mathbf{A}^k$ with $k \leq n^2$ representing the zero $n \times n$ matrix. The main question is that why this combination is given by the coefficients of the characteristic polynomial for $\mathbf{A}$. Instead of formula (3.69) we may use a closely related formula (3.59).

Theorem 3.39 (Cayley-Hamilton Theorem) *Let $\mathbf{A}$ be an $n \times n$ matrix with complex entries and let*

$$\det(\mathbf{A} - \lambda \mathbf{I_n}) = p_0 + p_1 \lambda + p_2 \lambda^2 + \cdots + p_n \lambda^n = P_{\mathbf{A}}(\lambda)$$

be the characteristic polynomial of $\mathbf{A}$. Then

$$P_{\mathbf{A}}(\mathbf{A}) = p_0 \mathbf{I}_n + p_1 \mathbf{A} + p_2 \mathbf{A}^2 + \cdots + p_n \mathbf{A}^n = \mathbf{0}_n.$$

Proof. By Cramer's formula (3.59) we have

$$(\text{adj}(\mathbf{A} - \lambda \mathbf{I_n}))\,(\mathbf{A} - \lambda \mathbf{I_n}) = (\det(\mathbf{A} - \lambda \mathbf{I_n}))\mathbf{I}_n.$$

It follows from the definition of the adjoint matrix that

$$\text{adj}(\mathbf{A} - \lambda \mathbf{I_n}) = \mathbf{Q}_0 + \mathbf{Q}_1 \lambda + \mathbf{Q}_2 \lambda^2 + \cdots + \mathbf{Q}_{n-1} \lambda^{n-1} = \mathbf{Q}(\lambda),$$

where $\mathbf{Q}_k$, $k = 1, 2, \ldots, n - 1$, are $n \times n$ matrices. Then

$$\mathbf{Q}(\lambda)(\mathbf{A} - \lambda \mathbf{I_n}) = \mathbf{Q}_0 \mathbf{A} + (\mathbf{Q}_1 \mathbf{A} - \mathbf{Q}_0)\lambda + (\mathbf{Q}_2 \mathbf{A} - \mathbf{Q}_1)\lambda^2 +$$
$$\cdots + (\mathbf{Q}_{n-1}\mathbf{A} - \mathbf{Q}_{n-2})\lambda^{n-1} - \mathbf{Q}_{n-1}\lambda^n.$$

Equating the matrix coefficients in the formula $P_A(\lambda) = Q(\lambda)(A - \lambda I_n)$, we obtain a series of equations:

$$
\begin{aligned}
Q_0 A &= p_0 I_n \\
Q_1 A - Q_0 &= p_1 I_n \\
Q_2 A - Q_1 &= p_2 I_n \\
&\vdots \\
Q_{n-1} A - Q_{n-2} &= p_{n-1} I_n \\
-Q_{n-1} &= p_n I_n
\end{aligned}
$$

Multiplying all equations starting from the second one by ascending powers of A, we obtain the system:

$$
\begin{aligned}
Q_0 A &= p_0 I_n \\
Q_1 A^2 - Q_0 A &= p_1 A \\
Q_2 A^3 - Q_1 A^2 &= p_2 A^2 \\
&\vdots \\
Q_{n-1} A^n - Q_{n-2} A^{n-1} &= p_{n-1} A^{n-1} \\
-Q_{n-1} A^n &= p_n A^n
\end{aligned}
$$

On adding all the equations together we obtain the zero matrix on the left and $P_A(A)$ on the right:

$$
0_n = p_0 I_n + p_1 A + p_2 A^2 + \cdots + p_n A^n. \qquad \square
$$

Let us put $\lambda = 0$ in the formula for the characteristic polynomials:

$$
\det(A - \lambda I_n) = P_0 + P_1 \lambda + P_2 \lambda^2 + \cdots + P_n \lambda^n = P_A(\lambda).
$$

Then we obtain that $P_0 = \det(A)$. By Cayley-Hamilton Theorem

$$
\det(A) I_n = -A \left(P_1 I_n + P_2 A + \cdots + P_n A^{n-1} \right) \Rightarrow
$$

$$
\boxed{A^{-1} = \frac{-1}{\det(A)} \left(P_1 I_n + P_2 A + \cdots + P_n A^{n-1} \right)}
$$

Problem 3.19 Using the Cayley-Hamilton Theorem, find the inverse of the following matrix

$$
A = \begin{pmatrix} 2 & 0 & 1 \\ -2 & 3 & 4 \\ -5 & 5 & 6 \end{pmatrix}.
$$

Solution: We have

$$-P_A(\lambda) = \lambda^3 - \text{Tr}(A)\lambda^2 + \underbrace{(C_{11} + C_{22} + C_{33})}_{\text{sum of the diagonal cofactors}} \lambda - \det(A).$$

Since
$$\text{Tr}(A) = a_{11} + a_{22} + a_{33} = 2 + 3 + 6 = 11,$$
$$(C_{11} + C_{22} + C_{33}) = -2 + 17 + 6 = 21,$$
$$\det(A) = 1,$$

we see that
$$-P_A(\lambda) = \lambda^3 - 11\lambda + 21\lambda - 1.$$

By the Cayley-Hamilton Theorem

$$A^3 - 11A^2 + 21A - I = 0 \Rightarrow A^2 - 11A + 21I - A^{-1} = 0 \Rightarrow$$
$$A^{-1} = A^2 - 11A + 21I =$$

$$\begin{pmatrix} -1 & 5 & 8 \\ -30 & 29 & 34 \\ -50 & 45 & 51 \end{pmatrix} - 11\begin{pmatrix} 2 & 0 & 1 \\ -2 & 3 & 4 \\ -5 & 5 & 6 \end{pmatrix} + 21\begin{pmatrix} 1 & 0 & 0 \\ 0 & 1 & 0 \\ 0 & 0 & 1 \end{pmatrix} = \boxed{\begin{pmatrix} -2 & 5 & -3 \\ -8 & 17 & -10 \\ 5 & -10 & 6 \end{pmatrix}}. \quad \Box$$

Matrix Exponents by the Cayley-Hamilton Theorem

$$f(\lambda) = \sum_{k=0}^{\infty} b_k \lambda^k.$$

Then
$$f(\lambda) = P(\lambda)Q(\lambda) + R(\lambda),$$

where $R(\lambda)$ is a polynomial of degree $n - 1$ or less, and n is the degree of the characteristic polynomial $P(\lambda)$ of the matrix A. Suppose that all the eigenvalues of A are different. Then A is diagonalisable. Putting $\lambda = \lambda_i$ in the formula results in the system

$$f(\lambda_i) = R(\lambda_i) = \sum_{k=0}^{n-1} \alpha_k \lambda_i^k, \quad i = 1, 2, \ldots, n.$$

By the Cayley-Hamilton Theorem $P(A) = 0$. This implies the formula for $f(A)$:

$$f(A) = R(A) = \sum_{k=0}^{n-1} \alpha_k A^k.$$

So, we first solve the system of linear equations and find α_k. Then we apply the above formula to find $f(A)$.

Problem 3.20 Let

$$A = \begin{pmatrix} 0 & 1 \\ -1 & 0 \end{pmatrix}.$$

Find e^{tA}.

Solution: We have

$$P_{\mathbf{A}}(\lambda) = \det \begin{pmatrix} -\lambda & 1 \\ -1 & -\lambda \end{pmatrix} = \lambda^2 + 1 \Rightarrow \lambda = \pm i.$$

$$\begin{aligned} e^{it} &= \cos t + i \sin t = \alpha_0 + \alpha_1 i \\ e^{-it} &= \cos t - +i \sin t = \alpha_0 - \alpha_1 i \end{aligned} \Rightarrow \begin{cases} \alpha_0 &= \cos t \\ \alpha_1 &= \sin t \end{cases} \Rightarrow$$

$$e^{\mathbf{A}t} = \cos t \mathbf{I} + \sin t \mathbf{A} = \begin{pmatrix} \cos t & \sin t \\ -\sin t & \cos t \end{pmatrix}. \quad \square$$

Theorem 3.40 *Any polynomial* $p(\lambda)$

$$p(\lambda) = (-1)^n \left(\lambda^n + a_{n-1}\lambda^{n-1} + \cdots + a_2\lambda^2 + a_1\lambda + a_0 \right),$$

is the characteristic polynomial of the matrix:

$$\mathbf{A} = \begin{pmatrix} 0 & 0 & 0 & \cdots & 0 & -a_0 \\ 1 & 0 & 0 & \cdots & 0 & -a_1 \\ 0 & 1 & 0 & \cdots & 0 & -a_2 \\ \vdots & \vdots & \vdots & \ddots & \vdots & \vdots \\ 0 & 0 & 0 & \cdots & 1 & -a_{n-1} \end{pmatrix}. \tag{3.73}$$

Proof. Let $\{\mathbf{e}_1, \mathbf{e}_2, \ldots, \mathbf{e}_n\}$ be the standard basis in $\mathbb{R}^n$. Then

$$\mathbf{A}\mathbf{e}_1 = \mathbf{e}_2, \ \mathbf{A}\mathbf{e}_2 = \mathbf{e}_3, \ \ldots \ \mathbf{A}\mathbf{e}_{n-1} = \mathbf{e}_n, \tag{3.74}$$

$$\mathbf{A}\mathbf{e}_n = -a_0\mathbf{e}_1 - a_2\mathbf{e}_2 - \ldots - a_{n-1}\mathbf{e}_{n-1}. \tag{3.75}$$

Substituting the formulas of (3.74) into (3.75), we obtain that

$$\mathbf{0} = a_0\mathbf{e}_1 + a_2\mathbf{e}_2 + \ldots + a_{n-1}\mathbf{e}_{n-1} + \mathbf{A}\mathbf{e}_n = \left(a_0 + a_1\mathbf{A} + \cdots + a_{n-1}\mathbf{A}^{n-1} + \mathbf{A}^n \right) \mathbf{e}_1 =$$

$$p(\mathbf{A})\mathbf{e}_1 \Rightarrow \mathbf{0} = A^k p(\mathbf{A})\mathbf{e}_1 = p(\mathbf{A})A^k \mathbf{e}_1 = p(\mathbf{A})e_{k+1}, \ \text{for} \ k = 0, \ldots, n - 1.$$

It follows that $p(\mathbf{A}) = \mathbf{0}$ as stated. $\quad \square$

A very special form of the matrix $\mathbf{A}$ in (3.73) shows that a characteristic polynomial cannot determine its matrix uniquely.

Problems

Prob. 109 — For any two 2×2 matrices $\mathbf{A}$, $\mathbf{B}$ with complex entries there is a complex number u such that

$$\det(\mathbf{A} + z\mathbf{B}) = \det \mathbf{A} + uz + \det \mathbf{B} \cdot z^2.$$

Prob. 110 — For any two 2×2 matrices $\mathbf{A}$, $\mathbf{B}$ with complex entries

$$\det(\mathbf{A} + \mathbf{B}) + \det(\mathbf{A} - \mathbf{B}) = 2 \det \mathbf{A} + 2 \det \mathbf{B}.$$

Prob. 111 — Let $\mathbf{A}$ be an invertible 2×2 matrix. Show that

$$\mathbf{A}^{-1} = \frac{1}{\det(\mathbf{A})} \left[\mathrm{tr}(\mathbf{A})\mathbf{I} - \mathbf{A} \right].$$

Prob. 112 — Show that the matrix

$$\mathbf{A} = \begin{pmatrix} 2 & 1 & 1 \\ 1 & 2 & 1 \\ 1 & 1 & 2 \end{pmatrix}$$

satisfies $\mathbf{A}^2 - 5\mathbf{A} + 4\mathbf{I} = \mathbf{0}$.

Prob. 113 — Using the Cayley-Hamilton Theorem find the inverse of the matrix

$$\mathbf{A} = \begin{pmatrix} 1 & 0 & 1 \\ 1 & 1 & 0 \\ 0 & 1 & 1 \end{pmatrix}.$$

3.13 Leontief's Input-Output Analysis

Wassily Leontief (1906-1999) graduated from Leningrad State University in 1925 and subsequently moved to Berlin to complete his Ph.D thesis in Economics. It is likely that Leontief received a solid education in Mathematics at Leningrad State University. In 1973 he was awarded the Nobel Prize in Economics for his input-output analysis of the US economy published in 1958.

The theory can be easily explained by 2×2 matrices.

Problem 3.21 A simple economy has only two sectors: the power industry and the coal industry. These sectors are interconnected as follows: producing 1 unit of power requires 0.4 units of power and 0.2 units of coal. Producing 1 unit of coal requires 0.5 unit of power and 0.3 unit of coal. If the surplus of power and coal required is

100 units of power and 100 units of coal, what would be the gross production of each industry?

Solution: The problem's data are conveniently organized into the Table 3.1. This

Inputs	Outputs Power	Outputs Coal
Power	0.4	0.5
Coal	0.2	0.3

Table 3.1: The Table of Problem 3.21.

table generates the **technology matrix**:

$$A = \begin{pmatrix} 0.4 & 0.5 \\ 0.2 & 0.3 \end{pmatrix}.$$

Then the first column Ae_1 of A specifies the amount of power and coal required to produce 1 unit of power. Similarly, the second column Ae_2 of A indicates the amount of power and coal utilized to produce one unit of coal. The **gross production matrix** for the economy is the column

$$X = \begin{pmatrix} x_1 \\ x_2 \end{pmatrix},$$

where x_1 is the gross production of power and x_2 is the gross production of coal. It follows that I_2X is the amount of the required production,

$$AX = x_1Ae_1 + x_2Ae_2 = \begin{pmatrix} 0.4x_1 + 0.5x_2 \\ 0.2x_1 + 0.3x_2 \end{pmatrix}$$

is the amount of production used for internal technological needs, and $(I_2 - A)X$ is the amount of surpluses, i.e. Y, which is also called **final demands**. The equation

$$(I_2 - A)X = Y$$

is called the **technology equation**. Then

$$X = (I_2 - A)^{-1}Y.$$

In our case $Y = \begin{pmatrix} 100 & 100 \end{pmatrix}^T$ implying that

$$X = \begin{pmatrix} 0.6 & -0.5 \\ -0.2 & 0.7 \end{pmatrix}^{-1} \begin{pmatrix} 100 \\ 100 \end{pmatrix} = \begin{pmatrix} 375 \\ 250 \end{pmatrix}. \quad \square$$

In general, we examine an economy – or even an enterprise – that encompasses n production processes, yielding commodities $C_1, C_2, \ldots, C_n$. Each production process requires inputs that may include the outputs from others within this system.

Additionally, there is external demand for each commodity. The challenge is to determine the level of production of these commodities that meets these demands.

> **Definition 3.15** Let
>
> $$a_{ij} \text{ be the amount of } C_i \text{ required to produce one unit of } C_j.$$
>
> Then $\mathbf{A} = (a_{ij})$ is called the **technology matrix**. An important matrix
>
> $$\mathbf{B} = (\mathbf{I}_n - \mathbf{A})$$
>
> is called the **Leontief matrix**.

It is not common to give a special name to the matrix $\mathbf{B}$, but this matrix has a noticeable role in calculations and it is useful to have a separate name for it.

We suppose that in a given time period

$$y_i = \text{ amount of } C_i \text{ required to satisfy the external demand;}$$
$$x_i = \text{ amount of } C_i \text{ required to satisfy all needs}$$

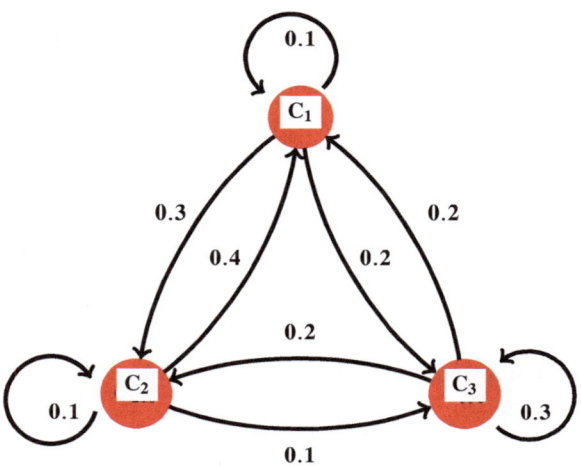

Fig. 3.7: The Graph for the technology matrix

Problem 3.22 The production process for three goods C_1, C_2, C_3 are interlinked. To produce one dollar's worth of C_1 requires inputs of $0.1 worth of C_1, $0.4 worth of C_2, and $0.2 worth of C_3. Producing one dollar's worth of C_2 requires inputs of $0.3 worth of C_1, $0.1 worth of C_2, and $0.2 worth of C_3. Producing one dollar's worth of C_3 requires inputs of $0.2 worth of C_1, $0.1 worth C_2, and $0.3 worth of C_3.

Assuming that, within a specific time frame, the external demands are y_1 dollars for C_1, y_2 dollars of C_2 and y_3 of C_3, what are the production levels x_1, x_2, x_3 of C_1, C_2, and C_3 respectively to meet all demands?

Solution: It is useful first to plot the graph of the problem shown in Figure 3.7. In contrast to the table, shown in Figure 3.1, the graph clearly indicates all inputs arrived to its vertices. For instance, the first column is made of inputs arrived to the vertex C_1. Therefore, the technology matrix $\mathbf{A}$ looks as follows:

$$\mathbf{A} = \begin{pmatrix} 0.1 & 0.3 & 0.2 \\ 0.4 & 0.1 & 0.1 \\ 0.2 & 0.2 & 0.3 \end{pmatrix} \Rightarrow \mathbf{B} = \mathbf{I}_3 - \mathbf{A} = \frac{1}{10}\begin{pmatrix} 9 & -3 & -2 \\ -4 & 9 & -1 \\ -2 & -2 & 7 \end{pmatrix} \Rightarrow \mathbf{B}^{-1} = \frac{10}{407}\begin{pmatrix} 61 & 25 & 21 \\ 30 & 59 & 17 \\ 26 & 24 & 69 \end{pmatrix}.$$

It follows that $\mathbf{X} = \mathbf{B}^{-1}\mathbf{Y}$.   □

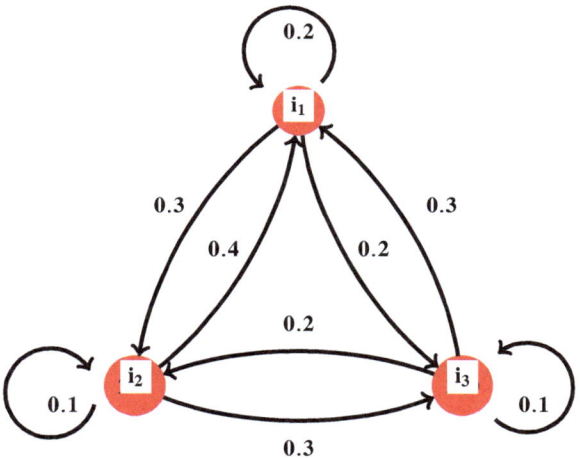

Fig. 3.8: The Graph for the technology matrix

The following problem illustrates applications of Leontief's Analysis to the economy of a city.

Problem 3.23 An utility economy can be split into three industries

$$i_1 : \text{ water } \quad i_2 : \text{ electricity } \quad i_3 : \text{ gas.}$$

These industries are interlinked as follows. To produce one dollar's worth of i_1 requires the input of $0.2 worth of i_1, of $0.4 worth of i_2 and of $0.3 worth of i_3. To produce one dollar's worth of i_2 requires the input of $0.3 worth of i_1, of $0.1

worth of i_2 and of $0.2 worth of i_3. To produce one dollar's worth of i_3 requires the input of $0.2 worth of i_1, $0.3 worth of i_2 and $0.1 worth of i_3.

Each week the external demands for water, electricity and gas are, respectively,

$$y_1 = \$40\,000 \quad y_2 = \$100\,000 \quad y_3 = \$72\,000.$$

(a) How much water, electricity and gas is needed to produce $1 worth of electricity?

(b) What should be the weekly production of each industry in order to satisfy all demands exactly?

(c) Suppose that the external demand on gas has reduced 10% whereas the external demand on water and electricity remain the same level. Find, how the internal productivity changed?

Solution: (a) The production of one dollar's worth of electricity (i_2) requires the input of $0.3 worth of water ($i_1$), of $0.1 worth of electricity (i_2) and of $0.2 worth of gas ($i_3$).

(b) The Graph of the problem is shown in Figure 3.8. It follows that the technology matrix is given by

$$\mathbf{A} = \begin{pmatrix} 0.2 & 0.3 & 0.2 \\ 0.4 & 0.1 & 0.3 \\ 0.3 & 0.2 & 0.1 \end{pmatrix} \Rightarrow \mathbf{B} = \mathbf{I}_3 - \mathbf{A} = \frac{1}{10}\begin{pmatrix} 8 & -3 & -2 \\ -4 & 9 & -3 \\ -3 & -2 & 9 \end{pmatrix} \Rightarrow \mathbf{B}^{-1} = \frac{2}{79}\begin{pmatrix} 75 & 31 & 27 \\ 45 & 66 & 32 \\ 35 & 25 & 60 \end{pmatrix}.$$

Therefore,

$$\mathbf{X} = \mathbf{B}^{-1}\mathbf{Y} = \frac{1}{79}\begin{pmatrix} 16088000 \\ 21408000 \\ 16440000 \end{pmatrix} \approx \begin{pmatrix} 203646. \\ 270987. \\ 208101. \end{pmatrix}.$$

(c) If the external demand on gas is reduced by 10%, then $y_3 = 0.9\times72000 = 64800$, implying that the gross product $\mathbf{Z}$ corresponding to this demand equals:

$$\mathbf{Z} = \mathbf{B}^{-1}\begin{pmatrix} y_1 \\ y_2 \\ 0.9y_3 \end{pmatrix} = \frac{2}{79}\begin{pmatrix} 75 & 31 & 27 \\ 45 & 66 & 32 \\ 35 & 25 & 60 \end{pmatrix}\begin{pmatrix} 40000 \\ 100000 \\ 64800 \end{pmatrix} = \frac{1}{79}\begin{pmatrix} 15699200 \\ 20947200 \\ 15576000 \end{pmatrix}.$$

In the first case the internal production is

$$\mathbf{AX} = \frac{1}{79}\begin{pmatrix} 0.2 & 0.3 & 0.2 \\ 0.4 & 0.1 & 0.3 \\ 0.3 & 0.2 & 0.1 \end{pmatrix}\begin{pmatrix} 16088000 \\ 21408000 \\ 16440000 \end{pmatrix} = \frac{1}{79}\begin{pmatrix} 12928000 \\ 13508000 \\ 10752000 \end{pmatrix}.$$

The internal production in the second case is

$$\mathbf{AZ} = \frac{1}{79}\begin{pmatrix} 0.2 & 0.3 & 0.2 \\ 0.4 & 0.1 & 0.3 \\ 0.3 & 0.2 & 0.1 \end{pmatrix}\begin{pmatrix} 15699200 \\ 20947200 \\ 15576000 \end{pmatrix} = \frac{1}{79}\begin{pmatrix} 12539200 \\ 13047200 \\ 10456800 \end{pmatrix}.$$

We see that

$$\frac{12539200}{12928000} \approx 0.969926, \quad \frac{13047200}{13508000} \approx 0.965887, \quad \frac{10456800}{10752000} \approx 0.972545.$$

Therefore, the internal productivity changed as shown in the table below.

	Water	Electricity	Gas
Drop Down	3%	3.4%	2.75%

□

Given an $n \times n$ technology matrix $\mathbf{A} = (a_{ij})$, the condition

$$a_{1j} + a_{2j} + \cdots + a_{nj} < 1 \tag{3.76}$$

expresses the fact that the total cost of producing a dollar's worth of the commodity C_j is less than one dollar. If it is required to make a profit in the production of each commodity, then this sum cannot exceed 1 in every column of the technology matrix $\mathbf{A}$.

Definition 3.16 If an $n \times n$ technology matrix $\mathbf{A}$ satisfies the condition

$$a_{1j} + a_{2j} + \cdots + a_{nj} < 1$$

for the commodity C_j, then $\mathbf{A}$ is called **productive at the commodity** C_j.
If

$$a_{1j} + a_{2j} + \cdots + a_{nj} = 1$$

for the commodity C_j, then $\mathbf{A}$ is called **balanced at the commodity** C_j.
Finally, if

$$a_{1j} + a_{2j} + \cdots + a_{nj} > 1$$

for the commodity C_j, then $\mathbf{A}$ is called **unproductive at the commodity** C_j.

Theorem 3.41 *Suppose that an $n \times n$ technology matrix $\mathbf{A}$ is productivity at each commodity. Then the Leontief matrix $\mathbf{B} = (\mathbf{I} - \mathbf{A})$ is invertible.*

Proof. By Theorem 3.4 the matrix $\mathbf{I} - \mathbf{A}$ is invertible if and only if its rank is n. By Corollary 2.7

$$r = \text{rank}(\mathbf{I} - \mathbf{A}) = \text{rank}(\mathbf{I} - \mathbf{A}^T).$$

If $r < n$ then the homogeneous system $(\mathbf{I} - \mathbf{A}^T)\mathbf{x} = \mathbf{0}$ has a non-zero solution $\mathbf{x}$:

$$\mathbf{x} = (x_1, x_2, \ldots, x_n),\; |x_k| = \max_{1 \le i \le n} |x_i| > 0.$$

By (3.76) we have

$$x_k = \sum_{i=1}^{n}(\mathbf{A}^T)_{ki}x_i = \sum_{i=1}^{n} a_{ik}x_i \Rightarrow |x_k| \le \sum_{i=1}^{n} a_{ik}|x_i| < |x_k|.$$

Since $|x_k| > 0$, we obtain a contradiction implying that $r = n$. $\square$

There is a simple formula for the inverse of the Leontief matrix in case the technology matrix is productive at each commodity.

Theorem 3.42 *Suppose that a technology matrix* $\mathbf{A}$ *is productive at each commodity. Then the matrix series*

$$\mathbf{B}^{-1} = (\mathbf{I} - \mathbf{A})^{-1} = \mathbf{I} + \mathbf{A} + \mathbf{A}^2 + \cdots + \mathbf{A}^k + \cdots \qquad (3.77)$$

converges absolutely at each entry to the inverse of the Leontief matrix .

Corollary 3.15 *Suppose that a technology matrix* $\mathbf{A}$ *is productive at each commodity. Then its Leontief matrix* $\mathbf{B}$ *is invertible and all entries of* $\mathbf{B}^{-1}$ *are non-negative. In particular, the technology problem*

$$(\mathbf{I}_n - \mathbf{A})\mathbf{X} = \mathbf{Y}$$

has a unique solution for any demand set of non-negative data

$$Y = \left(y_1\; y_2\; \cdots\; y_n\right)^T.$$

Lemma 3.12 *Let* $\mathbf{C}$ *and* $\mathbf{D}$ *be two technology matrices satisfying*

$$\sum_{i=1}^{n} c_{ij} < q < 1,\; \sum_{i=1}^{n} d_{ij} < q < 1,\; j = 1, 2, \ldots, n. \qquad (3.78)$$

Then

$$\sum_{i=1}^{n}(CD)_{ij} < q^2,\; j = 1, 2, \ldots, n. \qquad (3.79)$$

Proof. By the direct computation,

$$\sum_{i=1}^{n}(CD)_{ij} = \sum_{i=1}^{n}\sum_{k=1}^{n} c_{ik}d_{kj} = \sum_{k=1}^{n} d_{kj}\sum_{i=1}^{n} c_{ik} < \sum_{k=1}^{n} d_{kj}q < q^2.$$

□

Proof of Theorem 3.42. If a technology matrix $\mathbf{A}$ is productive at each commodity, then there is $0 < q < 1$ such that (3.78) holds for $\mathbf{C} = \mathbf{A}$. Applying Lemma 3.12 consequently to the powers of $\mathbf{A}$, we see that for every j

$$\sum_{i=1}^{n}(A^k)_{ij} < q^k \Rightarrow \max_{ij}(A^k)_{ij} < q^k.$$

By the comparison test, the positive series $\sum_{k=0}^{\infty}(A^k)_{ij}$ converge at any matrix entry $\{ij\}$. Since

$$(\mathbf{I} - \mathbf{A})\sum_{k=0}^{n} A^k = \mathbf{I} - \mathbf{A}^{n+1},$$

we conclude that the series in (3.77) converges to $(\mathbf{I} - \mathbf{A})^{-1}$ as stated in Theorem 3.42.

□

The following 3×3 technology matrix

$$\mathbf{A} = \begin{pmatrix} 0.2 & 0.2 & 0.1 \\ 0.2 & 0.2 & 0 \\ 0.8 & 0.8 & 0.2 \end{pmatrix}. \tag{3.80}$$

is productive only at the third commodity. However, the Leontief matrix is invertible

$$\mathbf{B}^{-1} = (\mathbf{I} - \mathbf{A})^{-1} = \frac{1}{20}\begin{pmatrix} 28 & 12 & 4 \\ 8 & 28 & 1 \\ 40 & 40 & 30 \end{pmatrix}$$

and all its entries are positive, implying that the economy described by $\mathbf{A}$ is **productive** as a whole.

Theorem 3.43 *Suppose that a matrix* $\mathbf{A} = (a_{ij})$ *with nonnegative entries* $a_{ij} \geq 0$ *satisfies the conditions*

$$\sum_{i=1}^{n} a_{ij} > 1$$

for every $j = 1, \ldots, n$. *Then the economy described by the technology matrix* $\mathbf{A}$ *is not productive.*

Proof. Suppose that for some vectors $\mathbf{y}$ and $\mathbf{x}$ with nonnegative coordinates such that at least one coordinate of $\mathbf{y}$ is positive, we have the equation of the productivity balance $\mathbf{y} = \mathbf{x} - \mathbf{A}\mathbf{x}$. This equation in the coordinates becomes a linear system of equations:

$$y_1 = x_1 - a_{11}x_1 - a_{12}x_2 - \cdots - a_{1n}x_n$$
$$y_2 = x_2 - a_{21}x_1 - a_{22}x_2 - \cdots - a_{2n}x_n$$
$$\vdots \quad \vdots \quad \vdots \quad \vdots \quad \vdots \quad \vdots \quad \vdots$$
$$y_n = x_n - a_{n1}x_1 - a_{n2}x_2 - \cdots - a_{nn}x_n$$

We sum up the above equations to obtain the identity:

$$\sum_{i=1}^{n} y_i = x_1 \left(1 - \sum_{i=1}^{n} a_{i1}\right) + x_2 \left(1 - \sum_{i=1}^{n} a_{i2}\right) + \cdots + x_n \left(1 - \sum_{i=1}^{n} a_{in}\right),$$

which shows that its left-hand side is positive, whereas the right-hand side cannot be positive. This contradiction shows that there is no production vectors, which implies that the matrix $\mathbf{A}$ is unproductive. $\square$

To clarify the mystery happened with the matrix $\mathbf{A}$ in (3.80), let us factor the Leontief matrix $\mathbf{B} = \mathbf{I}_3 - \mathbf{A}$ into a product of the following matrices:

$$\mathbf{B} = \begin{pmatrix} 0.1 & 0 & 0 \\ 0 & 0.1 & 0 \\ 0 & 0 & 0.4 \end{pmatrix} \begin{pmatrix} 1 & -0.25 & -0.5 \\ -0.25 & 1 & 0 \\ -0.25 & -0.25 & 1 \end{pmatrix} \begin{pmatrix} 8 & 0 & 0 \\ 0 & 8 & 0 \\ 0 & 0 & 2 \end{pmatrix}. \tag{3.81}$$

The matrix present at the middle of formula (3.81) has the form $\mathbf{I}_3 - \mathbf{A}_0$, where $\mathbf{A}_0$ is a productive technology matrix at each commodity. Two other factors are diagonal matrices with positive entries on the diagonal. By Theorem 3.77 the matrix $(\mathbf{I}_3 - \mathbf{A}_0)^{-1}$ exists and all its entries are non-negative. It follows that $(\mathbf{I}_3 - \mathbf{A})^{-1}$ exists and has also non-negative entries.

The problem of description of productive technology matrices is solved in terms of Leontief matrices $\mathbf{B} = \mathbf{I} - \mathbf{A}$ having the so-called dominant diagonal.

Definition 3.17 An $n \times n$ matrix $\mathbf{B} = (b_{ij})$ is said to have a **dominant diagonal** if there exist **positive** numbers $d_1, d_2, \ldots, d_n$ such that

$$\sum_{i \neq j} d_i |b_{ij}| < d_j |b_{jj}|, \text{ for } j = 1, 2, \ldots, n.$$

If $\mathbf{A}$ is productive at each commodity, then $d_1 = d_2 = \cdots = d_n = 1$, since by (3.76)

$$\sum_{i \neq j} |-a_{ij}| < 1 - a_{jj}, \ j = 1, 2, \ldots, n.$$

Definition 3.18 A vector $\mathbf{x}$ is called nonnegative (positive) if all its coordinates are nonnegative (positive).

The following simple lemma is crucial in what follows.

Lemma 3.13 *Let $\mathbf{B} = (b_{ij})$ be an $n \times n$ matrix with $b_{ii} > 0$ for every i and $b_{ij} \leq 0$ for $i \neq j$. If the system $\mathbf{Bx} = \mathbf{y}$ has a non-negative solution $\mathbf{x}$ for some positive vector $\mathbf{y}$, then the matrix $\mathbf{B}^T$ has a dominant diagonal.*

Proof. If $\mathbf{Bx} = \mathbf{y}$ for positive $\mathbf{y}$ and non-negative $\mathbf{x}$, then

$$\sum_{j=1}^{n} b_{ij} x_j = y_i > 0. \tag{3.82}$$

Since $b_{ij} \leq 0$ for $i \neq j$ and $x_j \geq 0$ for every j, we see that $x_i > 0$ and $b_{ii} > 0$ for every i. By (3.82)

$$\sum_{j \neq i}^{n} (-b_{ij}) x_j < b_{ii} x_i,$$

implying that $d_1 = x_1, d_1 = x_2, \ldots, d_n = x_n$ is a dominant diagonal for $\mathbf{B}^T$. $\square$

Now we prove a general result.

Theorem 3.44 *Let* $\mathbf{B} = (b_{ij})$ *be an* $n \times n$ *matrix with* $b_{ii} > 0$ *for every* i *and* $b_{ij} \leq 0$ *for* $i \neq j$. *Then for every nonnegative* $\mathbf{y}$ *the equation* $\mathbf{Bx} = \mathbf{y}$ *has a unique nonnegative solution* $\mathbf{x}$ *if and only if* $\mathbf{B}$ *has a dominant diagonal.*

Proof. Suppose that the matrix $\mathbf{B}$ has a dominant diagonal $d_1, \ldots, d_n$. Then

$$
\mathbf{B} = \begin{pmatrix} \frac{1}{d_1} & 0 & \cdots & 0 \\ 0 & \frac{1}{d_2} & \cdots & 0 \\ \vdots & \vdots & \ddots & \vdots \\ 0 & 0 & \cdots & \frac{1}{d_n} \end{pmatrix} \underbrace{\begin{pmatrix} 1 & \frac{d_1 b_{12}}{d_2 b_{22}} & \cdots & \frac{d_1 b_{1n}}{d_n b_{nn}} \\ \frac{d_2 b_{21}}{d_1 b_{11}} & 1 & \cdots & \frac{d_2 b_{2n}}{d_n b_{nn}} \\ \vdots & \vdots & \ddots & \vdots \\ \frac{d_n b_{n1}}{d_1 b_{11}} & \frac{d_n b_{n2}}{d_2 b_{22}} & \cdots & 1 \end{pmatrix}}_{\mathbf{E}} \begin{pmatrix} d_1 b_{11} & 0 & \cdots & 0 \\ 0 & d_2 b_{22} & \cdots & 0 \\ \vdots & \vdots & \ddots & \vdots \\ 0 & 0 & \cdots & d_n b_{nn} \end{pmatrix}.
$$

The matrix $\mathbf{E} = \mathbf{I} - \mathbf{A}_0$, where $\mathbf{A}_0$ is a technology matrix productive at each its commodity by the dominant diagonal property. Passing to the inverse matrices, we complete the proof by Theorem 3.42.

Suppose that the system $\mathbf{Bx} = \mathbf{y}$ has a unique non-negative solution $\mathbf{x}$ for any non-negative $\mathbf{y}$. Now, let us take $\mathbf{y}$ with all coordinates positive. Then by Lemma 3.13 the matrix $\mathbf{B}^T$ has a dominant diagonal. By the first part of the proof $\mathbf{B}^T \mathbf{u} = \mathbf{v}$ has a unique non-negative solution $\mathbf{u}$, $u_i \geq 0$ for any nonnegative $\mathbf{v}$. Let us take any $\mathbf{v}$ with positive coordinates. Then as above $\mathbf{B}^{TT} = \mathbf{B}$ has a dominant diagonal. $\qquad\square$

Theorem 3.45 (A description of productive technology matrices) *Let* $\mathbf{A}$ *be an* $n \times n$ *technology matrix. Then the following conditions are equivalent.*

(I) There exists an nonnegative vector $\mathbf{x}$ *such that* $(\mathbf{I} - \mathbf{A})\mathbf{x}$ *is positive.*
(II) For any nonnegative $\mathbf{y}$ *there exists a nonnegative* $\mathbf{x}$ *such that* $(\mathbf{I} - \mathbf{A})\mathbf{x} = \mathbf{y}$.
(III) The matrix $\mathbf{I} - \mathbf{A}$ *is nonsingular and all entries of* $(\mathbf{I} - \mathbf{A})^{-1}$ *are nonnegative.*

Proof. $(\mathbf{I}) \Rightarrow (\mathbf{III})$ Let $\mathbf{B} = \mathbf{I} - \mathbf{A}$. Since $b_{ij} \leq 0$ for $j \neq j$, the condition that $\mathbf{Bx}$ is positive implies that

$$
b_{ii} x_i > \sum_{j \neq i} (-b_{ij}) x_j \geq 0
$$

for every i. It follows that all b_{ii} and x_i are positive. Hence $\mathbf{B}^T$ has a dominant diagonal property with respect to this $\mathbf{x}$ implying that $\mathbf{B}^T$ is nonsingular. By Theorem 3.44, the vector $\left(\mathbf{B}^T\right)^{-1}\mathbf{y}$ is nonnegative for any nonnegative vector $\mathbf{y}$. Let $\mathbf{y} = \mathbf{e}_i$, where $\mathbf{e}_i$ is a vector of the standard basis in $\mathbb{R}^n$. Then vector $\mathbf{col}_i\left(\left(\mathbf{B}^T\right)^{-1}\right)$ is non-negative. Since $\left(\mathbf{B}^T\right)^{-1} = \left(\mathbf{B}^{-1}\right)^T$, we see that all rows $\mathbf{row}_i(\mathbf{B}^{-1})$ are nonnegative, which completes the proof of this part. The implications $\mathbf{(III)} \Rightarrow \mathbf{(II)}$ and $\mathbf{(II)} \Rightarrow \mathbf{(I)}$ are obvious. $\qquad\square$

One may think that condition $\mathbf{III}$ says that all $\mathbf{I}-\mathbf{A}$ can be obtained as the inverses to matrices with nonnegative entries. A simple calculation shows that

$$\begin{pmatrix} 1 & 2 & 3 & 4 \\ 4 & 3 & 2 & 1 \\ 2 & 4 & 1 & 3 \\ 2 & 3 & 4 & 1 \end{pmatrix}^{-1} = \begin{pmatrix} 0.1 & 0.45 & -0.2 & -0.25 \\ -0.3 & -0.25 & 0.4 & 0.25 \\ 0.1 & -0.05 & -0.2 & 0.25 \\ 0.3 & 0.05 & 0 & -0.25 \end{pmatrix}.$$

The reason for this is that the Leontief matrix $\mathbf{B} = \mathbf{I} - \mathbf{A}$ a'priory must satisfy the conditions $b_{ij} \leq 0$ for $i \neq j$. So, the question arises how to describe the matrices $\mathbf{I} - \mathbf{A}$ corresponding to productive technology matrices in terms of their entries? The answer is given by the **Hawkins-Simon Theorem**.

Theorem 3.46 *Let $\mathbf{B} = (b_{ij})$ be an $n \times n$ matrix such that $b_{ij} \leq 0$ for $i \neq j$. Then the following conditions are equivalent.*

(I) There exists an nonnegative vector $\mathbf{x}$ such that $\mathbf{Bx}$ is positive.
*(IV) All the successive **principal minors** of $\mathbf{B}$ are positive:*

$$b_{11} > 0, \quad \begin{vmatrix} b_{11} & b_{12} \\ b_{21} & b_{22} \end{vmatrix} > 0, \ldots, \quad \begin{vmatrix} b_{11} & b_{12} & \cdots & b_{1n} \\ b_{21} & b_{22} & \cdots & b_{2n} \\ \vdots & \vdots & \ddots & \vdots \\ b_{n1} & b_{n2} & \cdots & b_{nn} \end{vmatrix} > 0.$$

Proof of Theorem 3.46. $\mathbf{(I)} \Rightarrow \mathbf{(IV)}$ Condition $\mathbf{(I)}$ means that there exist $x_j \geq 0$ such that

$$\sum_{j=1}^{n} b_{ij}x_j > 0 \Leftrightarrow b_{ii}x_i > \sum_{j=1, j \neq i,}^{n} (-b_{ij})x_j \geq 0. \tag{3.83}$$

Since $b_{ij} \leq 0$ for $i \neq j$ we conclude that for every positive integer k the following conditions hold:

$$b_{ii}x_i > \sum_{j=1, j \neq i}^{k} (-b_{ij})x_j \geq 0, \; i = 1,\ldots, k \Leftrightarrow \sum_{j=1}^{k} b_{ij}x_j > 0, i = 1, \ldots k.$$

It follows that $b_{11} > 0$ and that every sub-matrix

$$\mathbf{B}_k = \begin{pmatrix} b_{11} & b_{12} & \cdots & b_{1k} \\ b_{21} & b_{22} & \cdots & b_{2k} \\ \vdots & \vdots & \ddots & \vdots \\ b_{k1} & b_{k2} & \cdots & b_{kk} \end{pmatrix}$$

of the matrix $\mathbf{B} = \mathbf{B}_n$ satisfies **(I)**. By Lemma 3.13 every matrix $\mathbf{B}_k^T$ has a dominant diagonal. By Theorem 3.44 for every non-negative $\mathbf{y}$ in $\mathbb{R}^k$ the equation $\mathbf{B}_k^T \mathbf{x} = \mathbf{y}$ has a unique non-negative solution $\mathbf{x} \in \mathbb{R}^k$. Since the solution $\mathbf{x}$ is unique, we see that nullity$(\mathbf{B}_k^T) = 0$. By Theorem 3.4 every matrix $\mathbf{B}_k^T$ is nonsingular. By Theorem 3.5 the matrix $\mathbf{B}_k$ is also non-singular.

If formulas (3.83) hold for some non-negative x, then for every t, $0 < t \le 1$ they hold for tx, implying that they also hold for the matrix $\mathbf{C}$ with $c_{ij} = tb_{ij}$ if $i \ne j$ and $c_{ii} = b_{ii}$ for every i and for the vector $\mathbf{x}$. It follows that all matrices $\mathbf{C}_k$ are non-singular, implying that $\det(\mathbf{C}_k(t)) \ne 0$ for any t. Since $\mathbf{C}_k(0)$ is a diagonal matrix with positive entries b_{ii} on the main diagonal, we conclude that $\det(\mathbf{C}_k)(0) > 0$. Since

$$\lim_{t \to 1^-} \det(\mathbf{C}_k)(t) = \det(\mathbf{C}_k(1)) = \det(\mathbf{B}_k) \ne 0,$$

and $\det(\mathbf{C}_k)(t) \ne 0$, we conclude that $\det(\mathbf{B}_k) > 0$. This proves **IV**.

(IV) $\Rightarrow$ **(II)**, were **II** is the condition in the statement of Theorem 3.45. We prove this by mathematical induction. For $n = 1$ we have $b_{11}x_1 = y_1$. By **IV** we see that $b_{11} > 0$. It follows that $x_1 \ge 0$ as soon as $y_1 \ge 0$. Suppose that **(IV)** $\Rightarrow$ **(II)** holds for $n - 1$ and prove that it holds for n. Consider the system

$$\sum_{j=1}^n b_{ij}x_j = y_i, \ i = 1, \dots n$$

and let us show that for a non-negative vector $\mathbf{y}$ the solution $\mathbf{x}$, which exists since $\det(\mathbf{B}_n) > 0$, is also non-negative.

Since $b_{11} > 0$ we can apply row operations to subtract multiples of the first row from other rows:

$$\begin{pmatrix} b_{11} & b_{12} & \cdots & b_{1n} \\ b_{21} & b_{22} & \cdots & b_{2n} \\ \vdots & \vdots & \ddots & \vdots \\ b_{n1} & b_{n2} & \cdots & b_{nn} \end{pmatrix} \sim \begin{pmatrix} b_{11} & b'_{12} & \cdots & b'_{1n} \\ 0 & b'_{22} & \cdots & b'_{2n} \\ \vdots & \vdots & \ddots & \vdots \\ 0 & b'_{n2} & \cdots & b'_{nn} \end{pmatrix}.$$

Since $\det(\mathbf{B}_k) > 0$ for $k = 1, \dots, n$, we obtain

$$\det \begin{pmatrix} b'_{22} & \cdots & b'_{2n} \\ \vdots & \ddots & \vdots \\ b'_{n2} & \cdots & b'_{nn} \end{pmatrix} = \frac{\det(\mathbf{B}_k)}{b_{11}} > 0, k = 2, \dots, n.$$

The row operations performed imply the formulas:

$$b'_{ij} = b_{ij} - b_{i1}\frac{b_{1j}}{b_{11}}, \quad i,j = 2,\ldots,n. \tag{3.84}$$

Since $i,j \geq 2$, we have $b_{i1} \leq 0$, $b_{1,j} \leq 0$. Hence the product $b_{i1}b_{1,j}$ is nonnegative. This implies that $b'_{ij} \leq 0$ for $i \neq j$, $i,j \geq 2$. Then by the induction hypothesis,

$$\sum_{j=2}^{n} b'_{ij}x_j = y'_i \tag{3.85}$$

has a nonnegative solution $(x_2,\ldots,x_n)$ for any nonnegative vector $(y'_2,\ldots,y'_n)$.

Let $(y_1,y_2,\ldots,y_n)$ be an arbitrary nonnegative vector. Define y'_i by the Gauss elimination formula for augmented matrices:

$$y'_i = y_i - y_1\frac{b_{i1}}{b_{11}} \geq 0 \text{ for } i = 2,3,\ldots,n, \tag{3.86}$$

since $b_{i1} \leq 0$ for $i \neq 1$. Substituting (3.84) and (3.86) into (3.85), we obtain:

$$y_i - y_1\frac{b_{i1}}{b_{11}} = \sum_{j=2}^{n}\left(b_{ij} - b_{i1}\frac{b_{1j}}{b_{11}}\right)x_j = \sum_{j=2}^{n}b_{ij}x_j - \frac{b_{i1}}{b_{11}}\sum_{j=2}^{n}b_{1j}x_j. \tag{3.87}$$

Let

$$x_1 = \frac{y_1 - \sum_{j=2}^{n}b_{1j}x_j}{b_{11}} \geq 0 \text{ since } b_{1j} \leq 0 \text{ for } j = 2,\ldots,n.$$

Then

$$\sum_{j=1}^{n}b_{1j}x_j = y_1,$$

$$\sum_{j=1}^{n}b_{ij}x_j = b_{i1}x_1 + \sum_{j=2}^{n}b_{ij}x_j = y_1\frac{b_{i1}}{b_{11}} - \frac{b_{i1}}{b_{11}}\sum_{j=2}^{n}b_{1j}x_j + \sum_{j=2}^{n}b_{ij}x_j = y_i$$

for $i = 2,\ldots,n$ by (3.87). $\qquad\square$

Problem 3.24 Find all values of x, $x \geq 0$ such that the technology matrix

$$A = \begin{pmatrix} 0.2 & 0.2 & 0.1 \\ 0.2 & 0.2 & x \\ 0.8 & 0.8 & 0.2 \end{pmatrix} \tag{3.88}$$

is productive.

Solution: Let us construct the Leontief matrix

$$\mathbf{B} = \mathbf{I} - \mathbf{A} = \begin{pmatrix} 0.8 & -0.2 & -0.1 \\ -0.2 & 0.8 & -x \\ -0.8 & -0.8 & 0.8 \end{pmatrix}$$

We apply the Hawkins-Simon Theorem. Elementary calculations show that

$$\Delta_1 = 0.8, \ \Delta_2 = 0.8^2 - (-0.2)^2 = 0.6 > 0, \ \Delta_3 = -0.8(x - 0.5) > 0 \Leftrightarrow 0 \le x < 0.5.$$

It follows that the technolgy matrix $\mathbf{A}$ is productive until the cost of production of $1 worth of the third commodity will be less than $0.8. □

Leontief used input-output analysis to study the 1958 U.S. economy. He divided the economy into 81 sectors and aggregated these sectors into six groups:

Sector		Examples
FN	Final nonmetal	Leather goods, furniture, foods
FM	Final metal	Construction mach'ry, household appliances
BM	Basic metal	Mining, machine shop products
BN	Basic nonmetal	Glass, wood, textile, livestock products
E	Energy	Coal, petroleum, electricity, gas
S	Services	Gov. services, transportation, real estate

	FN	FM	BM	BN	E	S
FN	0.170	0.004	0.000	0.029	0.000	0.008
FM	0.003	0.295	0.018	0.002	0.004	0.016
BM	0.025	0.173	0.460	0.007	0.011	0.007
BN	0.348	0.037	0.021	0.403	0.011	0.048
E	0.007	0.001	0.039	0.025	0.358	0.025
S	0.120	0.074	0.104	0.123	0.173	0.234

The units are millions of dollars. So the 0.173 in row 3 column 2 means that the production of $1 million worth of the final metal products requires the expenditure of $173 000 on basic metal.

External demands $\mathbf{d}$ for 1958 Economy in millions of dollars:

FN	99 640
FM	75 548
BM	14 444
BN	33 501
E	23 527
S	263 985

$$\mathbf{I} - \mathbf{A} = \begin{pmatrix} 0.830 & -0.004 & 0.000 & -0.029 & 0.000 & -0.008 \\ -0.003 & 0.705 & -0.018 & -0.002 & -0.004 & -0.016 \\ -0.025 & -0.173 & 0.540 & -0.007 & -0.011 & -0.007 \\ -0.348 & -0.037 & -0.021 & 0.597 & -0.011 & -0.048 \\ -0.007 & -0.001 & -0.039 & -0.025 & 0.642 & -0.025 \\ -0.120 & -0.074 & -0.104 & -0.123 & -0.173 & 0.766 \end{pmatrix} \Rightarrow$$

$$(\mathbf{I} - \mathbf{A})^{-1} = \begin{pmatrix} 1.234 & 0.014 & 0.006 & 0.064 & 0.007 & 0.018 \\ 0.017 & 1.436 & 0.057 & 0.012 & 0.020 & 0.032 \\ 0.071 & 0.465 & 1.877 & 0.019 & 0.045 & 0.031 \\ 0.751 & 0.134 & 0.100 & 1.740 & 0.066 & 0.124 \\ 0.060 & 0.045 & 0.130 & 0.082 & 1.578 & 0.059 \\ 0.339 & 0.236 & 0.307 & 0.312 & 0.376 & 1.349 \end{pmatrix} \Rightarrow$$

$$\begin{pmatrix} 1.234 & 0.014 & 0.006 & 0.064 & 0.007 & 0.018 \\ 0.017 & 1.436 & 0.057 & 0.012 & 0.020 & 0.032 \\ 0.071 & 0.465 & 1.877 & 0.019 & 0.045 & 0.031 \\ 0.751 & 0.134 & 0.100 & 1.740 & 0.066 & 0.124 \\ 0.060 & 0.045 & 0.130 & 0.082 & 1.578 & 0.059 \\ 0.339 & 0.236 & 0.307 & 0.312 & 0.376 & 1.349 \end{pmatrix} \begin{pmatrix} 99\,640 \\ 75\,548 \\ 14\,444 \\ 33\,501 \\ 23\,527 \\ 263\,985 \end{pmatrix} = \begin{pmatrix} 131\,161 \\ 120\,324 \\ 79\,194 \\ 178\,936 \\ 66\,703 \\ 426\,542 \end{pmatrix}.$$

It follows that it requires $131\,161$ million worth of final nonmetal products to meet both intermediate and final demands in the 1958 U.S. economy.

Problems

Prob. 114 — The city administration operates three utilities providing gas, electricity, and water to the population. The utilities' operational dependencies for supplying $1 worth of each service are as follows.
To supply $1 of gas, the gas utility purchases $0.25 worth of electricity and $0.25 worth of water. To supply $1 of electricity, the electricity utility consumes $0.65 worth of gas, $0.05 worth of its own electricity, and $0.05 worth of water. To supply $1 of water, the water utility purchases $0.55 worth of gas and $0.10 worth of electricity. In one week, the city receives orders for $60,000 worth of gas, $30,000 worth of electricity, and $10,000 worth of water to meet the needs of its residents

(a) Draw the graph of the problem.
(b) Construct the technology matrix.
(c) Construct the Leontief matrix for this problem. Check whether it satisfies the conditions of the Hawkins-Simon Theorem.
(d) How much must each of the three industries produce in that week to exactly satisfy their own demand and the outside demand?

Prob. 115 — An economy has three product-producing sectors: steel, agriculture, and electricity. To produce $1 of steel, this sector consumes $0.50 of its own products, $0.20 of agriculture and $0.10 of electricity. To produce $1 of agriculture, the agriculture sector requires $0.10 of steel, $0.50 of agriculture, and $0.30 of electricity. To provide $1 of electricity, this sector requires $0.10 of steel, $0.30 of agriculture, and $0.40 of electricity. Suppose that the open sector has a demand for $9,700 worth of steel, $5\,250 worth of agricultural products, and $1\,125 worth of electricity. Can the economy meet this demand? If so, find the production $\mathbf{x}$ that will meet it exactly.

(a) Draw the graph of the problem.

(b) Construct the technology matrix.
(c) Construct the Leontief matrix for this problem. Check whether this economy satisfies the conditions of the Hawkins-Simon Theorem.
(d) How much must each of the three industries produce to exactly satisfy their own demand and the outside demand?

Prob. 116 — An economy has three sectors: construction, food, and transport. To produce $1 of construction, the construction sector makes use of $0.10 of its own products, $0.30 of food and $0.40 of transportation. To produce $1 of food, the food sector requires $0.60 of construction, $0.20 of food, and $0.10 of transportation. To provide $1 of transportation, this sector requires $0.40 of construction, $0.30 of food, and $0.20 of transportation. Suppose that the open sector has a demand for $10 230 worth of construction, $4 250 worth of food, and $6 400 worth of trans- portatrion. Can the economy meet this demand? If so, find the production vector **x** that will meet it exactly.

(a) Draw the graph of the problem.
(b) Construct the technology matrix.
(c) Construct the Leontief matrix for this problem. Check whether it satisfies the conditions of the Hawkins-Simon Theorem.
(d) How much must each of the three industries produce to exactly satisfy their own demand and the outside demand?

Prob. 117 — Consider an open economy with the technology matrix

$$\mathbf{A} = \begin{pmatrix} a_{11} & a_{12} \\ a_{21} & 0 \end{pmatrix}.$$

Show by a direct computation that the Leontief equation $\mathbf{x} - \mathbf{Ax} = \mathbf{d}$ has a unique positive solution for every demand vector $\mathbf{d}$ if $a_{21}a_{12} < 1 - a_{11}$.

Prob. 118 — An economy has three sectors: agriculture, manufacturing, and en- ergy. To produce $1 of agriculture, the sector makes use of $0.50 of its own prod- ucts, $0.5 of manufacturing and $0.5 of energy. To produce $1 of manufacturing, the sector requires $0.25 of agriculture, $0.125 of manufacturing, and $0.25 of energy. To provide $1 of energy, the energy sector requires $0.25 of agriculture, $0.25 of manufacturing, and $0.125 of energy.

(a) Draw the graph of the problem.
(b) Construct the technology matrix.
(c) Construct the Leontief matrix for this problem. Check whether this economy satisfies the conditions of the Hawkins-Simon Theorem.
(d) Evaluate the inverse to the Leontief matrix. If the open sector demands the same dollar value from each product-producing sector, which such sector must produce the greatest dollar value to meet the demand? Is the economy productive?

Prob. 119 — The economy of a country has three sectors: agriculture, steel, and oil. An output of 1 ton of agricultural products requires inputs of 0.1 ton of agricultural products, 0.02 ton of steel, and 0.05 ton of oil. An output of 1 ton of steel requires inputs of 0.01 ton of agricultural products, 0.13 ton of steel, and 0.18 ton of oil. An output of 1 ton of oil requires inputs of 0.01 ton of agricultural products, 0.20 ton of steel, and 0.05 ton of oil.

(**a**) Write the technology table and matrix for this economy.
(**b**) Construct the Leontief matrix B for this problem and write it as $B = 0.01 \cdot D$, where all entries of D are integers.
(**c**) Using the cofactor method evaluate D^{-1} and show that all entries of B^{-1} are positive. Check whether the economy is productive.
(**d**) Production of which commodity is least dependent on the other two?
(**e**) If oil costs rise, which industry will be most affected?

Prob. 120 — An economy has three sectors: agricultural products, manufactured goods, and fuels. Suppose further that production of 10 units of agricultural products requires 5 units of agricultural products, 2 units of manufactured goods, and 1 unit of fuels; that production of 10 units of manufactured goods requires 1 unit of agricultural products, 5 units of manufactured products, and 3 units of fuels; and that production of 10 units of fuels requires 1 unit of agricultural products, 3 units of manufactured goods, and 4 units of fuels.

(**a**) Draw the graph of the problem.
(**b**) Construct the technology table and matrix.
(**c**) Construct the Leontief matrix B for this problem and write it as $B = 0.1 \cdot D$, where all entries of D are integers.
(**d**) Using the cofactor method evaluate D^{-1} and show that all entries of B^{-1} are positive. Check whether the economy is productive.
(**e**) How many units of agricultural products and of fuels are required to produce 100 units of manufactured goods?
(**f**) Production of which commodity is least dependent on the other two?
(**g**) If fuel costs rise, which two industries will be most affected?

Prob. 121 — The national economy of some country has four sectors: agricultural products, machinery, fuel, and steel. Producing 1 unit of agricultural products requires 0.2 unit of agricultural products, 0.3 unit of machinery, 0.2 unit of fuel, and 0.1 unit of steel. Producing 1 unit of machinery requires 0.1 unit of agricultural products, 0.2 unit of machinery, 0.2 unit of fuel, and 0.4 unit of steel. Producing 1 unit of fuel requires 0.1 unit of agricultural products, 0.2 unit of machinery, 0.3 unit of fuel, and 0.2 unit of steel. Producing 1 unit of steel requires 0.1 unit of agricultural products, 0.2 unit of machinery, 0.3 unit of fuel, and 0.2 unit of steel.

(**a**) Construct the technology table and matrix.

(b) Construct the Leontief matrix for this problem. Using Leontief's theory and the Hawkins-Simon Theorem show that all principal minors of Leontief's matrix are positive.

(c) Using Gauss elimination determine how many units of each product will give surpluses of 1600 units of agriculture products, 1200 units of machinery, 800 units of fuel, and 500 units of steel.

Prob. 122 — The economy of a country is divided into four sectors. The production of 1 unit of output of each sector requires the following:

(1) Gas requires 0.1 units of transportation, 0.5 units of chemicals, and 0.2 units of gas.

(2) Agriculture requires 0.3 units of gas, 0.2 units of agriculture, 0.15 units of transportation, and 0.2 units of chemicals.

(3) Transportation requires 0.4 units of gas, 0.2 units of transportation, and 0.15 units of chemicals.

(4) Chemicals requires 0.1 units of gas, 0.2 units of agriculture, 0.2 units of transportation, and 0.3 units of chemicals.

(a) Construct the technology matrix.

(b) Construct the Leontief matrix for this problem.

(c) If the economy produces 700 million dollars of gas, 400 million dollars of agriculture, 650 million dollars of transportation, and 800 million dollars of chemicals, how much of this production is internally consumed by the economy? How much is produced for the external demand?

(d) Find the inverse to Leontief matrix of the problem. Explain the economic meaning of the columns of the inverse matrix.

(e) Given the economy productions in (c), suppose that the demand for transportation decreased 10%, while other external demands remained unchanged. How did this change in the demand for transportation affect other sectors of economics, in percentage terms?.

Chapter 4
Vector Spaces

4.1 Introduction

One of the serious obstacles for students of Linear Algebra is the topic of Vector Spaces. Until now, our discussions have primarily focused on linear systems, matrices and vectors in $\mathbb{R}^n$. Therefore, a natural question arises: "Why does one need to consider such abstract notions?" There are several reasons for this. One of them is that such generalizations are often helpful for solutions of problems in other branches of Mathematics. A good example here is the solution of the ancient problem in Geometry on the arbitrary angle trisection using a compass and ruler. This problem was solved in negative using vector spaces over rational numbers and their quadratic extensions. Another reason to introduce vector spaces is their occurrence in forms distinct from the familiar vector spaces $\mathbb{R}^n$. Consider, for example, the vector space of all polynomials with real coefficients of degree less than or equal to a given positive number n. However, there is a very serious reason to study vector spaces even in this course. Let us consider a symmetric matrix:

$$\mathbf{A} = \frac{1}{3}\begin{pmatrix} 5 & -1 & -1 \\ -1 & 5 & -1 \\ -1 & -1 & 5 \end{pmatrix}.$$

It is easy to check that $\mathbf{A}\mathbf{v}_1 = 1 \cdot \mathbf{v}_1$, where $\mathbf{v}_1 = \begin{pmatrix} 1 & 1 & 1 \end{pmatrix}^T$ is called an eigenvector of $\mathbf{A}$ and 1 is called an eigenvalue. The space $\mathbb{R}^3$ is an inner product space with respect to the dot product of vectors:

$$\langle \mathbf{x}, \mathbf{y} \rangle \overset{\text{def}}{=} \mathbf{x}^T \mathbf{y}.$$

Symmetric matrices in terms of this inner product can be described as those matrices, which satisfy the equation

$$\langle \mathbf{A}\mathbf{x}, \mathbf{y} \rangle = \langle \mathbf{x}, \mathbf{A}\mathbf{y} \rangle \tag{4.1}$$

© The Author(s), under exclusive license to Springer Nature Switzerland AG 2024
S. Khrushchev, *Linear Algebra with Applications to Economics*, Classroom Companion:
Economics, https://doi.org/10.1007/978-3-031-68682-5_4

for every pair of vectors $\mathbf{x}$ and $\mathbf{y}$ in $\mathbb{R}^3$. For the vectors of the standard basis $\mathbf{x} = \mathbf{e}_i$ and $\mathbf{y} = \mathbf{e}_j$ it is nothing but the definition of a symmetric matrix. The equation (4.1) also shows that if $\mathbf{Ax} = \mathbf{x}$, then the matrix $\mathbf{A}$ maps the set of all vectors perpendicular to $\mathbf{x}$ into itself. In this particular case, this is nothing but the plane

$$x + y + z = 0. \tag{4.2}$$

This plane has the Gauss basis

$$B = \left\{ \left(-1\ 1\ 0 \right)^T,\ \left(-1\ 0\ 1 \right)^T \right\},$$

which unfortunately is not orthogonal. However, the plane (4.2) inherits the inner product from the larger vector space $\mathbb{R}^3$. Therefore, if we consider this problem in the setting of inner product vector spaces and symmetric matrices defined by (4.1), we can claim that this operator (this time not a matrix) must have an eigenvector in this smaller space. Then the above arguments can be repeated. As result, we obtain that every symmetric matrix is diagonalizable. Moreover, it is orthogonally diagonalizable. In this particular case we find that $\mathbf{v}_2 = \left(-1\ 1\ 0 \right)^T$ is the eigenvector of $\mathbf{A}$ corresponding to the eigenvalue 2: $\mathbf{Av}_2 = 2\mathbf{v}_2$. The third eigenvector corresponding to the eigenvalue 2 is

$$\mathbf{v}_3 = \begin{pmatrix} 1 \\ 1 \\ -2 \end{pmatrix} = \begin{pmatrix} -1 \\ 1 \\ 0 \end{pmatrix} - 2 \begin{pmatrix} -1 \\ 0 \\ 1 \end{pmatrix}.$$

To summarize, vector spaces with inner products can be used to show that every symmetric matrix has a basis of pairwise orthogonal eigenvectors arranging the process of finding them into mathematical induction arguments. At the same time, for a concrete matrix such as $\mathbf{A}$, we can follow this induction to evaluate all required eigenvectors.

Section 4.2 lists the axioms that formally define vector spaces, and even includes the proof of the related theorem. In Section 4.3 we develop a universal construction of vector spaces from $\mathbb{R}$, which explains the reasons why the axioms for vector spaces are nothing but the axioms of arithmetics for real numbers $\mathbb{R}$.

In Section 4.4 we present basic principles for treating operators on finite dimensional vector spaces as matrices. This section is crucial for understanding the topic of vector spaces as they are considered in this book.

Section 4.5 is an unusual introduction of complex numbers in the framework of 2×2 matrices.

In section 4.6 we study the main tool for work with bases in finite dimensional spaces. These are the so-called transition matrices. We present here a simple algorithm for calculating such matrices using the extended matrix method.

Section 4.8 considers skew projections. It includes useful and simple formulas to evaluate them.

In Section 4.7 we study matrices of linear transformations in finite dimensional vector spaces.

4.2 Finite Dimensional Vector Spaces

Using column notations, we can rewrite the system of linear equations (1.5) as follows:

$$x\mathbf{v}_1 + y\mathbf{v}_2 + z\mathbf{v}_3 = \mathbf{b}, \tag{4.3}$$

where

$$\mathbf{v}_1 = \begin{pmatrix} 6 \\ 2 \\ 1 \end{pmatrix}, \ \mathbf{v}_2 = \begin{pmatrix} 4 \\ 3 \\ 2 \end{pmatrix}, \ \mathbf{v}_3 = \begin{pmatrix} 5 \\ 3 \\ 3 \end{pmatrix}, \ \mathbf{b} = \begin{pmatrix} 31 \\ 15 \\ 10 \end{pmatrix}.$$

This means that the problem of solution of a linear system $\mathbf{Ax} = \mathbf{b}$ is equivalent to the problem of decomposition of a given vector $\mathbf{b}$ into a sum (4.3). In this particular case there is a unique solution $x = 3$, $y = 2$, $z = 1$.

The symbols $\mathbf{v}_1$, $\mathbf{v}_2$, $\mathbf{v}_3$, and $\mathbf{b}$ in (4.3) may denote any objects, which can be added and multiplied by numbers. For instance, they may be matrices of the form

$$\begin{pmatrix} a & b \\ c & 0 \end{pmatrix}.$$

If

$$\mathbf{v}_1 = \begin{pmatrix} 6 & 2 \\ 1 & 0 \end{pmatrix}, \ \mathbf{v}_2 = \begin{pmatrix} 4 & 3 \\ 2 & 0 \end{pmatrix}, \ \mathbf{v}_3 = \begin{pmatrix} 5 & 3 \\ 3 & 0 \end{pmatrix}, \ \mathbf{b} = \begin{pmatrix} 31 & 15 \\ 10 & 0 \end{pmatrix},$$

then the equation (4.3) also has the unique solution $x = 3$, $y = 2$, $z = 1$.

Another possibility for $\mathbf{v}_1$, $\mathbf{v}_2$, $\mathbf{v}_3$, and $\mathbf{b}$ in (4.3) are quadratic polynomials:

$$\mathbf{v}_1 = 6x^2 + 2x + 1, \ \mathbf{v}_2 = 4x^2 + 3x + 2, \ \mathbf{v}_3 = 5x^2 + 3x + 3, \ \mathbf{b} = 31x^2 + 15x + 10.$$

In this case the solution to (4.3) is the same.

These examples show that $\mathbb{R}^n$ is not the only space, where linear systems may be considered. Therefore, it is reasonable to extend the theory of Linear Systems in $\mathbb{R}^n$ to more general spaces, which are called **Vector Spaces**.

Roughly speaking a **Vector Space** is a set which elements are called vectors. Similar to the example of $\mathbb{R}^n$ vectors can be added and multiplied by **scalars**. In most cases, these scalars will be just real numbers. However, the scalars may be rational numbers, complex numbers and elements of the so-called finite fields. The list of axioms below sums up the properties of $\mathbb{R}^n$, which are necessary for the study of $\mathbb{R}^n$ from Linear Algebra point of view.

Definition 4.1 A (real) vector space V is a **non-empty** set of elements called **vectors** equipped with operations of **addition** and **scalar** multiplication such that for all $\alpha, \beta \in \mathbb{R}$ and all $\mathbf{u}, \mathbf{v}, \mathbf{w} \in V$:

1. $\mathbf{u} + \mathbf{v} \in V$ (**closure** under addition).
2. $\mathbf{u} + \mathbf{v} = \mathbf{v} + \mathbf{u}$ (the **commutative** law for addition).
3. $\mathbf{u} + (\mathbf{v} + \mathbf{w}) = (\mathbf{u} + \mathbf{v}) + \mathbf{w}$ (the **associative** law for addition).

4. there is a single member **0** of V, called the **zero vector**, such that $\mathbf{v} + \mathbf{0} = \mathbf{v}$ for all $\mathbf{v} \in V$.
5. for every $\mathbf{v} \in V$ there is a unique element $\mathbf{w}$ (usually written as $-\mathbf{v}$), called the **negative** of $\mathbf{v}$, such that $\mathbf{v} + \mathbf{w} = \mathbf{0}$.
6. $\alpha \mathbf{v} \in V$ (**closure** under scalar multiplication).
7. $\alpha(\mathbf{u} + \mathbf{v}) = \alpha\mathbf{u} + \alpha\mathbf{v}$ (**distributive** law).
8. $(\alpha + \beta)\mathbf{u} = \alpha\mathbf{u} + \alpha\mathbf{u}$ (**distributive** law).
9. $\alpha(\beta\mathbf{v}) = (\alpha\beta)\mathbf{v}$ (**associative** law for scalar multiplication).
10. $1\mathbf{v} = \mathbf{v}$.

Observe that the list of Axioms $1 - 10$ stated above is the list of arithmetic properties, which are used in arithmetic operations with real numbers.

The following theorem says that the negative of any vector is the result of the multiplication of the initial vector by -1 as well as establishes relations of the zero vector and the scalar multiplication.

Theorem 4.1 *Let V be a vector space, $\mathbf{u}$ a vector in V, and k a scalar. Then*

(a) $0\mathbf{u} = \mathbf{0}$
(b) $(-1)\mathbf{u} = -\mathbf{u}$
(c) $k\mathbf{0} = \mathbf{0}$
(d) *If $k\mathbf{u} = \mathbf{0}$, then $k = 0$ or $\mathbf{u} = \mathbf{0}$.*

Proof. **(a)** By Axiom 8 we have: $0\mathbf{u} + 0\mathbf{u} = (0 + 0)\mathbf{u} = 0\mathbf{u}$. By Axiom 5 the vector $0\mathbf{u}$ has a unique **negative** vector $-0\mathbf{u}$. Adding this negative to the both sides of the above equality yields $(0\mathbf{u} + 0\mathbf{u}) + (-0\mathbf{u}) = 0\mathbf{u} + (-0\mathbf{u})$. Now, by Axiom 3 we obtain

$$0\mathbf{u} + (0\mathbf{u} + (-0\mathbf{u})) = 0\mathbf{u} + (-0\mathbf{u}).$$

By Axiom 5 $0\mathbf{u} + (-0\mathbf{u}) = \mathbf{0}$, implying that $0\mathbf{u} + \mathbf{0} = \mathbf{0}$, and finally that $0\mathbf{u} = \mathbf{0}$ by Axiom 4.

(b) By Axiom 10 we have:

$$\mathbf{u} + (-1)\mathbf{u} = 1\mathbf{u} + (-1)\mathbf{u} \stackrel{\text{Axiom 8}}{=} (1 + (-1))\mathbf{u} = 0\mathbf{u} \stackrel{\text{(a)}}{=} \mathbf{0}.$$

Since the negative vector is unique, we conclude that $(-1)\mathbf{u} = -\mathbf{u}$.

(c) If $k = 0$ then we apply **(a)** for $\mathbf{u} = \mathbf{0}$. If $k \neq 0$, then for every vector $\mathbf{u}$ we have

$$\mathbf{u} + k\mathbf{0} \overset{\text{Axiom 10}}{=} 1\mathbf{u} + k\mathbf{0} = \left(k \cdot \frac{1}{k}\right)\mathbf{u} + k\mathbf{0} \overset{\text{Axiom 9}}{=} k\left(\frac{1}{k}\mathbf{u}\right) + k\mathbf{0}$$

$$\overset{\text{Axiom 7}}{=} k\left(\frac{1}{k}\mathbf{u} + \mathbf{0}\right) \overset{\text{Axiom 4}}{=} k\left(\frac{1}{k}\mathbf{u}\right) \overset{\text{Axiom 9}}{=} \left(k \cdot \frac{1}{k}\right)\mathbf{u} = 1\mathbf{u} = \mathbf{u}.$$

Since the zero vector is unique, we conclude that $k\mathbf{0} = \mathbf{0}$.

(d) If $k \neq 0$ then

$$\mathbf{0} = \frac{1}{k}\mathbf{0} = \frac{1}{k}(k\mathbf{u}) = \left(\frac{1}{k} \cdot k\right)\mathbf{u} = 1\mathbf{u} = \mathbf{u}. \qquad \square$$

Definition 4.2 Let $\mathbf{v}_1, \mathbf{v}_1, \ldots, \mathbf{v}_k$ be a set of vectors of a vector space V. The **linear span**

$$\text{Lin}(S) = \text{Lin}(\mathbf{v}_1, \mathbf{v}_2, \ldots, \mathbf{v}_k)$$

of $S = \{\mathbf{v}_1, \ldots, \mathbf{v}_k\}$ is the set of all linear combinations of $\mathbf{v}_1, \ldots, \mathbf{v}_k$,

$$\text{Lin}(S) = \{\alpha_1\mathbf{v}_1 + \cdots + \alpha_k\mathbf{v}_k \mid \alpha_1, \ldots, \alpha_k \in \mathbb{R}\}$$

Definition 4.3 A finite set $S = \{\mathbf{v}_1, \mathbf{v}_2, \ldots, \mathbf{v}_k\}$ of vectors of a vector space V is called **linearly independent** if the only linear combination of these vectors representing the zero vector

$$\mathbf{0} = \alpha_1\mathbf{v}_1 + \alpha_2\mathbf{v}_2 + \cdots + \alpha_k\mathbf{v}_k$$

is the combination with zero coefficients $\alpha_1 = \alpha_2 = \cdots = \alpha_k = 0$.

Lemma 4.1 *If a finite set $S = \{\mathbf{v}_1, \mathbf{v}_2, \ldots, \mathbf{v}_k\}$ of vectors of a vector space V is linearly independent, then every vector $\mathbf{v}$ in $\text{Lin}(S)$ has a unique representation as the sum*

$$\mathbf{v} = \alpha_1\mathbf{v}_1 + \alpha_2\mathbf{v}_2 + \cdots + \alpha_k\mathbf{v}_k. \tag{4.4}$$

Proof. The existence of representation (4.4) follows by the definition of the linear span. Suppose that

$$\mathbf{v} = \begin{cases} \alpha_1\mathbf{v}_1 + \alpha_2\mathbf{v}_2 + \cdots + \alpha_k\mathbf{v}_k, \\ \beta_1\mathbf{v}_1 + \beta_2\mathbf{v}_2 + \cdots + \beta_k\mathbf{v}_k. \end{cases}$$

By Axiom 5 there is a unique vector $-\mathbf{v}$ such that $\mathbf{v} + (-\mathbf{v}) = \mathbf{0}$. By (b) of Theorem 4.1 and Axiom 7

$$-\mathbf{v} = (-1)\,(\beta_1\mathbf{v}_1 + \beta_2\mathbf{v}_2 + \cdots + \beta_k\mathbf{v}_k) = -\beta_1\mathbf{v}_1 - \cdots - \beta_k\mathbf{v}_k.$$

It follows that

$$\mathbf{0} = \mathbf{v} - \mathbf{v} = \alpha_1\mathbf{v}_1 + \alpha_2\mathbf{v}_2 + \cdots + \alpha_k\mathbf{v}_k - \beta_1\mathbf{v}_1 - \beta_2\mathbf{v}_2 - \cdots - \beta_k\mathbf{v}_k \overset{\text{Axioms 2, 3}}{=}$$

$$\alpha_1\mathbf{v}_1 - \beta_1\mathbf{v}_1 + \alpha_2\mathbf{v}_2 - \beta_2\mathbf{v}_2 + \cdots + \alpha_k\mathbf{v}_k - \beta_k\mathbf{v}_k \overset{\text{Axiom 8}}{=}$$

$$(\alpha_1 - \beta_1)\mathbf{v}_1 + (\alpha_2 - \beta_2)\mathbf{v}_2 + \cdots + (\alpha_k - \beta_k)\mathbf{v}_k.$$

Since S is linearly independent, this implies that $\alpha_j = \beta_j$ for $j = 1, \ldots, k$.   □

Lemma 4.2 *Given a finite set* $\{\mathbf{v}_1, \mathbf{v}_2, \ldots, \mathbf{v}_k\}$ *of non-zero vectors in a vector space* V *it is linearly independent if and only if for every integer* i, $i = 2, \ldots, k$, *the vector* $\mathbf{v}_i$ *does not belong to* $\mathrm{Lin}\,(\mathbf{v}_1, \mathbf{v}_2, \ldots, \mathbf{v}_{i-1})$.

Proof. It follows the lines of the proof of Lemma 2.2.   □

Definition 4.4 A vector space V is called **finite dimensional** if either $V = \{\mathbf{0}\}$ or there is a finite linearly independent set $B = \{\mathbf{v}_1, \mathbf{v}_2, \ldots, \mathbf{v}_k\}$ of vectors in V such that $V = \mathrm{Lin}(B)$. The family B is called a **basis** of vector space V.

By Lemma 4.2 any vector space either coincide with $\mathrm{Lin}(S)$ for some finite linearly independent set $S = \{\mathbf{v}_1, \mathbf{v}_2, \ldots, \mathbf{v}_k\}$ of vectors in V, or there is an infinite increasing family of linear spans $\mathrm{Lin}(S)$ in V. In the later case V is called an **infinite dimensional** vector space.

Theorem 4.2 *A vector space* V *is finite dimensional if and only if either* $V = \{\mathbf{0}\}$ *or there exists a finite set of nonzero vectors* $S = \{\mathbf{v}_1, \mathbf{v}_2, \ldots, \mathbf{v}_k\}$ *in* V *such that* $V = \mathrm{Lin}(S)$.

Proof. If we have $V = \{\mathbf{0}\}$, then V is finite dimensional by definition. If S is linearly independent, then V is finite dimensional by definition. If S is linearly dependent, then by Lemma 4.2 there is the smallest index i, $i > 1$ such that $\mathbf{v}_i$ is a linear combination of $\mathbf{v}_k$ with $k < i$. If we delete $\mathbf{v}_i$ from the list S, then we obtain a smaller list S' such that $V = \mathrm{Lin}(S) = \mathrm{Lin}(S')$. If the set of vectors S' is linearly independent, then the proof is finished. Otherwise, we repeat the procedure described. In a finite number of steps, which does not exceed k, we obtain a subset of S making a basis for V.   □

4.3 Subspaces of Vector Spaces

It is a rare occasion in practice, when one verifies axioms $1 - 10$ in the definition of a vector space. Instead, a standard construction of a new vector space from a given one can be applied followed by the construction of a subspace.

> **Definition 4.5** A subspace W of a vector space V is a **non-empty** subset of V that is itself a vector space under the same operations of addition and scalar multiplication as V.

> **Theorem 4.3** *A non-empty subset W of a vector space is a subspace if and only if for all $\mathbf{u}, \mathbf{v} \in W$ and all $\alpha, \beta \in \mathbb{R}$, we have $\alpha\mathbf{u} + \beta\mathbf{v} \in W$.*

> *Proof.* Let W be a non-empty subset of a vector space V, which is a vector space with respect to the operations of addition and multiplication by scalars. If $\mathbf{u}, \mathbf{v}$ are vectors in W, then $\alpha\mathbf{u}$ and $\beta\mathbf{v}$ are vectors in W by Axiom 6. Similarly, $\alpha\mathbf{u} + \beta\mathbf{v}$ is a vector in W by Axiom 1.
>
> Suppose now that W is a non-empty subset of V such that $\alpha\mathbf{u} + \beta\mathbf{v}$ is a vector in W, as soon as $\mathbf{u}$ and $\mathbf{v}$ are vectors in W, and α, β are real numbers. It is clear that this condition implies that all formulas in Axioms $1 - 10$ make sense and show that the results of operations according to these formulas are in W. Hence, W is a vector space. □

> **Theorem 4.4** *Let X be an arbitrary set and V be a vector space. Then the set $\mathfrak{F}(X, V)$ of all functions from X to V with algebraic operations defined point-wise is a vector space.*

> *Proof.* The proof is obvious, since all operations are defined point-wise and V is a vector space. □

Corollary 4.1 $V = \mathbb{R}^n$ and $V = {}^n\mathbb{R}$ *are vector spaces.*

Proof. Let $X = \{1, 2, \ldots, n\}$. Then $\mathfrak{F}(X, \mathbb{R})$ is a vector space. Notice that $\mathbf{u} \in \mathfrak{F}(X, \mathbb{R})$ is uniquely defined by the set of values $\mathbf{u}(k) = u_k$, $k = 1, \ldots, n$. Such a function can be represented as a column (the case of $V = \mathbb{R}^n$), or as a row (the case of $V = {}^n\mathbb{R}$).   □

Corollary 4.2 *The set $\mathfrak{M}_{m \times n}$ of all matrices of size $m \times n$ is a vector space.*

Proof. We consider the set X of points on the plane with coordinates (i, j) with integer coordinates $1 \le i \le m$, $1 \le j \le n$. Then the set of all matrices is identified with $\mathfrak{F}(X, \mathbb{R})$.   □

Corollary 4.3 *The set of all polynomials of degree not exceeding a given nonnegative integer is a vector space with respect to the operations defined point-wise.*

Proof. The set of polynomials is a subset of $\mathfrak{F}(\mathbb{R}, \mathbb{R})$. Since a linear combination with real coefficients of two polynomials of degree $\le n$ is a polynomial of degree $\le n$, Theorem 4.3 implies the result.   □

Problems

Prob. 123 — Show that

$$V = \left\{ \left(x \ y \ z \right)^T \, \middle| \, x - 2y + z = 0, \ x, y, z \in \mathbb{R} \right\}$$

is a subspace of $\mathbb{R}^3$. Find the Gauss basis for this subspace.

Prob. 124 — Let $\mathcal{P}_n$ be the set of all polynomials with real coefficients of degree less or equal n. Show that the set

$$V = \left\{ p(x) \in \mathcal{P}_n \, \middle| \, p(1) = p(2) = 0 \right\}$$

is a subspace of the vector space $\mathcal{P}_n$. Find a basis for V.

Prob. 125 — Prove that the set

$$W = \left\{ \left(x \ y \ z \ u \right)^T \, \middle| \, \begin{matrix} x - y + z - u = 0 \\ x + y + z - u = 0 \end{matrix} \right\}$$

is a subspace of $\mathbb{R}^4$. Find the Gauss basis of this subspace. Find $\dim(W)$.

4.4 Linearly Isomorphic Vector Spaces

Definition 4.6 Let V and W be vector spaces. A mapping $T : V \to W$ is called **linear transformation** if for all $\mathbf{u}, \mathbf{v} \in V$ and all $\alpha \in \mathbb{R}$:

$$T(\alpha \mathbf{u} + \beta \mathbf{v}) = \alpha T(\mathbf{u}) + \beta T(\mathbf{v}).$$

A linear transformation $T : V \rightarrow V$ of a vector space V is called a **linear operator**.

Theorem 4.5 *Let V be a finite-dimensional vector space and let T be a linear transformation from V to a vector space W. Then T is completely determined by its values on the elements of a basis of V.*

Proof. Let $\{\mathbf{v}_1, \mathbf{v}_2, \ldots, \mathbf{v}_n,\}$ be a basis for V. Then every vector $\mathbf{v}$ in V can be uniquely represented as $\mathbf{v} = \alpha_1 \mathbf{v}_1 + \alpha_2 \mathbf{v}_2 + \cdots + \alpha_n \mathbf{v}_n$, where all coefficients α_k are real numbers. It follows that

$$T(\mathbf{v}) = \alpha_1 T(\mathbf{v}_1) + \alpha_2 T(\mathbf{v}_2) + \cdots + \alpha_n T(\mathbf{v}_n),$$

which proves the theorem. $\square$

Definition 4.7 Two finite dimensional vector spaces V and W are called **linearly isomorphic** if there exists a one-to-one linear transformation T of V onto W. In other words, T satisfies

(a)$T(\alpha \mathbf{x} + \beta \mathbf{y}) = \alpha T(\mathbf{x}) + \beta T(\mathbf{y})$ for every vectors $\mathbf{x}$, $\mathbf{y}$ and reals α, β;
(b)For every $\mathbf{w}$ in W there exists a unique $\mathbf{v}$ in V such that $T(\mathbf{v}) = \mathbf{w}$.

Any mapping T satisfying (a) and (b) above is called a **linear isomorphism** of vector spaces.

Theorem 4.6 *Let $T : V \rightarrow W$ be a linear isomorphism of finite-dimensional vector spaces V and W. Then there exists a unique linear isomorphism $T^{-1} : W \rightarrow V$ such that $T^{-1}(T(\mathbf{v})) = \mathbf{v}$ for every $\mathbf{v}$ in V.*

The linear isomorphism T^{-1} is called the **inverse** of the linear isomorphism T.

Proof. By Definition 4.7, (b), for every $\mathbf{w}$ in V there is a unique vector $\mathbf{v}$ such that $T(\mathbf{v}) = \mathbf{w}$. We define $T^{-1}(\mathbf{w}) = \mathbf{v}$. Since T is a one-to-one linear mapping,

$$T(T^{-1}(\alpha\mathbf{u} + \beta\mathbf{w})) = \alpha\mathbf{u} + \beta\mathbf{w} = \alpha T(T^{-1}(\mathbf{u})) + \beta T(T^{-1}(\mathbf{w})) =$$

$$T\left(\alpha T^{-1}(\mathbf{u}) + \beta T^{-1}(\mathbf{w})\right) \Rightarrow T^{-1}(\alpha\mathbf{u} + \beta\mathbf{w}) = \alpha T^{-1}(\mathbf{u}) + \beta T^{-1}(\mathbf{w}). \quad \square$$

Theorem 4.7 *A linear transformation $T : V \rightarrow W$ of finite dimensional vector spaces is a linear isomorphism of vector spaces if and only if there is a basis $B_1 = \{\mathbf{v}_1, \mathbf{v}_2, \ldots, \mathbf{v}_n\}$ for V such that $B_2 = \{T(\mathbf{v}_1), T(\mathbf{v}_2), \ldots, T(\mathbf{v}_n)\}$ is a basis for W.*

Proof. Since W is finite dimensional, there is a basis $B_2 = \{\mathbf{w}_1, \mathbf{w}_2, \ldots, \mathbf{w}_n\}$ for W. Suppose that T is a linear isomorphism. Then for every vector $\mathbf{w}$ in W there is a unique vector $\mathbf{v}$ in V such that $T(\mathbf{v}) = \mathbf{w}$. In particular, for $j = 1, \ldots, n$, there are vectors $\mathbf{v}_j$ in V satisfying $T(\mathbf{v}_j) = \mathbf{w}_j$. Let $\mathbf{v}$ be a vector in V. Since B_2 is a basis for W, we conclude that $\mathbf{w} = T(\mathbf{v})$ is a linear combination of vectors in B_2:

$$\mathbf{w} = \alpha_1\mathbf{w}_1 + \alpha_2\mathbf{w}_2 + \cdots + \alpha_n\mathbf{w}_n = T\left(\alpha_1\mathbf{v}_1 + \alpha_2\mathbf{v}_2 + \cdots + \alpha_n\mathbf{v}_n\right) = T(\mathbf{v}).$$

Since T is a one-to-one mapping, we conclude that

$$\mathbf{v} = \alpha_1\mathbf{v}_1 + \alpha_2\mathbf{v}_2 + \cdots + \alpha_n\mathbf{v}_n,$$

implying that $V = \mathrm{Lin}(B_1)$ and that B_1 is a linearly independent set of vectors in V. It follows that B_1 is a basis for V.

Suppose that for a linear transformation $T : V \rightarrow W$ the system B_2 is a basis for W and B_1 is a basis for V. Since T is linear,

$$\alpha_1 T(\mathbf{v}_1) + \alpha_2 T(\mathbf{v}_2) + \cdots + \alpha_n T(\mathbf{v}_n) = T(\alpha_1\mathbf{v}_1 + \alpha_2\mathbf{v}_2 + \cdots + \alpha_n\mathbf{v}_n). \quad (4.5)$$

Since B_2 is a basis for W, the linear mapping T is onto. Since B_1 is a basis for V, the linear mapping T is one-to-one. $\square$

Theorem 4.8 *Let V be a finite dimensional vector space, i.e. $V = \mathrm{Lin}(S)$, where $B = \{\mathbf{v}_1, \mathbf{v}_2, \ldots, \mathbf{v}_n\}$ is linearly independent. Then $\mathbb{R}^n$ and V are linearly isomorphic.*

Proof. We consider the standard basis $\{e_1, \ldots, e_n\}$ in $\mathbb{R}^n$. By Theorem 4.5 the linear transformation T is uniquely defined by the formulas $T(e_i) = v_i$ for $i = 1, \ldots, n$. By Theorem 4.7 this linear transformation is a linear isomorphism.

□

Corollary 4.4 *If V is a finite dimensional vector space, then the number of vectors in any of its bases is the same.*

Proof. By Theorem 2.3 any basis in $\mathbb{R}^n$ has n elements. By Theorem 4.8 the vector space V is linearly isomorphic to $\mathbb{R}^n$. Finally, we apply Lemma 4.7 to the case $V_1 = \mathbb{R}^n$, $V_2 = V$.

□

Definition 4.8 Let V be a finite dimensional vector space. Then $\dim(V)$ is the number of elements in any basis of V. It is called the **dimension** of V. If V is an infinite dimensional vector space, then we write $\dim(V) = \infty$.

Problems

Prob. 126 — Find a linear isomorphism of $\mathbb{R}^{n+1}$ onto the vector space $\mathcal{P}_n$ of all polynomials with real coefficients of degree less or equal n.

Prob. 127 — Find a linear isomorphism of $\mathbb{R}^4$ onto the vector space $\mathfrak{M}_{2,2}$ of all 2×2 real matrices.

Prob. 128 — Let $V = \{x + y + z = 0 \mid x, y, z \in \mathbb{R}\}$ be a subspace of $\mathbb{R}^3$. Find a linear isomorphism of $\mathbb{R}^2$ onto V.

4.5 Complex Numbers

Complex numbers were introduced in Algebra as imaginary solutions to algebraic equations, which have no real roots. For example, complex numbers $\pm i$ are solutions to quadratic equation $X^2 + 1 = 0$. In this section we give a simple construction of complex numbers based on the Matrix Algebra. First, we give a definition of a field.

Definition 4.9 A nonempty set F with two algebraic operations, addition $+$ and multiplication $\cdot$, is called a **field**, if these operations satisfy the following properties:

- $a + (b + c) = (a + b) + c$ and $a \cdot (b \cdot c) = (a \cdot b) \cdot c$;

- $a + b = b + a$ and $a \cdot b = b \cdot a$;
- there are two different elements 0 and 1 in F such that $a + 0 = a$, $a \cdot 1 = a$ for every a in F;
- for every a in F there is an element $-a$ such that $a + (-a) = 0$;
- for every $a \neq 0$ in F there is an element a^{-1} such that $a \cdot a^{-1}$;
- $a \cdot (b + c) = (a \cdot b) + (a \cdot c)$.

This definition summarizes basic algebraic properties of rational numbers $\mathbb{Q}$ and real numbers $\mathbb{R}$. So that both these sets are fields with respect to the naturally defined addition and multiplication. We are going now to define addition and multiplication of vectors in $\mathbb{R}^2$.

Vector $\mathbf{e}_1$ of the standard basis of $\mathbb{R}^2$ can be identified with number 1. Then $\mathrm{Lin}(\mathbf{e}_1) = \mathbb{R}$. Let us denote the second vector $\mathbf{e}_2$ of the standard basis of $\mathbb{R}^2$ by symbol $\mathbf{i}$. Then every vector $\mathbf{z}$ in $\mathbb{R}^2$ can be uniquely represented as $\mathbf{z} = x \cdot 1 + y \cdot \mathbf{i}$, where x and y are real numbers.

We define the addition on $\mathbb{R}^2$ as vector addition:

$$\mathbf{z}_1 + \mathbf{z}_2 = (x_1 + y_1 \mathbf{i}) + (x_2 + y_2) = (x_1 + x_2 \mathbf{i}) + (y_1 + y_2)\mathbf{i}. \tag{4.6}$$

By Corollary 4.2 the set $\mathcal{M}_{2 \times 2}$ of 2×2 matrices with real entries is a vector space over $\mathbb{R}$. The linear mapping of $\mathbb{R}^2$ to the vector space $\mathcal{M}_{2 \times 2}$ is defined by the formula:

$$\mathbf{z} \longmapsto \mathbf{A}_\mathbf{z} = \begin{pmatrix} x & -y \\ y & x \end{pmatrix} = x\mathbf{I}_2 + y\mathbf{J}_2, \quad \mathbf{I}_2 = \begin{pmatrix} 1 & 0 \\ 0 & 1 \end{pmatrix} \quad \mathbf{J}_2 = \begin{pmatrix} 0 & -1 \\ 1 & 0 \end{pmatrix}.$$

The mapping $\mathbf{z} \longmapsto \mathbf{A}_\mathbf{z}$ is linear:

$$\mathbf{A}_{\alpha\mathbf{z}} = \begin{pmatrix} \alpha x & -\alpha y \\ \alpha y & \alpha x \end{pmatrix} = \alpha\mathbf{A}_\mathbf{z},$$

$$\mathbf{A}_{\mathbf{z}_1 + \mathbf{z}_2} = \begin{pmatrix} x_1 + x_2 & -y_1 - y_2 \\ y_1 + y_2 & x_1 + x_2 \end{pmatrix} = \mathbf{A}_{\mathbf{z}_1} + \mathbf{A}_{\mathbf{z}_2}.$$

It follows that the range of $\mathbf{z} \longmapsto \mathbf{A}_\mathbf{z}$,

$$\left\{ \begin{pmatrix} x & -y \\ y & x \end{pmatrix} \,\middle|\, x, y \in \mathbb{R} \right\} \tag{4.7}$$

is a subspace of $\mathcal{M}_{2 \times 2}$. Since the null space of $\mathbf{z} \longmapsto \mathbf{A}_\mathbf{z}$ is zero, it is a linear isomorphism of $\mathbb{R}^2$ onto the vector space (4.7).

Since $\mathbf{J}_2^2 = -\mathbf{I}_2$, a simple calculation shows that the vector space (4.7) is closed under the matrix multiplication:

$$\mathbf{A}_{\mathbf{z}_1}\mathbf{A}_{\mathbf{z}_2} = (x_1\mathbf{I}_2 + y_1\mathbf{J}_2) \cdot (x_2\mathbf{I}_2 + y_2\mathbf{J}_2) = (x_1 x_2 - y_1 y_2)\mathbf{I}_2 + (x_1 y_2 + y_1 x_2)\mathbf{J}_2 = \mathbf{A}_{\mathbf{z}_3},$$

where we define

$$\mathbf{z}_3 = (x_1 x_2 - y_1 y_2) + (x_1 y_2 + y_1 x_2)\mathbf{i} \qquad (4.8)$$

as the product of two vectors $\mathbf{z}_1$ and $\mathbf{z}_2$ in $\mathbb{R}^2$.

Theorem 4.9 *The vector space $\mathbb{R}^2$ with addition defined by (4.6) and multiplication defined by (4.8) is a field.*

Proof. The zero element is the vector $0 + 0 \cdot \mathbf{i}$ and the unit element is the vector $1 + 0 \cdot i$. The linear isomorphism $\mathbf{z} \longmapsto \mathbf{A_z}$ is multiplicative: $\mathbf{A}_{\mathbf{z}_1 \cdot \mathbf{z}_2} = \mathbf{A}_{\mathbf{z}_1} \mathbf{A}_{\mathbf{z}_2}$. This allows us to drop a routine check of all necessary properties stated in Definition 4.9. We may just refer to the Matrix Algebra, see section 1.4, applied to the matrices $\mathbf{A_z}$. If $\mathbf{z} \neq 0$, then $x^2 + y^2 > 0$. By Problem 3.2

$$\begin{pmatrix} x & -y \\ y & x \end{pmatrix}^{-1} = \frac{1}{x^2 + y^2} \begin{pmatrix} x & y \\ -y & x \end{pmatrix}$$

is a matrix $\mathbf{A_w}$ with

$$\mathbf{w} = \frac{x}{x^2 + y^2} - \frac{y}{x^2 + y^2}\mathbf{i}.$$

By (4.8) we obtain that $\mathbf{z} \cdot \mathbf{w} = 1$. $\qquad \square$

Definition 4.10 The vector space $\mathbb{R}^2$ with addition defined by (4.6) and multiplication defined by (4.8) is called the **field of complex numbers**. It is denoted by $\mathbb{C}$.

Lemma 4.3 *Every complex number is a root of a quadratic polynomial with real coefficients.*

Proof. Since $\mathbf{A_i} = \mathbf{J}_2$ and $\mathbf{J}^2 = -\mathbf{I}_2$, we conclude that $\mathbf{i}$ is the root of the polynomial $X^2 + 1$.

If $\mathbf{z} = x + y \cdot \mathbf{i}$ is a vector in $\mathbb{R}^2$, then the polynomial in variable λ,

$$\det(\mathbf{A_z} - \lambda \mathbf{I}_2) = \det \begin{pmatrix} x - \lambda & -y \\ y & x - \lambda \end{pmatrix} = (\lambda - x)^2 + y^2$$

is a quadratic polynomial in λ vanishing at $\lambda = x + y\mathbf{i}$. $\qquad \square$

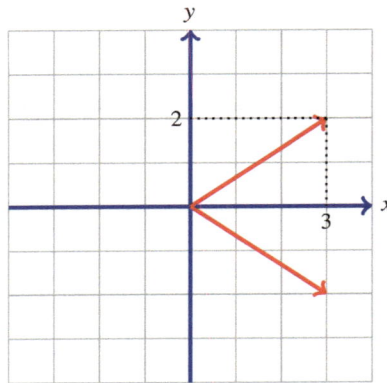

$$i = (0, 1)$$
$$z = x + yi$$
$$z = 3 + 2i$$
$$\bar{z} = 3 - 2i$$
$$\mathbb{C} = \{x + iy \; : \; x, y \in \mathbb{R}\}$$

Fig. 4.1: Complex number **z** in $\mathbb{C}$ as a vector.

Definition 4.11 The x coordinate of the complex number $z = x + yi$ is called the **real part** of **z**, and the y coordinate of the vector $z = x + yi$ is called the **imaginary part** of **z**. There is a special notation for these numbers in terms of **z**:

$$x = \text{Re}(z), \quad y = \text{Im}(z).$$

If $z = x + yi$ is a complex number, then the complex number $\bar{z} = x - yi$ is called the **complex conjugate** of **z**.

The set of all complex numbers is denoted by $\mathbb{C}$. It is clear that for every **z** in $\mathbb{C}$

$$z \cdot \bar{z} = (x + yi)(x - yi) = x^2 + y^2 = |z|^2; \qquad (4.9)$$

$$x = \text{Re}(z) = \frac{z + \bar{z}}{2}, \quad y = \text{Im}(z) = \frac{z - \bar{z}}{2}; \qquad (4.10)$$

$$(\lambda - z)(\lambda - \bar{z}) = \lambda^2 - 2\text{Re}(z)\lambda + |z|^2. \qquad (4.11)$$

The mapping $x + yi = z \rightarrow \bar{z} = x - yi$ can be realized by a matrix multiplication:

$$\bar{z} = \begin{pmatrix} x \\ -y \end{pmatrix} = S\begin{pmatrix} x \\ y \end{pmatrix}, \; S = \begin{pmatrix} 1 & 0 \\ 0 & -1 \end{pmatrix}.$$

It is clear that matrix **S** represents the reflection of the plane $\mathbb{C}$ through the x-axis.

Definition 4.12 The polar form of the complex number **z** is

$$z = r(\cos\theta + i\sin\theta). \qquad (4.12)$$

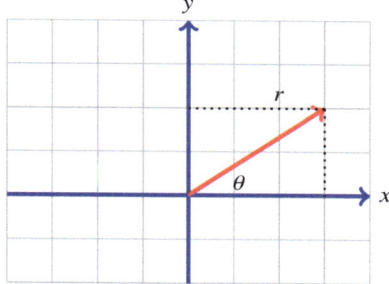

Fig. 4.2: The Polar Form of a Complex Number.

$$\mathbf{z} = x + y\mathbf{i}$$
$$x = r\cos\theta, \quad y = r\sin\theta$$
$$\mathbf{z} = r(\cos\theta + \mathbf{i}\cdot\sin\theta)$$
$$r = |z|$$

The length $r = \sqrt{x^2 + y^2}$ is called the **modulus** of $\mathbf{z}$, denoted by $|z|$, and the angle $\theta = \arg(z)$ is called the **argument** of $\mathbf{z}$. The **principal value** $\mathrm{Arg}(z)$ of $\arg(z)$ lies within the interval $(-\pi, \pi]$.

In terms of matrices, (4.12) can be rewritten as a product

$$\mathbf{A_z} = \begin{pmatrix} r & 0 \\ 0 & r \end{pmatrix}\begin{pmatrix} \cos\theta & -\sin\theta \\ \sin\theta & \cos\theta \end{pmatrix} = r\cdot\mathbf{R}_\theta. \tag{4.13}$$

of stretching (compression) by r and of rotation $\mathbf{R}_\theta$ through the angle θ, see Fig. 4.3.

Problem 4.1 Let l_1 and l_2 be two lines on the complex plane $\mathbb{C}$ passing through the origin. Let $\mathbf{S}_j$ be the reflection of $\mathbb{C}$ through l_j, $j = 1,2$. Suppose that l_2 is obtained from l_1 by the counterclockwise rotation through an angle θ. Show that the composition $\mathbf{S}_2\mathbf{S}_1$ of two reflections equals the rotation of $\mathbb{C}$ through the angle 2θ.

Solution: Without loss of generality we may assume that l_1 is the x-axis. Then the matrix of the linear reflection of $\mathbb{C}$ through l_2 is given by

$$\mathbf{R}_\theta\mathbf{SR}_{-\theta} = \begin{pmatrix} \cos\theta & -\sin\theta \\ \sin\theta & \cos\theta \end{pmatrix}\begin{pmatrix} 1 & 0 \\ 0 & -1 \end{pmatrix}\begin{pmatrix} \cos\theta & \sin\theta \\ -\sin\theta & \cos\theta \end{pmatrix} =$$
$$\begin{pmatrix} \cos^2\theta - \sin^2\theta & 2\cos\theta\sin\theta \\ 2\cos\theta\sin\theta & \sin^2\theta - \cos^2\theta \end{pmatrix} = \begin{pmatrix} \cos(2\theta) & \sin(2\theta) \\ \sin(2\theta) & -\cos(2\theta) \end{pmatrix}.$$

Finally,

$$\mathbf{R}_\theta\mathbf{SR}_{-\theta}\mathbf{S} = \begin{pmatrix} \cos(2\theta) & \sin(2\theta) \\ \sin(2\theta) & -\cos(2\theta) \end{pmatrix}\begin{pmatrix} 1 & 0 \\ 0 & -1 \end{pmatrix} = \begin{pmatrix} \cos(2\theta) & -\sin(2\theta) \\ \sin(2\theta) & \cos(2\theta) \end{pmatrix} = \mathbf{R}_{2\theta}. \quad \square$$

Problem 4.2 Using the fact that $\mathbf{R}_\theta\mathbf{R}_\phi = \mathbf{R}_{\theta+\phi}$ prove the addition formulas for sine and cosine:

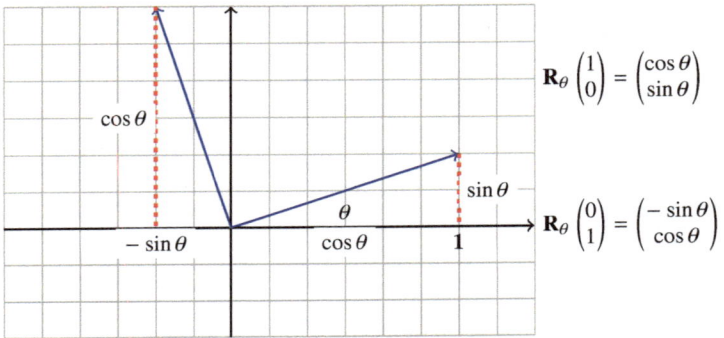

Fig. 4.3: The Rotation $\mathbf{R}_\theta$ of $\mathbb{C}$ through an angle θ.

$$\sin(\theta + \phi) = \sin\theta\cos\phi + \sin\phi\cos\theta, \; \cos(\theta + \phi) = \cos\theta\cos\phi - \sin\theta\sin\phi.$$

Solution:

$$\mathbf{R}_\theta\mathbf{R}_\phi = \begin{pmatrix} \cos\theta & -\sin\theta \\ \sin\theta & \cos\theta \end{pmatrix}\begin{pmatrix} \cos\phi & -\sin\phi \\ \sin\phi & \cos\phi \end{pmatrix} =$$

$$\begin{pmatrix} \cos\theta\cos\phi - \sin\theta\sin\phi & -\cos\theta\sin\phi - \cos\phi\sin\theta \\ \cos\theta\sin\phi + \cos\phi\sin\theta & \cos\theta\cos\phi - \sin\theta\sin\phi \end{pmatrix} = \mathbf{R}_{\theta+\phi}. \quad \square$$

Theorem 4.10 (de Moivre's Theorem) *For every real number θ and every integer n the following formula is true*

$$(\cos\theta + i\sin\theta)^n = \cos n\theta + i\sin n\theta.$$

Proof. Since

$$(\cos\theta + i\sin\theta)^{-1} = \frac{\cos\theta - i\sin\theta}{\cos^2\theta + \sin^2\theta} = \cos(-\theta) + i\sin(-\theta),$$

it is sufficient to prove the Theorem for positive integers n. Since $\mathbf{R}_\theta^n = \mathbf{R}_{n\theta}$, we obtain that

$$\begin{pmatrix} \cos\theta & -\sin\theta \\ \sin\theta & \cos\theta \end{pmatrix}^n = \begin{pmatrix} \cos n\theta & -\sin n\theta \\ \sin n\theta & \cos n\theta \end{pmatrix},$$

which proves the formula by (4.13). $\qquad\qquad\qquad\qquad\qquad\qquad\square$

Corollary 4.5 *The algebraic equation $\mathbf{z}^n = 1$ has n complex roots:*

$$\mathbf{z}_k = \cos \frac{2k\pi}{n} + i \cos \frac{2k\pi}{n}, \; k = 0, 1, \ldots, n - 1. \qquad (4.14)$$

The numbers $\mathbf{z}_k$ are called **roots of unity** or **de Moivre numbers**. It follows that

$$\mathbf{z}^n - 1 = (\mathbf{z} - \mathbf{z}_0)(\mathbf{z} - \mathbf{z}_1) \cdots (\mathbf{z} - \mathbf{z}_{n-1}).$$

Problem 4.3 Evaluate the powers $\mathbf{A}^n$, $n = \pm 1, \pm 2, \ldots$, of the matrix

$$\mathbf{A} = \begin{pmatrix} 1 & -1 \\ 1 & 1 \end{pmatrix}.$$

Solution: Since $\mathbf{A} = \mathbf{A}_{1+i}$, we conclude that $\mathbf{A}^n = \mathbf{A}_{(1+i)^n}$. The polar form of $1 + i$ is given by

$$1 + i = \sqrt{2} \left(\cos \frac{\pi}{4} + i \sin \frac{\pi}{4} \right).$$

By Theorem 4.10 we obtain that

$$(1 + i)^n = 2^{n/2} \left(\cos \frac{\pi n}{4} + i \sin \frac{\pi n}{4} \right).$$

It follows that

$$\mathbf{A}^n = 2^{n/2} \begin{pmatrix} \cos \frac{\pi n}{4} & -\sin \frac{\pi n}{4} \\ \sin \frac{\pi n}{4} & \cos \frac{\pi n}{4} \end{pmatrix}. \qquad \square$$

Theorem 4.11 (Fundamental Theorem of Algebra) *Every polynomial*

$$p(\mathbf{z}) = a_0 \mathbf{z}^n + a_1 \mathbf{z}^{n-1} + \cdots + a_{n-1} \mathbf{z} + a_n, \; a_0 \neq 0,$$

of degree n with complex coefficients $a_0, a_1, \ldots, a_{n-1}, a_n$, has at least one complex root $\mathbf{z} = c$, i.e. $p(c) = 0$.

This theorem was proved by Gauss. A very short proof can be found in Complex Analysis. Assuming that some polynomial $p(\mathbf{z})$ has no roots in $\mathbb{C}$, we consider the function $f(\mathbf{z}) = 1/p(\mathbf{z})$. Then $f(\mathbf{z})$ is differentiable in $\mathbf{z}$ everywhere on $\mathbb{C}$ and $\lim_{|\mathbf{z}| \to +\infty} f(\mathbf{z}) = 0$. By Liouville's Theorem this function must be identically zero, which is a contradiction.

A special case of Long Division of Polynomials, also known as the polynomial remainder theorem or little Bézout's theorem, says that any polynomial $p(\mathbf{z})$ of degree n with root $\mathbf{z} = c$ is a product $q(\mathbf{z})(\mathbf{z} - c)$, where q is a polynomial of degree $n - 1$. Applying little Bézout's theorem iteratively, we obtain that every polynomial p of degree n can be factored as a product of linear polynomials:

$$p(\mathbf{z}) = a_0(\mathbf{z} - \mathbf{z}_1)(\mathbf{z} - \mathbf{z}_2) \cdots (\mathbf{z} - \mathbf{z}_n),$$

where $\mathbf{z}_1, \mathbf{z}_2, \ldots, \mathbf{z}_n$, are complex numbers not necessarily different.

Problem 4.4 Solve the equation $\mathbf{z}^2 = a$ in $\mathbf{z}$, where a is an arbitrary complex number.

Solution: If $a = 0$, then $\mathbf{z} = 0$. If $a \neq 0$, then the complex number a can be written in the polar form: $a = |a|(\cos\theta + i\sin\theta)$, $\theta = \text{Arg}(a)$. We define

$$\mathbf{z}_1 = \sqrt{|a|}\left(\cos\left(\frac{\theta}{2}\right) + i\sin\left(\frac{\theta}{2}\right)\right), \quad \mathbf{z}_2 = -\sqrt{|a|}\left(\cos\left(\frac{\theta}{2}\right) + i\sin\left(\frac{\theta}{2}\right)\right).$$

Then by Theorem 4.10 we conclude that $\mathbf{z}_{1,2}^2 = a$.   □

Problem 4.5 Solve the equation $\mathbf{z}^2 = a + bi$ in $\mathbf{z}$, where a, b are real numbers and $b \neq 0$.

Solution: By Problem 4.4 there are two solutions of the form $x + iy$, where x and y are real numbers. Then

$$a + bi = (x + iy)^2 = x^2 + 2ixy - y^2 \Leftrightarrow \begin{cases} x^2 - y^2 & = a, \\ 2xy & = b \end{cases}.$$

Since $b \neq 0$, both x and y are nonzero real numbers. It follows that $y = b/2x$. Substituting this expression for y into the first equation, we get

$$a = x^2 - \frac{b^2}{4x^2} \Leftrightarrow 4x^4 - 4ax^2 - b^2 = 0.$$

Solving the quadratic equation in $x^2 > 0$ we obtain that

$$x^2 = \frac{2a + \sqrt{4a^2 + 4b^2}}{4} = \frac{a + \sqrt{a^2 + b^2}}{2}.$$

Finally,

$$x = \pm\sqrt{\frac{a + \sqrt{a^2 + b^2}}{2}}, \quad y = \pm\text{sign}(b)\sqrt{\frac{-a + \sqrt{a^2 + b^2}}{2}}, \qquad (4.15)$$

where $\text{sign}(b)$ is the sign of b and $\pm$ are equal signs in 4.15.   □

Definition 4.13 Every complex number $\mathbf{z}$ has a unique representation in the polar form:

$$\mathbf{z} = |\mathbf{z}|(\cos\theta + i\sin\theta), -\pi < \theta \leq \pi.$$

Then the **principal square root** $\sqrt{\mathbf{z}}$ of $\mathbf{z}$ of $\mathbf{z}$ is defined by the formula:

$$\sqrt{\mathbf{z}} = \sqrt{|\mathbf{z}|}\left(\cos\left(\frac{\theta}{2}\right) + i\sin\left(\frac{\theta}{2}\right)\right).$$

Fig. 4.4 shows the principal square roots of $-1, \mathbf{i}, -\mathbf{i}$.

$$\sqrt{-1} = \mathbf{i}$$

$$\sqrt{\mathbf{i}} = \frac{1+\mathbf{i}}{\sqrt{2}}$$

$$\sqrt{-\mathbf{i}} = -\frac{1+\mathbf{i}}{\sqrt{2}}$$

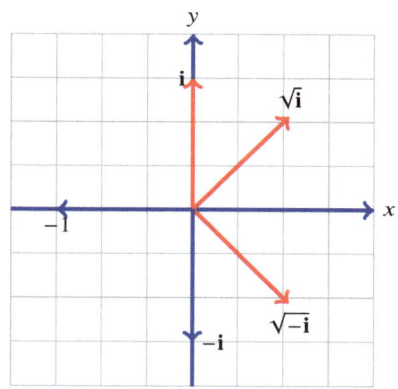

Fig. 4.4: Principal square root of z.

Problem 4.6 Find the square root of $1 + \mathbf{i}$.

Solution: By (4.15)

$$\sqrt{1 + \mathbf{i}} = \pm\left(\sqrt{\frac{1 + \sqrt{2}}{2}} + \mathbf{i}\sqrt{\frac{-1 + \sqrt{2}}{2}}\right). \quad \square$$

Theorem 4.12 *Let a, b, c be complex numbers and $a \neq 0$. Then the roots of the equation*

$$a\mathbf{z}^2 + b\mathbf{z} + c = 0$$

are given by the formula:

$$\mathbf{z}_{1,2} = \frac{-b \pm \sqrt{b^2 - 4ac}}{2a}.$$

Proof. As in the case of real numbers a, b, c, the proof follows from the identity:

$$a\mathbf{z}^2 + b\mathbf{z} + c = a\left(\mathbf{z} + \frac{b}{2a}\right)^2 + \frac{4ac - b^2}{4a}. \quad \square$$

Problems

Prob. 129 — Consider the complex numbers

$$z = \sqrt{3} - i, \quad w = 1 + i, \quad q = \frac{(\sqrt{3} - i)^6}{(1 + i)^{10}}.$$

Plot z and w as points in the complex plane. Express them in exponential form and hence evaluate q. Express q in the form $a + ib$.

4.6 Transition Matrices

If $B = \{v_1, v_2, \ldots, v_n\}$ is a basis for a vector space V, then every vector v in V can be uniquely represented as a linear combination

$$v = \alpha_1 v_1 + \alpha_2 v_2 + \cdots + \alpha_n v_n$$

with real coefficients $\alpha_1, \alpha_2, \ldots, \alpha_n$. In what follows

$$[v]_B = \left(\alpha_1 \ \alpha_2 \ \cdots \ \alpha_n\right)^T \tag{4.16}$$

denotes the column of these coefficients in $\mathbb{R}^n$. Then the mapping

$$P_B[v]_B = v$$

is a linear isomorphism of $\mathbb{R}^n$ onto V.

Problem 4.7 Let $\mathcal{P}_n$ be the vector space of polynomials with real coefficients in one variable x of degree not exceeding n. Show that it is linearly isomorphic to $\mathbb{R}^{n+1}$.

Solution: By Corollary 4.3 the set $\mathcal{P}_n$ is a vector space. If p is a polynomial in $\mathcal{P}_n$, then

$$p(x) = a_1 + a_2 x + \cdots + a_{n+1} x^n, \tag{4.17}$$

where $a_1, \ldots, a_{n+1}$ are the coefficients of p. Let $B = \{\mathbb{1}, x, x^2, \ldots, x^n\}$. Here $\mathbb{1}$ is the function $\mathbb{1}(x) = 1$ for every value of x, x is the identity function, and x^k are its powers, usually called monomials. Since every non-zero polynomial of degree n cannot have more that n zeros, the set B is a basis for $\mathcal{P}_n$. By (4.17) we have

$$[p]_B = \left(a_1 \ a_2 \ \cdots \ a_{n+1}\right)^T.$$

It follows that the mapping $P_B[p]_B = p$ is a linear isomorphism of $\mathbb{R}^{n+1}$ onto $\mathcal{P}_n$. $\square$

If $V = \mathbb{R}^n$ and $B = \{v_1, \ldots, v_n\}$ is a basis for $\mathbb{R}^n$, then P_B is defined by the matrix

$$\mathbf{P}_B = \left(v_1 \ \cdots \ v_n\right).$$

Indeed, $\mathrm{col}_i(\mathbf{P}_B) = \mathbf{v}_i$, $i = 1,\ldots,n$, i.e. the matrix $\mathbf{P}_B$ maps the standard basis $\{\mathbf{e}_1,\ldots,\mathbf{e}_n\}$ onto $\{\mathbf{v}_1,\ldots,\mathbf{v}_n\}$. The matrix $\mathbf{P}_B$ is called the **transition matrix**. It maps the column $[\mathbf{v}]_B$ of the coefficients of vector $\mathbf{v}$ in basis B to the coordinates of $\mathbf{v}$ in the standard basis of $\mathbb{R}^n$. Since B is a basis for $\mathbb{R}^n$, the null space of $\mathbf{P}_B$ is zero, implying that $\mathbf{P}_B$ is invertible. Then

$$\mathbf{P}_B[\mathbf{v}]_B = \mathbf{v} \Leftrightarrow [\mathbf{v}]_B = \mathbf{P}_B^{-1}\mathbf{v}. \tag{4.18}$$

It follows that $\mathbf{P}_B^{-1}$ maps a given vector $\mathbf{v}$ in $\mathbb{R}^n$ to the column $[\mathbf{v}]_B$ of its coefficients in basis B.

Problem 4.8 Let

$$B = \left\{ \begin{pmatrix} 1 \\ 2 \\ 1 \end{pmatrix}, \begin{pmatrix} 2 \\ 1 \\ 4 \end{pmatrix}, \begin{pmatrix} 3 \\ 2 \\ 1 \end{pmatrix} \right\}.$$

Show that B is a basis for $\mathbb{R}^3$. Find the transition matrix $\mathbf{P}_B$. Find the coefficients of vector $\mathbf{u} = \begin{pmatrix} 14 & 14 & 14 \end{pmatrix}^T$ in the basis B.

Solution: Let

$$\mathbf{P}_B = \begin{pmatrix} 1 & 2 & 3 \\ 2 & 1 & 2 \\ 1 & 4 & 1 \end{pmatrix} \Rightarrow \mathrm{rref}(\mathbf{P}_B) = \mathbf{I}_3.$$

It follows that $N(\mathbf{P}_B) = \mathbf{0}$. Hence the columns of $\mathbf{P}_B$ are linearly independent in $\mathbb{R}^3$. Since $\dim(\mathbb{R}^3) = 3$, the system B makes a basis for $\mathbb{R}^3$. It follows that $\mathbf{P}_B$ is the transition matrix. We have

$$\mathbf{P}_B^{-1} = \frac{1}{14}\begin{pmatrix} -7 & 10 & 1 \\ 0 & -2 & 4 \\ 7 & -2 & -3 \end{pmatrix} \Rightarrow [\mathbf{u}]_B = \mathbf{P}_B^{-1}\mathbf{u} = \begin{pmatrix} 4 \\ 2 \\ 2 \end{pmatrix}. \quad \square$$

Problem 4.9 Let B be the Gauss basis defined by the Gauss matrix $\mathbf{G}$ in Problem 2.5. Evaluate $\mathbf{P}_B$ and $\mathbf{P}_B^{-1}$. Evaluate $[\mathbf{u}]_B$ for the vector $\mathbf{u} = \begin{pmatrix} 1 & 1 & 1 & 1 & 1 & 1 \end{pmatrix}^T$.

Solution: Problem 2.5 shows that

$$\mathbf{P}_B = \mathbf{G} = \begin{pmatrix} 1 & 0 & -2 & 0 & -3 & -5 \\ 0 & 1 & -3 & 0 & -1 & -2 \\ 0 & 0 & 1 & 0 & 0 & 0 \\ 0 & 0 & 0 & 1 & -2 & -3 \\ 0 & 0 & 0 & 0 & 1 & 0 \\ 0 & 0 & 0 & 0 & 0 & 1 \end{pmatrix}.$$

By Theorem 3.3

$$\mathbf{P}_B^{-1} = \begin{pmatrix} 1 & 0 & 2 & 0 & 3 & 5 \\ 0 & 1 & 3 & 0 & 1 & 2 \\ 0 & 0 & 1 & 0 & 0 & 0 \\ 0 & 0 & 0 & 1 & 2 & 3 \\ 0 & 0 & 0 & 0 & 1 & 0 \\ 0 & 0 & 0 & 0 & 0 & 1 \end{pmatrix}.$$

Finally,

$$[\mathbf{u}]_B = \mathbf{P}_B^{-1}\mathbf{u} = \begin{pmatrix} 11 & 7 & 1 & 6 & 1 & 1 \end{pmatrix}^T. \quad \square$$

Problem 4.10 In the standard coordinates the equation of the curve C is given by the implicit equation

$$x^2 + \sqrt{3}xy + 2y^2 = 4.$$

Rotate the plane so that in the new basis the equation of C would be $AX^2 + BY^2 = C$. Sketch the curve C.

Solution: The matrix $\mathbf{R}_\theta$ of the rotation through an angle θ is the transition matrix from the new coordinates (X, Y) to the standard coordinates (x, y). By (4.13)

$$\begin{cases} x & = X\cos\theta - Y\sin\theta, \\ y & = X\sin\theta + Y\cos\theta. \end{cases}$$

It follows that

$$\begin{cases} x^2 & = X^2\cos^2\theta - 2XY\cos\theta\sin\theta + Y^2\sin^2\theta, \\ y^2 & = X^2\sin^2\theta + 2XY\sin\theta\cos\theta + Y^2\cos^2\theta, \\ xy & = X^2\cos\theta\sin\theta + XY(\cos^2\theta - \sin^2\theta) - Y^2\cos\theta\sin\theta. \end{cases}$$

Substituting these expression into the equation of C, we find that the coefficient at XY equals

$$2\cos\theta\sin\theta + \sqrt{3}(\cos^2\theta - \sin^2\theta) = \sin 2\theta + \sqrt{3}\cos 2\theta.$$

It is zero if $\tan 2\theta = -\sqrt{3}$, which happens if $\theta = -\pi/6$. Then $\sin\theta = -1/2$, $\cos\theta = \sqrt{3}/2$ and

$$x^2 + \sqrt{3}xy + 2y^2 = X^2\left(\cos^2\theta + 2\sin^2\theta + \sqrt{3}\cos\theta\sin\theta\right) +$$
$$Y^2\left(\sin^2\theta + 2\cos^2\theta - \sqrt{3}\cos\theta\sin\theta\right) = \frac{1}{2}X^2 + \frac{5}{2}Y^2.$$

The equation of C in new coordinates is $X^2 + 5Y^2 = 8$, see Fig. 4.5. $\quad \square$

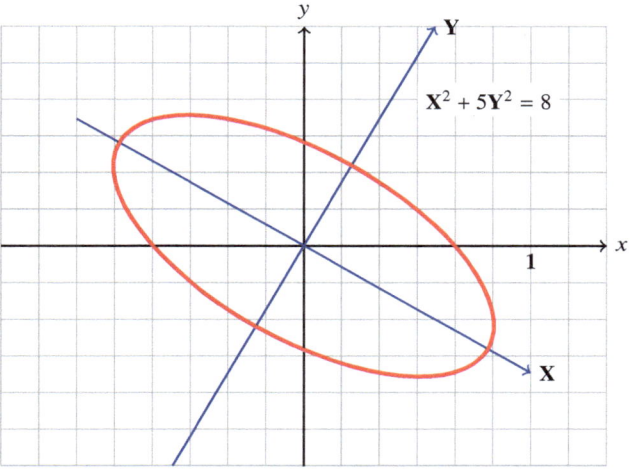

Fig. 4.5: The plot of $x^2 + \sqrt{3}xy + 2y^2 = 4$.

Theorem 4.13 *Let B and B′ be two bases in* $\mathbb{R}^n$*. Then for every* **v** *in* $\mathbb{R}^n$*:*

$$[\mathbf{v}]_{B'} = \mathbf{P}_{B'}^{-1}\mathbf{P}_B[\mathbf{v}]_B$$

Proof. By (4.18)
$$[\mathbf{v}]_{B'} = \mathbf{P}_{B'}^{-1}\mathbf{v} = \mathbf{P}_{B'}^{-1}\mathbf{P}_B[\mathbf{v}]_B. \qquad \square$$

Theorem 4.13 shows that the matrix $\mathbf{P}_{B \to B'}$, which sends the coordinates $[\mathbf{v}]_B$ of **v** in $\mathbb{R}^n$ to the coordinates $[\mathbf{v}]_{B'}$ of **v** in basis B' is given by the formula

$$\mathbf{P}_{B \to B'} = \mathbf{P}_{B'}^{-1}\mathbf{P}_B. \tag{4.19}$$

Problem 4.11 Let

$$\mathbf{A} = \begin{pmatrix} 2 & 1 & -1 & -1 & 1 \\ 1 & -1 & 1 & 1 & -2 \\ 3 & 3 & -3 & 3 & 4 \\ 4 & 5 & -5 & -5 & 7 \end{pmatrix}, \quad \mathbf{B} = \begin{pmatrix} 9 & 18 & 2 & 11 & 2 \\ 3 & 6 & 1 & 4 & 0 \\ 5 & 10 & 1 & 6 & 1 \\ 4 & 8 & 1 & 5 & 1 \end{pmatrix}$$

be two matrices. Find the transition matrix from the Gauss basis for solutions of matrix **A** to the Gauss basis for solutions of matrix **B**.

Solutions: We have

$$\text{rref}(\mathbf{A}) = \begin{pmatrix} 1 & 0 & 0 & 0 & -1/3 \\ 0 & 1 & -1 & 0 & 5/3 \\ 0 & 0 & 0 & 1 & 0 \\ 0 & 0 & 0 & 0 & 0 \end{pmatrix} \Rightarrow \mathbf{G_A} = \begin{pmatrix} 1 & 0 & 0 & 0 & 1/3 \\ 0 & 1 & 1 & 0 & -5/3 \\ 0 & 0 & 1 & 0 & 0 \\ 0 & 0 & 0 & 1 & 0 \\ 0 & 0 & 0 & 0 & 1 \end{pmatrix},$$

$$\text{rref}(\mathbf{B}) = \begin{pmatrix} 1 & 2 & 0 & 1 & 0 \\ 0 & 0 & 1 & 1 & 0 \\ 0 & 0 & 0 & 0 & 1 \\ 0 & 0 & 0 & 0 & 0 \end{pmatrix} \Rightarrow \mathbf{G_B}^{-1} = \begin{pmatrix} 1 & 2 & 0 & 1 & 0 \\ 0 & 1 & 0 & 0 & 0 \\ 0 & 0 & 1 & 1 & 0 \\ 0 & 0 & 0 & 1 & 0 \\ 0 & 0 & 0 & 0 & 1 \end{pmatrix}.$$

By (4.19) we find the transition matrix:

$$\mathbf{G_B}^{-1}\mathbf{G_A} = \begin{pmatrix} 1 & 2 & 2 & 1 & -3 \\ 0 & 1 & 1 & 0 & -5/3 \\ 0 & 0 & 1 & 1 & 0 \\ 0 & 0 & 0 & 1 & 0 \\ 0 & 0 & 0 & 0 & 1 \end{pmatrix}. \quad \square$$

An Efficient Method for Computing $\mathbf{P}_{B \to B'}$ in $\mathbb{R}^n$

Step 1: Form the matrix $(\mathbf{P}_{B'} \mid \mathbf{P}_B)$.

Step 2: Use elementary row operations to reduce the matrix in **Step 1** to reduced row echelon form.

Step 3: The resulting matrix will be $\left(\mathbf{I} \mid \mathbf{P}_{B \to B'} \right)$.

Step 4: Extract the matrix $\mathbf{P}_{B \to B'}$ from the right side of the matrix in **Step 3**. This procedure is captured in the following diagram.

$$\textbf{(new basis | old basis)} \xrightarrow{\text{row operations}} \begin{pmatrix} \mathbf{I} & \begin{array}{c} \textbf{transition from} \\ \textbf{old to new} \end{array} \end{pmatrix}. \tag{4.20}$$

Since B' is a basis for $\mathbb{R}^n$, the matrix $\mathbf{P}_{B'}$ is invertible. Since it is invertible, its reduced row echelon form is the identity matrix $\mathbf{I}$. Therefore, there is a chain of elementary row operations reducing $\mathbf{P}_{B'}$ to $\mathbf{I}$. We know that any elementary row operation is equivalent to a multiplication of the matrix by an elementary matrix on the left. It follows that

$$\mathbf{E}_r \mathbf{E}_{r-1} \cdots \mathbf{E}_1 \mathbf{P}_{B'} = \mathbf{I} \Rightarrow \mathbf{E}_r \mathbf{E}_{r-1} \cdots \mathbf{E}_1 = \mathbf{P}_{B'}^{-1}.$$

(new basis | old basis) = $(\mathbf{P}_{B'} \mid \mathbf{P}_B) \sim$

$\left(\mathbf{E}_r \mathbf{E}_{r-1} \cdots \mathbf{E}_1 \mathbf{P}_{B'} \mid \mathbf{E}_r \mathbf{E}_{r-1} \cdots \mathbf{E}_1 \mathbf{P}_B \right) \sim \left(\mathbf{I} \mid \mathbf{P}_{B'}^{-1} \mathbf{P}_B \right) = (\mathbf{I} \mid \mathbf{P}_{B \to B'}).$

Problem 4.12 Evaluate the transition matrix $\mathbf{P}_{B \to B'}$ for

$$B = \left\{ \begin{pmatrix} 2 \\ 1 \\ 0 \end{pmatrix}, \begin{pmatrix} 3 \\ 2 \\ 1 \end{pmatrix}, \begin{pmatrix} 2 \\ 1 \\ 1 \end{pmatrix} \right\}, \quad B' = \left\{ \begin{pmatrix} 2 \\ 3 \\ 2 \end{pmatrix}, \begin{pmatrix} 1 \\ 2 \\ 1 \end{pmatrix}, \begin{pmatrix} 0 \\ 1 \\ 1 \end{pmatrix} \right\}.$$

For $\mathbf{v} = \begin{pmatrix} 1 & 1 & 1 \end{pmatrix}^T$ find $[\mathbf{v}]_B$ and $[\mathbf{v}]_{B'}$.

Solution: First, we check that B is a basis:

$$\mathbf{P}_B^{-1} = \begin{pmatrix} 1 & -1 & -1 \\ -1 & 2 & 0 \\ 1 & -2 & 1 \end{pmatrix}.$$

The fact that B' is a basis follows automatically from both solutions given below. However, observing that $\mathbf{P}_{B'} = \mathbf{P}_B^T$ we obtain that

$$\mathbf{P}_{B'}^{-1} = \left(\mathbf{P}_B^T \right)^{-1} = \left(\mathbf{P}_B^{-1} \right)^T = \begin{pmatrix} 1 & -1 & 1 \\ -1 & 2 & -2 \\ -1 & 0 & 1 \end{pmatrix}.$$

By the formula (4.19) we have

$$\mathbf{P}_{B \to B'} = \mathbf{P}_{B'}^{-1} \mathbf{P}_B = \begin{pmatrix} 1 & -1 & 1 \\ -1 & 2 & -2 \\ -1 & 0 & 1 \end{pmatrix} \begin{pmatrix} 2 & 3 & 2 \\ 1 & 2 & 1 \\ 0 & 1 & 1 \end{pmatrix} = \begin{pmatrix} 1 & 2 & 2 \\ 0 & -1 & -2 \\ -2 & -2 & -1 \end{pmatrix}.$$

By (4.20)

(new basis | old basis) $= (\mathbf{P}_{B'} \mid \mathbf{P}_B) =$

$$\begin{pmatrix} 2 & 1 & 0 & 2 & 3 & 2 \\ 3 & 2 & 1 & 1 & 2 & 1 \\ 2 & 1 & 1 & 0 & 1 & 1 \end{pmatrix} \sim \begin{pmatrix} 1 & 0 & 0 & 1 & 2 & 2 \\ 0 & 1 & 0 & 0 & -1 & -2 \\ 0 & 0 & 1 & -2 & -2 & -1 \end{pmatrix} \Rightarrow \mathbf{P}_{B \to B'} = \begin{pmatrix} 1 & 2 & 2 \\ 0 & -1 & -2 \\ -2 & -2 & -1 \end{pmatrix}.$$

By (4.18)

$$[\mathbf{v}]_B = \mathbf{P}_B^{-1} \mathbf{v} = \begin{pmatrix} 1 & -1 & -1 \\ -1 & 2 & 0 \\ 1 & -2 & 1 \end{pmatrix} \begin{pmatrix} 1 \\ 1 \\ 1 \end{pmatrix} = \begin{pmatrix} -1 \\ 1 \\ 0 \end{pmatrix}.$$

Similarly,

$$[\mathbf{v}]_{B'} = \mathbf{P}_{B'}^{-1} \mathbf{v} = \begin{pmatrix} 1 & -1 & 1 \\ -1 & 2 & -2 \\ -1 & 0 & 1 \end{pmatrix} \begin{pmatrix} 1 \\ 1 \\ 1 \end{pmatrix} = \begin{pmatrix} 1 \\ -1 \\ 0 \end{pmatrix}.$$

It is easy to see that $\mathbf{P}_{B \to B'} [\mathbf{v}]_B = [\mathbf{v}]_{B'}$. $\qquad\qquad \square$

Problems

Prob. 130 — Show that the vectors

$$\mathbf{v}_1 = \begin{pmatrix} 2 \\ 1 \\ -1 \end{pmatrix}, \ \mathbf{v}_2 = \begin{pmatrix} 3 \\ 4 \\ 6 \end{pmatrix}, \ \mathbf{v}_3 = \begin{pmatrix} -2 \\ 3 \\ 2 \end{pmatrix}$$

form a basis $S = \{\mathbf{v}_1, \mathbf{v}_2, \mathbf{v}_3\}$ for $\mathbb{R}^3$ and find the coordinates $[\mathbf{v}]_S$ of the vector $(-5 \ 7 \ -2)^T$ in the basis S.

Prob. 131 — Let

$$\mathbf{v}_1 = \begin{pmatrix} 2 \\ 3 \\ 5 \end{pmatrix}, \ \mathbf{v}_2 = \begin{pmatrix} 1 \\ 1 \\ 2 \end{pmatrix}, \ \mathbf{v}_3 = \begin{pmatrix} a \\ b \\ c \end{pmatrix}.$$

Find a condition that a, b, c must satisfy for the set of vectors $\{\mathbf{v}_1, \mathbf{v}_2, \mathbf{v}_3\}$ to be linearly dependent.

Prob. 132 — Find a subset of $S = \{\mathbf{v}_1, \mathbf{v}_2, \mathbf{v}_3, \mathbf{v}_4\}$,

$$\mathbf{v}_1 = \begin{pmatrix} 1 \\ 2 \\ 1 \\ 2 \end{pmatrix}, \ \mathbf{v}_2 = \begin{pmatrix} 0 \\ -1 \\ 3 \\ 4 \end{pmatrix}, \ \mathbf{v}_3 = \begin{pmatrix} 4 \\ -11 \\ 5 \\ -1 \end{pmatrix}, \ \mathbf{v}_4 = \begin{pmatrix} 9 \\ 2 \\ 1 \\ -3 \end{pmatrix},$$

which is a basis for $\mathrm{Lin}\{\mathbf{v}_1, \mathbf{v}_2, \mathbf{v}_3, \mathbf{v}_4\}$.

Prob. 133 — Let $S = \{\mathbf{v}_1, \mathbf{v}_2, \mathbf{v}_3\}$, where

$$\mathbf{v}_1 = \begin{pmatrix} 1 \\ 1 \\ 0 \end{pmatrix}, \ \mathbf{v}_2 = \begin{pmatrix} -4 \\ 0 \\ 3 \end{pmatrix}, \ \mathbf{v}_3 = \begin{pmatrix} 3 \\ 5 \\ 1 \end{pmatrix}.$$

Show that S is a basis of $\mathbb{R}^3$. Let $\mathbf{w} = (-1 \ 7 \ 5)^T$ and $\mathbf{e}_1 = (1 \ 0 \ 0)^T$. Find the coordinates of $\mathbf{w}$ and $\mathbf{e}_1$ with respect of to the bases S.

Prob. 134 — Let

$$B = \{\mathbf{v}_1 = (1 \ 0 \ 1)^T, \ \mathbf{v}_2 = (1 \ 1 \ 3)^T, \ \mathbf{v}_3 = (0 \ 0 \ 1)^T\}$$
$$B' = \{\mathbf{v}'_1 = (1 \ 1 \ 1)^T, \ \mathbf{v}'_2 = (1 \ -1 \ 0)^T, \ \mathbf{v}'_3 = (0 \ 1 \ -1)^T\}$$

be two sets of vectors in $\mathbb{R}^3$. Show that both systems are bases in $\mathbb{R}^3$ and find the transition matrix from B to B'

Prob. 135 — Consider the vectors

$$\mathbf{v}_1 = \begin{pmatrix} 1 \\ 0 \\ 1 \end{pmatrix}, \quad \mathbf{v}_2 = \begin{pmatrix} 2 \\ 2 \\ -3 \end{pmatrix}, \quad \mathbf{v}_3 = \begin{pmatrix} 2 \\ -1 \\ 1 \end{pmatrix}, \quad \mathbf{u} = \begin{pmatrix} 1 \\ 2 \\ 3 \end{pmatrix}.$$

Show that $B = \{\mathbf{v}_1, \mathbf{v}_2, \mathbf{v}_3\}$ is a basis of $\mathbb{R}^3$.

Write down the transition matrix $\mathbf{P}$ from B coordinates to standard coordinates. Find $\mathbf{P}^{-1}$.

Find the B coordinates of $\mathbf{u}$, $[\mathbf{u}]_B$, and express $\mathbf{u}$ as a linear combination of the vectors $\mathbf{v}_1, \mathbf{v}_2, \mathbf{v}_3$.

4.7 Matrices of Linear Transformations

Let V and W be finite dimensional vector spaces satisfying $\dim(V) = n$, $\dim(W) = m$. Then there are linear isomorphisms $\mathrm{Iso}_1 : \mathbb{R}^n \to V$ and $\mathrm{Iso}_2 : \mathbb{R}^m \to W$. These isomorphisms define a basis $S = \{\mathbf{v}_1, \ldots, \mathbf{v}_n\}$ in V, and a basis $S' = \{\mathbf{w}_1, \ldots, \mathbf{w}_m\}$ in W, where

$$\mathbf{v}_i = \mathrm{Iso}_1(\mathbf{e}_i), \; i = 1, \ldots, n, \; \mathbf{w}_j = \mathrm{Iso}_2(\mathbf{e}_j), \; j = 1, \ldots, m. \quad (4.21)$$

We define an $m \times n$ matrix $\mathbf{A}_T$ by the formula

$$\mathbf{A}_T = \Big([T(\mathbf{v}_1)]_{S'} \; [T(\mathbf{v}_2)]_{S'} \; \cdots \; [T(\mathbf{v}_n)]_{S'} \Big). \quad (4.22)$$

Theorem 4.14 *The following diagram of mappings is commutative:*

In other words,

$$T \circ \mathrm{Iso}_1 = \mathrm{Iso}_2 \circ \mathbf{A}_T \iff \mathbf{A}_T = \mathrm{Iso}_2^{-1} \circ T \circ \mathrm{Iso}_1. \quad (4.23)$$

Proof. By Theorem 4.5 it is sufficient to show that for every $i = 1, \ldots, n$,

$$T \circ \mathrm{Iso}_1(\mathbf{e}_i) = \mathrm{Iso}_2 \circ \mathbf{A}_T(\mathbf{e}_i).$$

By (4.21) we have $T(\mathbf{v}_i) = T \circ \mathrm{Iso}_1(\mathbf{e}_i)$. Since $\mathrm{Col}_i(\mathbf{A}_T) = \mathbf{A}_T \mathbf{e}_i$, we conclude that $\mathbf{A}_T \mathbf{e}_i = [T(\mathbf{v}_i)]_{S'}$. By (4.16) we obtain that $T(\mathbf{v}_i) = [T(\mathbf{v}_i)]_{S'}$, which proves the Theorem.   □

Definition 4.14 The matrix $\mathbf{A}_T$ is called a **matrix of a linear transformation** T. By (4.21) it depends not only on T, but also on linear isomorphisms Iso_1 and Iso_2.

Problem 4.13 Let $V = \mathbb{R}^n$, $W = \mathbb{R}^m$, $\mathrm{Iso}_1 = \mathbf{I}_n$, $\mathrm{Iso}_2 = \mathbf{I}_m$, $T : \mathbb{R}^n \to \mathbb{R}^m$ be a linear transformation. Then

$$\mathbf{A}_T = \big([T(\mathbf{e}_1)]_{S'} \ [T(\mathbf{e}_2)]_{S'} \ \cdots \ [T(\mathbf{e}_n)]_{S'}\big),$$

where $S' = \{\mathbf{e}_1, \ldots, \mathbf{e}_m\}$ in $\mathbb{R}^m$.

This problem shows that every linear transformation $T : \mathbb{R}^n \to \mathbb{R}^m$ is defined by the formula

$$T(\mathbf{x}) = \mathbf{A}_T \mathbf{x},$$

where $\mathbf{A}_T$ is the $m \times n$ matrix.

If $\mathrm{Iso}_1 : \mathbb{R}^n \to V$ and $\mathrm{Iso}_2 : \mathbb{R}^n \to W$ are linear isomorphisms, then the matrix $\mathbf{A}_T$ is defined: $\mathbf{A}_T = \mathrm{Iso}_2^{-1} \circ T \circ \mathrm{Iso}_1$. Being a composition of linear isomorphisms the linear mapping defined by the matrix $\mathbf{A}_T$ is one-to-one. Hence the matrix $\mathbf{A}_T$ is invertible and

$$\mathbf{A}_T^{-1} = \big(\mathrm{Iso}_2^{-1} \circ T \circ \mathrm{Iso}_1\big)^{-1} = \mathrm{Iso}_1^{-1} \circ T^{-1} \circ \mathrm{Iso}_2 \Rightarrow (\mathbf{A}_T)^{-1} = \mathbf{A}_{T^{-1}}. \quad (4.24)$$

Problem 4.14 The linear operator on $\mathbb{R}^3$ is defined by the formula:

$$T\begin{pmatrix} x \\ y \\ z \end{pmatrix} = \begin{pmatrix} 3x + 2y + z \\ x - y - 2z \\ x + 2y + 3z \end{pmatrix}.$$

Evaluate the matrix $\mathbf{A}_T$ for the identity isomorphisms $\mathrm{Iso}_1 = \mathbf{I}_3$ and $\mathrm{Iso}_2 = \mathbf{I}_3$.

Solution: We have

$$\begin{pmatrix} 3x + 2y + z \\ x - y - 2z \\ x + 2y + 3z \end{pmatrix} = \begin{pmatrix} 3 & 2 & 1 \\ 1 & -1 & -2 \\ 1 & 2 & 3 \end{pmatrix}\begin{pmatrix} x \\ y \\ z \end{pmatrix} \Rightarrow \mathbf{A}_T = \begin{pmatrix} 3 & 2 & 1 \\ 1 & -1 & -2 \\ 1 & 2 & 3 \end{pmatrix}. \quad □$$

Definition 4.15 Suppose that T is a linear transformation from a vector space V to a vector space W. Then the **range**, $R(T)$, of T is

$$R(T) = \{T(\mathbf{v}) \mid \mathbf{v} \in V\},$$

and the **null space**, $N(T)$, of T is

$$N(T) = \{\mathbf{v} \in V \mid T(\mathbf{v}) = \mathbf{0}\}.$$

The null space is also called the **kernel** and is denoted as $\ker(T)$.

Definition 4.16 The **rank**, $\text{rank}(T)$, of a linear transformation T is $\dim(R(T))$. The **nullity** of a linear transformation T is $\text{nullity}(T) = \dim(N(T))$.

Theorem 4.15 (Rank-Nullity Theorem for linear transformations) *Let T be a linear transformation from the finite dimensional vector space V to the vector space W. Then*

$$\text{rank}(T) + \text{nullity}(T) = \dim(V). \tag{4.25}$$

Proof. Since W does not enter the formula (4.25), we may assume that $W = R(T)$. Since V is finite dimensional, there is a basis $\{\mathbf{v}_1, \mathbf{v}_2, \ldots, \mathbf{v}_n\}$ for V. Then

$$W = R(T) = \text{Lin}\{T(\mathbf{v}_1), T(\mathbf{v}_2), \ldots, T(\mathbf{v}_n)\}.$$

By Theorem 4.2 the vector space W is finite dimensional. By (4.23) we see that

$$\mathbf{A}_T\mathbf{x} = \mathbf{0} \Leftrightarrow \text{Iso}_2^{-1} \circ T \circ \text{Iso}_1\mathbf{x} = \mathbf{0} \Leftrightarrow \mathbf{v} = \text{Iso}_1\mathbf{x} \text{ is a vector in } N(T).$$

Since Iso_1 is a linear isomorphism, we conclude by Lemma 4.7 that

$$\text{nullity}(T) = \text{nullity}(\mathbf{A}_T).$$

Since Iso_2 is a linear isomorphism, we conclude by Lemma 4.7 that

$$\text{rank}(T) = \text{rank}(\mathbf{A}_T).$$

Since Iso_1 is a linear isomorphism, we conclude by Lemma 4.7 that

$$\dim(V) = \dim(\mathbb{R}^n) = n.$$

The result now follows by Theorem 2.14.   □

Remark. Note that this result holds even if W is not finite-dimensional.

Problem 4.15 Consider the linear transformations $T : \mathcal{P}_1 \to \mathcal{P}_2$ and $S : \mathcal{P}_2 \to \mathcal{P}_2$

$$\begin{cases} T\,(a + bx) & = a - 2b + (-2a + 4b)x + (a - 2b)x^2, \\ S\,(a + bx + cx^2) & = a + c + (a - b)x + (a + b + 2c)x^2. \end{cases}$$

(a) Find the null space $N(T)$ of T and the range $R(T)$ of T. Check the Rank-Nullity Theorem for T. Do the same for the operator S.
(b) Which linear transformation is defined: ST or TS?
(c) Use the rank-nullity theorem to find the dimensions of the range and the null space of the composed linear transformation.

Solution: Let $B' = \{1, x\}$ be a basis in $\mathcal{P}_1$ and $B = \{1, x, x^2\}$ be a basis in $\mathcal{P}_1$. Then

$$\mathbf{P}_{B'} \begin{pmatrix} a \\ b \end{pmatrix} = a + bx, \quad \mathbf{P}_B \begin{pmatrix} a \\ b \\ c \end{pmatrix} = a + bx + cx^2,$$

are isomorphisms of vector spaces $\mathbb{R}^2$ and $\mathcal{P}_1$, $\mathbb{R}^3$ and $\mathcal{P}_2$. Then the matrix $\mathbf{A}_T$ for operator T satisfies

$$\textbf{(a) } \mathbf{A}_T = \begin{pmatrix} 1 & -2 \\ -2 & 4 \\ 1 & -2 \end{pmatrix} \sim \begin{pmatrix} 1 & -2 \\ 0 & 0 \\ 0 & 0 \end{pmatrix} = \mathbf{rref}(\mathbf{A}_T) \Rightarrow \begin{cases} x & = 2s \\ y & = s \\ J & = \{1\} \end{cases} \Rightarrow$$

$$N(T) = \{2s + sx^2 | s \in \mathbb{R}\}, \quad R(T) = \{s - 2sx + sx^2 | s \in \mathbb{R}\}.$$

We conclude that

$$\dim(\mathcal{P}_1) = 2 = 1 + 1 = \text{rank}(T) + \text{nullity}(T).$$

The matrix $\mathbf{A}_S$ for operator S satisfies

$$\mathbf{A}_S = \begin{pmatrix} 1 & 0 & 1 \\ 1 & -1 & 0 \\ 1 & 1 & 2 \end{pmatrix} \sim \begin{pmatrix} 1 & 0 & 1 \\ 0 & 1 & 1 \\ 0 & 0 & 0 \end{pmatrix} = \text{rref}(\mathbf{A}_S) \Rightarrow \begin{cases} x & = -s \\ y & = -s \\ z & = s \\ J & = \{1, 2\} \end{cases} \Rightarrow$$

$$N(S) = \{-s - sx + sx^2 | s \in \mathbb{R}\}, \quad R(S) = \text{Lin}\left(1 + x + x^2, -x + x^2\right).$$

We conclude that

$$\dim(\mathcal{P}_2) = 3 = 2 + 1 = \text{rank}(S) + \text{nullity}(S).$$

(**b**) The diagram

$$ST : \mathbf{R}^2 \xrightarrow{T} \mathbf{R}^3 \xrightarrow{S} \mathbf{R}^3$$

exists, implying that ST is defined. This follows also from the fact that the size of $\mathbf{A}_S$ is 3×3 and of $\mathbf{A}_T$ is 3×2. Hence the matrix product $\mathbf{A}_S \mathbf{A}_T$ is defined and equals $\mathbf{A}_{ST}$. Similarly, the product $\mathbf{A}_T \mathbf{A}_S$ does not exist, implying that TS is not defined.

(**c**) By (**a**), we have $\text{nullity}(T) = 1$ and $\text{rank}(T) = 1$. It follows that $\text{nullity}(ST) \geq 1$ and $\text{rank}(ST) \leq 1$. Since $1 - 2x + x^2 \in R(T)$, a simple calculation shows that

$$\mathbf{A}_S \begin{pmatrix} 1 \\ -2 \\ 1 \end{pmatrix} = \begin{pmatrix} 1 & 0 & 1 \\ 1 & -1 & 0 \\ 1 & 1 & 2 \end{pmatrix} \begin{pmatrix} 1 \\ -2 \\ 1 \end{pmatrix} = \begin{pmatrix} 2 \\ 3 \\ 1 \end{pmatrix} \neq \mathbf{0}_3, \tag{4.26}$$

implying that $\dim(R(ST)) \geq 1$. By the rank-nullity theorem

$$2 = \dim(\mathbb{R}^2) = \dim(R(ST)) + \text{nullity}(ST).$$

Since the both summands in the right hand part are nonzero, we conclude that

$$\dim(R(ST)) = \text{nullity}(ST) = 1. \quad \square$$

Theorem 4.16 (The Inverse Rank-Nullity Theorem for linear transformations) *Let V be a finite dimensional vector space and W be a vector space. Given a subspace N of V and R of W, there exists a linear mapping $T : V \to W$ with $N(T) = N$ and $R(T) = R$ if and only if*

$$\dim(N) + \dim(R) = \dim(V). \tag{4.27}$$

Proof. By Theorem 4.15 the condition (4.27) is necessary. If (4.27) holds, then $\dim(R) \leq \dim(V)$ is finite. Likewise in the proof of Theorem 4.15 we may assume that $\dim(W) < +\infty$. Then T exists if and only if the corresponding matrix $\mathbf{A}_T$ exists. Therefore, we may assume that $V = \mathbb{R}^n$ and $W = \mathbb{R}^m$. Then N is a subspace of $\mathbb{R}^n$ and R is a subspace of $\mathbb{R}^m$. The result follows by Corollary 3.3. $\square$

In practice, solving such problems, we find first the reduced row echelon form $\text{rref}(N)$ for the subspace N. Next, we find the Gauss basis for the subspace R and arrange the columns of this Gauss basis to make the matrix B. Then $B \cdot \text{rref}(N)$ is the required matrix.

Problem 4.16 Let $V = \text{Lin}(\mathbf{v}_1, \mathbf{v}_2, \mathbf{v}_3, \mathbf{v}_4)$ be the subspace of $\mathbb{R}^4$ defined by the columns

$$\mathbf{v}_1 = \begin{pmatrix} -1 \\ -1 \\ 0 \\ 1 \end{pmatrix}, \quad \mathbf{v}_2 = \begin{pmatrix} 4 \\ 1 \\ 3 \\ 2 \end{pmatrix}, \quad \mathbf{v}_3 = \begin{pmatrix} 2 \\ 1 \\ 1 \\ 0 \end{pmatrix}, \quad \mathbf{v}_4 = \begin{pmatrix} -5 \\ -2 \\ -3 \\ -1 \end{pmatrix}.$$

Let $W = \text{Lin}(\mathbf{w}_1, \mathbf{w}_1, \mathbf{w}_3, \mathbf{w}_4)$ be the subspace of $\mathbb{R}^5$ defined by the columns

$$\mathbf{w}_1 = \begin{pmatrix} 1 \\ -1 \\ 3 \\ 0 \\ 2 \end{pmatrix}, \quad \mathbf{w}_2 = \begin{pmatrix} 2 \\ 1 \\ -2 \\ 1 \\ 1 \end{pmatrix}, \quad \mathbf{w}_3 = \begin{pmatrix} 0 \\ 3 \\ -8 \\ 1 \\ -3 \end{pmatrix}, \quad \mathbf{w}_4 = \begin{pmatrix} 5 \\ 1 \\ -1 \\ 2 \\ 4 \end{pmatrix}.$$

Does there exist a linear transformation $T : \mathbb{R}^4 \to \mathbb{R}^5$ such that $N(T) = V$ and $R(T) = W$? Find such a T if it exists.

Solution: Let $\mathbf{A} = \begin{pmatrix} \mathbf{v}_1 & \mathbf{v}_2 & \mathbf{v}_3 & \mathbf{v}_4 \end{pmatrix}$. Then

$$\text{rref}\left(\mathbf{A}^T\right) = \begin{pmatrix} 1 & 0 & 1 & 1 \\ 0 & 1 & -1 & -2 \\ 0 & 0 & 0 & 0 \\ 0 & 0 & 0 & 0 \end{pmatrix} \Rightarrow \mathbf{G} = \begin{pmatrix} 1 & 0 & -1 & -1 \\ 0 & 1 & 1 & 2 \\ 0 & 0 & 1 & 0 \\ 0 & 0 & 0 & 1 \end{pmatrix} \Rightarrow N(\mathbf{A}^T) = \text{Lin}\left\{ \begin{pmatrix} -1 \\ 1 \\ 1 \\ 0 \end{pmatrix}, \begin{pmatrix} -1 \\ 2 \\ 0 \\ 1 \end{pmatrix} \right\}.$$

It follows that

$$\text{rref}(V) = \text{rref}\begin{pmatrix} -1 & 1 & 1 & 0 \\ -1 & 2 & 0 & 1 \end{pmatrix} = \begin{pmatrix} 1 & 0 & -2 & 1 \\ 0 & 1 & -1 & 1 \end{pmatrix}.$$

Since

$$\text{rref}\begin{pmatrix} 1 & 2 & 0 & 5 \\ -1 & 1 & 3 & 1 \\ 3 & -2 & -8 & -1 \\ 0 & 1 & 1 & 2 \\ 2 & 1 & -3 & 4 \end{pmatrix} = \begin{pmatrix} 1 & 0 & -2 & 1 \\ 0 & 1 & 1 & 2 \\ 0 & 0 & 0 & 0 \\ 0 & 0 & 0 & 0 \\ 0 & 0 & 0 & 0 \end{pmatrix},$$

we conclude that

$$\dim(V) + \dim(W) = 2 + 2 = \dim(\mathbb{R}^4).$$

By Theorem 4.16 such a transformation exists and

$$\mathbf{A}_T = \begin{pmatrix} 1 & 2 \\ -1 & 1 \\ 3 & -2 \\ 0 & 1 \\ 2 & 1 \end{pmatrix} \begin{pmatrix} 1 & 0 & -2 & 1 \\ 0 & 1 & -1 & 1 \end{pmatrix} = \begin{pmatrix} 1 & 2 & -4 & 3 \\ -1 & 1 & 1 & 0 \\ 3 & -2 & -4 & 1 \\ 0 & 1 & -1 & 1 \\ 2 & 1 & -5 & 3 \end{pmatrix}. \quad \square$$

Problem 4.17 Let $V = \text{Lin}\{u_1, u_2\}$ be a subspace of $\mathbb{R}^3$, where

$$\mathbf{u}_1 = \begin{pmatrix} 1 & 2 & -1 \end{pmatrix}^T, \quad \mathbf{u}_2 = \begin{pmatrix} 1 & 1 & 1 \end{pmatrix}^T.$$

Find the matrix of the operator $T : V \rightarrow V$ defined by

$$T\mathbf{u}_1 = \begin{pmatrix} 2 & 3 & 0 \end{pmatrix}^T, \quad T\mathbf{u}_2 = \begin{pmatrix} 0 & 1 & -2 \end{pmatrix}^T,$$

in the Gauss basis of V.

Solution: To find the Gauss basis of V, we apply Corollary 3.11, which says that the cross-product

$$\mathbf{u}_1 \times \mathbf{u}_2 = \begin{vmatrix} \mathbf{i} & \mathbf{j} & \mathbf{k} \\ 1 & 2 & -1 \\ 1 & 1 & 1 \end{vmatrix} = 3\mathbf{i} - 2\mathbf{j} - \mathbf{k}.$$

It follows that $V = N(A)$, where $\mathbf{A} = \begin{pmatrix} 3 & -2 & -1 \end{pmatrix}$. We have

$$\text{rref}(\mathbf{A}) = \begin{pmatrix} 1 & -2/3 & -1/3 \end{pmatrix} \Rightarrow \mathbf{v} \in N(\mathbf{A}) \Leftrightarrow \mathbf{x} = s \begin{pmatrix} 2/3 \\ 1 \\ 0 \end{pmatrix} + t \begin{pmatrix} 1/3 \\ 0 \\ 1 \end{pmatrix}, \quad s, t \in \mathbb{R}.$$

Hence the Gauss basis of V is given by the two following vectors:

$$\mathbf{v}_1 = s \begin{pmatrix} 2/3 \\ 1 \\ 0 \end{pmatrix}, \quad \mathbf{v}_2 = \begin{pmatrix} 1/3 \\ 0 \\ 1 \end{pmatrix}.$$

To find the coefficients of u_1 and u_2 in the basis $S = \{v_1, v_2\}$, we solve two systems simultaneously by the method of extended matrix:

$$\text{rref} \begin{pmatrix} 2/3 & 1/3 & | & 1 & 1 \\ 1 & 0 & | & 2 & 1 \\ 0 & 1 & | & -1 & 1 \end{pmatrix} = \begin{pmatrix} 1 & 0 & | & 2 & 1 \\ 0 & 1 & | & -1 & 1 \\ 0 & 0 & | & 0 & 0 \end{pmatrix} \Rightarrow \begin{cases} \mathbf{u}_1 &= 2\mathbf{v}_1 - \mathbf{v}_2, \\ \mathbf{u}_2 &= \mathbf{v}_1 + \mathbf{v}_2. \end{cases} \tag{4.28}$$

Similarly,

$$\text{rref} \begin{pmatrix} 2/3 & 1/3 & | & 2 & 0 \\ 1 & 0 & | & 3 & 1 \\ 0 & 1 & | & 0 & -2 \end{pmatrix} = \begin{pmatrix} 1 & 0 & | & 3 & 1 \\ 0 & 1 & | & 0 & -2 \\ 0 & 0 & | & 0 & 0 \end{pmatrix} \Rightarrow \begin{cases} T\mathbf{u}_1 &= 3\mathbf{v}_1, \\ T\mathbf{u}_2 &= \mathbf{v}_1 - 2\mathbf{v}_2. \end{cases} \tag{4.29}$$

By (4.28) and (4.29),

$$2T\mathbf{v}_1 - T\mathbf{v}_2 = T\mathbf{u}_1 = 3\mathbf{v}_1,$$
$$T\mathbf{v}_1 + T\mathbf{v}_2 = T\mathbf{u}_2 = \mathbf{v}_1 - 2\mathbf{v}_2.$$

It follows that

$$\begin{pmatrix} 2 & -1 \\ 1 & 1 \end{pmatrix} \begin{pmatrix} T\mathbf{v}_1 \\ T\mathbf{v}_2 \end{pmatrix} = \begin{pmatrix} 3 & 0 \\ 1 & -2 \end{pmatrix} \begin{pmatrix} \mathbf{v}_1 \\ \mathbf{v}_2 \end{pmatrix} \Rightarrow \begin{pmatrix} 2 & -1 \\ 1 & 1 \end{pmatrix}^{-1} \begin{pmatrix} 3 & 0 \\ 1 & -2 \end{pmatrix} \begin{pmatrix} \mathbf{v}_1 \\ \mathbf{v}_2 \end{pmatrix} = \frac{1}{3} \begin{pmatrix} 4 & -2 \\ -1 & -4 \end{pmatrix} \begin{pmatrix} \mathbf{v}_1 \\ \mathbf{v}_2 \end{pmatrix},$$

implying that the matrix of the operator T in the Gauss basis of V is given by

$$\frac{1}{3} \begin{pmatrix} 4 & -2 \\ -1 & -4 \end{pmatrix}. \quad \square$$

If $T : V \to V$ is a linear operator on a finite dimensional vector space V, then by the definition there is a finite basis $B = \{\mathbf{v}_1, \ldots, \mathbf{v}_n\}$, where $n = \dim(V)$, in V. By Theorem 4.14, the basis B defines a unique matrix $\mathbf{A}_T$ for the operator T. This matrix has a unique characteristic polynomial $p_\mathbf{A}(\lambda)$, which can be considered as a candidate for the position of the characteristic polynomial $p_T(\lambda)$ of the operator T. The only problem is to prove that $p_T(\lambda)$ does not depend on the choice of the basis B.

Suppose that B' is another basis for V. By (4.23) and (4.19) we have

$$\begin{cases} \mathbf{A}_T & = \mathbf{P}_B^{-1} T \mathbf{P}_B, \\ \mathbf{A}'_T & = \mathbf{P}_{B'}^{-1} T \mathbf{P}_{B'} \end{cases} \Rightarrow \mathbf{A}'_T = P_{B \to B'} \mathbf{A}_T P_{B \to B'}^{-1}.$$

Since $\mathbf{R} = P_{B \to B'}$ is an invertible matrix in $\mathbb{R}^n$, we obtain by Theorem 3.29

$$\det(\mathbf{A}'_T - \lambda \mathbf{I}_n) = \det(\mathbf{R}(\mathbf{A}_T - \lambda \mathbf{I}_n)\mathbf{R}^{-1}) = \det(\mathbf{R})\det(\mathbf{A}_T - \lambda \mathbf{I}_n)\det(\mathbf{R})^{-1} = \det(\mathbf{A}_T - \lambda \mathbf{I}_n$$

> **Definition 4.17** Let $T : V \to V$ be a linear operator on a finite dimensional vector space V and let $\mathbf{A}_T$ be its matrix in any basis $B = \{\mathbf{v}_1, \ldots, \mathbf{v}_n\}$ for V. Then the **characteristic polynomial** $p_T(\lambda)$ of T is defined by the formula
>
> $$p_T(\lambda) = \det(\mathbf{A}_T - \lambda \mathbf{I}_n) \qquad (4.30)$$

Notice that Theorem 3.38 allows one to define the **determinant** $\det(T)$ and **trace** $\mathrm{Tr}(T)$ for linear transformations $T : V \to V$ on a finite dimensional vector space V.

Problems

Prob. 136 — Is there an inverse to the linear transformation

$$T \begin{pmatrix} x \\ y \\ z \end{pmatrix} = \begin{pmatrix} x + y + z \\ x - y \\ x + 2y - 3z \end{pmatrix}?$$

Prob. 137 — Find the null space and range of the linear transformation $S : \mathbb{R}^2 \to \mathbb{R}^4$ given by the formula

$$S\begin{pmatrix} x \\ y \end{pmatrix} = \begin{pmatrix} x+y \\ x \\ x-y \\ y \end{pmatrix}$$

Prob. 138 — Is it possible to construct a linear transformation $T : \mathbb{R}^3 \to \mathbb{R}^3$ with

$$N(T) = \left\{ t \begin{pmatrix} 1 \\ 2 \\ 3 \end{pmatrix} : t \in \mathbb{R} \right\}, \quad R(T) = \mathbb{R}^2 \subset \mathbb{R}^3?$$

Prob. 139 — Consider the vectors

$$\mathbf{v}_1 = \begin{pmatrix} 1 \\ 0 \\ 1 \end{pmatrix}, \quad \mathbf{v}_2 = \begin{pmatrix} 2 \\ 2 \\ -3 \end{pmatrix}, \quad \mathbf{v}_3 = \begin{pmatrix} 2 \\ -1 \\ 1 \end{pmatrix}, \quad \mathbf{u} = \begin{pmatrix} 1 \\ 2 \\ 3 \end{pmatrix}.$$

Let T be a linear transformation, $T : \mathbb{R}^3 \to \mathbb{R}^3$ whose null space, $N(T)$, is the plane through two vectors $\mathbf{v}_1$ and $\mathbf{u}$ and whose range, $R(T)$, contains $\mathbf{v}_2$. Find a basis of $N(T)$. State the rank-nullity theorem for linear transformations and verify that T satisfies it. Find a basis of $R(T)$.

4.8 Skew Projections

Definition 4.18 Let W be a vector space, U and V its subspaces such that

$$U \cap V = \{\mathbf{0}_W\}, \quad \mathrm{Lin}(U,V) = W.$$

Then we say that $W = U \dotplus V$ is a direct sum of the subspaces U and W.

Since $\mathrm{Lin}(U,V) = W$, every vector $\mathbf{w}$ can be represented as a sum $\mathbf{w}_U + \mathbf{w}_V$. Since $U \cap V = \{\mathbf{0}_W\}$, this representation is unique. It follows that the mapping $\mathbf{w} \to \mathbf{w}_U$ is a linear operator $P_{U\|V}$ on W, which keeps all vectors in U and vanishes on the subspace V. The fact that $P_{U\|V}$ is linear follows from the following simple arguments. If $\mathbf{w} = \mathbf{w}_U + \mathbf{w}_V$ and $\mathbf{z} = \mathbf{z}_U + \mathbf{z}_V$, then for any real α and β

$$\alpha\mathbf{w} + \beta\mathbf{z} = (\alpha\mathbf{w}_U + \beta\mathbf{z}_U) + (\alpha\mathbf{w}_V + \beta\mathbf{z}_V),$$

implying that $P_{U\|V}(\alpha\mathbf{w} + \beta\mathbf{z}) = \alpha\mathbf{w}_U + \beta\mathbf{z}_U$. It is clear that the operator $P = P_{U\|V}$ satisfies the equation

$$P^2 = P. \tag{4.31}$$

Definition 4.19 An operator P on a vector space W is called **idempotent** if it satisfies the equation (4.31).

Theorem 4.17 *An operator P on a finite dimensional vector space W is idempotent if and only if it is the projection onto $\mathrm{R}(P)$ parallel to its null space $\mathrm{N}(P)$.*

Proof. If P is a projection then $P_{U\|V}\mathbf{w} = \mathbf{w}_U$ for every $\mathbf{w}$ in W. Since $\{\mathbf{w}_U\}_U = \mathbf{w}_U$, we obtain that $P_{U\|V}^2 = P_{U\|V}$, which shows that $P_{U\|V}$ is an idempotent operator. If $P^2 = P$ then $P(P\mathbf{w}) = P\mathbf{w}$ implying that P acts as the identity operator I on $\mathrm{R}(P)$. It follows that $\mathrm{R}(P)$ and $\mathrm{N}(P)$ have only the zero vector as a common element. By the Rank-Nullity theorem $\dim(\mathrm{R}(P)) + \dim(\mathrm{N}(P)) = \dim W$ implying that $\mathrm{R}(P) \dotplus \mathrm{N}(P) = W$ is a direct sum.   □

Given a direct sum $W = U \dotplus V$ the operator $P_{U\|V}$ is called the **skew projection** onto U parallel to V. Skew projections appear naturally if one has a basis which is split into two blocks so that the first spans the subspace U and the second spans the subspace V.

Let $\mathbb{R}^n = U \dotplus V$ be a direct sum. Then the matrix

$$\mathbf{P}_{U\|V} = \begin{pmatrix} \mathbf{e}_{1U} & \mathbf{e}_{2U} & \cdots & \mathbf{e}_{nU} \end{pmatrix}.$$

is a skew projection onto U parallel to V.

Definition 4.20 An $n \times n$ matrix $\mathbf{P}$ is called **idempotent** if $\mathbf{P}^2 = \mathbf{P}$.

Problem 4.18 Let

$$U = \mathrm{Lin}\left\{ \begin{pmatrix} 1 \\ 0 \\ 1 \\ 1 \end{pmatrix}, \begin{pmatrix} 1 \\ 1 \\ 0 \\ 0 \end{pmatrix} \right\}, \quad V = \mathrm{Lin}\left\{ \begin{pmatrix} 3 \\ -1 \\ 1 \\ -1 \end{pmatrix}, \begin{pmatrix} 2 \\ 0 \\ 1 \\ -1 \end{pmatrix} \right\}.$$

Show that $U \dotplus V = \mathbb{R}^4$ and find the matrix $\mathbf{P}_{U\|V}$.

Solution: The reduced row echelon form rref($\mathbf{B}$) of the matrix

$$\mathbf{B} = \begin{pmatrix} \mathbf{u}_1 & \mathbf{u}_2 & \mathbf{v}_1 & \mathbf{v}_2 \end{pmatrix} = \begin{pmatrix} 1 & 1 & 3 & 2 \\ 0 & 1 & -1 & 0 \\ 1 & 0 & 1 & 1 \\ 1 & 0 & -1 & -1 \end{pmatrix}$$

is $\mathbf{I}_4$. Hence $N(\mathbf{B}) = \{\mathbf{0}\}$. It follows that the system $\{\mathbf{u}_1, \mathbf{u}_2, \mathbf{v}_1, \mathbf{v}_2\}$ makes a basis for $\mathbb{R}^4$. This implies that $U \dotplus V = \mathbb{R}^4$. Applying the method of extended matrix we find that

$$\mathbf{B}^{-1} = \frac{1}{4} \begin{pmatrix} 0 & 0 & 2 & 2 \\ 2 & 2 & -3 & 1 \\ 2 & -2 & -3 & 1 \\ -2 & 2 & 5 & -3 \end{pmatrix}.$$

Then

$$\mathbf{Be}_1 = \mathbf{u}_1, \quad \mathbf{Be}_2 = \mathbf{u}_2, \quad \mathbf{Be}_3 = \mathbf{v}_1, \quad \mathbf{Be}_4 = \mathbf{v}_2,$$

$$\mathbf{B}^{-1}\mathbf{u}_1 = \mathbf{e}_1, \quad \mathbf{B}^{-1}\mathbf{u}_2 = \mathbf{e}_2, \quad \mathbf{B}^{-1}\mathbf{v}_1 = \mathbf{e}_3, \quad \mathbf{B}^{-1}\mathbf{v}_2 = \mathbf{e}_4.$$

Therefore, the matrix

$$\mathbf{C} = \begin{pmatrix} \mathbf{u}_1 & \mathbf{u}_2 & \mathbf{0} & \mathbf{0} \end{pmatrix} \mathbf{B}^{-1} = \frac{1}{4} \begin{pmatrix} 2 & 2 & -1 & 3 \\ 2 & 2 & -3 & 1 \\ 0 & 0 & 2 & 2 \\ 0 & 0 & 2 & 2 \end{pmatrix}$$

satisfies $\mathbf{Cu}_1 = \mathbf{u}_1$, $\mathbf{Cu}_2 = \mathbf{u}_2$, $\mathbf{Cv}_1 = \mathbf{Cv}_2 = \mathbf{0}$. It follows that $\mathbf{Cx}_U = \mathbf{x}_U$, $\mathbf{Cx}_V = \mathbf{0}$ for every $\mathbf{x}$. Hence $\mathbf{C} = \mathbf{P}_{U\|V}$. □

In general, let $\{\mathbf{u}_1, \ldots, \mathbf{u}_r\}$ be a basis for a vector subspace U in $\mathbb{R}^n$, $\{\mathbf{v}_1, \ldots, \mathbf{v}_s\}$ be a basis for a vector subspace V in $\mathbb{R}^n$. If U and V have only zero vector in common and $r + s = n$ then $U \dotplus V = \mathbb{R}^n$. In this case the union of these two bases makes a basis for $\mathbb{R}^n$. Therefore, the $n \times n$ matrix with the columns equal to the vectors of this basis is invertible. The formula for $\mathbf{P}_{U\|V}$ is given by

$$\mathbf{P}_{U\|V} = \begin{pmatrix} \mathbf{u}_1 & \ldots & \mathbf{u}_r & \mathbf{0} & \ldots & \mathbf{0} \end{pmatrix} \begin{pmatrix} \mathbf{u}_1 & \ldots & \mathbf{u}_r & \mathbf{v}_1 & \ldots & \mathbf{v}_s \end{pmatrix}^{-1}. \tag{4.32}$$

Problems

Prob. 140 — If V is a finite-dimensional vector space and if A is a non-zero subspace of V prove that there exists a subspace B of V such that $V = A \dotplus B$.

Prob. 141 — Let V be a finite-dimensional vector space and A, B be subspaces of V such that $\text{Lin}(A, B) = V$ and $\dim(A) + \dim(B) = \dim V$. Prove that $V = A \dotplus B$.

Prob. 142 — Let V be a finite-dimensional vector space and $T : V \to V$ be a linear operator on V. Prove that $V = R(T) \dotplus N(T)$ if and only if $R(T) = R(T^2)$.

Prob. 143 — Consider the subspaces V and W of $\mathbb{R}^3$, where

$$V = \text{Lin}\left\{\begin{pmatrix}1\\0\\1\end{pmatrix}, \begin{pmatrix}1\\1\\0\end{pmatrix}\right\} \quad \text{and} \quad W = \text{Lin}\left\{\begin{pmatrix}0\\1\\1\end{pmatrix}\right\}$$

(**a**) Show that $\mathbb{R}^3$ is the direct sum of V and W.
(**b**) Find a basis of $W^\perp$, the orthogonal complement of W in $\mathbb{R}^3$.
(**c**) Write down a 3×2 matrix $\mathbf{A}$ such that the range $R(\mathbf{A})$ of $\mathbf{A}$ is V.
(**d**) Write down a 2×3 matrix $\mathbf{B}$ such that the null space $N(\mathbf{B})$ is W.
(**e**) Show that $\mathbf{BA}$, where $\mathbf{A}$ and $\mathbf{B}$ are defined in (**c**) and (**d**), is an invertible matrix and consider the matrix

$$\mathbf{Q} = \mathbf{A}(\mathbf{BA})^{-1}\mathbf{B}$$

(**1**) Show that $\mathbf{Q}$ is idempotent.
(**2**) Find $\mathbf{Q}$.
(**3**) Show that the linear transformation $T : \mathbb{R}^3 \to \mathbb{R}^3$ defined by $T(\mathbf{x}) = \mathbf{Qx}$ is a projection of $\mathbb{R}^3$ onto V parallel to W.

Chapter 5
Diagonalization

5.1 Introduction

Let us consider the arithmetic-geometric progression defined by the following recurrence equation, which has applications to elementary finance theory.

$$y_{n+1} = qy_n + d, \ n = 0,1,\ldots, y_0 = a. \tag{5.1}$$

If $q = 1$, then we obtain the arithmetic progression. If $d = 0$ and $q \neq 1$, we obtain the geometric progression. To get a formula for the general term of the sequence (5.1), when $q \neq 1$, we rewrite (5.1) in a matrix form:

$$\begin{pmatrix} y_{n+1} \\ 1 \end{pmatrix} = \begin{pmatrix} q & d \\ 0 & 1 \end{pmatrix} \begin{pmatrix} y_n \\ 1 \end{pmatrix}.$$

Easy calculations show that

$$\begin{pmatrix} q & d \\ 0 & 1 \end{pmatrix} \begin{pmatrix} 1 \\ 0 \end{pmatrix} = q \begin{pmatrix} 1 \\ 0 \end{pmatrix}, \quad \begin{pmatrix} q & d \\ 0 & 1 \end{pmatrix} \begin{pmatrix} \frac{d}{1-q} \\ 1 \end{pmatrix} = \begin{pmatrix} \frac{d}{1-q} \\ 1 \end{pmatrix}. \tag{5.2}$$

To simplify notations, let

$$\mathbf{A} = \begin{pmatrix} q & d \\ 0 & 1 \end{pmatrix}, \ \mathbf{v}_1 = \begin{pmatrix} 1 \\ 0 \end{pmatrix}, \ \mathbf{v}_2 = \begin{pmatrix} \frac{d}{1-q} \\ 1 \end{pmatrix}.$$

Then formulas (5.2) can be rewritten as follows:

$$\mathbf{A}\mathbf{v}_1 = q\mathbf{v}_1, \ \mathbf{A}\mathbf{v}_2 = \mathbf{v}_2. \tag{5.3}$$

Since $\mathbf{v}_1$ and $\mathbf{v}_2$ are not proportional, they make a basis in $\mathbb{R}^2$, see Lemma 2.1. Since

$$\begin{pmatrix} y_0 \\ 1 \end{pmatrix} = \left(y_0 - \frac{d}{1-q} \right) \mathbf{v}_1 + \mathbf{v}_2,$$

we obtain that

$$\binom{y_n}{1} = \mathbf{A}^n \binom{y_0}{1} = \left(y_0 - \frac{d}{1-q}\right)\mathbf{A}^n\mathbf{v}_1 + \mathbf{A}^n\mathbf{v}_2 = \left(y_0 - \frac{d}{1-q}\right)q^n\mathbf{v}_1 + \mathbf{v}_2,$$

implying the formula

$$y_n = \left(y_0 - \frac{d}{1-q}\right)q^n + \frac{d}{1-q}, \quad n = 0, 1, 2, \ldots. \tag{5.4}$$

Problem 5.1 Each week a company produces 1,000 LCD panels and sells 20% of its warehouse stock, which is empty initially. What is the minimum warehouse size needed to accommodate the company's production?

Solution: Let y_n be the number of panels stored in the warehouse at the end of the nth week. Then $y_0=0$ and $y_{n+1} = (1 - 0.2)y_n + 1000 = 0.8y_n + 1000$. By (5.1) we see that $q = 0.8$ and $d = 1,000$. Then by (5.4) we obtain that

$$y_n = \left(0 - \frac{1000}{1 - 0.8}\right)0.8^n + \frac{1000}{1 - 0.8} = 5000(1 - 0.8^n) < 5000. \tag{5.5}$$

Thus the warehouse must accommodate at least $5,000$ LCD panels.   □

Problem 5.2 Determine the monthly repayments needed to repay a $100,000 loan that is paid back over 25 years with an annual compounded interest rate of 8%

Solution: In this problem, the time interval between consecutive repayments is one month, whereas the period during which interest is charged is one year. So, if the monthly installment is x dollars, then the outstanding debt must be decreased by $12x$ at the end of each year.

Let y_n be the balance of the loan by the end of the nth year, Then

$$y_n = 1.08y_{n-1} - 12x, \quad y_0 = 0.$$

By the formula (5.1) we have $y_0 = 100,000$, $q = 1.08$, $d = -12x$. By (5.4), we obtain that

$$y_n = \left(100,000 - \frac{-12x}{1 - 1.08}\right)1.08^n + \frac{-12x}{1 - 1.08} = (100,000 - 150x)1.08^n + 150x.$$

By the loan contract we must have $y_{25} = 0$, implying the equation for x:

$$100,000 \times 1.08^{25} = 150x\left(1.08^{25} - 1\right) \Rightarrow x = 780.66. \quad □$$

Let us consider the invertible matrices

$$\mathbf{P} = \begin{pmatrix} \mathbf{v}_1 & \mathbf{v}_2 \end{pmatrix}, \quad \mathbf{D} = \begin{pmatrix} q & 0 \\ 0 & 1 \end{pmatrix}$$

Then

$$\mathbf{AP} = \begin{pmatrix} \mathbf{Av}_1 & \mathbf{Av}_2 \end{pmatrix} = \begin{pmatrix} q\mathbf{v}_1 & \mathbf{v}_2 \end{pmatrix} = \mathbf{PD} \Leftrightarrow \mathbf{A} = \mathbf{PDP}^{-1}. \tag{5.6}$$

The formula on the right-hand side of (5.6) shows that

$$\mathbf{A}^n = \underbrace{\mathbf{PDP}^{-1}\mathbf{PDP}^{-1}\cdots\mathbf{PDP}^{-1}}_{n} = \mathbf{PD}^n\mathbf{P}^{-1}, \tag{5.7}$$

where

$$\mathbf{D}^n = \begin{pmatrix} q^n & 0 \\ 0 & 1 \end{pmatrix}.$$

Using the above formula and the formula (5.7), one can easily evaluate any power of the matrix $\mathbf{A}$ multiplying only three matrices as it ia shown in (5.7).

The method described by the formulas (5.6) and (5.7) is called "Diagonalization." It was successfully applied to solution of equations described by (5.1), but it can also be applied to solution of the second order recurrences

$$y_{n+1} = ay_n + by_{n-1}, \ n = 1, 2, \ldots, y_0 = c_0, y_1 = c_1, \tag{5.8}$$

where a, b are some constants and c_0 and c_1 are the initial values for the dynamical system described by (5.8).

Let

$$\mathbf{A} = \begin{pmatrix} a & b \\ 1 & 0 \end{pmatrix}. \tag{5.9}$$

Then

$$\begin{pmatrix} y_{n+1} \\ y_n \end{pmatrix} = \begin{pmatrix} a & b \\ 1 & 0 \end{pmatrix} \begin{pmatrix} y_n \\ y_{n-1} \end{pmatrix}, \ n = 1, 2, \ldots, \tag{5.10}$$

is equivalent to (5.8). To find the vectors $\mathbf{v}_1$ and $\mathbf{v}_2$ for the matrix $\mathbf{A}$ defined in (5.9), we find the characteristic polynomial of $\mathbf{A}$:

$$\det(\mathbf{A} - \lambda \mathbf{I}_2) = \det \begin{pmatrix} a - \lambda & b \\ 1 & -\lambda \end{pmatrix} = \lambda^2 - a\lambda - b. \tag{5.11}$$

Problem 5.3 Find a formula for the sequence $\{y_n\}$ defined by

$$y_{n+1} = y_n + 2y_{n-1}, \ n = 1, 2, \ldots, \ y_0 = c_0, y_1 = c_1.$$

Solution: By (5.9)

$$\mathbf{A} = \begin{pmatrix} 1 & 2 \\ 1 & 0 \end{pmatrix}.$$

Then $p_{\mathbf{A}}(\lambda) = \lambda^2 - \lambda - 2 = (\lambda + 1)(\lambda - 2)$, implying that

$$\begin{cases} \mathbf{A} + \mathbf{I}_2 &= \begin{pmatrix} 2 & 2 \\ 1 & 1 \end{pmatrix} \Rightarrow \mathbf{Av}_1 = -\mathbf{v}_1, \ \mathbf{v}_1 = \begin{pmatrix} 1 \\ -1 \end{pmatrix} \\ \mathbf{A} - 2\mathbf{I}_2 &= \begin{pmatrix} -1 & 2 \\ 1 & -2 \end{pmatrix} \Rightarrow \mathbf{Av}_2 = 2\mathbf{v}_2, \ \mathbf{v}_2 = \begin{pmatrix} 2 \\ 1 \end{pmatrix}. \end{cases}$$

The vectors $\mathbf{v}_1$ and $\mathbf{v}_2$ being not proportional make a basis for $\mathbb{R}^2$. Let $\mathbf{P} = \begin{pmatrix} \mathbf{v}_1 & \mathbf{v}_2 \end{pmatrix}$ be the matrix made by the columns $\mathbf{v}_1$ and $\mathbf{v}_2$. Then

$$\mathbf{P}^{-1} \begin{pmatrix} c_1 \\ c_0 \end{pmatrix} = \frac{1}{3} \begin{pmatrix} 1 & -2 \\ 1 & 1 \end{pmatrix} \begin{pmatrix} c_1 \\ c_0 \end{pmatrix} = \frac{1}{3} \begin{pmatrix} c_1 - 2c_0 \\ c_1 + c_0 \end{pmatrix} \tag{5.12}$$

Multiplying both sides of (5.12) by $\mathbf{P}$, we obtain that

$$\begin{pmatrix} c_1 \\ c_0 \end{pmatrix} = \frac{c_1 - 2c_0}{3} \mathbf{v}_1 + \frac{c_1 + c_0}{3} \mathbf{v}_2. \tag{5.13}$$

Then

$$\begin{pmatrix} y_{n+1} \\ y_n \end{pmatrix} = \mathbf{A} \begin{pmatrix} c_1 \\ c_0 \end{pmatrix} = \frac{c_1 - 2c_0}{3} \mathbf{A}^n \mathbf{v}_1 + \frac{c_1 + c_0}{3} \mathbf{A}^n \mathbf{v}_2 =$$

$$(-1)^n \frac{c_1 - 2c_0}{3} \mathbf{v}_1 + 2^n \frac{c_1 + c_0}{3} \mathbf{v}_2 \Rightarrow \boxed{y_n = (-1)^{n+1} \frac{c_1 - 2c_0}{3} + \frac{c_1 + c_0}{3}.} \quad \square$$

$$\tag{5.14}$$

Problem 5.4 Find a formula for the sequence $\{y_n\}$ defined by

$$y_{n+1} = 2y_n - y_{n-1}, \ n = 1, 2, \ldots, \ y_0 = c_0, y_1 = c_1.$$

Solution: By (5.9) and (5.11)

$$\mathbf{A} = \begin{pmatrix} 2 & -1 \\ 1 & 0 \end{pmatrix}, \ \det(\mathbf{A} - \lambda \mathbf{I}_2) = (\lambda - 1)^2.$$

Then

$$\mathbf{A} - \mathbf{I}_2 \sim \begin{pmatrix} 1 & -1 \\ 0 & 0 \end{pmatrix} \Rightarrow \mathbf{A}\mathbf{v}_1 = \mathbf{v}_1, \ \mathbf{v}_1 = \begin{pmatrix} 1 \\ 1 \end{pmatrix}.$$

In Problem 5.3 we introduced the matrix $\mathbf{P}$ which is a transition matrix between the standard basis and the basis $\{\mathbf{v}_1, \mathbf{v}_2\}$ in which the matrix $\mathbf{A}$ acts as a simple multiplication by real numbers. As result, we could express the powers of $\mathbf{A}$ via the powers of a diagonal matrix. This schema does not work in this case since we have only one vector $\mathbf{v}_1$. Let us search the second vector among the solutions to the system

$$(\mathbf{A} - \mathbf{I}_2)\mathbf{x} = \mathbf{v}_1 \Leftrightarrow \mathbf{x} = \begin{pmatrix} 1 \\ 0 \end{pmatrix} + s \begin{pmatrix} 1 \\ 1 \end{pmatrix}, \ s \in \mathbb{R}.$$

Then

$$\mathbf{v}_2 = \begin{pmatrix} 1 \\ 0 \end{pmatrix} \Rightarrow \mathbf{A}\mathbf{v}_2 = \mathbf{v}_1 + \mathbf{v}_2.$$

We define

$$\mathbf{P} = \begin{pmatrix} \mathbf{v}_1 & \mathbf{v}_2 \end{pmatrix} = \begin{pmatrix} 1 & 1 \\ 1 & 0 \end{pmatrix} \Rightarrow \mathbf{P}^{-1} = \begin{pmatrix} 0 & 1 \\ 1 & -1 \end{pmatrix}.$$

The action of $\mathbf{A}$ in the basis $\{\mathbf{v}_1, \mathbf{v}_2\}$ can be represented by the matrix

$$\mathbf{J} = \begin{pmatrix} 1 & 1 \\ 0 & 1 \end{pmatrix} \Rightarrow \mathbf{AP} = \begin{pmatrix} 1 & 2 \\ 1 & 1 \end{pmatrix} = \mathbf{PJ} \Leftrightarrow \mathbf{A} = \mathbf{PJP}^{-1}.$$

Then $\mathbf{A}^n = \mathbf{PJ}^n\mathbf{P}^{-1}$. Using mathematical induction, one can easily check that

$$\mathbf{J}^n = \begin{pmatrix} 1 & n \\ 0 & 1 \end{pmatrix}.$$

It follows that

$$\mathbf{A}^n = \begin{pmatrix} 1 & 1 \\ 1 & 0 \end{pmatrix} \begin{pmatrix} 1 & n \\ 0 & 1 \end{pmatrix} \begin{pmatrix} 0 & 1 \\ 1 & -1 \end{pmatrix} = \begin{pmatrix} n+1 & -n \\ n & -(n-1) \end{pmatrix}.$$

Applying formula (5.10), we obtain that

$$y_{n+1} = (n+1)c_1 - nc_0, \ n = 1, 2, \ldots . \quad \square$$

Problems

Prob. 144 — A person saves \$300 in a bank account at the beginning of each month. The bank offers a return of 2% compounded monthly.

(a) Determine the total amount saved after 12 months.
(b) After how many months does the amount saved first exceed \$5,000?

Prob. 145 — A gentleman opens a savings account with a bank. The interest rate is fixed at 3% per annum, compounded annually, for 10 years after he opens the account. He opens the savings account with a payment of \$500 on 1 January 2023, and makes further contributions of \$500 annually, on 1 January each year, beginning on 1 January 2024. Find the value of his savings in 10 years after he opens the account.

Prob. 146 — A savings account pays interest annually at a rate of 6%. An investor deposits an amount P which is large enough to ensure that each year for the next 12 years, this investor can withdraw \$1,000 from the account at the end of the year, maintaining a non-negative balance. Find the minimal value of P.

Prob. 147 — A customer invests P dollars into a financial institution under a fixed annual interest rate $100 \cdot r\%$. How much capital I this customer can withdraw from the account at the beginning of each year so that in N years the capital would be used up?

Prob. 148 — Find a formula for the sequence $\{y_n\}$ defined by

$$y_{n+1} = 4y_n - 4y_{n-1}, \ n = 1, 2, 3, \ldots, \ y_0 = c_0, y_1 = c_1.$$

Prob. 149 — Find a formula for the sequence $\{y_n\}$ defined by

$$y_{n+1} = 6y_n - 11y_{n-1} + 6y_{n-2}, \ n = 2,3,\ldots, \ y_0 = y_1 = y_2 = 1.$$

Hint: Consider the matrix

$$\mathbf{A} = \begin{pmatrix} 6 & -11 & 6 \\ 1 & 0 & 0 \\ 0 & 1 & 0 \end{pmatrix}$$

and apply the proof of Theorem 3.40 to prove that

$$p_{\mathbf{A}}(\lambda) = -\lambda^3 + 6\lambda^2 - 11\lambda + 6 = -(\lambda - 1)(\lambda - 2)(\lambda - 3).$$

5.2 Diagonalization

Definition 5.1 Suppose that $\mathbf{A}$ is a **square matrix**. The number λ is said to be an **eigenvalue** of $\mathbf{A}$ if for some non-zero vector $\mathbf{x}$,

$$\mathbf{Ax} = \lambda\mathbf{x}.$$

A **non-zero** vector $\mathbf{x}$ for which this equation holds is called an **eigenvector** for eigenvalue λ or an eigenvector of $\mathbf{A}$ corresponding to eigenvalue λ.

The equation $\mathbf{Ax} = \lambda\mathbf{x}$ has a nonzero solution $\mathbf{x}$ if and only if the homogeneous equation $(\mathbf{A} - \lambda\mathbf{I})\mathbf{x} = \mathbf{0}$ has a nonzero solution, which happens if and only if the matrix $\mathbf{A} - \lambda\mathbf{I}$ is singular or equivalently $\det(\mathbf{A} - \lambda\mathbf{I}) = 0$. Hence all eigenvalues are roots of the characteristic polynomial of matrix $\mathbf{A}$.

Problem 5.5 Find the characteristic polynomial, eigenvalues and eigenvectors of the matrix

$$\mathbf{A} = \begin{pmatrix} 7 & -15 \\ 2 & -4 \end{pmatrix}.$$

Solution: Step 1. First, we evaluate the matrix

$$\mathbf{A} - \lambda\mathbf{I} = \begin{pmatrix} 7 & -15 \\ 2 & -4 \end{pmatrix} - \lambda\begin{pmatrix} 1 & 0 \\ 0 & 1 \end{pmatrix} = \begin{pmatrix} 7 - \lambda & -15 \\ 2 & -4 - \lambda \end{pmatrix}.$$

Step 2. Next, we evaluate the characteristic polynomial:

$$|\mathbf{A} - \lambda\mathbf{I}| = \begin{vmatrix} 7 - \lambda & -15 \\ 2 & -4 - \lambda \end{vmatrix} = (7 - \lambda)(-4 - \lambda) + 30 =$$

$$\lambda^2 - 3\lambda + 2 = (\lambda - 1)(\lambda - 2).$$

Step 3. The eigenvalues of $\mathbf{A}$ are the roots of the characteristic polynomial:

$$\lambda_1 = 1 \quad \lambda_2 = 2.$$

Step 4. To find the eigenvector corresponding to the eigenvalue $\lambda_1 = 1$ we find the reduced row echelon form of the matrix

$$\mathbf{A} - \mathbf{I} = \begin{pmatrix} 6 & -15 \\ 2 & -5 \end{pmatrix} \sim \begin{pmatrix} 0 & 0 \\ 2 & -5 \end{pmatrix} \sim \begin{pmatrix} 1 & \frac{-5}{2} \\ 0 & 0 \end{pmatrix}.$$

We see that $x_2 = 2s$ is the non-leading variable. It follows that the solutions to the equation $(\mathbf{A} - \mathbf{I})\mathbf{x} = \mathbf{0}$ are given by

$$\mathbf{x} = \begin{pmatrix} 5s \\ 2s \end{pmatrix} = s \begin{pmatrix} 5 \\ 2 \end{pmatrix}.$$

To find the eigenvector corresponding to the eigenvalue $\lambda_1 = 2$ we find the reduced row echelon form of the matrix

$$\mathbf{A} - 2\mathbf{I} = \begin{pmatrix} 5 & -15 \\ 2 & -6 \end{pmatrix} \sim \begin{pmatrix} 0 & 0 \\ 2 & -6 \end{pmatrix} \sim \begin{pmatrix} 1 & -3 \\ 0 & 0 \end{pmatrix}.$$

We see that $x_2 = s$ is the non-leading variable. It follows that the solutions to the equation $(\mathbf{A} - 2\mathbf{I})\mathbf{x} = \mathbf{0}$ are given by

$$\mathbf{x} = \begin{pmatrix} 3s \\ 1s \end{pmatrix} = s \begin{pmatrix} 3 \\ 1 \end{pmatrix}. \quad \square$$

Problem 5.6 Find the characteristic polynomial, eigenvalues and eigenvectors of the matrix

$$\mathbf{A} = \begin{pmatrix} 3 & 0 & 4 \\ 0 & 3 & 4 \\ 4 & 4 & 7 \end{pmatrix}.$$

Solution: Step 1. First, we evaluate the matrix

$$\mathbf{A} - \lambda\mathbf{I} = \begin{pmatrix} 3 & 0 & 4 \\ 0 & 3 & 4 \\ 4 & 4 & 7 \end{pmatrix} - \lambda\mathbf{I} = \begin{pmatrix} 3 - \lambda & 0 & 4 \\ 0 & 3 - \lambda & 4 \\ 4 & 4 & 7 - \lambda \end{pmatrix}.$$

Step 2. Next, we evaluate the characteristic polynomial of $\mathbf{A}$:

$$p_\mathbf{A}(\lambda) = \begin{vmatrix} 3 - \lambda & 0 & 4 \\ 0 & 3 - \lambda & 4 \\ 4 & 4 & 7 - \lambda \end{vmatrix} = \boxed{-\lambda^3 + 13\lambda^2 - 19\lambda - 33}.$$

Step 3. We find the roots of $p_A(\lambda)$. By the Rational Zeros Theorem the rational roots are of the form:

$$\frac{\textbf{factor of } -33}{\textbf{factor of } 1} = \textbf{factor of } 33.$$

$$
\begin{array}{r|rrrr}
1 & -1 & 13 & -19 & -33 \\
  &   & -1 & 12 & -7 \\
\hline
  & -1 & 12 & -7 & \mathbf{-40}
\end{array}
\qquad\qquad
\begin{array}{r|rrrr}
-1 & -1 & 13 & -19 & -33 \\
   &   & 1 & -14 & 33 \\
\hline
   & -1 & 14 & -33 & \mathbf{0}
\end{array}
$$

By the Synthetic Division

$$-\lambda^3 + 13\lambda^2 - 19\lambda - 33 = (\lambda + 1)(-\lambda^2 + 14\lambda - 33).$$

By the formula for the roots of quadratic equations in a reduced form we find

$$\lambda_1 = -1, \quad \lambda_2 = 7 - 4 = 3, \quad \lambda_3 = 7 + 4 = 11.$$

Step 4. We find the eigenvector corresponding to the eigenvalue $\lambda_1 = -1$:

$$
\mathbf{A} + \mathbf{I} = \begin{pmatrix} 4 & 0 & 4 \\ 0 & 4 & 4 \\ 4 & 4 & 8 \end{pmatrix} \sim \begin{pmatrix} 1 & 0 & 1 \\ 0 & 1 & 1 \\ 1 & 1 & 2 \end{pmatrix} \sim \begin{pmatrix} 1 & 0 & 1 \\ 0 & 1 & 1 \\ 0 & 0 & 0 \end{pmatrix} \Rightarrow \mathbf{x} = \begin{pmatrix} -s \\ -s \\ s \end{pmatrix} = s \begin{pmatrix} -1 \\ -1 \\ 1 \end{pmatrix}.
$$

We find the eigenvector corresponding to the eigenvalue $\lambda_2 = 3$:

$$
\mathbf{A} - 3\mathbf{I} = \begin{pmatrix} 0 & 0 & 4 \\ 0 & 0 & 4 \\ 4 & 4 & 4 \end{pmatrix} \sim \begin{pmatrix} 0 & 0 & 1 \\ 0 & 0 & 1 \\ 1 & 1 & 1 \end{pmatrix} \sim \begin{pmatrix} 1 & 1 & 0 \\ 0 & 0 & 1 \\ 0 & 0 & 0 \end{pmatrix} \Rightarrow \mathbf{x} = \begin{pmatrix} -s \\ s \\ 0 \end{pmatrix} = s \begin{pmatrix} -1 \\ 1 \\ 0 \end{pmatrix}.
$$

We find the eigenvector corresponding to the eigenvalue $\lambda_3 = 11$:

$$
\mathbf{A} - 11 \cdot \mathbf{I} = \begin{pmatrix} -8 & 0 & 4 \\ 0 & -8 & 4 \\ 4 & 4 & -4 \end{pmatrix} \sim \begin{pmatrix} 2 & 0 & -1 \\ 0 & 2 & -1 \\ 1 & 1 & -1 \end{pmatrix} \sim \begin{pmatrix} 0 & -2 & 1 \\ 0 & 2 & -1 \\ 1 & 1 & -1 \end{pmatrix} \sim
$$

$$
\begin{pmatrix} 1 & 0 & -1/2 \\ 0 & 1 & -1/2 \\ 0 & 0 & 0 \end{pmatrix} \Rightarrow \mathbf{x} = \begin{pmatrix} s \\ s \\ 2s \end{pmatrix} = s \begin{pmatrix} 1 \\ 1 \\ 2 \end{pmatrix}.
$$

Problem 5.7 Find the characteristic polynomial, eigenvalues and eigenvectors of the matrix

$$A = \begin{pmatrix} -3 & -1 & -2 \\ 1 & -1 & 1 \\ 1 & 1 & 0 \end{pmatrix}. \tag{5.15}$$

Solution: **Step 1.** First, we evaluate the matrix

$$A - \lambda I = \begin{pmatrix} -3 & -1 & -2 \\ 1 & -1 & 1 \\ 1 & 1 & 0 \end{pmatrix} - \lambda I = \begin{pmatrix} -3 - \lambda & -1 & -2 \\ 1 & -1 - \lambda & 1 \\ 1 & 1 & -\lambda \end{pmatrix}.$$

Step 2. Next, we evaluate the characteristic polynomial:

$$p_A(\lambda) = \begin{vmatrix} -3 - \lambda & -1 & -2 \\ 1 & -1 - \lambda & 1 \\ 1 & 1 & -\lambda \end{vmatrix} = \boxed{-(\lambda + 2)(\lambda + 1)^2}.$$

Step 3. We find the eigenvector corresponding to the eigenvalue $\lambda_1 = -2$:

$$A + 2I = \begin{pmatrix} -1 & -1 & -2 \\ 1 & 1 & 1 \\ 1 & 1 & 2 \end{pmatrix} \sim \begin{pmatrix} 1 & 1 & 2 \\ 1 & 1 & 1 \\ 0 & 0 & 0 \end{pmatrix} \sim \begin{pmatrix} 1 & 1 & 0 \\ 0 & 0 & 1 \\ 0 & 0 & 0 \end{pmatrix} \Rightarrow v_1 = \boxed{\begin{pmatrix} -1 \\ 1 \\ 0 \end{pmatrix}}.$$

Step 4. We find the eigenvector(s) corresponding to the eigenvalue $\lambda_1 = -1$:

$$A + I = \begin{pmatrix} -2 & -1 & -2 \\ 1 & 0 & 1 \\ 1 & 1 & 1 \end{pmatrix} \sim \begin{pmatrix} 0 & 1 & 0 \\ 1 & 0 & 1 \\ 1 & 1 & 1 \end{pmatrix} \sim \begin{pmatrix} 1 & 0 & 1 \\ 0 & 1 & 0 \\ 0 & 0 & 0 \end{pmatrix} \Rightarrow v_2 = \boxed{\begin{pmatrix} -1 \\ 0 \\ 1 \end{pmatrix}}.$$

Notice that in Problem 5.7 there are only TWO eigenvectors. However, in all previous problems the eigenvectors corresponding to different eigenvalues are linearly independent. It turns out that it is a general result.

Theorem 5.1 *Eigenvectors corresponding to different eigenvalues of a square matrix are linearly independent.*

Proof. Let $S = \{\mathbf{v}_1, \ldots, \mathbf{v}_k\}$ be any set of eigenvectors of $\mathbf{A}$ corresponding to **different** eigenvalues $\lambda_1, \ldots, \lambda_k$. By definition, the set S is independent set of vectors if the formula

$$c_1 \mathbf{v}_1 + \ldots + c_k \mathbf{v}_k = \mathbf{0}, \tag{5.16}$$

where all numbers c_i are real, can hold only if $c_1 = \cdots = c_k = 0$. Applying $\mathbf{A}$, $\mathbf{A}^2, \ldots, \mathbf{A}^{k-1}$ to (5.16), we obtain the system:

$$\begin{cases} c_1 \mathbf{v}_1 + \ldots + c_k \mathbf{v}_k & = 0 \\ c_1 \lambda_1 \mathbf{v}_1 + \ldots + c_k \lambda_k \mathbf{v}_k & = 0 \\ \ldots\ldots\ldots\ldots\ldots\ldots\ldots\ldots & = 0 \\ c_1 \lambda_1^{k-1} \mathbf{v}_1 + \ldots + c_k \lambda_k^{k-1} \mathbf{v}_k & = 0. \end{cases} \tag{5.17}$$

Multiplying the first equation in (5.17) by a real number x_1, the second equation by x_2 etc, and adding them together, we obtain

$$(x_1 + \lambda_1 x_2 + \lambda_1^2 x_3 + \cdots \lambda_1^{k-1} x_k) c_1 \mathbf{v}_1 +$$
$$(x_1 + \lambda_2 x_2 + \lambda_2^2 x_3 + \cdots \lambda_2^{k-1} x_k) c_2 \mathbf{v}_2 + \cdots +$$
$$(x_1 + \lambda_k x_2 + \lambda_k^2 x_3 + \cdots \lambda_k^{k-1} x_k) c_k \mathbf{v}_k = \mathbf{0}. \tag{5.18}$$

Let

$$\mathbf{V}_k = \begin{pmatrix} 1 & \lambda_1 & \lambda_1^2 & \cdots & \lambda_1^{k-1} \\ 1 & \lambda_2 & \lambda_2^2 & \cdots & \lambda_2^{k-1} \\ \vdots & \vdots & \vdots & \ddots & \vdots \\ 1 & \lambda_k & \lambda_k^2 & \cdots & \lambda_k^{k-1} \end{pmatrix}, \quad \mathbf{x} = \begin{pmatrix} x_1 \\ x_2 \\ \vdots \\ x_k \end{pmatrix}.$$

By Theorem 3.34

$$\det(\mathbf{V}_k) = \prod_{1 \le i < j \le k} (\lambda_i - \lambda_j) \ne 0,$$

implying that the matrix $\mathbf{V}_k$ is invertible. It follows that $\mathbf{V}_k^{-1}$ exists and for every j, $j = 1, \ldots, k$ there is $\mathbf{x}$ such that $\mathbf{V}_k \mathbf{x} = \mathbf{e}_j$. Substituting this $\mathbf{x}$ into (5.18), we obtain that $c_j \mathbf{v}_j = \mathbf{0}$. But $\mathbf{v}_j$ is an eigenvector corresponding to the eigenvalue λ_j. Hence $\mathbf{v}_j \ne \mathbf{0}$, implying that $c_j = 0$. $\qquad\square$

Theorem 5.2 *If an $n \times n$ matrix $\mathbf{A}$ has n different eigenvalues, then it has a set of eigenvectors which make a basis for $\mathbb{R}^n$.*

Proof. By Theorem 5.1 any set of eigenvectors corresponding to different eigenvalues is linearly independent. Any n linearly independent vectors in $\mathbb{R}^n$ make a basis.   □

Let $\{v_1, \ldots, v_n\}$ be n linearly independent eigenvectors of an $n \times n$ matrix A. Then they make a basis in $\mathbb{R}^n$ and we may define the invertible matrix

$$P = \begin{pmatrix} v_1 & v_2 & \cdots & v_n \end{pmatrix}. \tag{5.19}$$

Let λ_k be the eigenvalue corresponding to v_k, $k = 1, \ldots, n$. We denote by $D = D(\lambda_1, \lambda_2, \ldots, \lambda_n)$ the diagonal matrix with eigenvalues on the diagonal placed in the order of the eigenvectors defining the matrix P in (5.19). Then

$$AP = \begin{pmatrix} Av_1 & Av_2 & \cdots & Av_n \end{pmatrix} = \begin{pmatrix} \lambda_1 v_1 & \lambda_2 v_2 & \cdots & \lambda_n v_n \end{pmatrix} = PD.$$

It follows that
$$AP = PD. \tag{5.20}$$

Theorem 5.3 *Let A be an $n \times n$ matrix, P an $n \times n$ invertible matrix and $D = D(\lambda_1, \lambda_2, \ldots, \lambda_n)$ a diagonal matrix satisfying (5.20). Then λ_k is the eigenvalue of A corresponding to the eigenvector $v_k = \mathrm{Col}_k(P)$.*

Proof. Under conditions of the theorem, we have

$$\begin{cases} AP &= \begin{pmatrix} Av_1 & Av_2 & \cdots & Av_n \end{pmatrix} \\ PD &= \begin{pmatrix} \lambda_1 v_1 & \lambda_2 v_2 & \cdots & \lambda_n v_n \end{pmatrix} \end{cases} \Rightarrow Av_k = \lambda_k v_k, \; k = 1, 2, \ldots, n.$$

□

Definition 5.2 An $n \times n$ matrix is called **diagonalizable** if there are a diagonal matrix $D = D(\lambda_1, \ldots, \lambda_n)$ and an invertible $n \times n$ matrix P such that $AP = DP$.

If A is a diagonalizable matrix, then by (5.7)

$$A^m = \underbrace{PDP^{-1}PDP^{-1}\cdots PDP^{-1}}_{n} = PD(\lambda_1^m, \ldots, \lambda_n^m)P^{-1}, \tag{5.21}$$

The formula (5.21) is the main purpose of diagonalization. It allows one to study the asymptotic behavior of the powers of a diagonalizable matrix.

By definition and Theorem 5.3 a matrix is diagonalizable if and only if it has a basis of eigenvectors. The matrix in Problem 5.7 does not have such a basis and therefore is not diagonalizable.

Definition 5.3 An $n \times n$ matrix $\mathbf{A}$ is called **nilpotent** if $\mathbf{A}^k = \mathbf{0}$ for some positive integer k.

Theorem 5.4 *Any non-zero nilpotent matrix is not diagonalizable.*

Proof. If $\mathbf{A}$ is diagonalizable, then there are an invertible matrix $\mathbf{P}$ and a diagonal matrix $\mathbf{D} = \mathbf{D}(\lambda_1 \ldots, \lambda_n)\mathbf{P}^{-1}$ such that $\mathbf{A} = \mathbf{PD}(\lambda_1 \ldots, \lambda_n)\mathbf{P}^{-1}$, implying that $\mathbf{0} = \mathbf{A}^k = \mathbf{PD}(\lambda_1^k \ldots, \lambda_n^k)\mathbf{P}^{-1}$. It follows that $\mathbf{0} = \mathbf{P}^{-1}\mathbf{0P} = \mathbf{D}(\lambda_1^k \ldots, \lambda_n^k)$. Then $\lambda_j^k = 0$ for every j. Hence $\lambda_1 = \cdots = \lambda_n = 0$, implying that $\mathbf{A} = \mathbf{0}$, which is a contradiction.   □

Here are examples of nilpotent matrices:

$$\begin{pmatrix} 0 & 1 \\ 0 & 0 \end{pmatrix}, \begin{pmatrix} 0 & 1 & 0 \\ 0 & 0 & 1 \\ 0 & 0 & 0 \end{pmatrix}, \begin{pmatrix} 0 & 1 & 0 & 0 \\ 0 & 0 & 1 & 0 \\ 0 & 0 & 0 & 1 \\ 0 & 0 & 0 & 0 \end{pmatrix}, \begin{pmatrix} 0 & 1 & 0 & 0 & 0 \\ 0 & 0 & 1 & 0 & 0 \\ 0 & 0 & 0 & 1 & 0 \\ 0 & 0 & 0 & 0 & 1 \\ 0 & 0 & 0 & 0 & 0 \end{pmatrix}. \tag{5.22}$$

Theorem 5.5 *Let $\mathbf{A}$ be an $n \times n$ nilpotent matrix. Then its characteristic polynomial $p_\mathbf{A}(\lambda)$ equals $(-1)^n \lambda^n$.*

Proof. If $\mathbf{A}$ is nilpotent, then $\mathbf{A}^k = \mathbf{0}$ for some positive integer k. If $\lambda \neq 0$ is an eigenvalue of $\mathbf{A}$, then there is a non-zero eigenvector $\mathbf{v}$ for $\mathbf{A}$. Since $\mathbf{A}^k = \mathbf{0}$, we obtain that $\mathbf{0} = \mathbf{A}^k\mathbf{v} = \lambda^k\mathbf{v}$, implying that $\lambda = 0$, which is a contradiction.   □

Corollary 5.1 *If* $\mathbf{A}$ *is an* $n \times n$ *nilpotent matrix, then* $\mathbf{A}^n = \mathbf{0}$.

Proof. Apply the Cayley-Hamilton Theorem (see Theorem 3.39).   □

Definition 5.4 If $\mathbf{A}$ is an $n \times n$ matrix and λ is an eigenvalue of $\mathbf{A}$, then the eigenspace of the eigenvalue λ is the subspace $N(\mathbf{A} - \lambda \mathbf{I})$ of $\mathbb{R}^n$.

The **eigenspace** $E(\lambda, \mathbf{A})$ of an eigenvalue λ can also be described as the set S, where $S = \{\mathbf{x} \mid \mathbf{Ax} = \lambda \mathbf{x}\}$. Notice that the eigenspace of an eigenvalue λ is the set of all eigenvectors belonging to the eigenvalue λ and the zero vector $\mathbf{0}$.

In what follows, we denote by $\mathbf{geo_A}(\mu)$ the dimension of $N(\mathbf{A} - \mu \mathbf{I})$. The number $\mathbf{geo_A}(\mu)$ is called the **geometric multiplicity** of an eigenvalue μ of a matrix $\mathbf{A}$.

The number $\mathbf{alg_A}(\mu)$ is the largest integer k such that $(\lambda - \mu)^k$ is a factor of the characteristic polynomial $p_\mathbf{A}(\lambda)$ of $\mathbf{A}$. The number $\mathbf{alg_A}(\mu)$ is called the **algebraic multiplicity** of an eigenvalue μ of a matrix $\mathbf{A}$.

For the matrix $\mathbf{A}$ in Problem 5.7 we have

$$\mathbf{geo_A}(-2) = \mathbf{alg_A}(-2), \ \mathbf{geo_A}(-1) < \mathbf{alg_A}(-1).$$

It follows that the main obstacle for diagonalization of a matrix $\mathbf{A}$ is the existence of eigenvalues μ for this matrix such that $\mathbf{geo_A}(\mu) < \mathbf{alg_A}(\mu)$.

Theorem 5.6 *Let* $\mathbf{A}$ *be an* $n \times n$ *matrix. Then* $\mathbf{A}$ *is diagonalizable if and only if*

$$\mathbf{geo_A}(\lambda) = \mathbf{alg_A}(\lambda)$$

for every eigenvalue λ *of* $\mathbf{A}$.

Proof. Since

$$p_\mathbf{A}(x) = \prod_{p_\mathbf{A}(\lambda)=0} (x - \lambda)^{\mathbf{alg_A}(\lambda)} \Rightarrow \sum_{p_\mathbf{A}(\lambda)=0} \mathbf{alg_A}(\lambda) = n,$$

for every λ with $p_\mathbf{A}(\lambda) = 0$ we can find a basis for $N(\mathbf{A} - \lambda \mathbf{I})$ and combine the obtained vectors in one system S. Since $\mathbf{geo_A}(\lambda) = \mathbf{alg_A}(\lambda)$, the total number of vectors in this system is n. If a linear combination of vectors in S represents a zero vector, then it can be split into a number of blocks so that the sums of vectors in each block be in one and the same null-space $N(\mathbf{A} - \lambda\mathbf{I})$. By Theorem 5.1 these sums are linearly independent, implying that each sum is zero. But vectors included in one block make a basis for $N(\mathbf{A} - \lambda\mathbf{I})$, which completes the proof. $\square$

Theorem 5.6 shows that the only reason for violation of the diagonalization property is the lack of eigenvectors to complete the construction of the matrix $\mathbf{P}$. There are two main reasons for this. The matrix

$$\mathbf{A} = \begin{pmatrix} 0 & 1 \\ -1 & 0 \end{pmatrix}$$

is not diagonalizable since $p_{\mathbf{A}}(\lambda) = \lambda^2 + 1$ has no real roots. But $p_{\mathbf{A}}(\lambda) = (\lambda - i)(\lambda + i)$ has two different complex roots. The proof of Theorem (5.2) is the same if instead of $\mathbb{R}^n$ we consider the complex vector space $\mathbb{C}^n$, implying that the matrix $\mathbf{A}$ is diagonalizable which can be established directly:

$$\begin{pmatrix} 0 & 1 \\ -1 & 0 \end{pmatrix}\begin{pmatrix} i \\ 1 \end{pmatrix} = (-i)\begin{pmatrix} i \\ 1 \end{pmatrix}, \begin{pmatrix} 0 & 1 \\ -1 & 0 \end{pmatrix}\begin{pmatrix} -i \\ 1 \end{pmatrix} = (i)\begin{pmatrix} -i \\ 1 \end{pmatrix} \Rightarrow \mathbf{P} = \begin{pmatrix} i & -i \\ 1 & 1 \end{pmatrix}, \mathbf{D} = \begin{pmatrix} -i & 0 \\ 0 & i \end{pmatrix},$$

$$\mathbf{AP} = \begin{pmatrix} 1 & 1 \\ -i & i \end{pmatrix} = \mathbf{PD}.$$

The second reason for the diagonalization violation are nilpotent matrices. Let us demonstrate how to resolve this issue on the example of the matrix (5.15). Since in Problem 5.7 we could find only two vectors $\mathbf{v}_1$ and $\mathbf{v}_2$, let us add the third vector $\mathbf{v}_3$, which satisfies the equation $(\mathbf{A} + \mathbf{I})\mathbf{v}_3 = \mathbf{v}_2$. It is easy to see that we may put $\mathbf{v}_3 = \mathbf{e}_2$. Then $\mathbf{Av}_3 = -\mathbf{v}_3 + \mathbf{v}_2$. Now, we put

$$\mathbf{P} = \begin{pmatrix} \mathbf{v}_1 & \mathbf{v}_2 & \mathbf{v}_3 \end{pmatrix}, \mathbf{J} = \begin{pmatrix} -2 & 0 & 0 \\ 0 & -1 & 1 \\ 0 & 0 & -1 \end{pmatrix}. \tag{5.23}$$

We have

$$\mathbf{AP} = \begin{pmatrix} \mathbf{Av}_1 & \mathbf{Av}_2 & \mathbf{Av}_3 \end{pmatrix}\begin{pmatrix} -2\mathbf{v}_1 & -\mathbf{v}_2 & -\mathbf{v}_3 + \mathbf{v}_2 \end{pmatrix} = \mathbf{PJ}.$$

It follows that

$$\mathbf{A}^n = \mathbf{PJ}^n\mathbf{P}^{-1}, \quad n = 1, 2, \ldots . \tag{5.24}$$

Since $\mathbf{J}$ is a block diagonal matrix made by a 1×1 block and 2×2 blocks, we see that

$$\mathbf{J}^n = \begin{pmatrix} (-2)^n & \begin{pmatrix} 0 & 0 \end{pmatrix} \\ \begin{pmatrix} 0 \\ 0 \end{pmatrix} & \begin{pmatrix} -1 & 1 \\ 0 & -1 \end{pmatrix}^n \end{pmatrix}$$

Using mathematical induction, one can easily prove that

$$\begin{pmatrix} -1 & 1 \\ 0 & -1 \end{pmatrix}^n = \begin{pmatrix} (-1)^n & n(-1)^{n-1} \\ 0 & (-1)^n \end{pmatrix}, \ n = 1,2,\dots , \tag{5.25}$$

which gives a formula for the powers of $\mathbf{A}^n$:

$$\mathbf{A}^n = \begin{pmatrix} -1 & -1 & 0 \\ 1 & 0 & 1 \\ 0 & 1 & 0 \end{pmatrix} \begin{pmatrix} (-2)^n & 0 & 0 \\ 0 & (-1)^n & n(-1)^{n-1} \\ 0 & 0 & (-1)^n \end{pmatrix} \begin{pmatrix} -1 & 0 & -1 \\ 0 & 0 & 1 \\ 1 & 1 & 1 \end{pmatrix} =$$

$$(-1)^n \begin{pmatrix} 2^n + n & n & 2^n + n - 1 \\ -2^n + 1 & 1 & -2^n + 1 \\ -n & -n & 1 - n \end{pmatrix}.$$

Matrix Exponents by the Cayley-Hamilton Theorem. If

$$f(\lambda) = \sum_{k=0}^{\infty} b_k \lambda^k,$$

then

$$f(\lambda) = P(\lambda)Q(\lambda) + R(\lambda),$$

where $R(\lambda)$ is a polynomial of degree $n-1$ or less, and n is the degree of the characteristic polynomial $P(\lambda)$ of the matrix $\mathbf{A}$. Suppose that all the eigenvalues of $\mathbf{A}$ are different. Then $\mathbf{A}$ is diagonalizable. Putting $\lambda = \lambda_i$ in the formula results in the system

$$f(\lambda_i) = R(\lambda_i) = \sum_{k=0}^{n-1} \alpha_k \lambda_i^k, \ i = 1,2,\dots,n.$$

By the Cayley-Hamilton Theorem $P(\mathbf{A}) = \mathbf{0}$. This implies the formula for $f(\mathbf{A})$:

$$f(\mathbf{A}) = R(\mathbf{A}) = \sum_{k=0}^{n-1} \alpha_k \mathbf{A}^k.$$

So, we first solve the system of linear equations and find α_k. Then we apply the above formula to find $f(\mathbf{A})$.

Problems

Prob. 150 — Check if the matrix

$$\begin{pmatrix} 7 & -15 \\ 2 & -4 \end{pmatrix}$$

is diagonalizable.

Prob. 151 — Check if the matrix

$$\mathbf{A} = \begin{pmatrix} 4 & 3 & -7 \\ 1 & 2 & 1 \\ 2 & 2 & -3 \end{pmatrix}$$

is diagonalizable.

Prob. 152 — Check that the 2×2 matrix

$$\mathbf{A} = \begin{pmatrix} 4 & 1 \\ -1 & 2 \end{pmatrix}$$

is not diagonalisable. Find a matrix $\mathbf{J}$ for it following the directions given in Section 5.2. Find the formula for $\mathbf{A}^n$, $n = 1, 2, \ldots$.

Prob. 153 — Investigate if the following matrix

$$\mathbf{A} = \begin{pmatrix} -10 & 11 & -6 \\ -15 & 16 & -10 \\ -3 & 3 & -2 \end{pmatrix}$$

is diagonalisable?

Prob. 154 — Investigate if the following matrix

$$\mathbf{A} = \begin{pmatrix} -13 & -60 & -60 \\ 10 & 42 & 40 \\ -5 & -20 & -18 \end{pmatrix}$$

is diagonalisable?

Prob. 155 — Let $\mathbf{A}$ be a real 2×2 matrix with complex eigenvalues $\lambda = a \pm bi$, where $b \neq 0$. If $\mathbf{x}$ is an eigenvector of $\mathbf{A}$ corresponding to $\lambda = a - bi$, then the matrix $\mathbf{P} = \big(\mathrm{Re}(\mathbf{x})\ \mathrm{Im}(\mathbf{x})\big)$ is invertible and

$$\mathbf{A} = \mathbf{P} \begin{pmatrix} a & -b \\ b & a \end{pmatrix} \mathbf{P}^{-1}.$$

Prob. 156 — The eigenvalues of the real matrix

$$\mathbf{C} = \begin{pmatrix} a & -b \\ b & a \end{pmatrix}$$

are $\lambda = a \pm bi$. If a and b are not both zero, then the matrix can be factored as

$$\begin{pmatrix} a & -b \\ b & a \end{pmatrix} = \begin{pmatrix} |\lambda| & 0 \\ 0 & |\lambda| \end{pmatrix} \begin{pmatrix} \cos\phi & -\sin\phi \\ \sin\phi & \cos\phi \end{pmatrix},$$

where ϕ is is the angle from the positive x-axis to the ray that joins the origin to the point (a, b).

5.3 Similar Matrices

Definition 5.5 An $n \times n$ matrix $\mathbf{A}$ is called **similar** to an $n \times n$ matrix $\mathbf{B}$ if there is an $n \times n$ invertible matrix $\mathbf{P}$ such that $\mathbf{A} = \mathbf{PBP}^{-1}$. In this case we write $\mathbf{A} \cong \mathbf{B}$.

As in the case of row operations introduced in Section 1.3, the similarity relation is an **equivalence relation**. It is an easy exercise to check that it satisfies the following equivalence relation properties.

SIM1 $\mathbf{A} \cong \mathbf{A}$.
SIM2 $\mathbf{A} \cong \mathbf{B} \Rightarrow \mathbf{B} \cong \mathbf{A}$.
SIM3 $\mathbf{A} \cong \mathbf{B}$ and $\mathbf{B} \cong \mathbf{C}$ imply that $\mathbf{A} \cong \mathbf{C}$.

The set of all row equivalent matrices to a given matrix $\mathbf{A}$ contains a unique matrix called rref$(\mathbf{A})$. The set of all matrices similar to a diagonalizable matrix $\mathbf{A}$ contains a diagonal matrix $\mathbf{D}(\lambda_1, \ldots, \lambda_2)$ which is also unique if we assume that $\lambda_1 \leq \lambda_2 \leq \cdots \leq \lambda_n$.

Theorem 5.7 *If* $\mathbf{A} \cong \mathbf{B}$ *then the characteristic polynomial* $p_{\mathbf{A}}(\lambda)$ *of* $\mathbf{A}$ *equals the characteristic polynomial* $p_{\mathbf{B}}(\lambda)$ *of* $\mathbf{B}$.

Proof. By the properties of determinants we have:

$$p_{\mathbf{B}}(\lambda) = |\mathbf{B} - \lambda\mathbf{I}| = |\mathbf{P}^{-1}\mathbf{AP} - \lambda\mathbf{I}| = |\mathbf{P}^{-1}\mathbf{AP} - \mathbf{P}^{-1}(\lambda\mathbf{I})\mathbf{P}| =$$
$$|\mathbf{P}^{-1}(\mathbf{A} - \lambda\mathbf{I})\mathbf{P}| = |\mathbf{P}^{-1}| \cdot |\mathbf{A} - \lambda\mathbf{I}| \cdot |\mathbf{P}| = |\mathbf{A} - \lambda\mathbf{I}| = p_{\mathbf{A}}(\lambda).$$

□

Corollary 5.2 *If* $\mathbf{A} \cong \mathbf{B}$ *then* $\det(\mathbf{A}) = \det(\mathbf{B})$ *and* $\mathrm{Tr}(\mathbf{A}) = \mathrm{Tr}(\mathbf{B})$

Proof. Apply Theorems 3.38 and 5.7.   □

Theorem 5.8 *The determinant of an $n \times n$ matrix $\mathbf{A}$ is equal to the product of its eigenvalues.*

Proof.

$$|\mathbf{A} - \lambda \cdot \mathbf{I}| = p_{\mathbf{A}}(\lambda) = (-1)^n(\lambda - \lambda_1)\cdots(\lambda - \lambda_n) \overset{\lambda=0}{\Rightarrow} |\mathbf{A}| = \lambda_1 \cdots \lambda_n.$$

$\square$

Theorem 5.9 *The trace of an $n \times n$ matrix $\mathbf{A}$ is equal to the sum of its eigenvalues.*

Proof.

$$p_{\mathbf{A}}(\lambda) = (-1)^n(\lambda - \lambda_1)\cdots(\lambda - \lambda_n) =$$
$$(-1)^n\lambda^n + (-1)^n(-\lambda_1 - \lambda_2 - \cdots - \lambda_n)\lambda^{n-1} + \cdots + \lambda_1\lambda_2\cdots\lambda_n =$$
$$(-1)^n\lambda^n + (-1)^{n-1}(\lambda_1 + \lambda_2 + \cdots + \lambda_n)\lambda^{n-1} + \cdots + \lambda_1\lambda_2\cdots\lambda_n$$

The result follows from the formula

$$|\mathbf{A} - \lambda \cdot \mathbf{I}| = (-1)^n\lambda^n + (-1)^{n-1}(a_{11} + a_{22} + \cdots + a_{nn})\lambda^{n-1} + \cdots + \det(\mathbf{A}),$$

which is proved in Theorem 3.38 and immediately implies the result. $\square$

It is not true that the equality of characteristic polynomials implies that their matrices are similar. In the following problem the matrices $\mathbf{A}$ and $\mathbf{B}$ have equal characteristic polynomials, differ only at one entry but are not similar.

Problem 5.8 Investigate two matrices on the property of diagonalization

$$\mathbf{A} = \begin{pmatrix} 4 & 0 & -3 \\ 0 & 7 & 0 \\ -3 & 0 & 4 \end{pmatrix}, \mathbf{B} = \begin{pmatrix} 4 & 0 & -3 \\ 0 & 7 & 0 \\ -3 & 1 & 4 \end{pmatrix}.$$

Solution: We have

$$p_{\mathbf{A}}(\lambda) = p_{\mathbf{B}}(\lambda) = (\lambda - 1)(\lambda - 7)^2.$$

We consider the case of the matrix $\mathbf{A}$ first.

$$\mathrm{rref}(\mathbf{A} - \mathbf{I}) = \begin{pmatrix} 1 & 0 & -1 \\ 0 & 1 & 0 \\ 0 & 0 & 0 \end{pmatrix} \Rightarrow \mathbf{v}_1 = \begin{pmatrix} 1 \\ 0 \\ 1 \end{pmatrix}.$$

$$\mathrm{rref}(\mathbf{A} - 7\mathbf{I}) = \begin{pmatrix} 1 & 0 & 1 \\ 0 & 0 & 0 \\ 0 & 0 & 0 \end{pmatrix} \Rightarrow \mathbf{v}_2 = \begin{pmatrix} -1 \\ 0 \\ 1 \end{pmatrix}, \; \mathbf{v}_3 = \begin{pmatrix} 0 \\ 1 \\ 0 \end{pmatrix}.$$

It follows that

$$\mathbf{P} = \begin{pmatrix} \mathbf{v}_1 & \mathbf{v}_2 & \mathbf{v}_3 \end{pmatrix} = \begin{pmatrix} 1 & -1 & 0 \\ 0 & 0 & 1 \\ 1 & 1 & 0 \end{pmatrix}.$$

Then

$$\mathbf{AP} = \begin{pmatrix} 1 & -7 & 0 \\ 0 & 0 & 7 \\ 1 & 7 & 0 \end{pmatrix} = \mathbf{PD}(1,7,7),$$

implying that $\mathbf{A}$ is diagonalizable matrix.

Let us consider the case of matrix $\mathbf{B}$.

$$\mathrm{rref}(\mathbf{B} - \mathbf{I}) = \begin{pmatrix} 1 & 0 & -1 \\ 0 & 1 & 0 \\ 0 & 0 & 0 \end{pmatrix} \Rightarrow \mathbf{v}_1 = \begin{pmatrix} 1 \\ 0 \\ 1 \end{pmatrix}.$$

$$\mathrm{rref}(\mathbf{B} - 7\mathbf{I}) = \begin{pmatrix} 1 & 0 & 1 \\ 0 & 1 & 0 \\ 0 & 0 & 0 \end{pmatrix} \Rightarrow \mathbf{v}_2 = \begin{pmatrix} -1 \\ 0 \\ 1 \end{pmatrix}.$$

Since $\mathbf{B}$ has only two eigenvectors, it is not diagonalizable. □

If $\mathbf{AP} = \mathbf{PD}$ for an invertible $\mathbf{P}$ and some $\mathbf{D} = \mathbf{D}(\lambda_1, \ldots, \lambda_n)$ then one may consider $\mathbf{P}$ is the transition matrix from the coordinates in the basis $B = \{\mathbf{v}_1, \mathbf{v}_2, \ldots, \mathbf{v}_n\}$ to the standard coordinates in the basis $\{\mathbf{e}_1, \ldots, \mathbf{e}_n\}$. The formula relating $\mathbf{A}$ and $\mathbf{D}$ is $\mathbf{A} = \mathbf{PDP}^{-1}$ and can be interpreted as follows.

The action of $\mathbf{A}$ on a vector $\mathbf{v}$ can be evaluated in three steps. First, we find the coordinates of $\mathbf{v}$ in the basis B. Then we apply the diagonal matrix $\mathbf{D}$ to the result, and finally we return to the standard coordinates.

Theorem 5.10 *For any eigenvalue of a square matrix, the geometric multiplicity is less or equal than the algebraic multiplicity.*

Proof. Let $\{v_1, \ldots, v_k\}$ be the basis of the eigenspace of $\mathbf{A}$ corresponding to the eigenvalue μ. We can extend it to a basis $B = \{v_1, \ldots, v_k, \ldots, v_n\}$ of $\mathbb{R}^n$. Since the matrix $\mathbf{P} = \mathbf{P}_B = \begin{pmatrix} v_1 & \cdots & v_k & \cdots & v_n \end{pmatrix}$ is a transition matrix from basis B to the standard basis in $\mathbb{R}^n$, we obtain the formula

$$\mathbf{B} \overset{def}{=} \mathbf{P}^{-1}\mathbf{A}\mathbf{P} = \begin{pmatrix} \mu\mathbf{I}_k & \mathbf{X} \\ \mathbf{0} & \mathbf{C} \end{pmatrix}.$$

Since $\mathbf{B} \cong \mathbf{A}$, these matrices have equal characteristic polynomials:

$$p_{\mathbf{A}}(\lambda) = p_{\mathbf{B}}(\lambda) = \begin{vmatrix} (\mu - \lambda)\mathbf{I}_k & \mathbf{X} \\ \mathbf{0} & \mathbf{C} - \lambda\mathbf{I}_{n-k} \end{vmatrix} = (\mu - \lambda)^k p_{\mathbf{C}}(\lambda).$$

$\square$

Theorem 5.11 *If $\mathbf{A}$ is a real matrix with an eigenvalue $\lambda \in \mathbb{R}$, then λ is an eigenvalue of $\mathbf{A}^T$ and its geometric multiplicities for both matrices coincide.*

Proof. By Theorem 2.10

$$\text{nullity}(\mathbf{A}) = \text{nullity}(\mathbf{A}^T).$$

Then

$$\mathbf{geo}_{\mathbf{A}}(\lambda) = \dim\left(\mathbf{N}\left(\mathbf{A} - \lambda\mathbf{I}\right)\right) = \text{nullity}\left(\mathbf{A} - \lambda\mathbf{I}\right) =$$
$$\text{nullity}\left(\mathbf{A}^T - \lambda\mathbf{I}\right) = \dim\left(\mathbf{N}\left(\mathbf{A}^T - \lambda\mathbf{I}\right)\right) = \mathbf{alg}_{\mathbf{A}^T}(\lambda).$$

$\square$

Theorem 5.12 *If $\mathbf{A}$ is a square matrix then the characteristic polynomials of $\mathbf{A}$ and $\mathbf{A}^T$ coincide: $p_{\mathbf{A}}(\lambda) = p_{\mathbf{A}^T}(\lambda)$.*

Proof. We have

$$p_{\mathbf{A}}(\lambda) = \det(\mathbf{A} - \lambda\mathbf{I}) = \det\left(\mathbf{A}^T - \lambda\mathbf{I}\right) = p_{\mathbf{A}^T}(\lambda).$$

□

Theorem 5.13 *A matrix $\mathbf{A}$ is diagonalisable if and only if $\mathbf{A}^T$ is diagonalisable.*

Proof. Apply Theorems 5.6 and 5.11. □

Problems

Prob. 157 — Show that one of the following two matrices can be diagonalized and the other cannot.

$$\mathbf{A} = \begin{pmatrix} 2 & 3 & 0 \\ 3 & 2 & 0 \\ 1 & 1 & 5 \end{pmatrix}, \quad \mathbf{B} = \begin{pmatrix} 2 & 3 & 0 \\ 3 & 2 & 0 \\ 1 & -1 & 5 \end{pmatrix}.$$

Diagonalize the matrix which can be diagonalized and diagonalize with $\mathbf{J}$ the matrix which cannot be diagonalized.

Prob. 158 — Diagonalize the matrix

$$\mathbf{B} = \begin{pmatrix} 0 & 2 & 1 \\ 16 & 4 & -6 \\ -16 & 4 & 10 \end{pmatrix}.$$

Prob. 159 — Check whether the following matrices are similar:

$$\begin{pmatrix} 0 & 1 \\ 0 & 0 \end{pmatrix}, \quad \begin{pmatrix} 0 & 2 \\ 0 & 0 \end{pmatrix}.$$

5.4 Jordan Normal Form

The following problem gives an interesting example of what one can face in the diagonalization process. Theorem 3.14 allows one to obtain many similar examples.

Problem 5.9 Investigate the matrix on the possibility of diagonalization.

$$A = \begin{pmatrix} 1 & 3 & -2 \\ 0 & 7 & -4 \\ 0 & 9 & -5 \end{pmatrix}.$$

Solution: **Step 1.** The characteristic polynomial of A is

$$\det(A - \lambda I) = \begin{vmatrix} 1 - \lambda & 3 & -2 \\ 0 & 7 - \lambda & -4 \\ 0 & 9 & -5 - \lambda \end{vmatrix} = -(\lambda - 1)^3.$$

Step 2. We evaluate $E(1, A)$

$$A - I = \begin{pmatrix} 1 - 1 & 3 & -2 \\ 0 & 7 - 1 & -4 \\ 0 & 9 & -5 - 1 \end{pmatrix} \sim \begin{pmatrix} 0 & 1 & -\frac{2}{3} \\ 0 & 0 & 0 \\ 0 & 0 & 0 \end{pmatrix} \Rightarrow \begin{cases} x_1 & = s \\ x_2 & = \frac{2}{3}t \\ x_3 & = t \end{cases} \Rightarrow$$

$$v_1 = \begin{pmatrix} 1 \\ 0 \\ 0 \end{pmatrix}, \quad v_2 = \begin{pmatrix} 0 \\ 2 \\ 3 \end{pmatrix} \Rightarrow \begin{cases} \dim E(1, A) & = 2 \\ R(A - I) & = \mathrm{Lin}\{(1\ 2\ 3)^T\}. \\ v_1, v_2 & \notin R(A - I) \end{cases}$$

The problem with $E(1, A)$ is that its dimension is clearly two, but neither of obtained vectors v_1, v_2 belong to $R(A - I)$. Therefore, we cannot apply the extension, which was used in Problems 5.7 and 5.4. However, it is easy to see that

$$u_1 := v_1 + v_2 \in R(A - I).$$

Step 3. For the vector $u_1 = (1\ 2\ 3)^T$ we find a vector u_2 satisfying : $u_1 = (A - I)u_2$:

$$(A - I \mid u_1) = \begin{pmatrix} 0 & 3 & -2 & 1 \\ 0 & 6 & -4 & 2 \\ 0 & 9 & -6 & 3 \end{pmatrix} \sim \begin{pmatrix} 0 & 3 & -2 & 1 \\ 0 & 0 & 0 & 0 \\ 0 & 0 & 0 & 0 \end{pmatrix} \Rightarrow u_2 = \begin{pmatrix} 0 \\ 1 \\ 1 \end{pmatrix}.$$

Then we put:

$$J = \begin{pmatrix} 1 & 0 & 0 \\ 0 & 1 & 1 \\ 0 & 0 & 1 \end{pmatrix}, \quad P = \begin{pmatrix} v_1 & u_1 & u_2 \end{pmatrix} = \begin{pmatrix} 1 & 1 & 0 \\ 0 & 2 & 1 \\ 0 & 3 & 1 \end{pmatrix}.$$

Step 4. Easy calculations show that $AP = PJ \Rightarrow A = PJP^{-1}$. Then

$$A^n = \begin{pmatrix} 1 & 1 & 0 \\ 0 & 2 & 1 \\ 0 & 3 & 1 \end{pmatrix} \begin{pmatrix} 1 & 0 & 0 \\ 0 & 1 & n \\ 0 & 0 & 1 \end{pmatrix} \begin{pmatrix} 1 & 1 & -1 \\ 0 & -1 & 1 \\ 0 & 3 & -2 \end{pmatrix} = \begin{pmatrix} 1 & 3n & -2n \\ 9 & 6n + 1 & -4n \\ 0 & 9n & -6n + 1 \end{pmatrix}. \quad \square$$

In cases when diagonalization was not possible, the transition matrix P was completed to a square invertible matrix with vectors v_2 satisfying the equation

$$\mathbf{v}_1 = (\mathbf{A} - \lambda\mathbf{I})\mathbf{v}_2, \ \mathbf{v}_1 \in E(\lambda, \mathbf{A}).$$

Since $\mathbf{v}_1 \in E(\lambda, \mathbf{A})$, we see that $(\mathbf{A} - \lambda\mathbf{I})^2\mathbf{v}_2 = \mathbf{0}$. These vectors, as well as more general one, are important for constructions of matrices $\mathbf{J}$ and we consider this topic here.

Definition 5.6 Let T be an operator on V and λ be its eigenvalue. A **non-zero** vector $\mathbf{v} \in V$ is called a **generalized eigenvector** of T corresponding to λ if there is a positive integer k such that $(T - \lambda\mathbf{I})^k\mathbf{v} = \mathbf{0}$.

Definition 5.7 Let T be an operator on a finite dimensional vector space V and λ be its eigenvalue. A **generalized eigenspace** $G(\lambda, T)$ of T corresponding to λ is the set of all generalized eigenvectors $\mathbf{v}$ of T, corresponding to λ, along with the $\mathbf{0}$ vector.

Definition 5.8 Let T be an operator on a finite dimensional space V. Then a subspace V_1 is called an **invariant subspace** for T if $T\mathbf{v} \in V_1$ for every $\mathbf{v} \in V_1$.

Every eigenspace $E(\lambda, T)$ of an operator T on a finite dimensional vector space is invariant. Indeed, if $(T - \lambda I)\mathbf{v} = \mathbf{0}$, then $(T - \lambda I)T\mathbf{v} = T(T - \lambda I)\mathbf{v} = \mathbf{0}$. Similarly, every generalized eigenspace is invariant under the action of T.

Lemma 5.1 *If T is a linear operator on a finite dimensional space V and $\lambda \in \mathbb{C}$, then*

$$G(\lambda, T) = N\left((T - \lambda\mathbf{I})^{\dim V}\right)$$

is a subspace of V.

Proof. We have the following increasing chain of subspaces:

$$N\left((T - \lambda\mathbf{I})\right) \subseteq N\left((T - \lambda\mathbf{I})^2\right) \subseteq \cdots \subseteq V, \tag{5.26}$$

implying that

$$G(\lambda, T) = \bigcup_{k=1}^{\infty} N\left((T - \lambda\mathbf{I})^k\right)$$

is a subspace. Since $\dim(V) < +\infty$, the sequence of subspaces in (5.26) must stabilize at some index $n \leq \dim(V)$. $\qquad\square$

As we have already seen on the example of nilpotent operators, it is not true that

$$V = N(T) \dotplus R(T).$$

However, it is true for some particular operators.

Lemma 5.2 *Let T be a linear operator on a complex finite-dimensional vector space V, dim $V = n$. Then $V = N(T^n) \dotplus R(T^n)$.*

Proof. First we show that $N(T^n) \cap R(T^n) = \{0\}$. Suppose that $\mathbf{v} \in N(T^n) \cap R(T^n)$. Then

$$\begin{cases} T^n\mathbf{v} &= \mathbf{0} \\ \mathbf{v} &= T^n\mathbf{u} \end{cases} \Rightarrow \begin{cases} T^n\mathbf{v} &= \mathbf{0} \\ T^n\mathbf{v} &= T^{2n}\mathbf{u} \end{cases} \Rightarrow T^{2n}\mathbf{u} = \mathbf{0} \Rightarrow \mathbf{u} \in N(T^{2n}).$$

But Lemma 5.1 with $\lambda = 0$, $N(T^n) = N(T^{n+1}) = \cdots = N(T^{2n})$. It follows that $\{0\} = T^n\mathbf{u} = \mathbf{v}$. By the Rank-Nullity Theorem $\dim N(T^n) + \dim R(T^n) = \dim V$. This shows that $N(T^n) \dotplus R(T^n) = V$.   □

Theorem 5.14 *Let T be an operator on a finite-dimensional space V. Let $\lambda_1, \lambda_2, \ldots, \lambda_m$ be the list of all distinct eigenvalues of T. Then the vector space V decomposes into a direct sum of generalized eigenspaces:*

$$V = G(\lambda_1, T) \dotplus G(\lambda_2, T) \dotplus \cdots \dotplus G(\lambda_m, T). \tag{5.27}$$

Proof. Every linear operator T defined on a vector space V with $\dim V = n$ has a characteristic polynomial (4.30) of degree n. If $n = 0$, then there is nothing to prove. If $n > 1$, then by the Gauss theorem on the roots of polynomials, see Theorem 4.11, $p_T(\lambda_1) = 0$ at least for one complex number λ_1, implying that $E(\lambda_1, T) = N(T - \lambda_1 I) \neq \{0\}$. Lemma 5.2 applied to $T := T - \lambda I$ implies that the vector space V decomposes into the following direct sum of its subspaces:

$$V = G(\lambda_1, T) \dotplus R\left((T - \lambda_1 I)^n\right). \tag{5.28}$$

Let us consider the subspace $V_1 = R\left((T - \lambda_1 I)^n\right)$ of V. This subspace is invariant under operator T. Indeed, if $\mathbf{v} = (T - \lambda_1 I)(\mathbf{w})$, then $T(\mathbf{v}) = (T - \lambda_1 I)(T(\mathbf{w})) \in V_1$. Since $G(\lambda_1, T) \neq \mathbf{0}$, we see that $\dim V_1 < \dim V$. The operator T being considered on subspace V_1 has a characteristic polynomial which has at least one root λ_2. This root must be different from λ_1. Indeed, otherwise there will be an eigenvector $\mathbf{u}$ in V_1 satisfying $T(\mathbf{u}) = \lambda_2\mathbf{u} = \lambda_1\mathbf{u}$. Then $\mathbf{u} \in G(\lambda_1, T)$, which is impossible since the sum in (5.28) is direct. It is clear that this process stops in m steps. Since in each step the decomposition is made inside the last subspace in the direct sum, the sum obtained is also direct.   □

Theorem 5.14 shows that if one combines bases in components of the direct sum in (5.27) in one basis, then the matrix $\mathbf{A}_T$ of the operator T is similar to the block matrix

$$\mathbf{A}_T \cong \begin{pmatrix} \mathbf{A}_1 & \mathbf{0} & \cdots & \mathbf{0} \\ \mathbf{0} & \mathbf{A}_2 & \cdots & \mathbf{0} \\ \vdots & \vdots & \ddots & \vdots \\ \mathbf{0} & \mathbf{0} & \cdots & \mathbf{A}_m \end{pmatrix}, \tag{5.29}$$

where each $\mathbf{A}_k - \lambda_k \mathbf{I}$ is a nilpotent matrix. It follows that to complete the study of non-diagonalizable matrices, we must understand the structure of nilpotent matrices in $\mathbb{R}^n$. Typical examples of nilpotent matrices are shown in (5.22). These matrices can be described as matrices $\mathbf{A}$ with all zero entries except for the entries $a_{i,i+1} = 1$. For every integer n, $n > 1$, there is only one $n \times n$ matrix satisfying such a condition. We denote this nilpotent matrix by $\mathbf{J}_n$.

Definition 5.9 Given a complex number λ, we denote by $\mathbf{J}_n(\lambda) = \lambda \mathbf{I}_n + \mathbf{J}_n$ the Jordan block of size $n \times n$.

Theorem 5.15 *An operator T on a finite dimensional vector space V is similar to a Jordan block $\mathbf{J}_n(\lambda)$ if and only if there exists a basis $S = \{\mathbf{v}_1, \ldots, \mathbf{v}_n\}$ in V such that*

$$\mathbf{0} = (T - \lambda I)\mathbf{v}_1, \ \mathbf{v}_1 = (T - \lambda I)\mathbf{v}_2, \ \ldots, \mathbf{v}_{n-1} = (T - \lambda I)\mathbf{v}_n. \tag{5.30}$$

Proof. If $T - \lambda \mathbf{I}_n = \mathbf{J}_n$, then $\mathbf{J}_n \mathbf{e}_1 = \mathbf{0}$ and $\mathbf{J}_n \mathbf{e}_k = \mathbf{e}_{k-1}$ for $k > 1$. Therefore, we may put $\mathbf{v}_k = \mathbf{e}_k$.

Let $T \cong \mathbf{J}_n(\lambda)$. Then there is an invertible Transformation $P : \mathbb{R}^n \to V$ such that

$$T = P(\mathbf{J}_n + \lambda I)P^{-1} = P\mathbf{J}_n P^{-1} + \lambda I \Rightarrow T - \lambda I = P\mathbf{J}_n P^{-1}.$$

Hence, if $\mathbf{v}_k = P\mathbf{e}_k$, then $S = \{\mathbf{v}_1, \ldots, \mathbf{v}_n\}$ is a basis in V (notice that P is an invertible operator), and

$$(T - \lambda I)\mathbf{v}_k = P\mathbf{J}_n P^{-1}\mathbf{v}_k = P\mathbf{J}_n \mathbf{e}_k = P\mathbf{e}_{k-1} = \mathbf{v}_{k-1}.$$

If there is a basis $S = \{\mathbf{v}_1, \ldots, \mathbf{v}_n\}$ satisfying (5.30), then we define the invertible transformation $P :\to \mathbb{R}^n$ by $P\mathbf{e}_k = \mathbf{v}_k$, $k = 1, \ldots, n$. It is easy to see that $P^{-1}(T - \lambda I)P$ satisfies

$$P^{-1}(T - \lambda I)P e_k = P^{-1}(T - \lambda I)\mathbf{v}_k = P^{-1}\mathbf{v}_{k-1} = \mathbf{e}_{k-1},$$

which shows that $P^{-1}(T - \lambda I)P = \mathbf{J}_n$.   □

Definition 5.10 Let $\lambda \in \mathbb{C}$ and T be an operator on a finite dimensional space V. A finite sequence $\{\mathbf{v}_k\}_{k=1}^n$ of nonzero vectors in V is called a Jordan chain of length n if

$$\mathbf{0} = (T - \lambda \mathbf{I}_n)\mathbf{v}_1, \ \mathbf{v}_1 = (T - \lambda \mathbf{I}_n)\mathbf{v}_2, \ \ldots, \mathbf{v}_{n-1} = (T - \lambda \mathbf{I}_n)\mathbf{v}_n.$$

Notice that any Jordan chain is linearly independent. Indeed, it consists of nonzero vectors by the definition. If any linear combination of these vectors is zero, then we apply the operator $(T - \lambda I)^k$ to it, where k is the greatest integer such that $\mathbf{v}_{k+1}$ enters this linear combination with a nonzero coefficient to obtain a contradiction.

Similarly, if we have a finite number of Jordan chains with linearly independent first vectors in $E(\lambda, T)$, then the combined system of vectors in these chains is also linearly independent.

Theorem 5.16 *A linear operator T on a finite dimensional vector space is nilpotent if and only if the vector space V is decomposed into a direct sum*

$$V = V_1 \dotplus V_2 \dotplus \cdots \dotplus V_m$$

of invariant subspaces for T such that each restriction $T|V_k$ of T onto V_k is similar to a Jordan block $\mathbf{J}_n$ with $n = \dim(V_k)$.

Proof. We run the proof of the theorem by induction in $\dim(V)$. If $\dim V = 1$, then every operator on V is a multiplication by a constant, implying that the only nilpotent operator on V is the zero operator.

The case of $\dim(V) = 2$ was illustrated earlier on the example of Problem 5.4. Since every operator T on V is similar to its matrix $\mathbf{A}_T$, we may assume that $T = \mathbf{A}$ is a 2×2 matrix. Since $\mathbf{A}$ is nilpotent, its characteristic polynomial is λ^2 by Theorem 5.5. If there are two linearly independent vectors $\mathbf{v}_1$ and $\mathbf{v}_2$ such that $\mathbf{A}\mathbf{v}_1 = \mathbf{A}\mathbf{v}_2 = \mathbf{0}$, then $\mathbf{A} = \mathbf{0}$ and there is nothing to prove. Otherwise, there is only one eigenvector $\mathbf{v}_1$ such that $\mathbf{A}\mathbf{v}_1 = \mathbf{0}$. If $\mathbf{v}_1 \notin R(\mathbf{A}$, then $\mathbf{A}^2$ cannot be a zero matrix. Therefore, the equation $\mathbf{v}_1 = \mathbf{A}\mathbf{v}_2$ has a solution $\mathbf{v}_2$ implying that $S = \{\mathbf{v}_1, \mathbf{v}_2\}$ is a Jordan chain of length 2 for $\lambda = 0$. By Theorem 5.15 the matrix $\mathbf{A}$ is similar to $\mathbf{J}_2$.

Suppose now that the theorem is proved for every vector space V, $\dim(V) \leq n$ and every nilpotent operator T on V and consider a nilpotent operator T on a vector space V, $\dim(V) = n+1$. By Theorem 5.5 we see that $p_T(\lambda) = (-1)^{n+1}\lambda^{n+1}$, which implies that $\dim(N(T)) \geq 1$. By the Rank-Nullity Theorem, see Theorewm 4.15,

$$n + 1 = \dim(N(T)) + \dim(R(T)) \Rightarrow \dim(R(T)) = n + 1 - \dim(N(T)) \le n.$$

By induction hypothesis, the restriction of T onto its invariant subspace $R(T)$ satisfies the conclusion of the theorem, implying that

$$R(T) = V_1 \dotplus V_2 \dotplus \cdots \dotplus V_m, \; T|V_k \cong J_{n_k}, \; \dim(V_k) = n_k.$$

By Theorewm 5.30, each V_k is $\mathrm{Lin}\{\mathbf{v}_{k1}, \mathbf{v}_{k2}, \dots, \mathbf{v}_{kn_k}\}$, where

$$\mathbf{0} = T\mathbf{v}_{k1}, \; \mathbf{v}_{k1} = T\mathbf{v}_{k2}, \; \cdots, \; \mathbf{v}_{kn_k-1} = T\mathbf{v}_{kn_k}$$

is the Jordan chain of order n_k. Since $\mathbf{v}_{kn_k} \in R(T)$, there is a vector $\mathbf{v}_{kn_k+1} \in V$ such that $\mathbf{v}_{kn_k} = T\mathbf{v}_{kn_k+1}$. Since $\mathbf{v}_{k1} \in V_k$ and the subspaces V_k make a direct decomposition of $R(T)$ into their sum, this system of vectors is linearly independent, implying the linear independence of the extended system of vectors obtained by adding the vectors $\mathbf{v}_{kn_k+1}$ to it. Let

$$U = \mathrm{Lin}\{\mathbf{v}_{1n_1+1}, \mathbf{v}_{2n_2+1}, \cdots, \mathbf{v}_{mn_m+1}\} \Rightarrow \dim(U) = m.$$

It is easy to see that

$$N(T) \cap R(T) = \mathrm{Lin}\{\mathbf{v}_{11}, \mathbf{v}_{21}, \dots, \mathbf{v}_{m1}\} \Rightarrow \dim(N(T) \cap R(T)) = m = \dim(U).$$

We complete a linearly independent system $\{\mathbf{v}_{11}, \mathbf{v}_{21}, \dots, \mathbf{v}_{m1}\}$ to a basis of $N(T)$ by adding a system $\{\mathbf{w}_1, \dots, \mathbf{w}_s\}$ so that $\dim(N(T)) = m + s$. We observe that

$$\dim(N(T)) + \dim(R(T)) = n + 1$$
$$\dim(N(T) + R(T)) = n + 1 - m$$

Since $\dim(U) = m$ and since U intersects both $N(T)$ and $R(T)$ by a zero vector, we obtain that $V = (N(T) + R(T)) \dotplus U$.

Let $W_k = \mathrm{Lin}\{V_k, \mathbf{v}_{kn_k+1}\}$ and $W = \mathrm{Lin}\{\mathbf{w}_1, \dots, \mathbf{w}_s\}$. Then

$$V = W \dotplus W_1 \dotplus W_2 \dotplus \cdots \dotplus W_m.$$

It follows that the restriction of T onto W is diagonalizable and the restriction to W_k is similar to the Jordan block J_{n_k+1}. $\qquad\square$

Both theorems, Theorem 5.14 and Theorem 5.16, are of theoretical value. They guarantee that every operator on a finite dimensional vector space is similar to a block matrix (5.29), whereas each block is a direct sum of Jordan blocks $J_n(\lambda_k)$ of different sizes n. In practice, one first finds a block decomposition (5.29), and then for each eigenvalue λ_k evaluates a representation of this block as a direct sum of Jordan blocks.

Theorem 5.17 *Let* $\mathbf{A}$ *be an* $n \times n$ *matrix with complex entries. Then*

$$\mathbf{A}^T \cong \mathbf{A}.$$

Proof. Since every complex square matrix is similar to its Jordan normal form, it is sufficient to prove this theorem for Jordan blocks $\mathbf{J}_n(\lambda)$. Let $\mathbf{C}$ be the matrix with zero entries except for the entries of the auxiliary diagonal wihich are all 1. Then $\mathbf{C}^2 = \mathbf{I}_n$ and the matrices $\mathbf{C}$ and $\mathbf{J}_n(\lambda)^T \mathbf{C}$ is symmetric. It follows that

$$\mathbf{J}_n(\lambda)^T \mathbf{C} = \left(\mathbf{J}_n(\lambda)^T \mathbf{C} \right)^T = \mathbf{C} \mathbf{J}_n(\lambda) \tag{5.31}$$

Multiplying both sides of the equation (5.31) on the right, we obtain that $\mathbf{J}_n(\lambda)^T = \mathbf{C} \mathbf{J}_n(\lambda) \mathbf{C}$. □

The following formulas illustrate calculations made in the proof of Theorem 5.17:

$$\mathbf{J}_3(\lambda)^T \mathbf{C} = \begin{pmatrix} 1 & 0 & 0 \\ \lambda & 1 & 0 \\ 0 & \lambda & 1 \end{pmatrix} \begin{pmatrix} 0 & 0 & 1 \\ 0 & 1 & 0 \\ 1 & 0 & 0 \end{pmatrix} = \begin{pmatrix} 0 & 0 & 1 \\ 0 & 1 & \lambda \\ 1 & \lambda & 0 \end{pmatrix}.$$

We complete this section with a formula for $\mathbf{J}_m(\lambda)^n$.

Theorem 5.18 *For every polynomial* f

$$f(\mathbf{J}_m(\lambda)) = \begin{pmatrix} f(\lambda) & \frac{f^{(1)}(\lambda)}{1!} & \frac{f^{(2)}(\lambda)}{2!} & \cdots & \frac{f^{(m-2)}(\lambda)}{(m-2)!} & \frac{f^{(m-1)}(\lambda)}{(m-1)!} \\ 0 & f(\lambda) & \frac{f^{(1)}(\lambda)}{1!} & \cdots & \frac{f^{(m-3)}(\lambda)}{(m-3)!} & \frac{f^{(m-2)}(\lambda)}{(m-2)!} \\ 0 & 0 & f(\lambda) & \cdots & \frac{f^{(m-4)}(\lambda)}{(m-4)!} & \frac{f^{(m-3)}(\lambda)}{(m-3)!} \\ \vdots & \vdots & \vdots & \ddots & \vdots & \vdots \\ 0 & 0 & 0 & \cdots & f(\lambda) & \frac{f^{(1)}(\lambda)}{1!} \\ 0 & 0 & 0 & \cdots & 0 & f(\lambda) \end{pmatrix} \tag{5.32}$$

Proof. Since any polynomial is a linear combination of monomials λ^n and both sides of the formula are linear in f, it is sufficient to prove (5.32) only for monomials $f(\lambda) = \lambda^n$. By the Binomial Theorem we have

$$\mathbf{J}_m(\lambda)^n = (\lambda\mathbf{I}_m+\mathbf{J}_m)^n = \lambda^n\mathbf{I}_m + \binom{n}{1}\lambda^{n-1}\mathbf{J}_m + \binom{n}{2}\lambda^{n-2}\mathbf{J}_m^2 + \binom{n}{3}\lambda^{n-3}\mathbf{J}_m^3 + \cdots$$

$$f(\lambda)\mathbf{I}_m + \frac{f^{(1)}(\lambda)}{1!}\mathbf{J}_m + \frac{f^{(2)}(\lambda)}{2!}\mathbf{J}_m^2 + \frac{f^{(3)}(\lambda)}{3!}\mathbf{J}_m^3 + \cdots . \quad \square$$

Notice that the formula (5.25) follows by Theorem (5.18) with $m = 2$.

Problems

Prob. 160 — Determine the Jordan normal forms of the matrices

$$\mathbf{A} = \begin{pmatrix} 0 & 2 & 0 & -2 \\ -2 & 4 & 0 & -2 \\ -3 & 3 & 2 & -2 \\ -2 & 0 & 0 & 2 \end{pmatrix}, \quad \mathbf{B} = \begin{pmatrix} 1 & -1 & 0 & -1 \\ 0 & 2 & 0 & 1 \\ -2 & 1 & -1 & 1 \\ 2 & -1 & 2 & 0 \end{pmatrix}.$$

Prob. 161 — Determine which of the two matrices below is not a Jordan matrix.

$$\mathbf{J}_1 = \begin{pmatrix} 2 & 0 & 0 & 0 \\ 0 & 2 & 1 & 0 \\ 0 & 0 & 2 & 0 \\ 0 & 0 & 0 & 3 \end{pmatrix} \quad \mathbf{J}_2 = \begin{pmatrix} 2 & 1 & 0 & 0 \\ 0 & 2 & 1 & 0 \\ 0 & 0 & 2 & 1 \\ 0 & 0 & 0 & 3 \end{pmatrix}.$$

Let **J** be the one which is a Jordan matrix and answer the following questions.

(a) Let **P** be an invertible matrix and let **A** be a matrix such that $\mathbf{J} = \mathbf{P}^{-1}\mathbf{AP}$. Write down the characteristic polynomial of **A** as a product of linear terms.

(b) Let $\mathbf{v}_1$, $\mathbf{v}_2$, $\mathbf{v}_3$, $\mathbf{v}_4$ denote the column vectors of the matrix **P** in part (a). Express each vector $\mathbf{Av}_i$ as a linear combination of the vectors $\mathbf{v}_1$, $\mathbf{v}_2$, $\mathbf{v}_3$, $\mathbf{v}_4$.

Prob. 162 — Let

$$\mathbf{A} = \begin{pmatrix} 0 & 1 & 0 & -1 \\ -2 & 3 & 0 & -1 \\ -2 & 1 & 2 & -1 \\ 2 & -1 & 0 & 3 \end{pmatrix}.$$

(a) Evaluate the Jordan Normal Form **J** for the matrix **A**. Find a matrix **P** such that $\mathbf{A} = \mathbf{PJP}^{-1}$.

(b) Let $\mathbf{v}_1, \mathbf{v}_2, \mathbf{v}_3, \mathbf{v}_4$ denote the columns of **P**. Express each vector $\mathbf{Av}_i$ as a linear combination of $\mathbf{v}_1, \mathbf{v}_2, \mathbf{v}_3, \mathbf{v}_4$.

Prob. 163 — Using direct sums and properties of Jordan blocks, find $\mathbf{B}^{-1}$ for

$$\mathbf{B} = \begin{pmatrix} 2 & 1 & 0 & 0 & 0 \\ 0 & 2 & 1 & 0 & 0 \\ 0 & 0 & 2 & 0 & 0 \\ 0 & 0 & 0 & 1 & 1 \\ 0 & 0 & 0 & 0 & 1 \end{pmatrix}.$$

Prob. 164 — Let

$$\mathbf{A} = \begin{pmatrix} -13 & 8 & 1 & 2 \\ -22 & 13 & 0 & 3 \\ 8 & -5 & 0 & -1 \\ -22 & 13 & 5 & 5 \end{pmatrix}.$$

(a) Evaluate the Jordan Normal Form **J** for the matrix **A**. Find a matrix **P** such that
   $\mathbf{A} = \mathbf{PJP}^{-1}$.
(b) Let $\mathbf{v}_1, \mathbf{v}_2, \mathbf{v}_3, \mathbf{v}_4$ denote the columns of **P**. Express each vector $\mathbf{Av}_i$ as a linear
   combination of $\mathbf{v}_1, \mathbf{v}_2, \mathbf{v}_3, \mathbf{v}_4$.

Prob. 165 — Let **A** be an invertible $n \times n$ matrix with complex entries, and let k
be a positive integer. Then $\mathbf{A}^k$ is diagonalizable if and only if **A** is diagonalizable.
Verify whether this statement is true if **A** is not invertible.

Prob. 166 — The null space V and the matrix **A** are given by the formulas:

$$V = \text{Lin} \left\{ \begin{pmatrix} 5 \\ 3 \\ 2 \\ -1 \end{pmatrix}, \begin{pmatrix} -2 \\ 0 \\ 1 \\ 1 \end{pmatrix} \right\}, \ \mathbf{A} = \begin{pmatrix} 5 & -2 & * & * \\ 3 & 0 & * & * \\ 2 & 1 & * & * \\ -1 & 1 & * & * \end{pmatrix}.$$

Explain why **A** cannot be diagonalized. Find a Jordan matrix **J** and invertible matrix
P such that $\mathbf{P}^{-1}\mathbf{AP} = \mathbf{J}$.

5.5 Difference Equations

Diagonalization is the main tool in solving difference equations and systems of difference equations as it was shown in Section 5.1. We illustrate this approach with
the solution of the following problem.

Problem 5.10 Find the sequences x_t, y_t, z_t, which satisfy the difference equations

$$\begin{cases} x_{t+1} &= 6x_t + 13y_t - 8z_t \\ y_{t+1} &= 2x_t + 5y_t - 2z_t \\ z_{t+1} &= 7x_t + 17y_t - 9z_t \end{cases}$$

and the initial conditions $x_0 = 1, y_0 = 1, z_0 = 0$.

Solution: Let

$$\mathbf{A} = \begin{pmatrix} 6 & 13 & -8 \\ 2 & 5 & -2 \\ 7 & 17 & -9 \end{pmatrix}, \quad \mathbf{x}_t = \begin{pmatrix} x_t \\ y_t \\ z_t \end{pmatrix}.$$

Then the system can be rewritten in the matrix form:

$$\mathbf{x}_{t+1} = \mathbf{A}\mathbf{x}_t, \quad \mathbf{x}_0 = \begin{pmatrix} 1 & 1 & 0 \end{pmatrix}^T.$$

The solution is given by the formula: $\mathbf{x}_t = \mathbf{A}^t\mathbf{x}_0$. To evaluate $\mathbf{A}^t$ we diagonalize the matrix $\mathbf{A}$.

$$p_\mathbf{A}(\lambda) = |\det(\mathbf{A} - \lambda\mathbf{I})| = \begin{vmatrix} 6-\lambda & 13 & -8 \\ 2 & 5-\lambda & -2 \\ 7 & 17 & -9-\lambda \end{vmatrix} = \begin{vmatrix} 6-\lambda & 13 & -2-\lambda \\ 2 & 5-\lambda & 0 \\ 7 & 17 & -2-\lambda \end{vmatrix} =$$

$$\begin{vmatrix} -1-\lambda & -4 & 0 \\ 2 & 5-\lambda & 0 \\ 7 & 17 & -2-\lambda \end{vmatrix} = \begin{vmatrix} 1+\lambda & 4 & 0 \\ 2 & 5-\lambda & 0 \\ -7 & -17 & 2+\lambda \end{vmatrix} = (\lambda+2)\begin{vmatrix} 1+\lambda & 4 \\ 2 & 5-\lambda \end{vmatrix} =$$

$$(\lambda+2)(-\lambda^2 + 4\lambda - 3) = -(\lambda-3)(\lambda-1)(\lambda+2) \Rightarrow$$

$$\boxed{\lambda_1 = 3, \ \lambda_2 = 1, \ \lambda_3 = -2}.$$

$$\mathbf{A} - 3\cdot\mathbf{I} = \begin{pmatrix} 3 & 13 & -8 \\ 2 & 2 & -2 \\ 7 & 17 & -12 \end{pmatrix} \sim \begin{pmatrix} 1 & 0 & -0.5 \\ 0 & 1 & -0.5 \\ 0 & 0 & 0 \end{pmatrix} \Rightarrow \boxed{\mathbf{v}_1 = \begin{pmatrix} 1 \\ 1 \\ 2 \end{pmatrix}}.$$

$$\mathbf{A} - \mathbf{I} = \begin{pmatrix} 5 & 13 & -8 \\ 2 & 4 & -2 \\ 7 & 17 & -10 \end{pmatrix} \sim \begin{pmatrix} 1 & 0 & 1 \\ 0 & 1 & -1 \\ 0 & 0 & 0 \end{pmatrix} \Rightarrow \boxed{\mathbf{v}_2 = \begin{pmatrix} -1 \\ 1 \\ 1 \end{pmatrix}}.$$

$$\mathbf{A} + 2\mathbf{I} = \begin{pmatrix} 8 & 13 & -8 \\ 2 & 7 & -2 \\ 7 & 17 & -7 \end{pmatrix} \sim \begin{pmatrix} 1 & 0 & -1 \\ 0 & 1 & 0 \\ 0 & 0 & 0 \end{pmatrix} \Rightarrow \boxed{\mathbf{v}_3 = \begin{pmatrix} 1 \\ 0 \\ 1 \end{pmatrix}}.$$

$$(\mathbf{P}|\mathbf{I}) = \begin{pmatrix} 1 & -1 & 1 & | & 1 & 0 & 0 \\ 1 & 1 & 0 & | & 0 & 1 & 0 \\ 2 & 1 & 1 & | & 0 & 0 & 1 \end{pmatrix} \sim \begin{pmatrix} 1 & 0 & 0 & | & 1 & 2 & -1 \\ 0 & 1 & 0 & | & -1 & -1 & 1 \\ 0 & 0 & 1 & | & -1 & -3 & 2 \end{pmatrix} \Rightarrow \mathbf{P}^{-1} = \begin{pmatrix} 1 & 2 & -1 \\ -1 & -1 & 1 \\ -1 & -3 & 2 \end{pmatrix};$$

$$\mathbf{A}^t\mathbf{x}_0 = \mathbf{P}\mathbf{D}^t\mathbf{P}^{-1}\mathbf{x}_0 = \begin{pmatrix} 1 & -1 & 1 \\ 1 & 1 & 0 \\ 2 & 1 & 1 \end{pmatrix}\begin{pmatrix} 3^t & 0 & 0 \\ 0 & 1 & 0 \\ 0 & 0 & (-2)^t \end{pmatrix}\begin{pmatrix} 1 & 2 & -1 \\ -1 & -1 & 1 \\ -1 & -3 & 2 \end{pmatrix}\begin{pmatrix} 1 \\ 1 \\ 0 \end{pmatrix} =$$

$$\begin{pmatrix} 1 & -1 & 1 \\ 1 & 1 & 0 \\ 2 & 1 & 1 \end{pmatrix} \begin{pmatrix} 3^t & 0 & 0 \\ 0 & 1 & 0 \\ 0 & 0 & (-2)^t \end{pmatrix} \begin{pmatrix} 3 \\ -2 \\ -4 \end{pmatrix} = \begin{pmatrix} 1 & -1 & 1 \\ 1 & 1 & 0 \\ 2 & 1 & 1 \end{pmatrix} \begin{pmatrix} 3^{t+1} \\ -2 \\ -(-2)^{t+2} \end{pmatrix} =$$

$$\begin{pmatrix} 3^{t+1} + 2 - (-2)^{t+2} \\ 3^{t+1} - 2 \\ 2 \cdot 3^{t+1} - 2 - (-2)^{t+2} \end{pmatrix} \Rightarrow \begin{cases} x_t & = 3^{t+1} + 2 - (-2)^{t+2} \\ y_t & = 3^{t+1} - 2 \\ z_t & = 2 \cdot 3^{t+1} - 2 - (-2)^{t+2} \end{cases} . \quad \square$$

Problem 5.10 by Change of Variables. Since $\mathbf{P} = \begin{pmatrix} \mathbf{v}_1 & \mathbf{v}_2 & \mathbf{v}_3 \end{pmatrix}$ diagonalizes $\mathbf{A}$, we have $\mathbf{P}^{-1}\mathbf{AP} = \mathbf{D}$. If we define new variables $\mathbf{u}_t$ by $\mathbf{x}_t = \mathbf{Pu}_t$, then

$$\mathbf{Pu}_{t+1} = \mathbf{x}_{t+1} = \mathbf{Ax}_t = \mathbf{APu}_t \Rightarrow \mathbf{u}_{t+1} = \mathbf{P}^{-1}\mathbf{APu}_t = \mathbf{Du}_t.$$

Since $\mathbf{D}$ is diagonal, there is a very simple formula for $\mathbf{D}^t$. So, $\mathbf{u}_t = \mathbf{D}^t \mathbf{u}_0$. Then $\mathbf{x}_t = \mathbf{Pu}_t$. The advantage of this approach is that it is not necessary to evaluate $\mathbf{P}^{-1}$. Instead, using Gauss'elimination we solve the linear system $\mathbf{x}_0 = \mathbf{Pu}_0$ to find $\mathbf{u}_0$. To show how this works we consider Problem 5.10 again.

$$\mathbf{P} = \begin{pmatrix} 1 & -1 & 1 \\ 1 & 1 & 0 \\ 2 & 1 & 1 \end{pmatrix}, \quad \mathbf{D} = \begin{pmatrix} 3 & 0 & 0 \\ 0 & 1 & 0 \\ 0 & 0 & -2 \end{pmatrix}.$$

Let $\mathbf{u}_t = \begin{pmatrix} u_t & v_t & w_t \end{pmatrix}^T$. Then $\mathbf{Pu}_0 = \mathbf{x}_0$. We find the reduced row echelon form for the augmented matrix of this system:

$$\begin{pmatrix} 1 & -1 & 1 & 1 \\ 1 & 1 & 0 & 1 \\ 2 & 1 & 1 & 0 \end{pmatrix} \sim \begin{pmatrix} 1 & 0 & 0 & 3 \\ 0 & 1 & 0 & -2 \\ 0 & 0 & 1 & -4 \end{pmatrix} \Rightarrow \mathbf{u}_0 = \begin{pmatrix} 3 \\ -2 \\ -4 \end{pmatrix}.$$

$$\mathbf{x}_t = \mathbf{Pu}_t = \mathbf{PD}^t \mathbf{u}_0 = \begin{pmatrix} 1 & -1 & 1 \\ 1 & 1 & 0 \\ 2 & 1 & 1 \end{pmatrix} \begin{pmatrix} 3^t & 0 & 0 \\ 0 & 1 & 0 \\ 0 & 0 & (-2)^t \end{pmatrix} \begin{pmatrix} 3 \\ -2 \\ -4 \end{pmatrix} = \begin{pmatrix} 1 & -1 & 1 \\ 1 & 1 & 0 \\ 2 & 1 & 1 \end{pmatrix} \begin{pmatrix} 3^{t+1} \\ -2 \\ -(-2)^{t+2} \end{pmatrix} =$$

$$\begin{pmatrix} 3^{t+1} + 2 - (-2)^{t+2} \\ 3^{t+1} - 2 \\ 2 \cdot 3^{t+1} - 2 - (-2)^{t+2} \end{pmatrix} \Rightarrow \begin{cases} x_t & = 3^{t+1} + 2 - (-2)^{t+2} \\ y_t & = 3^{t+1} - 2 \\ z_t & = 2 \cdot 3^{t+1} - 2 - (-2)^{t+2} \end{cases} .$$

The following theorem gives the formula for the solutions of recurrence equations, which have negative discriminant. Since the discriminant is negative, the corresponding characteristic polynomial has two different complex conjugate roots, which implies that the diagonalization of the matrix $\mathbf{A}$ is possible over the field $\mathbb{C}$ of complex numbers.

Theorem 5.19 *The solution to the recurrence equation*

$$y_n = ay_{n-1} - by_{n-2}, \ n = 2,3,\ldots, \ y_0 = c_0, y_1 = c_1, \qquad (5.33)$$

where $a,b \in \mathbb{R}$ and $D = a^2 - 4b < 0$, is given by the formula

$$y_n = r^n \left(\frac{2c_1 - ac_0}{\sqrt{|D|}} \sin n\theta + c_0 \cos n\theta \right), n = 0,1,2\ldots . \qquad (5.34)$$

Proof. Formula (5.10) shows that the difference equation (5.33) is equivalent to the system

$$\binom{y_{n+1}}{y_n} = A \binom{y_n}{y_{n-1}}, \ A = \begin{pmatrix} a & -b \\ 1 & 0 \end{pmatrix}, \ p_A(\lambda) = \lambda^2 - a\lambda + b. \qquad (5.35)$$

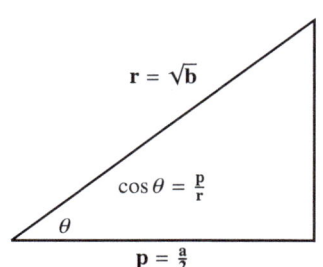

$$y_{n+1} - ay_n + by_{n-1} = 0$$
$$\lambda^2 - a\lambda + b = 0$$

$$D = a^2 - 4b < 0 \Rightarrow$$
$$b > 0, \ \frac{|a|}{2} < \sqrt{b}$$

$$\lambda_1 = r(\cos\theta + i\sin\theta)$$
$$\lambda_2 = r(\cos\theta - i\sin\theta)$$

General Solution:
$$y_t = Er^t \cos\theta t + Fr^t \sin\theta t$$

$$r = \sqrt{b}$$
$$\cos\theta = \frac{p}{r}$$
$$p = \frac{a}{2}$$

Fig. 5.1: The Difference Equation.

Since $D = a^2 - 4b < 0$, the equation $p_A(\lambda) = 0$ has two complex roots, defined by the formulas

$$\lambda_1 = \frac{a + i\sqrt{|D|}}{2}, \quad \lambda_2 = \frac{a - i\sqrt{|D|}}{2}.$$

Hence the matrix A in (5.35) is diagonalizable over the field of complex numbers $\mathbb{C}$. The geometrical sense of the parameters involved is pictured in Figure 5.1. By Vièta's formulas

$$\lambda_1 + \lambda_2 = a, \ \lambda_1\lambda_2 = b \qquad (5.36)$$

we obtain that

$$A = \begin{pmatrix} \lambda_1 + \lambda_2 & -\lambda_1\lambda_2 \\ 1 & 0 \end{pmatrix} \Rightarrow A\binom{\lambda_1}{1} = \lambda_1\binom{\lambda_1}{1}, \ A\binom{\lambda_2}{1} = \lambda_2\binom{\lambda_2}{1}.$$

Then

$$\mathbf{P} = \begin{pmatrix} \lambda_1 & \lambda_2 \\ 1 & 1 \end{pmatrix}, \ \mathbf{P}^{-1} = \frac{1}{\lambda_1 - \lambda_2} \begin{pmatrix} 1 & -\lambda_2 \\ -1 & \lambda_1 \end{pmatrix}, \ \mathbf{D} = \begin{pmatrix} \lambda_1 & 0 \\ 0 & \lambda_2 \end{pmatrix}. \tag{5.37}$$

It follows that

$$\mathbf{A}^n = \mathbf{PD}^n\mathbf{P}^{-1} = \begin{pmatrix} \lambda_1 & \lambda_2 \\ 1 & 1 \end{pmatrix} \begin{pmatrix} \lambda_1^n & 0 \\ 0 & \lambda_2^n \end{pmatrix} \frac{1}{\lambda_1 - \lambda_2} \begin{pmatrix} 1 & -\lambda_2 \\ -1 & \lambda_1 \end{pmatrix} =$$

$$\frac{1}{\lambda_1 - \lambda_2} \begin{pmatrix} \lambda_1^{n+1} & \lambda_2^{n+1} \\ \lambda_1^n & \lambda_2^n \end{pmatrix} \begin{pmatrix} 1 & -\lambda_2 \\ -1 & \lambda_1 \end{pmatrix} = \frac{1}{\lambda_1 - \lambda_2} \begin{pmatrix} \lambda_1^{n+1} - \lambda_2^{n+1} & -b(\lambda_1^n - \lambda_2^n) \\ \lambda_1^n - \lambda_2^n & -b(\lambda_1^{n-1} - \lambda_2^{n-1}) \end{pmatrix}.$$

Therefore, by (5.35)

$$\begin{pmatrix} y_{n+1} \\ y_n \end{pmatrix} = \mathbf{A}^n \begin{pmatrix} y_1 \\ y_0 \end{pmatrix} \Rightarrow y_n = \frac{\lambda_1^n - \lambda_2^n}{\lambda_1 - \lambda_2} c_1 - bc_0 \frac{\lambda_1^{n-1} - \lambda_2^{n-1}}{\lambda_1 - \lambda_2}, \ n = 1, 2, \dots . \tag{5.38}$$

Using the trigonometric formula

$$\sin(n - 1)\theta = \sin(n\theta - \theta) = \sin n\theta \cos\theta - \cos n\theta \sin\theta$$

and the de Moivre's Theorem (see Theorem 4.10), we obtain

$$\lambda_1^n - \lambda_2^n = 2ir^n(\sin n\theta), \ \lambda_1^{n-1} - \lambda_2^{n-1} = 2ir^{n-1}(\sin(n - 1)\theta), \lambda_1 - \lambda_2 = 2ir\sin\theta,$$

and

$$y_n = \frac{\lambda_1^n - \lambda_2^n}{\lambda_1 - \lambda_2} c_1 - bc_0 \frac{\lambda_1^{n-1} - \lambda_2^{n-1}}{\lambda_1 - \lambda_2} = r^{n-1} \frac{\sin n\theta}{\sin\theta} c_1 - r^{n-2} \frac{\sin(n - 1)\theta}{\sin\theta} bc_0 =$$

$$r^n \frac{\sin n\theta}{r\sin\theta} c_1 - r^n \frac{bc_0}{r^2} \cot\theta \sin n\theta + r^n \frac{bc_0}{r^2} \cos n\theta =$$

$$r^n \left(\frac{c_1}{r\sin\theta} - \frac{bc_0 \cot\theta}{r^2} \right) \sin n\theta + r^n \frac{bc_0}{r^2} \cos n\theta = \begin{cases} b = r^2 \\ r\sin\theta = 0.5\sqrt{|D|} \\ \cot\theta = \frac{a}{\sqrt{|D|}} \end{cases}$$

$$r^n \left(\frac{2c_1 - ac_0}{\sqrt{|D|}} \sin n\theta + c_0 \cos n\theta \right).$$

□

Problems

Prob. 167 — Suppose that, for non-negative integers n, the sequences x_n and y_n satisfy the system of difference equations

$$x_{n+1} = x_n + 2y_n$$
$$y_{n+1} = 2x_n + y_n$$

and the initial conditions $x_0 = 0$ and $y_0 = 1$. Find formulas for both sequences.

Prob. 168 — Find the sequences x_t, y_t, z_t, which satisfy the difference equations

$$\begin{cases} x_{t+1} & = -9x_t - 17y_t + 7z_t \\ y_{t+1} & = 8x_t + 14y_t - 5z_t \\ z_{t+1} & = 2x_t + 2y_t + z_t \end{cases}$$

and the initial conditions: $x_0 = -1$, $y_0 = 1$, $z_0 = 2$.

Prob. 169 — Let

$$A = \begin{pmatrix} -1 & 7 & -1 \\ 0 & 1 & 0 \\ 0 & 15 & -2 \end{pmatrix}, \quad P = \begin{pmatrix} 1 & 1 & 1 \\ 0 & 0 & 1 \\ 1 & 0 & 5 \end{pmatrix}.$$

Evaluate A^{10}.

Prob. 170 — Sequences x_t, y_t, z_t are defined by $x_0 = -1$, $y_0 = 2$, $z_0 = 1$ and the recurrence system

$$x_{t+1} = 7x_t - 3z_t$$
$$y_{t+1} = x_t + 6y_t + 5z_t$$
$$z_{t+1} = 5x_t - z_t.$$

Find explicit formulas for x_t, y_t, and z_t by diagonalization.

5.6 Difference Equations in Economics

Figure 5.2 demonstrates the graphs of the inverse demand function $p = p^D(q)$ and inverse supply function $p = p^S(q)$. Both graphs intersect at the equilibrium point (q^*, p^*). The cobweb model is used to investigate the stability of the market about the equilibrium point. The process starts from some price $p_0 \neq p^*$ and generates a sequence of points on the graphs of the both functions as it is shown in Figure 5.2. The crucial equation is boxed in Figure 5.2. Since the both demand and supply curves are usually smooth, they can easily be replaced with their tangents at (q^*, p^*). We are going to study the sequence

$$p_t = p^D(q^S(p_{t-1}))$$

in the case of linear supply and demand:

$$S = \{(q,p) : q = bp - a\}, \quad D = \{(q,p) : q = c - dp\}.$$

We assume that a, b, c, d are positive, which correspond to the natural assumptions that $q^S(p)$ increases and $p^D(q)$ decreases. Then

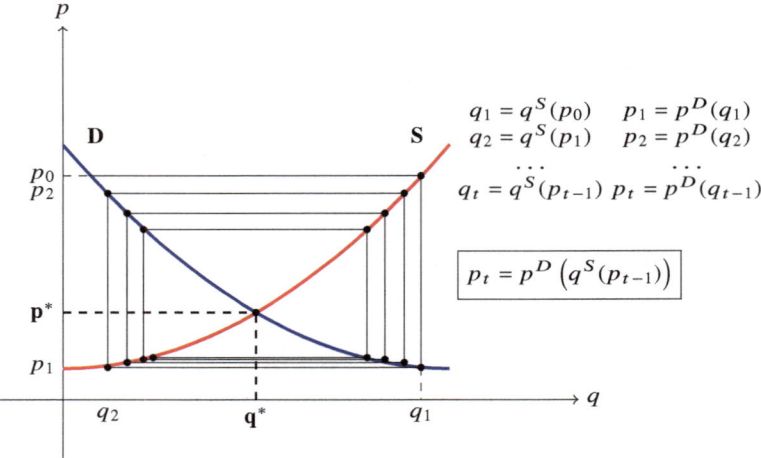

Fig. 5.2: The Cobweb Model.

$$q^S(p) = bp - a, \ p^D(q) = \frac{c-q}{d} \Rightarrow p^D(q^S(p)) = \frac{c-(bp-a)}{d},$$

implying the recurrence relation:

$$p_t = \left(\frac{-b}{d}\right) p_{t-1} + \frac{c+a}{d}. \tag{5.39}$$

We observe that the equilibrium price p^*

$$bp^* - a = c - dp^* \Leftrightarrow p^* = \frac{a+c}{b+d}$$

gives the time-independent solution to (5.39). The general solution to (5.39) is given by (5.4), where

$$q = \frac{-b}{d}, \ \frac{d}{1-q} = \frac{a+c}{b+d} = p^*,$$

implying that the solution to (5.39) is given by

$$p_t = p^* + (p_0 - p^*) \left(\frac{-b}{d}\right)^t.$$

If $b < d$ then $p_t \to p^*$ as $t \to +\infty$. It is clear that then the market equilibrium is **stable**. If $b > d$ then the p_t's oscillate with increasing magnitude. The market equilibrium is called **unstable** in this case. If $b = d$ then the price oscillates about p^*.

Notice that the stability condition **b** < **d** means that the supply line is steeper than the demand line when they both are considered as the graphs of functions in *p*.

Corollary 5.3 *If a change in the quantity affects the suppliers price more than the consumers price, then the equilibrium will be stable.*

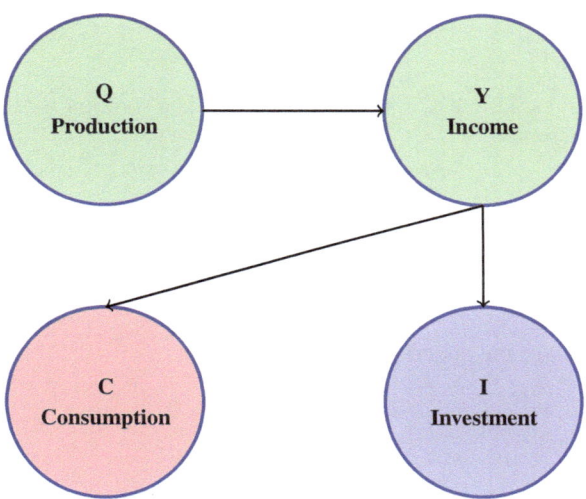

Fig. 5.3: A Simple Model of the National Economy.

A Simple Model of the National Economy.

In this model, all yearly production is used for the income Y, whereas the investment Y is split between the consumption C and the investment I so that

$$Q = Y, \quad Y = C + I,$$

see Figure 5.3. The first natural condition is that the consumption C_t for the tth year is the sum of necessary constant consumption $c > 0$ and a proportion bY_{t-1} of the previous year income Y_{t-1}, where $0 < b < 1$.

1. $C_t = c + bY_{t-1}$, $c > 0$, $b > 0$,
2. $I_t = i + v(Q_{t-1} - Q_{t-2})$, $i > 0$, $v > 0$.

The second equation says that the investment for the tth year is made by a constant and absolutely necessary investment i and a part of the increment of the production for two previous years. Excluding C_t, I_t, Q_t we obtain the following difference equation

$$Y_t = C_t + I_t = (c + bY_{t-1}) + (i + v(Q_{t-1} - Q_{t-2})) = c + bY_{t-1} + i + v(Y_{t-1} - Y_{t-2}) =$$

$$(c + i) + (b + v)Y_{t-1} - vY_{t-2} \Rightarrow \boxed{Y_t - (b + v)Y_{t-1} + vY_{t-2} = c + i}.$$

The constant particular solution $Y_t \equiv Y^*$ is

$$Y^* - (b + v)Y^* + vY^* = c + i \Rightarrow \boxed{\mathbf{Y^* = \frac{c + i}{1 - b} > 0}}.$$

To find a general solution $Y_t = Y^* + y_t$, where y_t is a solution for the second-order recurrence equation

$$y_t - (b + v)y_{t-1} + vy_{t-2} = 0, \tag{5.40}$$

following (5.35) we diagonalize the matrix

$$\mathbf{A} = \begin{pmatrix} b + v & -v \\ 1 & 0 \end{pmatrix} \Rightarrow p_{\mathbf{A}}(\lambda) = \lambda^2 - (b + v)\lambda + v.$$

The discriminant of the quadratic equation in λ

$$\lambda^2 - (b + v)\lambda + v = 0 \tag{5.41}$$

equals

$$D = (b + v)^2 - 4v = v^2 + 2(b - 2)v + b^2.$$

Equation (5.41) has two different real solutions if and only if $D > 0$. This happens if and only if

either $0 \le v < v_1(b) = (2 - b) - 2\sqrt{1 - b}$ **or** $(2 - b) + 2\sqrt{1 - b} = v_2(b) < v.$
$$\tag{5.42}$$

Indeed, the quadratic equation $D = 0$ in v has two roots

$$v_1(b) = 2 - b - 2\sqrt{1 - b} < v_2(b) = 2 - b + 2\sqrt{1 - b}$$

and $D = D(v) > 0$ if and only if (5.42) is satisfied. It follows that the domain $G = \{(b, v) \mid 0 \le b \le 1, 0 \le v\}$ shown in Figure 5.4 splits into four sub-domains

$$\begin{aligned}
G_1 &= \{(b, v) \mid 0 \le v < v_1(b)\}, \\
G_2 &= \{(b, v) \mid v_1(b) < v < 1\}, \\
G_3 &= \{(b, v) \mid 1 < v < v_2(b)\}, \\
G_4 &= \{(b, v) \mid v_2(b) < v\}.
\end{aligned} \tag{5.43}$$

The solutions to the equation (5.41) are positive in G_1 and G_2:

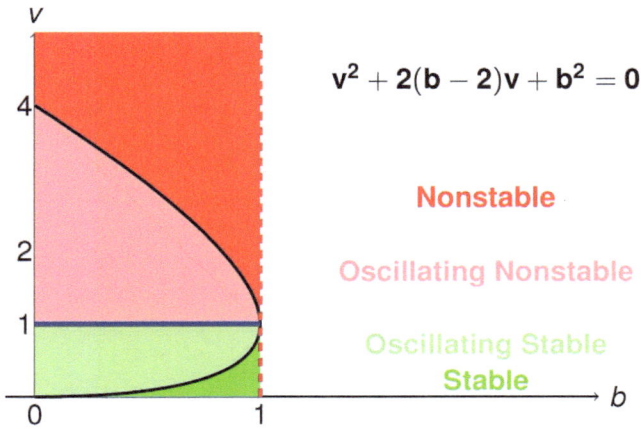

Fig. 5.4: Simple National Economy Analysis

$$0 < \lambda_1 = \frac{(b+v) - \sqrt{(b+v)^2 - 4v}}{2} < \lambda_2 = \frac{(b+v) + \sqrt{(b+v)^2 - 4v}}{2}.$$

Notice that the domain G_1 is placed below the line $b+v = 2$, implying that $b+v < 2$ for any point (b,v) in G_1. It follows that for any $(b,v) \in G_1$

$$\lambda_2 < 1 \Leftrightarrow \sqrt{(b+v)^2 - 4v} < 2 - (b+v) \Leftrightarrow -4v < 4 - 4(b+v) \Leftrightarrow 0 < 4(1-b).$$

We conclude that the case $(b,v) \in G_1$ corresponds to the case of the stable economy:

$$Y_t = Y^* + A\lambda_1^t + B\lambda_2^t \to Y^* \text{ if } t \to +\infty.$$

To the contrary, the domain G_2 is placed above the line $b + v = 2$, implying that

$$1 < \lambda_1 \Leftrightarrow (b+v) - 2 > \sqrt{(b+v)^2 - 4v} \Leftrightarrow 4 - 4(b+v) > -4v \Leftrightarrow 0 < 4(1-b).$$

We conclude that the case $(b,v) \in G_2$ corresponds to an extremely unstable economy. Since in our model $Q_t = Y_t$, this growth is a result of the overproduction. If $(b,v) \in G_2 \cup G_3$, then the characteristic polynomial has two complex roots and the economy behavior is controlled by Theorem 5.19. In our case $b = -v$, implying the exponentially decaying oscillation if $v < 1$, oscillation about the equilibrium state if $v = 1$, and exponentially increasing oscillation if $1 < v$.

There is an interesting case of points (b,v) on the curve $v_1 = v_1(b)$ and on $v_2 = v_2(b)$. In both cases $D = 0$ and therefore there is only one eigenvalue $\lambda_1 = 1 - \sqrt{1-b}$ for points on the curve $v_1 = v_1(b)$ and $\lambda_1 = 1 + \sqrt{1-b}$ for points on $v_2 = v_2(b)$. The dimension of the eigenspace corresponding to $\lambda_1 = 1 - \sqrt{1-b}$ is one. The vector

$$\mathbf{v}_1 = \begin{pmatrix} b \\ 1 + \sqrt{1 - b} \end{pmatrix}$$

is the eigenvector of $\mathbf{A}$ corresponding to λ_1. It is easy to check that

$$(\mathbf{A} - \lambda_1 \mathbf{I}_2)\mathbf{v}_2 = \mathbf{v}_1, \quad \text{where} \quad \mathbf{v}_2 = \begin{pmatrix} 1 + \sqrt{1 - b} \\ 0 \end{pmatrix},$$

see Problem 5.3. Then

$$\mathbf{A}^n = \mathbf{P}\mathbf{J}^n\mathbf{P}^{-1}, \quad \mathbf{J} = \begin{pmatrix} \lambda_1 & 1 \\ 0 & \lambda_1 \end{pmatrix}.$$

It is easy to check by induction that

$$\mathbf{J}^n = \begin{pmatrix} \lambda_1^n & n\lambda_1^{n-1} \\ 0 & \lambda_1^n \end{pmatrix}.$$

Since $0 < \lambda_1 < 1$ and $\lim_n n\lambda_1^{n-1} = 0$, we see that $\lim \mathbf{A}^n = 0$, implying that if the parameters b and v are placed on this curve, then the income approaches the equilibrium income exponentially fast. For points on the curve $v = v_2(b)$ we get a catastrophic growth of the production.

Problems

Prob. 171 — Find the general solution of the recurrence:

$$y_t - 5y_{t-1} + 6y_{t-2} = 12.$$

Prob. 172 — Find the general solution of the recurrence:

$$y_t - 2y_{t-1} + y_{t-2} = a, \quad a \neq 0.$$

Prob. 173 — Find the solution of the recurrence:

$$y_t - 2y_{t-1} - 3y_{t-2} = t, \quad y_0 = 0, y_1 = 1.$$

Hint: Find a particular solution in the form $at + b$.

Prob. 174 — Find the general solution of the recurrence equation

$$y_t - 6y_{t-1} + 9y_{t-2} = 21.$$

Prob. 175 — Find the solution of the recurrence equation

$$y_t - 2y_{t-1} + 2y_{t-2} = 0, \quad y_0 = 2, \ y_1 = 1.$$

Prob. 176 — Determine whether the cobweb model predicts stable or unstable equilibrium for the market with

$$q^S(p) = 4p - 3, \quad q^D(p) = 18 - 3p.$$

Prob. 177 — Consider a simple national economy satisfying

$$C_t = \frac{3}{4}Y_{t-1}, \quad \textbf{2.} \; I_t = 40 + \frac{1}{4}(Q_{t-1} - Q_{t-2}); \; Y_0 = 35, \; Y_1 = 30.$$

Classify this economy according to the picture shown in Figure 5.4.

5.7 Markov's Theory

A **Markov Chain (Process)** is a closed system consisting of a population distributed across n different states and which is observed at scheduled times. The population changes with time from one distribution to another. It is assumed that the probability of a member transition from one state to another is known. This probability depends on the state occupied at the previous observation. This information is used to predict the system's distribution at a future time t.

The probabilities are represented in an $n \times n$ matrix $\mathbf{A} = (a_{ij})$, where the entry a_{ij} denotes the probability of a member transitioning from state **j** to state **i**. Such a matrix, called a **transition matrix**, has the following two properties:

(**i**) The entries of $\mathbf{A}$ are all non-negative.
(**ii**)The sum of the entries in each **column** of $\mathbf{A}$ is equal to 1:

$$a_{1j} + a_{2j} + \cdots + a_{nj} = 1.$$

Property (**ii**) follows from the assumption that all members of the population must be in one of the n states at any given time.

In Markov chains, p_i ($i = 1, 2, \ldots, n$) equals the probability that the system is in state i. The entry a_{ij} of the transition matrix $\mathbf{A}$ is the conditional probability for the system to be in state i if it is in the state j. Then, the probability that the system arrives to state i equals

$$\sum_{j=1}^{n} a_{ij}p_j,$$

where a_{ij} is the conditional probability of the transition to state i and p_j is the probability to be in state j.

It follows that

$$\mathbf{Ap} = \begin{pmatrix} a_{11} & a_{12} & \cdots & a_{1n} \\ a_{21} & a_{22} & \cdots & a_{2n} \\ \vdots & \vdots & \ddots & \vdots \\ a_{n1} & a_{n2} & \cdots & a_{nn} \end{pmatrix} \begin{pmatrix} p_1 \\ p_2 \\ \vdots \\ p_n \end{pmatrix}$$

gives the probability distribution in one step of the dynamical system described by the transition matrix $\mathbf{A}$.

Problem 5.11 (Unemployment Model) A city with a population of $1,100,000$ people who are either employed or seeking employment has steady-state records of unemployment. Being employed, the chances to keep this status in the next month are 30%. Being unemployed, the chances to find employment during the next month are 40%. If initially $600,000$ of citizens are employed and $500,000$ are unemployed, then what will be the proportion of employed and unemployed people in a long run?

Solution: Let x_t be the number of employed people and y_t the number of unemployed people in the tth month. Then

$$\begin{cases} x_{t+1} & = 0.3x_t + 0.4y_t \\ y_{t+1} & = 0.7x_t + 0.6y_t \end{cases}.$$

If $\mathbf{x}_t = \begin{pmatrix} x_t & y_t \end{pmatrix}^T$, then this system can be rewritten in the matrix form:

$$\mathbf{x}_{t+1} = \mathbf{A}\mathbf{x}_t, \quad \mathbf{A} = \begin{pmatrix} 0.3 & 0.4 \\ 0.7 & 0.6 \end{pmatrix}.$$

Using diagonalization, we can solve this system of recurrence equations assuming that initially there were $x_0 = 600,000$ employed people and $y_0 = 500,000$ unemployed people.

The characteristic polynomial of $\mathbf{A}$ equals

$$\det(\mathbf{A} - \lambda \mathbf{I}_2) = (\lambda - 1)(\lambda + 0.1).$$

For $\lambda = 1$ we have

$$\mathbf{A} - \mathbf{I}_2 \sim \begin{pmatrix} 1 & -4/7 \\ 0 & 0 \end{pmatrix} \Rightarrow \mathbf{v}_1 = \begin{pmatrix} 4 \\ 7 \end{pmatrix}.$$

For $\lambda = -1$ we have

$$\mathbf{A} + 0.1\mathbf{I}_2 \sim \begin{pmatrix} 1 & 1 \\ 0 & 0 \end{pmatrix} \Rightarrow \mathbf{v}_2 = \begin{pmatrix} 1 \\ -1 \end{pmatrix}.$$

It follows that

$$\mathbf{P} = \begin{pmatrix} \mathbf{v}_1 & \mathbf{v}_2 \end{pmatrix} = \begin{pmatrix} 4 & 1 \\ 7 & -1 \end{pmatrix} \Rightarrow \mathbf{P}^{-1} = \frac{1}{11} \begin{pmatrix} 1 & 1 \\ 7 & -4 \end{pmatrix}, \quad \mathbf{D} = \begin{pmatrix} 1 & 0 \\ 0 & -0.1 \end{pmatrix}.$$

Then

$$\mathbf{A}^t = \mathbf{P}\mathbf{D}^t\mathbf{P}^{-1} = \frac{1}{11}\begin{pmatrix} 4+7(-0.1)^t & 4-4(-0.1)^t \\ 7-7(-0.1)^t & 7+4(-0.1)^t \end{pmatrix}.$$

Passing to the limit, we find that

$$\lim_{t\to+\infty} \mathbf{A}^t = \begin{pmatrix} 4/11 & 4/11 \\ 7/11 & 7/11 \end{pmatrix} = \frac{1}{11}\begin{pmatrix} \mathbf{v}_1 & \mathbf{v}_1 \end{pmatrix}.$$

A conclusion from these calculations is that we found $\mathbf{v}_1$, normalized it so that a new eigenvector $\mathbf{u}$ would be a probability vector, and wrote that $\lim_{t\to+\infty} \mathbf{A}^t = \begin{pmatrix} \mathbf{u} & \mathbf{u} \end{pmatrix}$. This means that the limit distribution $\mathbf{u} = \begin{pmatrix} 4/11 & 7/11 \end{pmatrix}^T$ does not depend on the initial conditions. In the case of this problem, there will be 400,000 employed and 700,000 unemployed citizens in a long run. □

Problem 5.12 [Rent a car company again] A rent a car company owns 900 cars. Customers can either rent a car with a fixed day of return (the scheduled rent) or can leave this free. The later rent is more expensive. Each day, customers rent 40% of the cars in the garage of which 10% are not fixed. The same day, 50% of cars with fixed date and 50% of free date cars return. At present, 500 cars wait for customers in the garage, whereas 300 cars are on a scheduled rent. How many cars will be in the garage, rented on schedule and free, in the long run?

Solution: In Problem 1.9, we have introduced appropriate notation and, in particular, constructed the matrices

$$\mathbf{A} = \begin{pmatrix} 0.6 & 0.5 & 0.5 \\ 0.3 & 0.5 & 0 \\ 0.1 & 0 & 0.5 \end{pmatrix}, \quad \mathbf{x}_t = \begin{pmatrix} x_t \\ y_t \\ z_t \end{pmatrix}$$

To simplify calculations, we introduce a matrix $\mathbf{B}$ with integer entries:

$$\mathbf{B} = 10\mathbf{A} = \begin{pmatrix} 6 & 5 & 5 \\ 3 & 5 & 0 \\ 1 & 0 & 5 \end{pmatrix} \Rightarrow p_{\mathbf{B}}(\lambda) = -\lambda^3 + 16\lambda^2 - 65\lambda + 50 = (10-\lambda)(5-\lambda)(1-\lambda).$$

The eigenvalues of $\mathbf{B}$ and $\mathbf{A}$ are related by obvious formulas $\lambda_i(\mathbf{B}) = 10\lambda_i(\mathbf{A})$, $i = 1,2,3$, whereas the corresponding eigenvectors are the same. For $\lambda_1(\mathbf{B}) = 10$, $\lambda_1(\mathbf{A}) = 1$, we have

$$\mathbf{B} - 10\mathbf{I}_3 \sim \begin{pmatrix} 1 & 0 & 5 \\ 0 & 1 & -3 \\ 0 & 0 & 0 \end{pmatrix} \Rightarrow \mathbf{v}_1 = \begin{pmatrix} 5 \\ 3 \\ 1 \end{pmatrix}.$$

For $\lambda_1(\mathbf{B}) = 5$, $\lambda_1(\mathbf{A}) = 0.5$, we have

$$\mathbf{B} - 5\mathbf{I}_3 \sim \begin{pmatrix} 1 & 0 & 0 \\ 0 & 1 & 1 \\ 0 & 0 & 0 \end{pmatrix} \Rightarrow \mathbf{v}_2 = \begin{pmatrix} 0 \\ 1 \\ -1 \end{pmatrix}.$$

For $\lambda_1(\mathbf{B}) = 1$, $\lambda_1(\mathbf{A}) = 0.1$, we have

$$\mathbf{B} - \mathbf{I}_3 \sim \begin{pmatrix} 1 & 0 & 4 \\ 0 & 1 & -3 \\ 0 & 0 & 0 \end{pmatrix} \Rightarrow \mathbf{v}_3 = \begin{pmatrix} -4 \\ 3 \\ 1 \end{pmatrix}.$$

It follows that

$$\mathbf{P} = \begin{pmatrix} 5 & 0 & -4 \\ 3 & 1 & 3 \\ 1 & -1 & 1 \end{pmatrix}, \ \mathbf{P}^{-1} = \begin{pmatrix} 1/9 & 1/9 & 1/9 \\ 0 & 1/4 & -3/4 \\ -1/9 & 5/36 & 5/36 \end{pmatrix}, \ \mathbf{D} = \begin{pmatrix} 1 & 0 & 0 \\ 0 & 0.5 & 0 \\ 0 & 0 & 0.1 \end{pmatrix}.$$

Then

$$\lim_{t \to +\infty} \mathbf{A}^t = \mathbf{P} \lim_{t \to +\infty} \mathbf{D}^t \mathbf{P}^{-1} = \mathbf{P} \begin{pmatrix} 1 & 0 & 0 \\ 0 & 0 & 0 \\ 0 & 0 & 0 \end{pmatrix} \mathbf{P}^{-1}. \tag{5.44}$$

Rather than to compute the limit directly, we are going to use some theory.

> **Theorem 5.20** *If $\mathbf{A}$ and $\mathbf{B}$ are transition matrices of Markov chains, then $\mathbf{AB}$ is also the transition matrix of a matrix chain. In particular, the powers $\mathbf{A}^k$, $k = 2, 3, \ldots$ are transition matrices of Markov chains.*

> *Proof.* Let $\mathbf{C} = \mathbf{AB}$. Then
>
> $$c_{ij} = \sum_{k=1}^{n} a_{ik} b_{kj} \Rightarrow \sum_{i=1}^{n} c_{ij} = \sum_{i=1}^{n} \sum_{k=1}^{n} a_{ik} b_{kj} = \sum_{k=1}^{n} b_{kj} \left(\sum_{i=1}^{n} a_{ik} \right) = \sum_{k=1}^{n} b_{kj} = 1.$$
>
> $\square$

Theorem 5.20 shows that the columns of $\mathbf{A}^t$ being probability vectors converge to probability vectors making the columns of the matrix on the right-hand side of (5.44). If $\mathrm{row}(\mathbf{P}^{-1}) = \begin{pmatrix} c_{11} & c_{12} & c_{13} \end{pmatrix}$, and $\mathbf{P} = \begin{pmatrix} \mathbf{v}_1 & \mathbf{v}_2 & \mathbf{v}_3 \end{pmatrix}$, then

$$\begin{pmatrix} 1 & 0 & 0 \\ 0 & 0 & 0 \\ 0 & 0 & 0 \end{pmatrix} \mathbf{P}^{-1} = \begin{pmatrix} c_{11} & c_{12} & c_{13} \\ 0 & 0 & 0 \\ 0 & 0 & 0 \end{pmatrix} \Rightarrow \lim_{t \to +\infty} \mathbf{A}^t = \begin{pmatrix} c_{11}\mathbf{v}_1 & c_{12}\mathbf{v}_1 & c_{13}\mathbf{v}_1 \end{pmatrix}.$$

Hence $c_{11}\mathbf{v}_1 = c_{12}\mathbf{v}_1 = c_{13}\mathbf{v}_1$ is the long run distribution, which equals $\begin{pmatrix} 5/9 & 3/9 & 1/9 \end{pmatrix}$ in the case of our problem. Therefore, there will be 500 cars in the garage, 300 cars rented on schedule, and 100 cars rented with a free return date. Please, notice that if $\mathbf{v}_1$ is initially a probability vector, then $c_{11} = c_{12} = c_{13} = 1$. $\square$

The example of the matrix

$$A = \begin{pmatrix} 0 & 1 \\ 1 & 0 \end{pmatrix}$$

shows that things are not so simple with Markov matrices in general:

$$A^{2k} = \begin{pmatrix} 1 & 0 \\ 0 & 1 \end{pmatrix}, \quad A^{2k+1} = \begin{pmatrix} 0 & 1 \\ 1 & 0 \end{pmatrix}.$$

Definition 5.11 A Markov chain with a transition matrix A is called **regular** if all entries of A^k are positive for some $k = 1, 2, \ldots$.

The transition matrix A in Problem 5.12 is regular since

$$A^2 = \begin{pmatrix} 0.56 & 0.55 & 0.55 \\ 0.33 & 0.4 & 0.15 \\ 0.11 & 0.05 & 0.3 \end{pmatrix}.$$

Stochastic matrices explain clearly why Markov transition matrices always have eigenvalue $\lambda_1 = 1$.

Definition 5.12 A matrix C is called a **stochastic matrix** if it has the following two properties:

(i) The entries of C are all non-negative.
(ii) The sum of the entries in each **row** of C is equal to 1:

$$c_{i1} + c_{i2} + \cdots + c_{in} = 1.$$

It is clear that a matrix A is a transition matrix of a Markov process if and only if $C = A^T$ is a stochastic matrix.

Theorem 5.21 *If C is a stochastic matrix, then:*

(i) $v = \begin{pmatrix} 1 & 1 & \cdots & 1 \end{pmatrix}^T$ *is an eigenvector of C with eigenvalue $\lambda = 1$.*
(ii) *If λ is an eigenvalue of C, then $|\lambda| \leq 1$.*

Proof.

$$
\textbf{(i)} \begin{pmatrix} c_{11} & c_{12} & \cdots & c_{1n} \\ c_{21} & c_{22} & \cdots & c_{2n} \\ \vdots & \vdots & \ddots & \vdots \\ c_{n1} & c_{n2} & \cdots & c_{nn} \end{pmatrix} \begin{pmatrix} 1 \\ 1 \\ \vdots \\ 1 \end{pmatrix} = \begin{pmatrix} c_{11} + c_{12} + \cdots + c_{1n} \\ c_{21} + c_{22} + \cdots + c_{2n} \\ \vdots \\ c_{n1} + c_{n2} + \cdots + c_{nn} \end{pmatrix} = \begin{pmatrix} 1 \\ 1 \\ \vdots \\ 1 \end{pmatrix}.
$$

(ii) Let λ be an eigenvalue of $\mathbf{C}$ and $\mathbf{u} \neq 0$ its eigenvector. Let u_i be the largest component of the vector $\mathbf{u}$. Then $\mathbf{w} = \mathbf{u}/u_i$, $\mathbf{Cw} = \lambda\mathbf{w}$. It follows that

$$
|\lambda| = |\lambda||w_i| = |c_{i1}w_1 + \cdots + c_{in}w_n| \leq c_{i1} + \cdots + c_{in} = 1 \Rightarrow |\lambda| \leq 1. \quad \square
$$

Corollary 5.4 *If $\mathbf{A}$ is a transition matrix of a Markov chain, then:*

(i) $\lambda = 1$ *is an eigenvalue of $\mathbf{A}$;*
(ii) *If λ_i is an eigenvalue of $\mathbf{A}$, then $|\lambda_i| \leq 1$.*

Proof. By Theorem 2.10:

$$
\text{nullity}(\mathbf{A} - \lambda\mathbf{I}) = \text{nullity}(\mathbf{A}^T - \lambda\mathbf{I}) = \text{nullity}(\mathbf{C} - \lambda\mathbf{I}). \qquad \square
$$

Theorem 5.22 (Fundamental Limit Theorem for Markov Matrices) *If $\mathbf{A}$ is a regular transition Markov matrix of size $n \times n$, then*

$$
\lim_{k \to +\infty} \mathbf{A}^k = \mathbf{U} = \begin{pmatrix} \mathbf{v}_1 & \cdots & \mathbf{v}_1 \end{pmatrix}, \tag{5.45}
$$

where $\mathbf{Av}_1 = \mathbf{v}_1$ and $\mathbf{v}_1$ is positive $n \times 1$ eigenvector representing the long run probability distribution.

We split the proof of this important theorem (see Kennedy J. (Snell J.)) into a number of simple steps. Given a vector $\mathbf{x} = \begin{pmatrix} x_1 & x_2 & \cdots & x_n \end{pmatrix}^T$ in $\mathbb{R}^n$, we put

$$
m(\mathbf{x}) = \min_{1 \leq i \leq n} x_i, \quad M(\mathbf{x}) = \max_{1 \leq i \leq n} x_i
$$

Lemma 5.3 *Let* **C** *be an* $n \times n$ *stochastic matrix with non-negative entries and* $\mathbf{x} \in \mathbb{R}^n$. *Then*

$$m(\mathbf{x}) \leq m(\mathbf{Cx}) \leq M(\mathbf{Cx}) \leq M(\mathbf{x}).$$

Proof. We have

$$\begin{cases} (\mathbf{Cx})_i & = \sum_{j=1}^n c_{ij} x_j \geq m(\mathbf{x}) \sum_{j=1}^n c_{ij} = m(\mathbf{x}) \Rightarrow m(\mathbf{Cx}) \geq m(\mathbf{x}); \\ (\mathbf{Cx})_i & = \sum_{j=1}^n c_{ij} x_j \leq M(\mathbf{x}) \sum_{j=1}^n c_{ij} = M(\mathbf{x}) \Rightarrow M(\mathbf{Cx}) \leq M(\mathbf{x}). \end{cases} \qquad \square$$

Lemma 5.4 *Let* **C** *be an* $n \times n$ *stochastic matrix with positive entries and* $\mathbf{x} \in \mathbb{R}^n$. *Let* q *be the minimal entry of* **C**. *Then* $0 < q \leq 0.5$ *and*

$$M(\mathbf{Cx}) - m(\mathbf{Cx}) \leq (1 - 2q)(M(\mathbf{x}) - m(\mathbf{x})). \qquad (5.46)$$

Proof. Let us consider $\text{row}_i(\mathbf{C})$ which contains the entry q. The entry q is the minimal in this row. Since **C** is a stochastic matrix, the sum of all entries in $\text{row}_i(\mathbf{C})$ is one, implying that $qn \leq 1$. It follows that $0 < q \leq 1/n \leq 0.5$.

For any index k we replace all x_j in the sum of formula (5.47) with $M(\mathbf{x})$ except for $j = r$ satisfying $x_r = m(\mathbf{x})$:

$$(\mathbf{Cx})_k = \sum_{j=1}^n c_{kj} x_j \leq c_{kr} m(\mathbf{x}) + M(\mathbf{x}) \sum_{j \neq r} c_{kj} =$$

$$M(\mathbf{x}) - c_{kr}(M(\mathbf{x}) - m(\mathbf{x})) \leq M(\mathbf{x}) - q(M(\mathbf{x}) - m(\mathbf{x})), \quad (5.47)$$

since $q \leq c_{kr}$. Replacing $\mathbf{x}$ with $-\mathbf{x}$ in the obtained inequality

$$M(\mathbf{Cx}) \leq M(\mathbf{x}) - q(M(\mathbf{x}) - m(\mathbf{x})), \qquad (5.48)$$

we conclude that

$$-m(\mathbf{Cx}) \leq -m(\mathbf{x}) - q(-m(\mathbf{x}) + M(\mathbf{x})) \qquad (5.49)$$

Adding (5.48) and (5.49), we obtain (5.46). $\qquad \square$

Proof of Theorem 5.22. Suppose first that all entries of the initial transition matrix are positive. Then Lemma 5.4 applied to $\mathbf{C} = \mathbf{A}^T$ says that for every $\mathbf{x} \in \mathbb{R}^n$ the inequalities (5.46) hold. Iterating them for $\mathbf{x} = \mathbf{e}_j$ shows that

$$M(\mathbf{C}^k \mathbf{e}_j) - m(\mathbf{C}^k \mathbf{e}_j) \le (1 - 2q)^k \to 0 \text{ as } k \to +\infty. \qquad (5.50)$$

It follows that the j-th columns of $\mathbf{C}^k$ converge to a column with a constant component a_j. By Lemma 5.3 the sequence $\{m(\mathbf{C}^k \mathbf{e}_j)\}$ is non-decreasing. Since all entries of $\mathbf{C}$ are positive, we conclude that $a_j > 0$ for every j. In terms of the transition matrix $\mathbf{A}$ this means that the limit in (5.45) exists and equals to the matrix $\mathbf{U}$ with equal columns $\mathbf{v}_1$. Applying $\mathbf{A}$ to the both sides of (5.45), we conclude that $\mathbf{A}\mathbf{v}_1 = \mathbf{v}_1$. Since the columns of $\mathbf{A}^k$ are probability distributions, The vector $\mathbf{v}_1$ is a long-run probability distribution as well.

If there is a number l such that $\mathbf{A}^l$ has all positive entries, then the increasing subsequence $\{m(\mathbf{A}^{lk}\mathbf{x})\}_{k \ge 0}$ and decreasing subsequence $\{M(\mathbf{A}^{lk}\mathbf{x})\}_{k \ge 0}$ of increasing sequence $\{m(\mathbf{A}^k \mathbf{x})\}_{k \ge 0}$ and decreasing sequence $\{M(\mathbf{A}^k \mathbf{x})\}_{k \ge 0}$ (see Lemma 5.3) converge to a common limit implying the convergence to this very limit of the larger sequences.   □

Corollary 5.5 *If $\mathbf{A}$ is the transition matrix of a regular Markov chain, then any eigenvector corresponding to the eigenvalue $\lambda = 1$ is proportional to $\mathrm{col}_1(\mathbf{U})$, all other eigenvalues λ satisfy $|\lambda| < 1$, and the sum of the coordinates of any eigenvector corresponding to any eigenvalue $\lambda \ne 1$ equals 0.*

Proof. If $\mathbf{y}$ is an eigenvector corresponding to an eigenvalue λ, then

$$\sum_{i=1}^{n} y_i \cdot \mathrm{col}_1(\mathbf{U}) = \lim_{k \to +\infty} \mathbf{A}^k \mathbf{y} = \left(\lim_{k \to +\infty} \lambda^k \right) \cdot \mathbf{y}.$$

It follows that $\lim_{k \to +\infty} \lambda^k$ exists. Hence, either $\lambda = 1$ or $|\lambda| < 1$. It follows from this formula that any eigenvector corresponding to $\lambda = 1$ is proportional to $\mathrm{col}_1(\mathbf{U})$ and $\sum_{i=1}^{n} y_i = 0$ otherwise.   □

Problem 5.13 A carpenter, an electrician, and a plumber agreed to repair their three homes, each working a total of ten days according to the following schedule:

	Carpenter	Electrician	Plumber
Carpenter's Home	2 days	1 day	6 days
Electrician's Home	4 days	5 days	1 day
Plumber's Home	4 days	4 days	3 days

They agreed that their daily wages must be such that each homeowner will break even. Can this agreement be realized?

Solution:

$$
\begin{aligned}
x_1 &= \textbf{daily wage of carpenter} \\
x_2 &= \textbf{daily wage of electrician} \\
x_3 &= \textbf{daily wage of plumber}
\end{aligned}
\Rightarrow
\begin{cases}
2x_1 + x_2 + 6x_3 &= 10x_1 \\
4x_1 + 5x_2 + x_3 &= 10x_2 \\
4x_1 + 4x_2 + 3x_3 &= 10x_3
\end{cases} \tag{5.51}
$$

The first equation of the system (5.51) states that the carpenter pays a total of $2x_1 + x_2 + 6x_3$ for the repairs in his own home and receives a total income of $10x_1$. If

$$
\mathbf{B} = \begin{pmatrix} 2 & 1 & 6 \\ 4 & 5 & 1 \\ 4 & 4 & 3 \end{pmatrix},
$$

then $\mathbf{A} = 0.1\mathbf{B}$ is the transition matrix of a Markov chain.

Since all entries of $\mathbf{A}$ are positive, the equation $\mathbf{Ax} = \mathbf{x}$ has a unique solution and all coordinates of $\mathbf{x}$ are positive. We can check this by the direct calculation:

$$
\mathbf{B} - 10 \cdot \mathbf{I} = \begin{pmatrix} -8 & 1 & 6 \\ 4 & -5 & 1 \\ 4 & 4 & -7 \end{pmatrix} \sim \begin{pmatrix} 0 & -9 & 8 \\ 4 & -5 & 1 \\ 0 & 9 & -8 \end{pmatrix} \sim \begin{pmatrix} 4 & -5 & 1 \\ 0 & 9 & -8 \\ 0 & 0 & 0 \end{pmatrix} \sim
$$

$$
\begin{pmatrix} 4 & -5 & 1 \\ 0 & 1 & \frac{-8}{9} \\ 0 & 0 & 0 \end{pmatrix} \sim \begin{pmatrix} 4 & 0 & \frac{-31}{9} \\ 0 & 1 & \frac{-8}{9} \\ 0 & 0 & 0 \end{pmatrix} \sim \begin{pmatrix} 1 & 0 & \frac{-31}{36} \\ 0 & 1 & \frac{-8}{9} \\ 0 & 0 & 0 \end{pmatrix} \Rightarrow \mathbf{x} = s \begin{pmatrix} 31 \\ 32 \\ 36 \end{pmatrix}. \quad \square
$$

Problem 5.14 Three lawyers — specializing in civil law (CI), criminal law (CR) and notary law (NL) — each owns a legal consulting firm. Their consulting services are multidisciplinary, requiring them to purchase portions of each others' services. For every \$1 of consulting the CI firm does, it purchases \$0.10 of the CR's services and \$0.30 of the NL's services. For each \$1 of consulting provided by the CR firm, it buys \$0.20 of the CI's services and \$0.40 of the NL's services. Similarly, for every \$1 of consulting that the NL firm delivers, it purchases \$0.30 of the CI's services and \$0.40 of the CR's services. In a specific week, the consulting firms each receive outside consulting orders totaling \$347.

(a) Construct Leontief matrix, find its inverse and determine the dollar amount of consulting each lawyer does in this week.

(b) Using Leontief's equation of balance, derive a general formula for the profit of each lawyer. Assuming that only the lawyer specializing in notary law received an order valued at $347, what would be the profit of each lawyer in this scenario?

(c) Under the conditions where each of the lawyer receives an order valued at $347, what will be the profit for each lawyer?

Solution: (a) The graph of the problem is shown on Fig. 5.5.

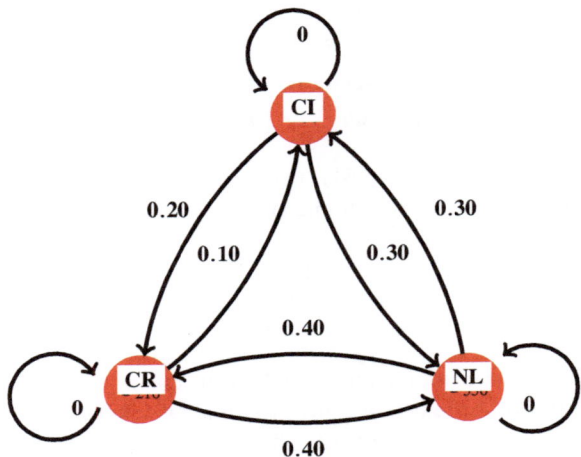

Fig. 5.5: The graph of Problem 5.14.

The technology table is given by:

	$1 of CI	$1 of CR	$1 of NL
CI	0	0.20	0.30
CR	0.10	0	0.40
NL	0.30	0.40	0

The technology matrix looks as follows:

$$\mathbf{A} = \begin{pmatrix} 0 & 0.2 & 0.3 \\ 0.1 & 0 & 0.4 \\ 0.3 & 0.4 & 0 \end{pmatrix}$$

The Leontief matrix is

$$\mathbf{B} = \mathbf{I} - \mathbf{A} = \begin{pmatrix} 1 & -0.2 & -0.3 \\ -0.1 & 1 & -0.4 \\ -0.3 & -0.4 & 1 \end{pmatrix} = \frac{1}{10} \begin{pmatrix} 10 & -2 & -3 \\ -1 & 10 & -4 \\ -3 & -4 & 10 \end{pmatrix}.$$

$$\begin{pmatrix} 10 & -2 & -3 & 1 & 0 & 0 \\ -1 & 10 & -4 & 0 & 1 & 0 \\ -3 & -4 & 10 & 0 & 0 & 1 \end{pmatrix} \sim \begin{pmatrix} 1 & 0 & 0 & 42/347 & 16/347 & 19/347 \\ 0 & 1 & 0 & 11/347 & 91/694 & 43/694 \\ 0 & 0 & 1 & 17/347 & 23/347 & 49/347 \end{pmatrix} \Rightarrow$$

$$\mathbf{B}^{-1} = \begin{pmatrix} \frac{420}{347} & \frac{160}{347} & \frac{190}{347} \\ \frac{110}{347} & \frac{455}{347} & \frac{215}{347} \\ \frac{170}{347} & \frac{230}{347} & \frac{490}{347} \end{pmatrix}.$$

It follows that the dollar amounts of consulting each lawyer did in this week equal

$$\begin{pmatrix} x_1 \\ x_2 \\ x_3 \end{pmatrix} = \begin{pmatrix} \frac{420}{347} & \frac{160}{347} & \frac{190}{347} \\ \frac{110}{347} & \frac{455}{347} & \frac{215}{347} \\ \frac{170}{347} & \frac{230}{347} & \frac{490}{347} \end{pmatrix} \begin{pmatrix} 347 \\ 347 \\ 347 \end{pmatrix} = \begin{pmatrix} 770 \\ 780 \\ 890 \end{pmatrix} \Rightarrow \begin{cases} x_1 & = \mathbf{770} \\ x_2 & = \mathbf{780} \\ x_3 & = \mathbf{890} \end{cases}.$$

(b) We find a general formula for the profit using Leontief Equation:

$$\begin{array}{cccc} \mathbf{CI} & \mathbf{CR} & \mathbf{NL} & \mathbf{D} \end{array}$$
$$\mathbf{CI}\ a_{11}x_1 + a_{12}x_2 + a_{13}x_3 + d_1 = x_1$$
$$\mathbf{CR}\ a_{21}x_1 + a_{22}x_2 + a_{23}x_3 + d_2 = x_2$$
$$\mathbf{NL}\ a_{31}x_1 + a_{32}x_2 + a_{33}x_3 + d_3 = x_3$$

The amount of work in dollars made by a CI lawyer for others, including two other lawyers, equals

$$a_{12}x_2 + a_{13}x_3 + d_1 \Rightarrow \mathbf{revenue(CI)} = a_{12}x_2 + a_{13}x_3 + d_1 = (1 - a_{11})x_1.$$

The internal expanses of the CI lawyer are $(a_{21} + a_{31})x_1$. Considering that the CI lawyer must compensate for the service provided by two other lawyers, we obtain the formula for the profit of the CI firm:

$$\mathbf{profit(CI)} = (1 - a_{11} - a_{21} - a_{31})x_1 = \mathbf{0.6x_1}.$$

Similarly,
$$\mathbf{profit(CR)} = 0.1(1 - a_{12} - a_{22} - a_{32})x_2 = \mathbf{0.4x_2},$$
$$\mathbf{profit(NL)} = 0.1(1 - a_{13} - a_{23} - a_{33})x_3 = \mathbf{0.3x_3}.$$

In case **(b)** we have

$$\begin{pmatrix} x_1 \\ x_2 \\ x_3 \end{pmatrix} = \mathbf{B}^{-1} \begin{pmatrix} 0 \\ 0 \\ 347 \end{pmatrix} = \begin{pmatrix} 190 \\ 215 \\ 490 \end{pmatrix}.$$

Then the profits of the lawyers for this week are

$$\begin{pmatrix} \text{profit}(\mathbf{CI}) \\ \text{profit}(\mathbf{CR}) \\ \text{profit}(\mathbf{NL}) \end{pmatrix} = \begin{pmatrix} 190 \cdot 0.6 \\ 215 \cdot 0.4 \\ 490 \cdot 0.3 \end{pmatrix} = \begin{pmatrix} 114 \\ 86 \\ 147 \end{pmatrix}.$$

Notice that

$$114 + 86 + 147 = 347.$$

In other words, the sum of \$347 obtained from the external order of the NL firm, is distributed among the three lawyers in the proportions specified above.

(c) In this case

$$\mathbf{x} = \begin{pmatrix} \frac{420}{347} & \frac{160}{347} & \frac{190}{347} \\ \frac{110}{347} & \frac{455}{347} & \frac{215}{347} \\ \frac{170}{347} & \frac{230}{347} & \frac{490}{347} \end{pmatrix} \begin{pmatrix} 347 \\ 347 \\ 347 \end{pmatrix} = \begin{pmatrix} 420 + 160 + 190 \\ 110 + 455 + 215 \\ 170 + 230 + 490 \end{pmatrix} = \begin{pmatrix} 770 \\ 780 \\ 890 \end{pmatrix}.$$

It follows that

$$\begin{pmatrix} \text{profit}(\mathbf{CI}) \\ \text{profit}(\mathbf{CR}) \\ \text{profit}(\mathbf{NL}) \end{pmatrix} = \begin{pmatrix} 770 \cdot 0.6 \\ 780 \cdot 0.4 \\ 890 \cdot 0.3 \end{pmatrix} = \begin{pmatrix} 462 \\ 312 \\ 267 \end{pmatrix}.$$

Notice that

$$462 + 312 + 267 = 1041 = 3 \times 347,$$

which confirms that the total profit amounts to \$1,041, exactly three times the individual orders of \$347.

Problems

Prob. 178 — Let $\mathbf{v}_i$ be three columns in $\mathbb{R}^3$, $i = 1, 2, 3$.

$$\begin{pmatrix} \mathbf{v}_1 & \mathbf{v}_2 & \mathbf{v}_3 \end{pmatrix} \begin{pmatrix} 1 & 0 & 0 \\ 0 & 0 & 0 \\ 0 & 0 & 0 \end{pmatrix} =? \quad \begin{pmatrix} 1 & 0 & 0 \\ 0 & 0 & 0 \\ 0 & 0 & 0 \end{pmatrix} \begin{pmatrix} \mathbf{v}_1 & \mathbf{v}_2 & \mathbf{v}_3 \end{pmatrix} =?$$

Prob. 179 — A car rental company has two locations at the airport and at the city. Customer rent cars from any of the two locations and return cars to any of them. The cars are rented and returned according to the following table of probabilities:

	Rented from Airport	Rented from City
Returned to Airport	0.6	0.2
Returned to City	0.4	0.8

(i) Assuming that a car is rented from the airport, what is the probability that it will be at the airport in two rentals?

(ii) Assuming that this dynamical system can be modeled as a Markov chain, find the long-run behavior of this system.

(iii) If the rental agency owns 120 cars, how many parking spaces should it allocate at each location to be reasonably certain that it will have enough spaces for the cars over the long term? Explain your reasoning.

Prob. 180 — It is known that when the air quality in a city is good on one day, then there is a 80% chance that it will be good the next day. When the air quality is bad on one day, then there is a 40% chance that it will remain bad the next day.

(**a**) Find a transition matrix for this process.
(**b**) If the air quality is good today, what is the probability that it will be good three days from now?
(**c**) If the air quality is bad today, what is the probability that it will be bad three days from now?
(**d**) What is the air quality distribution between good and bad days in the long run?

Prob. 181 — Consider a matrix chain with the transition matrix:

$$A = \begin{pmatrix} 0.5 & 0.4 & 0.6 \\ 0.2 & 0.2 & 0.3 \\ 0.3 & 0.4 & 0.1 \end{pmatrix}.$$

Evaluate the limit $\lim_n A^n$.

Prob. 182 — Consider a Matrix chain with the transition matrix:

$$A = \begin{pmatrix} 0.7 & 0.2 & 0.2 \\ 0 & 0.2 & 0.4 \\ 0.3 & 0.6 & 0.4 \end{pmatrix}.$$

Evaluate the limit $\lim_n A^n$.

Prob. 183 — The total population of animals in a forest amounts to $100\,000$ animals consisting of rabbits, wolves, foxes, bears, and rats. Initially the animals are located in five different parts of the forest. Being in a particular part of the forest each spring each animal stays in the same part with probability $3/7$ and can migrate to each of the other parts with equal probability $1/7$. Find the long-term distribution of this population.

Prob. 184 — National television channels, channel 1 and channel 2, each have 50% of the viewer market initially. During each one-month period channel 1 captures 10% of channel 2's share, and channel 1 captures 20% of channel 1's share. Determine each channel's market share in the long run.

Prob. 185 — A tagged bear migrates over three forests, forest 1, forest 2, and forest 3. The monthly migration of the bear is modeled by a Markov chain with transition matrix

$$A = \begin{pmatrix} 0.5 & 0.4 & 0.6 \\ 0.2 & 0.2 & 0.3 \\ 0.3 & 0.4 & 0.1 \end{pmatrix}.$$

Determine the long term behavior of the tagged bear.

Prob. 186 — Show that

$$A = \begin{pmatrix} 1/2 & 1/2 & 0 \\ 1/4 & 1/2 & 1/3 \\ 1/4 & 0 & 2/3 \end{pmatrix}$$

is a regular Markov matrix and find $\lim_n A^n$.

Prob. 187 — By the end of the t-th year, the population of wolves on a certain island totals x_t animals and the population of rabbits totals y_t individuals. The populations are related by the following difference equation

$$\mathbf{x}_{t+1} = \begin{pmatrix} x_{t+1} \\ y_{t+1} \end{pmatrix} = \begin{pmatrix} 0.4 & 0.1 \\ 1.2 & 0.8 \end{pmatrix} \begin{pmatrix} x_t \\ y_t \end{pmatrix}, \quad \mathbf{x}_0 = \begin{pmatrix} 40 \\ 800 \end{pmatrix}.$$

Show that populations will reach steady states as $t \to +\infty$. Determine the long-term populations of wolves and rabbits.

Prob. 188 — The forest on the island is divided into three parts: a section covered with coniferous trees (C), a section with deciduous trees (D), and a section with mixed vegetation (V). Initially, there were 480 wild animals distributed in this forest with 120 in section C, 200 in section D, and 160 in section V. The animals migrate between these three areas on a daily basis. The probability that an animal in section C remains in the same place is 90%, moves to D is 5%, and moves to V is 5%. The probability that an animal in section D remains in the same place is 75%, moves to C is 15%, and moves to V is 10%. The probability that an animal in section V remains in the same place is 85%, moves to C is 10%, and moves to D is 5%. Determine the long-term distribution of animals across the three sections.

Chapter 6
Inner Product Spaces

6.1 Introduction

The notion of orthogonality of two vectors in $\mathbb{R}^n$ appears at the very beginning of Linear Algebra theory. For example, the rows of any real $m \times n$ matrix $\mathbf{A}$ are perpendicular to any vector in the null space $N(\mathbf{A})$. If $\mathbf{A}$ is an $n \times n$ square matrix $\mathbf{A} = \begin{pmatrix} \mathbf{v}_1 & \cdots & \mathbf{v}_n \end{pmatrix}$ with pairwise orthogonal non-zero columns $\mathbf{v}_j$, then the system $\mathbf{A}\mathbf{x} = \mathbf{b}$ has a very simple solution. We rewrite it as

$$x_1\mathbf{v}_1 + x_2\mathbf{v}_2 + \cdot + x_n\mathbf{v}_n = \mathbf{b}. \tag{6.1}$$

Since $\mathbf{v}_i \perp \mathbf{v}_j$ if $i \neq j$, we can evaluate both sides of (6.1) by Theorem 1.12:

$$x_j\langle \mathbf{v}_j, \mathbf{v}_j \rangle = \langle \mathbf{b}, \mathbf{v}_j \rangle \Rightarrow x_j = \frac{\langle \mathbf{b}, \mathbf{v}_j \rangle}{\langle \mathbf{v}_j, \mathbf{v}_j \rangle}.$$

These simple arguments hint that orthogonal systems may be very useful for Linear Algebra problems.

This chapter includes the Gram-Schmidt orthogonalization process, formulas for orthogonal projections and the orthogonal diagonalization of matrices. The theory developed is applied to the theory of quadratic forms. At the end of the chapter a couple of applications of quadratic forms to optimization of the profit are considered.

6.2 Inner Product Spaces

The Euclidean inner product on $\mathbb{R}^n$ is defined by formula (1.48), whereas Theorem 1.12 lists its basic properties. These properties $(\mathbf{a} - \mathbf{c})$ can be taken to extend the notion of the inner product to general linear spaces.

© The Author(s), under exclusive license to Springer Nature Switzerland AG 2024
S. Khrushchev, *Linear Algebra with Applications to Economics*, Classroom Companion:
Economics, https://doi.org/10.1007/978-3-031-68682-5_6

Definition 6.1 A mapping $\langle *, * \rangle : V \times V \longrightarrow \mathbb{R}$ is called an **inner product** on a real vector space V if for all vectors $\mathbf{x}$, $\mathbf{y}$, and $\mathbf{z}$ in V and all scalars $\alpha, \beta \in \mathbb{R}$ the following axioms are satisfied:

(i) $\langle \mathbf{x}, \mathbf{y} \rangle = \langle \mathbf{y}, \mathbf{x} \rangle$;
(ii) $\langle \alpha \mathbf{x} + \beta \mathbf{y}, \mathbf{z} \rangle = \alpha \langle \mathbf{x}, \mathbf{z} \rangle + \beta \langle \mathbf{y}, \mathbf{z} \rangle$;
(iii) $\langle \mathbf{x}, \mathbf{x} \rangle \geq 0$ and $\langle \mathbf{x}, \mathbf{x} \rangle = 0$ if and only if $\mathbf{x} = \mathbf{0}$.

Problem 6.1 Let $x_0 < x_1 < \cdots < x_n$ be real numbers and $\mathcal{P}_n$ be the real vector space of all polynomials $\mathbf{p}(x) = a_0 + a_1 x + a_2 x^2 + \cdots + a_n x^n$ with real coefficients $a_0, \ldots, a_n$. Then

$$\langle \mathbf{p}, \mathbf{q} \rangle = \sum_{i=0}^{n} \mathbf{p}(x_i) \mathbf{q}(x_i)$$

is an inner product on $\mathcal{P}_n$.

Solution: Properties (**i** – **ii**) are obvious. If $\langle \mathbf{p}, \mathbf{p} \rangle = 0$, then the polynomial $\mathbf{p}$ vanishes at $n + 1$ different real numbers, which is only possible if $a_0 = \cdots = a_n$, since its degree cannot exceed n. $\square$

Problem 6.2 Let $\mathbf{A}$ be an invertible $n \times n$ matrix. Then

$$\langle \mathbf{u}, \mathbf{v} \rangle_{\mathbf{A}} = (\mathbf{A}^T \mathbf{A} \mathbf{u}, \mathbf{v}) = (\mathbf{A}\mathbf{u}, \mathbf{A}\mathbf{v}) = (\mathbf{A}\mathbf{u})^T \mathbf{A}\mathbf{v} = \mathbf{u}^T \mathbf{A}^T \mathbf{A}\mathbf{v} = \mathbf{u}^T \left(\mathbf{A}^T \mathbf{A} \right) \mathbf{v}$$

is an inner product on $\mathbb{R}^n$.

Problem 6.3 Let

$$\mathbf{A} = \mathrm{diag} \left(\sqrt{w_1}, \sqrt{w_2}, \ldots, \sqrt{w_n} \right)$$

be a diagonal matrix with positive weights $w_1, \ldots, w_n$. Then

$$\langle \mathbf{u}, \mathbf{v} \rangle_{\mathbf{A}} = \mathbf{u}^T \mathbf{A}^T \mathbf{A}\mathbf{v} = \mathbf{u}^T \mathrm{diag}(w_1, w_2, \ldots, w_n) \mathbf{v} = \sum_{i=1}^{n} u_i v_i w_i.$$

is a **weighted inner product** on $\mathbb{R}^n$.

Problem 6.4 If $\mathbf{A}$ and $\mathbf{B}$ are $m \times n$ matrices, then $\mathbf{A}^T$ is an $n \times m$ matrix, so that the product $\mathbf{A}^T \mathbf{B}$ exists and is an $n \times n$ matrix. Then

$$\langle \mathbf{A}, \mathbf{B} \rangle = \mathrm{Tr} \left(\mathbf{A}^T \mathbf{B} \right)$$

is an inner product on $\mathfrak{M}_{m \times n}$.

Remark. Notice that for $m = 1$ this inner product is the standard inner product in $^n\mathbb{R}$.

Solution: All axioms are trivial except for the last:

$$\left(\mathbf{A}^T\mathbf{A}\right)_{ii} = \text{row}_i(\mathbf{A}^T)\cdot\text{col}_i(\mathbf{A}) = \text{col}_i(\mathbf{A})\cdot\text{col}_i(\mathbf{A}) \Rightarrow$$

$$\text{Tr}\left(\mathbf{A}^T\mathbf{A}\right) = \sum_{i=1}^{n}(\text{col}_i(\mathbf{A}),\text{col}_i(\mathbf{A})) \geq 0 \text{ and } \text{Tr}\left(\mathbf{A}^T\mathbf{A}\right) = 0 \Leftrightarrow \mathbf{A} = 0. \quad \Box$$

Now we pay attention to an extension of the property (**d**) of Euclidean inner product (see Theorem 1.12) to general inner product vector spaces.

Definition 6.2 Two vectors $\mathbf{v}_1$ and $\mathbf{v}_2$ in an inner product space V are called **orthogonal** if $\langle\mathbf{v}_1,\mathbf{v}_2\rangle = 0$.

Lemma 6.1 *Suppose that $S = \{\mathbf{v}_1,\ldots,\mathbf{v}_k\}$ is a set of nonzero pairwise orthogonal vectors in an inner product space V. Then the set S is linearly independent.*

Proof. For $j = 1,\ldots,k$, we have

$$0 = \langle\alpha_1\mathbf{v}_1 + \cdots + \alpha_k\mathbf{v}_k,\mathbf{v}_j\rangle = \alpha_j(\mathbf{v}_j,\mathbf{v}_j) \Rightarrow \alpha_j = 0. \quad \Box$$

Definition 6.3 A basis $S = \{\mathbf{v}_1,\ldots,\mathbf{v}_n\}$ of an inner product space V is called **orthogonal** if $\langle\mathbf{v}_i,\mathbf{v}_j\rangle = 0$ for $i \neq j$. It is called **orthonormal** if in addition $\|\mathbf{v}_i\| = 1, i = 1,\ldots,n$.

The following theorem says that every nonzero finite dimensional vector space with an inner product has an orthonormal basis.

Theorem 6.1 *Let V be a nonzero inner product space with a basis $S = \{\mathbf{v}_1,\ldots,\mathbf{v}_n\}$. Then there exists an orthonormal basis $OS = \{\mathbf{u}_1,\ldots,\mathbf{u}_n\}$ for V such that*

$$\text{Lin}(\mathbf{v}_1,\ldots,\mathbf{v}_k) = \text{Lin}(\mathbf{u}_1,\ldots,\mathbf{u}_k)$$

for $k = 1,\ldots,n$.

Proof. First, we construct an orthogonal basis $\{\mathbf{w}_1,\ldots,\mathbf{w}_n\}$.

Step 1. Let $\mathbf{w}_1 = \mathbf{v}_1$. Since S is a basis, we see that $\mathbf{w}_1 \neq \mathbf{0}$. It is clear that $\text{Lin}(\mathbf{v}_1) = \text{Lin}(\mathbf{w}_1)$

Step 2. If nonzero vectors $\{\mathbf{w}_1,\mathbf{w}_2,\ldots,\mathbf{w}_{k-1}\}$ are constructed, are orthogonal, and

$$\text{Lin}(\mathbf{v}_1,\ldots,\mathbf{v}_{k-1}) = \text{Lin}(\mathbf{w}_1,\ldots,\mathbf{w}_{k-1}), \tag{6.2}$$

then we define $\mathbf{w}_k$ by the formula

$$\mathbf{w}_k = \mathbf{v}_k - \frac{\langle \mathbf{v}_k, \mathbf{w}_1 \rangle}{\langle \mathbf{w}_1, \mathbf{w}_1 \rangle} \mathbf{w}_1 - \frac{\langle \mathbf{v}_k, \mathbf{w}_2 \rangle}{\langle \mathbf{w}_2, \mathbf{w}_2 \rangle} \mathbf{w}_2 - \cdots - \frac{\langle \mathbf{v}_k, \mathbf{w}_{k-1} \rangle}{\langle \mathbf{w}_{k-1}, \mathbf{w}_{k-1} \rangle} \mathbf{w}_{k-1}. \quad (6.3)$$

Formula (6.3) shows that $\mathbf{w}_k \perp \mathbf{w}_j$ for $j \leq k - 1$. Vector $\mathbf{w}_k$ cannot be zero, since otherwise $\mathbf{v}_k$ is in $\mathrm{Lin}(\mathbf{v}_1, \ldots, \mathbf{v}_{k-1})$ by (6.2), which contradict to our assumption that S is a basis for V. Finally, we see that $\mathbf{w}_k$ is in $\mathrm{Lin}(\mathbf{v}_1, \ldots, \mathbf{v}_k)$ and $\mathbf{v}_k$ is in $\mathrm{Lin}(\mathbf{w}_1, \ldots, \mathbf{w}_k)$, implying that

$$\mathrm{Lin}(\mathbf{v}_1, \ldots, \mathbf{v}_k) = \mathrm{Lin}(\mathbf{w}_1, \ldots, \mathbf{w}_k).$$

The construction can be continued by induction on k until $k = n$.
Step 3.

$$\mathbf{u}_k = \frac{1}{\|\mathbf{w}_k\|} \mathbf{w}_k, \ k = 1, \ldots, n. \qquad \square$$

Definition 6.4 The transformation of any linearly independent system of vectors $S = \{\mathbf{v}_1, \ldots, \mathbf{v}_n\}$ to the orthonormal system $OS = \{\mathbf{u}_1, \ldots, \mathbf{u}_n\}$ described by **Steps 1 – 3** above is called the **Gram-Schmidt process.**

Since any subspace of an inner product space is an inner product space itself, the Gram-Schmidt process can be applied to any linearly independent system of vectors.

Problem 6.5 Use the Gram-Schmidt process to find an orthonormal basis for the subspace of $\mathbb{R}^4$ with basis

$$S = \{(1, 1, 1, 0)^T, (0, 1, 0, 1)^T, (1, 0, 0, 1)^T\}.$$

Solution:
Step 1. $\mathbf{w}_1 = \mathbf{v}_1 = (1, 1, 1, 0)^T.$
Step 2.

$$\mathbf{w}_2 = \mathbf{v}_2 - \frac{(\mathbf{v}_2, \mathbf{w}_1)}{(\mathbf{w}_1, \mathbf{w}_1)} \mathbf{w}_1 = (0, 1, 0, 1)^T - \frac{1}{3}(1, 1, 1, 0)^T = \left(-\frac{1}{3}, \frac{2}{3}, -\frac{1}{3}, 1\right)^T;$$

$$\mathbf{w}_3 = \mathbf{v}_3 - \frac{(\mathbf{v}_3, \mathbf{w}_1)}{(\mathbf{w}_1, \mathbf{w}_1)} \mathbf{w}_1 - \frac{(\mathbf{v}_3, \mathbf{w}_2)}{(\mathbf{w}_2, \mathbf{w}_2)} \mathbf{w}_2 =$$

$$(1, 0, 0, 1)^T - \frac{1}{3}(1, 1, 1, 0)^T - \frac{2}{5}\left(-\frac{1}{3}, \frac{2}{3}, -\frac{1}{3}, 1\right)^T = \left(\frac{4}{5}, -\frac{3}{5}, -\frac{1}{5}, \frac{3}{5}\right)^T.$$

Step 3.

$$\mathbf{u}_1 = \frac{1}{\sqrt{3}}\begin{pmatrix} 1 \\ 1 \\ 1 \\ 0 \end{pmatrix}, \; \mathbf{u}_2 = \frac{1}{\sqrt{15}}\begin{pmatrix} -1 \\ 2 \\ -1 \\ 3 \end{pmatrix}, \; \mathbf{u}_3 = \frac{1}{\sqrt{35}}\begin{pmatrix} 4 \\ -3 \\ -1 \\ 3 \end{pmatrix}. \qquad \Box$$

Lemma 6.2 *Let V be a finite dimensional inner product space with an orthonormal basis $\{\mathbf{u}_1, \ldots, \mathbf{u}_n\}$. Then every vector $\mathbf{v}$ in V is uniquely represented as the sum*

$$\mathbf{v} = \langle \mathbf{v}, \mathbf{u}_1 \rangle \mathbf{u}_1 + \cdots + \langle \mathbf{v}, \mathbf{u}_n \rangle \mathbf{u}_n. \tag{6.4}$$

For every pair $\mathbf{v}$, $\mathbf{w}$ of vectors in V

$$\langle \mathbf{v}, \mathbf{w} \rangle = \sum_{k=1}^{n} \langle \mathbf{v}, \mathbf{u}_k \rangle \langle \mathbf{w}, \mathbf{u}_k \rangle. \tag{6.5}$$

In particular,

$$||\mathbf{v}||^2 = \sum_{k=1}^{n} |\langle \mathbf{v}, \mathbf{u}_k \rangle|^2. \tag{6.6}$$

Proof. Since $\{\mathbf{u}_1, \ldots, \mathbf{u}_n\}$ is a basis for V, any vector $\mathbf{v}$ in V can be uniquely represented as the sum

$$\mathbf{v} = \alpha_1 \mathbf{u}_1 + \cdots + \alpha_n \mathbf{u}_n$$

with real coefficients $\alpha_1, \ldots, \alpha_n$. Using orthogonality of the basis, we find that $\langle \mathbf{v}, \mathbf{u}_j \rangle = \alpha_j$ for $j = 1, \ldots, n$. Using the representation (6.4) for $\mathbf{v}$ and $\mathbf{w}$, and the property **(ii)** of the definition of an inner product, we immediately obtain (6.5). Formula (6.6) is a partial case of (6.5) with $\mathbf{w} = \mathbf{v}$. $\qquad \Box$

It is interesting that there is a formula for vectors $\mathbf{w}_k$ obtained in the steps of the Gram-Schmidt process:

$$\mathbf{w}_k = \begin{vmatrix} (\mathbf{v}_1, \mathbf{v}_1) & (\mathbf{v}_1, \mathbf{v}_2) & \cdots & (\mathbf{v}_1, \mathbf{v}_k) \\ (\mathbf{v}_2, \mathbf{v}_1) & (\mathbf{v}_2, \mathbf{v}_2) & \cdots & (\mathbf{v}_2, \mathbf{v}_k) \\ \vdots & \vdots & \ddots & \vdots \\ (\mathbf{v}_{k-1}, \mathbf{v}_1) & (\mathbf{v}_{k-1}, \mathbf{v}_2) & \cdots & (\mathbf{v}_{k-1}, \mathbf{v}_k) \\ \mathbf{v}_1 & \mathbf{v}_2 & \cdots & \mathbf{v}_k \end{vmatrix}$$

The above determinant is understood in Cramer's sense. If $j < k$, then $(\mathbf{v}_j, \mathbf{w}_k)$ is the determinant with two equal rows at places j and k. Hence $\mathbf{w}_k \perp \mathbf{v}_j$ for $j = 1, \ldots, k-1$ and

$$(\mathbf{v}_k, \mathbf{w}_k) = \Delta_k = \begin{vmatrix} (\mathbf{v}_1, \mathbf{v}_1) & (\mathbf{v}_1, \mathbf{v}_2) & \cdots & (\mathbf{v}_1, \mathbf{v}_k) \\ (\mathbf{v}_2, \mathbf{v}_1) & (\mathbf{v}_2, \mathbf{v}_2) & \cdots & (\mathbf{v}_2, \mathbf{v}_k) \\ \vdots & \vdots & \ddots & \vdots \\ (\mathbf{v}_{k-1}, \mathbf{v}_1) & (\mathbf{v}_{k-1}, \mathbf{v}_2) & \cdots & (\mathbf{v}_{k-1}, \mathbf{v}_k) \\ (\mathbf{v}_k, \mathbf{v}_1) & (\mathbf{v}_k, \mathbf{v}_2) & \cdots & (\mathbf{v}_k, \mathbf{v}_k) \end{vmatrix}.$$

We have

$$
(\mathbf{w}_k, \mathbf{w}_k) =
\begin{vmatrix}
(\mathbf{v}_1,\mathbf{v}_1) & (\mathbf{v}_1,\mathbf{v}_2) & \cdots & (\mathbf{v}_1,\mathbf{v}_k) \\
(\mathbf{v}_2,\mathbf{v}_1) & (\mathbf{v}_2,\mathbf{v}_2) & \cdots & (\mathbf{v}_2,\mathbf{v}_k) \\
\vdots & \vdots & \ddots & \vdots \\
(\mathbf{v}_{k-1},\mathbf{v}_1) & (\mathbf{v}_{k-1},\mathbf{v}_2) & \cdots & (\mathbf{v}_{k-1},\mathbf{v}_k) \\
(\mathbf{w}_k,\mathbf{v}_1) & (\mathbf{w}_k,\mathbf{v}_2) & \cdots & (\mathbf{w}_k,\mathbf{v}_k)
\end{vmatrix} =
$$

$$
\begin{vmatrix}
(\mathbf{v}_1,\mathbf{v}_1) & (\mathbf{v}_1,\mathbf{v}_2) & \cdots & (\mathbf{v}_1,\mathbf{v}_k) \\
(\mathbf{v}_2,\mathbf{v}_1) & (\mathbf{v}_2,\mathbf{v}_2) & \cdots & (\mathbf{v}_2,\mathbf{v}_k) \\
\vdots & \vdots & \ddots & \vdots \\
(\mathbf{v}_{k-1},\mathbf{v}_1) & (\mathbf{v}_{k-1},\mathbf{v}_2) & \cdots & (\mathbf{v}_{k-1},\mathbf{v}_k) \\
0 & 0 & \cdots & (\mathbf{w}_k,\mathbf{v}_k)
\end{vmatrix} =
$$

$$
(\mathbf{v}_k,\mathbf{w}_k)
\begin{vmatrix}
(\mathbf{v}_1,\mathbf{v}_1) & (\mathbf{v}_1,\mathbf{v}_2) & \cdots & (\mathbf{v}_1,\mathbf{v}_{k-1}) \\
(\mathbf{v}_2,\mathbf{v}_1) & (\mathbf{v}_2,\mathbf{v}_2) & \cdots & (\mathbf{v}_2,\mathbf{v}_{k-1}) \\
\vdots & \vdots & \ddots & \vdots \\
(\mathbf{v}_{k-1},\mathbf{v}_1) & (\mathbf{v}_{k-1},\mathbf{v}_2) & \cdots & (\mathbf{v}_{k-1},\mathbf{v}_{k-1})
\end{vmatrix} = \Delta_k \Delta_{k-1}.
$$

It follows that

$$
\mathbf{u}_k = \frac{1}{\sqrt{\Delta_k \Delta_{k-1}}}
\begin{vmatrix}
(\mathbf{v}_1,\mathbf{v}_1) & (\mathbf{v}_1,\mathbf{v}_2) & \cdots & (\mathbf{v}_1,\mathbf{v}_k) \\
(\mathbf{v}_2,\mathbf{v}_1) & (\mathbf{v}_2,\mathbf{v}_2) & \cdots & (\mathbf{v}_2,\mathbf{v}_k) \\
\vdots & \vdots & \ddots & \vdots \\
(\mathbf{v}_{k-1},\mathbf{v}_1) & (\mathbf{v}_{k-1},\mathbf{v}_2) & \cdots & (\mathbf{v}_{k-1},\mathbf{v}_k) \\
\mathbf{v}_1 & \mathbf{v}_2 & \cdots & \mathbf{v}_k
\end{vmatrix}. \tag{6.7}
$$

The explicit formulas for orthogonalization obtained in (6.7) are not suitable for practical calculations. However, they have a great theoretical value. Formula (6.4) shows that the transition matrix $P_{S \to OS}$ is given by:

$$
P_{S \to OS} =
\begin{pmatrix}
\langle \mathbf{v}_1,\mathbf{u}_1 \rangle & \langle \mathbf{v}_2,\mathbf{u}_1 \rangle & \cdots & \langle \mathbf{v}_n,\mathbf{u}_1 \rangle \\
0 & \langle \mathbf{v}_2,\mathbf{u}_2 \rangle & \cdots & \langle \mathbf{v}_n,\mathbf{u}_2 \rangle \\
0 & 0 & \cdots & \langle \mathbf{v}_n,\mathbf{u}_3 \rangle \\
\vdots & \vdots & \ddots & \vdots \\
0 & 0 & \cdots & \langle \mathbf{v}_n,\mathbf{u}_n \rangle
\end{pmatrix}.
$$

Definition 6.5 A linear isomorphism $T : V_1 \to V_2$ of inner product spaces V_1 and V_2 is called an **isometry** if $||T(\mathbf{v})||^2 = ||\mathbf{v}||^2$ for every vector $\mathbf{v}$ in V_1.

Elementary formula

$$||T(\mathbf{v} + \mathbf{w})||^2 = ||T(\mathbf{v})||^2 + 2\langle T(\mathbf{v}), T(\mathbf{w})\rangle + ||T(\mathbf{w})||^2$$

shows that

$$\langle T(\mathbf{v}), T(\mathbf{w})\rangle = \langle \mathbf{v}, \mathbf{w}\rangle \tag{6.8}$$

for any vectors $\mathbf{v}$ and $\mathbf{w}$ in V_1 and any isometry T.

Theorem 6.2 *If V is a finite dimensional inner product space with $\dim(V) = n > 0$, then there exists an isometry $T : V \to \mathbb{R}^n$ of V onto $\mathbb{R}^n$ with the standard inner product.*

Proof. Since $\dim(V) = n > 0$, there is a basis $S = \{\mathbf{v}_1, \ldots, \mathbf{v}_n\}$ for V. By Theorem 6.1 there is an orthonormal basis $OS = \{\mathbf{u}_1, \ldots, \mathbf{u}_n\}$ for V. By Theorem 4.5 a linear transformation $T : V \to \mathbb{R}^n$ is uniquely determined by its values on the elements of the basis OS. We define T to satisfy $T(\mathbf{u}_k) = \mathbf{e}_k$ for $k = 1, \ldots, n$. By (6.4)

$$T(\mathbf{v}) = \langle \mathbf{v}, \mathbf{u}_1\rangle \mathbf{e}_1 + \ldots + \langle \mathbf{v}, \mathbf{u}_n\rangle \mathbf{e}_n.$$

If $T(\mathbf{v}) = \mathbf{0}$, then $\langle \mathbf{v}, \mathbf{u}_1\rangle = \cdots = \langle \mathbf{v}, \mathbf{u}_n\rangle = 0$, since $\{\mathbf{e}_1, \ldots, \mathbf{e}_n\}$ is the standard basis for $\mathbb{R}^n$. By (6.4) we conclude that $\mathbf{v} = \mathbf{0}$. It follows that T is a linear isomorphism. By (6.6) we obtain that

$$||T(\mathbf{v})||^2 = \sum_{k=1}^n |\langle \mathbf{v}, \mathbf{u}_k\rangle|^2 = ||\mathbf{v}||^2,$$

implying that T is an isometry. $\square$

Corollary 6.1 *A mapping $\langle *, *\rangle$ is an inner product on $\mathbb{R}^n$ if and only if there is an invertible $n \times n$ matrix $\mathbf{A}$ such that $\langle \mathbf{x}, \mathbf{y}\rangle = (\mathbf{A}^T\mathbf{A}\mathbf{x}, \mathbf{y})$ for any vectors $\mathbf{x}$, $\mathbf{y}$ in $\mathbb{R}^n$.*

Proof. Example 6.2 shows that

$$\langle \mathbf{x}, \mathbf{y}\rangle_{\mathbf{A}} = (\mathbf{A}\mathbf{x}, \mathbf{A}\mathbf{y}) = (\mathbf{A}^T\mathbf{A}\mathbf{x}, \mathbf{y})$$

is an inner product on $\mathbb{R}^n$ for every invertible matrix $\mathbf{A}$. Let V be $\mathbb{R}^n$ with some inner product $\langle *, *\rangle$. By Theorem 6.2 there is an isometry T of $\mathbb{R}^n$ with inner product $\langle *, *\rangle$ onto $\mathbb{R}^n$ with the standard inner product. By (6.8)

$$\langle \mathbf{x}, \mathbf{y}\rangle = (T(\mathbf{x}), T(\mathbf{y})) = (\mathbf{A}_T\mathbf{x}, \mathbf{A}_T\mathbf{y}) = (\mathbf{A}_T^T\mathbf{A}_T\mathbf{x}, \mathbf{y}). \qquad \square$$

6.3 Linear Operators on Inner Product Spaces

Definition 6.6 If W is a subspace of a real inner product space V, then the set of all vectors in V that are orthogonal to every vector in W is called the **orthogonal complement** of W and is denoted by the symbol $W^\perp$.

Theorem 6.3 *If W is a subspace of a real inner product space V, then:*

(i) $W^\perp$ *is a subspace of V.*
(ii) $W \cap W^\perp = \{0\}$.

Proof. (i) If $\mathbf{x}, \mathbf{y} \perp W$ and $\alpha, \beta \in \mathbb{R}$, then by property (ii) of the inner product

$$\langle \alpha \mathbf{x} + \beta \mathbf{y}, \mathbf{z} \rangle = \alpha \langle \mathbf{x}, \mathbf{z} \rangle + \beta \langle \mathbf{y}, \mathbf{z} \rangle = 0$$

for every $\mathbf{z} \in W$. Hence $\alpha \mathbf{x} + \beta \mathbf{y} \in W^\perp$. By Theorem 4.3 the set $W^\perp$ is a subspace of V.

(ii) If $\mathbf{x} \in W \cap W^\perp$, then $\mathbf{x} \perp \mathbf{x}$, implying that $\langle \mathbf{x}, \mathbf{x} \rangle = 0$. Hence $\mathbf{x} = \mathbf{0}$. □

Theorem 6.4 *If W is a subspace of a finite dimensional inner product space V, then*
$$V = W \oplus W^\perp, \text{ and } \dim(V) = \dim(W) + \dim(W^\perp).$$

Proof. Suppose first that $V = \mathbb{R}^n$ with the standard inner product and that $\dim(W) = n - r$. By Theorem 2.16 there is a unique $r \times n$ matrix $\mathbf{A}$ in reduced row echelon form such that $N(\mathbf{A}) = W$. By Theorem 2.15

$$N(\mathbf{A}) \oplus \mathbf{row}(\mathbf{A})^T = \mathbb{R}^n,$$

implying that $W^\perp = \mathbf{row}(\mathbf{A})^T$.

For V with $\dim(V) = 0$ the Theorem is obvious. If $\dim(V) = n > 0$, then by Theorem 6.2 there is an isometry $T : V \to \mathbb{R}^n$. Since T is a linear isomorphism of V onto $\mathbb{R}^n$, the image $T(W)$ of the subspace W is a subspace of $\mathbb{R}^n$, $\dim(V) = n$, and $\dim(T(W)) = \dim(W) = n - r$. Since $T(W)$ is a

subspace of $\mathbb{R}^n$,
$$T(W) \oplus T(W)^\perp = \mathbb{R}^n.$$

By (6.8) the isometry T keeps orthogonality and maps V onto $\mathbb{R}^n$. Hence $W \oplus W^\perp = V$ as stated.   □

Corollary 6.2 *If W is a subspace of a real finite dimensional inner product space V, then*
$$\left(W^\perp\right)^\perp = W.$$

Proof. It follows from the formula $W \oplus W^\perp = W^\perp \oplus W$.   □

Let V be a finite dimensional inner product vector space. By Theorem 6.4 $V = W \oplus W^\perp$, which implies that every vector $\mathbf{v}$ in V can be uniquely decomposed as the sum of two vectors:

$$\mathbf{v} = \mathbf{w} + \mathbf{w}^\perp, \quad \mathbf{w} = \mathrm{Proj}_W \mathbf{v} \in W, \quad \mathbf{w}^\perp = \mathrm{Proj}_{W^\perp} \mathbf{v} \in W^\perp. \tag{6.9}$$

Definition 6.7 The vectors $\mathrm{Proj}_W \mathbf{v}$ and $\mathrm{Proj}_{W^\perp} \mathbf{v}$ are called the **orthogonal projection** of $\mathbf{v}$ on W and the **orthogonal projection** of $\mathbf{v}$ on $W^\perp$, respectively.

Since every subspace W of a finite dimensional inner product space V is itself a finite dimensional inner product space, Theorem 6.1 says that there exists an orthonormal basis $\{\mathbf{u}_1, \ldots, \mathbf{u}_r\}$ for W.

Theorem 6.5 *Let W be a subspace of a finite dimensional inner product space V and $S = \{\mathbf{u}_1, \ldots, \mathbf{u}_r\}$ be an orthonormal basis for W. Then for every vector v in V*
$$\begin{aligned} \mathrm{Proj}_W \mathbf{v} &= \langle \mathbf{v}, \mathbf{u}_1 \rangle \mathbf{u}_1 + \langle \mathbf{v}, \mathbf{u}_2 \rangle \mathbf{u}_2 + \cdots + \langle \mathbf{v}, \mathbf{u}_r \rangle \mathbf{u}_r, \\ \mathrm{Proj}_{W^\perp} \mathbf{v} &= \mathbf{I} - \mathrm{Proj}_W \mathbf{v}, \end{aligned} \tag{6.10}$$

where $\mathbf{I}$ is the identity transformation on V.

Proof. By (6.9) we have $\mathbf{v} = \mathbf{w} + \mathbf{w}^\perp$. Since $\mathbf{w}$ is a vector in W and S is an orthonormal basis for W, we obtain that

$$\mathrm{Proj}_W \mathbf{v} = \mathbf{w} = \langle \mathbf{w}, \mathbf{u}_1 \rangle \mathbf{u}_1 + \langle \mathbf{w}, \mathbf{u}_2 \rangle \mathbf{u}_2 + \cdots + \langle \mathbf{w}, \mathbf{u}_r \rangle \mathbf{u}_r$$

by (6.4). Since $\mathbf{w}^\perp$ is a vector in $W^\perp$, it is orthogonal to any vector of the basis S. In other words, $\langle \mathbf{w}, \mathbf{u}_j \rangle = \langle \mathbf{v}, \mathbf{u}_j \rangle$ for $j = 1, \ldots, r$. This proves the first formula in (6.10). The second formula follows from the obvious identity $\mathbf{w}^\perp = \mathbf{v} - \mathbf{w}$. $\square$

Corollary 6.3 *Let W be a subspace of a finite dimensional inner product space V. Then* $P(\mathbf{v}) = \text{Proj}_W \mathbf{v}$ *is a linear operator on V.*

Problem 6.6 Let $\mathbf{u}$ be a unit vector in $\mathbb{R}^n$. Evaluate the matrix $\mathbf{A}_T$ of the linear operator $T(\mathbf{v}) = \langle \mathbf{v}, \mathbf{u} \rangle \mathbf{u}$.

Solution: Since $T(\mathbf{e}_j) = \langle \mathbf{e}_j, \mathbf{u} \rangle \mathbf{u} = u_j \mathbf{u}$, we obtain that

$$\mathbf{A}_T = \left(u_1 \mathbf{u} \; u_2 \mathbf{u} \cdots u_n \mathbf{u} \right). \quad \square \qquad\qquad (6.11)$$

Definition 6.8 A linear transformation $L : V \to \mathbb{R}$ is called a **linear functional** on V.

If $\mathbf{y}$ is a vector in an inner product space V, then $L(\mathbf{v}) = \langle \mathbf{v}, \mathbf{y} \rangle$ is a linear functional on V. Indeed,

$$L(\alpha \mathbf{u} + \beta \mathbf{v}) = \langle \alpha \mathbf{u} + \beta \mathbf{v}, \mathbf{y} \rangle = \alpha \langle \mathbf{u}, \mathbf{y} \rangle + \beta \langle \mathbf{v}, \mathbf{y} \rangle = \alpha L(\mathbf{u}) + \beta L(\mathbf{v})$$

by property (**ii**) of Definition 6.1.

Theorem 6.6 *Any linear functional L on a finite dimensional inner product space V is defined by the formula*

$$L(\mathbf{v}) = \langle \mathbf{v}, \mathbf{y}_L \rangle,$$

where $\mathbf{y}_L$ *is a vector in V uniquely defined by L.*

Proof. Since either $R(L) = \mathbf{0}$ or $R(L) = \mathbb{R}$, we see that either $\dim(R(L)) = 0$ or $\dim(R(L)) = 1$. By Theorem 4.15,

$$\dim(N(L)) + \dim(R(L)) = \dim(V).$$

If $R(L) = \mathbf{0}$, then $\dim(N(L)) = \dim(V)$, which implies that $N(L) = V$. In this case we may put $\mathbf{y}_L = \mathbf{0}$.

If $R(L) = \mathbb{R}$, then $\dim(N(L)) = \dim(V) - 1$. By Theorem 6.4, $V = N(L) \oplus N(L)^\perp$, where $\dim(N(L)^\perp) = 1$. It follows that there exists a vector $\mathbf{n}$ in V of unit length such that $N(L)^\perp = \text{Lin}(\mathbf{n})$.

Since $V = N(L) \oplus N(L)^{\perp}$, every vector $\mathbf{v}$ in V can be uniquely represented as $\mathbf{v} = \mathbf{w} + \alpha \cdot \mathbf{n}$, where α is a real number. If we put $\mathbf{y}_L = L(\mathbf{n}) \cdot \mathbf{n}$, then

$$\langle \mathbf{v}, \mathbf{y}_L \rangle = \langle \mathbf{w} + \alpha \cdot \mathbf{n}, L(\mathbf{n}) \cdot \mathbf{n} \rangle = \alpha L(\mathbf{n}) \langle \mathbf{n}, \mathbf{n} \rangle = L(\alpha \mathbf{n}) = L(\mathbf{w} + \alpha \cdot \mathbf{n}) = L(\mathbf{v}).$$

To prove the uniqueness of $\mathbf{y}_L$ we assume that $\langle \mathbf{v}, \mathbf{y}_L \rangle = \langle \mathbf{v}, \mathbf{y}^* \rangle$ for every vector $\mathbf{v}$ in V and some vector $\mathbf{y}^*$. Then $\langle \mathbf{v}, \mathbf{y}_L - \mathbf{y}^* \rangle = 0$ for every $\mathbf{v}$ in V. In particular, this holds for $\mathbf{v} = \mathbf{y}_L - \mathbf{y}^*$, implying that $\mathbf{y}_L = \mathbf{y}^*$. $\qquad\square$

If T is a linear operator on a finite dimensional inner product space V, then it is uniquely defined by the bilinear form $\langle T(\mathbf{x}), \mathbf{y} \rangle$, where $\mathbf{x}$ and $\mathbf{y}$ are vectors in V. Indeed,

$$\langle T_1(\mathbf{x}), \mathbf{y} \rangle = \langle T_2(\mathbf{x}), \mathbf{y} \rangle \Leftrightarrow \langle T_1(\mathbf{x}) - T_2(\mathbf{x}), \mathbf{y} \rangle = 0.$$

Since the above identity holds for any choice of $\mathbf{x}$ and $\mathbf{y}$ in V, we may put $\mathbf{y} = T_1(\mathbf{x}) - T_2(\mathbf{x})$, implying that $||T_1(\mathbf{x}) - T_2(\mathbf{x})||^2 = 0$. Hence $T_1(\mathbf{x}) = T_2(\mathbf{x})$ for every $\mathbf{x}$ as stated.

For every $\mathbf{y}$ in V the functional $L_{\mathbf{y}}(\mathbf{x}) = \langle T(\mathbf{x}), \mathbf{y} \rangle$ is a linear functional in $\mathbf{x}$ on V. By Theorem 6.6 there is a unique vector $\mathbf{y}_L$ in V such that

$$L_{\mathbf{y}}(\mathbf{x}) = \langle T(\mathbf{x}), \mathbf{y} \rangle = \langle \mathbf{x}, \mathbf{y}_L \rangle. \tag{6.12}$$

Then

$$L_{\alpha \mathbf{y}_1 + \beta \mathbf{y}_2}(\mathbf{x}) = \langle T(\mathbf{x}), \alpha \mathbf{y}_1 + \beta \mathbf{y}_2 \rangle = \alpha \langle T(\mathbf{x}), \mathbf{y}_1 \rangle + \beta \langle T(\mathbf{x}), \mathbf{y}_2 \rangle = \langle \mathbf{x}, \alpha (\mathbf{y}_1)_L + \beta (\mathbf{y}_2)_L \rangle.$$

It follows that the mapping $T^*(\mathbf{y}) = \mathbf{y}_L$ is a linear operator on V. It is called the **adjoint operator** for T. It is clear that the adjoint operator is determined by the formula

$$\langle T(\mathbf{x}), \mathbf{y} \rangle = \langle \mathbf{x}, T^*(\mathbf{y}) \rangle, \tag{6.13}$$

where $\mathbf{x}$ and $\mathbf{y}$ are vectors in V.

In case $V = \mathbb{R}^n$ with the standard inner product any linear operator is represented by an $n \times n$ matrix $\mathbf{A}_T$. To evaluate the matrix $\mathbf{A}_{T^*}$ we use identity (6.13):

$$\langle \mathbf{A}_T \mathbf{x}, \mathbf{y} \rangle = \langle \mathbf{x}, \mathbf{A}_{T^*} \mathbf{y} \rangle. \tag{6.14}$$

If we apply (6.14) with $\mathbf{e}_j$ and $\mathbf{y} = \mathbf{e}_i$, then we obtain that the matrix of the adjoint operator is the transpose matrix of the given operator:

$$a_{ij} = \langle \mathbf{A}_T \mathbf{e}_j, \mathbf{e}_i \rangle = \langle \mathbf{e}_j, \mathbf{A}_{T^*} \mathbf{e}_i \rangle = (\mathbf{A}_{T^*})_{ji} \Rightarrow \boxed{\mathbf{A}_{T^*} = \mathbf{A}_T^T}.$$

The first formula in (6.10) shows that the projection $P = \mathrm{Proj}_W$ is the linear operator, which satisfies two properties:

$$P(\mathbf{v}) = \begin{cases} \mathbf{v} & \text{if } \mathbf{v} \text{ is in } W \\ \mathbf{0} & \text{if } \mathbf{v} \text{ is in } W^{\perp} \end{cases}. \tag{6.15}$$

The following Theorem describes orthogonal projections in terms of operator algebra.

Theorem 6.7 *A linear operator P on a finite dimensional inner product vector space V is an orthogonal projection if and only if*

$$\textbf{(a) } P^2 = P, \quad \textbf{(b) } P^* = P.$$

Proof. By Theorem 4.17 a linear operator P is a skew projection if and only if it is an idempotent (see condition **(a)** above). Therefore, it remains to prove that the null space $N(P)$ of P is orthogonal to its range $R(P)$ if and only if $P^* = P$.

If $N(P) \perp R(P)$, then by the Rank-Nullity Theorem (see Theorem 4.15), $N(P) \oplus R(P) = V$. It follows that any $\mathbf{x} \in V$ can be uniquely represented as a sum of $\mathbf{w} \in N(P)$ and $\mathbf{w}^{\perp} \in R(P)$. Similarly, $\mathbf{y} = \mathbf{z} + \mathbf{z}^{\perp}$. Then

$$\langle P(\mathbf{x}), \mathbf{y} \rangle = \langle P(\mathbf{w}+\mathbf{w}^{\perp}), \mathbf{z}+\mathbf{z}^{\perp} \rangle = \langle \mathbf{w}^{\perp}, \mathbf{z}^{\perp} \rangle = \langle \mathbf{w}+\mathbf{w}^{\perp}, P(\mathbf{z}+\mathbf{z}^{\perp}) \rangle = \langle \mathbf{x}, P(\mathbf{y}) \rangle,$$

which implies **(b)**.

Suppose now that P is a skew projection onto $R(P)$ satisfying condition **(b)**. If $\mathbf{v} \in R(P)$, then $P(\mathbf{v}) = \mathbf{v}$ by property **(a)**. Let $\mathbf{v} \in N(P)$ and $\mathbf{z} \in V$. Then

$$\langle P(\mathbf{z}), \mathbf{v} \rangle = \langle \mathbf{z}, P^*(\mathbf{v}) \rangle = \langle \mathbf{z}, P(\mathbf{v}) \rangle = 0,$$

see **(b)**. Since $\mathbf{z}$ is an arbitrary vector in V, we conclude that $N(P) \perp R(P)$. □

In Linear Algebra subspaces are usually given as linear spans of some linearly independent vectors. To find a formula for the orthogonal projection onto such a subspaces one can apply the Gram-Schmidt process to obtain the representation (6.10).

For orthogonal projections there is another important formula, which is a corollary of Theorem 7.3.

Theorem 6.8 *Let $W = \mathrm{Lin}\{\mathbf{v}_1, \ldots, \mathbf{v}_r\}$, where $\{\mathbf{v}_1, \ldots, \mathbf{v}_r\}$ is an arbitrary basis for a subspace W of $\mathbb{R}^n$. Let $\mathbf{A} = \left(\mathbf{v}_1 \cdots \mathbf{v}_r \right)$ be an $n \times r$ matrix. Then*

$$\mathrm{Proj}_W = \mathbf{A} \left(\mathbf{A}^T \mathbf{A} \right)^{-1} \mathbf{A}^T. \tag{6.16}$$

Problem 6.7 Find the orthogonal projection in $\mathbb{R}^4$ onto the subspace

$$W = \mathrm{Lin}\left\{ \begin{pmatrix} 1 \\ 0 \\ 1 \\ 1 \end{pmatrix}, \begin{pmatrix} 1 \\ 1 \\ 0 \\ 0 \end{pmatrix} \right\}.$$

Solution: We apply Theorem 6.8:

$$A = \begin{pmatrix} 1 & 1 \\ 0 & 1 \\ 1 & 0 \\ 1 & 0 \end{pmatrix} \Rightarrow A^T A = \begin{pmatrix} 3 & 1 \\ 1 & 2 \end{pmatrix} \Rightarrow (A^T A)^{-1} = \frac{1}{5}\begin{pmatrix} 2 & -1 \\ -1 & 3 \end{pmatrix} \Rightarrow A(A^T A)^{-1} A^T =$$

$$\frac{1}{5}\begin{pmatrix} 1 & 1 \\ 0 & 1 \\ 1 & 0 \\ 1 & 0 \end{pmatrix}\begin{pmatrix} 2 & -1 \\ -1 & 3 \end{pmatrix}\begin{pmatrix} 1 & 0 & 1 & 1 \\ 1 & 1 & 0 & 0 \end{pmatrix} = \frac{1}{5}\begin{pmatrix} 3 & 2 & 1 & 1 \\ 2 & 3 & -1 & -1 \\ 1 & -1 & 2 & 2 \\ 1 & -1 & 2 & 2 \end{pmatrix}. \qquad \square$$

Using formulas (6.10) and (6.11), we can also find the matrix of the orthogonal projection onto W. Ferst, we apply the Gram-Schmidt process to obtain the orthonormal basis $\{\mathbf{u}_1, \mathbf{u}_2\}$:

$$\mathbf{u}_1 = \frac{1}{\sqrt{3}}\begin{pmatrix} 1 \\ 0 \\ 1 \\ 1 \end{pmatrix}, \quad \mathbf{u}_2 = \frac{1}{\sqrt{15}}\begin{pmatrix} 2 \\ 3 \\ -1 \\ -1 \end{pmatrix}.$$

By (6.11) the operators $T_1(\mathbf{v}) = \langle \mathbf{v}, \mathbf{u}_1 \rangle \mathbf{u}_1$ and $T_2(\mathbf{v}) = \langle \mathbf{v}, \mathbf{u}_2 \rangle \mathbf{u}_2$ have the matrices:

$$A_{T_1} = \frac{1}{3}\begin{pmatrix} 1 & 0 & 1 & 1 \\ 0 & 0 & 0 & 0 \\ 1 & 0 & 1 & 1 \\ 1 & 0 & 1 & 1 \end{pmatrix}, \quad A_{T_2} = \frac{1}{15}\begin{pmatrix} 4 & 6 & -2 & -2 \\ 6 & 9 & -3 & -3 \\ -2 & -3 & 1 & 1 \\ -2 & -3 & 1 & 1 \end{pmatrix}.$$

Then

$$A_{T_1} + A_{T_2} = \frac{1}{5}\begin{pmatrix} 3 & 2 & 1 & 1 \\ 2 & 3 & -1 & -1 \\ 1 & -1 & 2 & 2 \\ 1 & -1 & 2 & 2 \end{pmatrix}.$$

The second algorithm requires more calculations than the first. It is also possible to apply the formulas obtained for skew projections in this case too. We define

$$A = \begin{pmatrix} 1 & 0 & 1 & 1 \\ 1 & 1 & 0 & 0 \\ 0 & 0 & 0 & 0 \\ 0 & 0 & 0 & 0 \end{pmatrix} \sim \begin{pmatrix} 1 & 0 & 1 & 1 \\ 0 & 1 & -1 & -1 \\ 0 & 0 & 0 & 0 \\ 0 & 0 & 0 & 0 \end{pmatrix}.$$

Then $j_1 = 1$, $j_2 = 2$, $x_3 = s$, $x_4 = t$. We obtain that $\mathbf{x}$ is in $N(\mathbf{A})$ if and only if

$$\mathbf{x} = \begin{pmatrix} -s - t \\ s + t \\ s \\ t \end{pmatrix} = s \begin{pmatrix} -1 \\ 1 \\ 1 \\ 0 \end{pmatrix} + t \begin{pmatrix} -1 \\ 1 \\ 0 \\ 1 \end{pmatrix}.$$

Let

$$\mathbf{B} = \begin{pmatrix} 1 & 1 & -1 & -1 \\ 0 & 1 & 1 & 1 \\ 1 & 0 & 1 & 0 \\ 1 & 0 & 0 & 1 \end{pmatrix} \Rightarrow \mathbf{B}^{-1} = \frac{1}{5} \begin{pmatrix} 1 & -1 & 2 & 2 \\ 2 & 3 & -1 & -1 \\ -1 & 1 & 3 & -2 \\ -1 & 1 & -2 & 3 \end{pmatrix}.$$

By (4.32)

$$\mathbf{P}_U = \begin{pmatrix} \mathbf{u}_1 & \mathbf{u}_2 & 0 & 0 \end{pmatrix} \mathbf{B}^{-1} = \frac{1}{5} \begin{pmatrix} 1 & 1 & 0 & 0 \\ 0 & 1 & 0 & 0 \\ 1 & 0 & 0 & 0 \\ 1 & 0 & 0 & 0 \end{pmatrix} \begin{pmatrix} 1 & -1 & 2 & 2 \\ 2 & 3 & -1 & -1 \\ -1 & 1 & 3 & -2 \\ -1 & 1 & -2 & 3 \end{pmatrix} = \frac{1}{5} \begin{pmatrix} 3 & 2 & 1 & 1 \\ 2 & 3 & -1 & -1 \\ 1 & -1 & 2 & 2 \\ 1 & -1 & 2 & 2 \end{pmatrix}. \quad \square$$

6.4 Orthogonal diagonalization

Definition 6.9 An $n \times n$ matrix $\mathbf{P}$ is called **orthogonal** if

$$\mathbf{P}^T \mathbf{P} = \mathbf{P} \mathbf{P}^T = \mathbf{I}.$$

Theorem 6.9 *A matrix $\mathbf{P}$ is orthogonal if and only if its columns are pairwise orthogonal, and each has length 1.*

Proof. Let $\mathbf{P} = \begin{pmatrix} \mathbf{x}_1 & \mathbf{x}_2 & \cdots & \mathbf{x}_n \end{pmatrix}$. Then

$$\mathbf{I} = \mathbf{P}^T \mathbf{P} = \begin{pmatrix} \mathbf{x}_1^T \mathbf{x}_1 & \mathbf{x}_1^T \mathbf{x}_2 & \cdots & \mathbf{x}_1^T \mathbf{x}_n \\ \mathbf{x}_2^T \mathbf{x}_1 & \mathbf{x}_2^T \mathbf{x}_2 & \cdots & \mathbf{x}_2^T \mathbf{x}_n \\ \vdots & \vdots & \ddots & \vdots \\ \mathbf{x}_n^T \mathbf{x}_1 & \mathbf{x}_n^T \mathbf{x}_2 & \cdots & \mathbf{x}_n^T \mathbf{x}_n \end{pmatrix} \Leftrightarrow \mathbf{x}_i^T \mathbf{x}_j = \begin{cases} 1 \text{ if } i = j \\ 0 \text{ if } i \neq j \end{cases}.$$

If the columns of $\mathbf{P}$ are pairwise orthogonal, and each has length 1, then the above formula shows that $\mathbf{P}^T \mathbf{P} = \mathbf{I}$. It follows that

$$1 = \det(\mathbf{I}) = \det(\mathbf{P}^T)\det(\mathbf{P}) = \det(\mathbf{P})^2 \Rightarrow \det(\mathbf{P}) = \pm 1 \neq 0.$$

Hence $\mathbf{P}$ is not singular. Hence $\mathbf{P}$ is invertible and therefore the left inverse $\mathbf{P}^T$ is at the same time the right inverse: $\mathbf{PP}^T = \mathbf{I}$.

□

Definition 6.10 A system $S = \{\mathbf{v}_1, \ldots, \mathbf{v}_n\}$ is called a **complete orthonormal system** in $\mathbb{R}^n$ if $\|\mathbf{v}_i\| = 1$ for every i and $\mathbf{v}_i \perp \mathbf{v}_j = 0 \Leftrightarrow \mathbf{v}_i^T\mathbf{v}_j = 0$ if $i \neq j$.

Since any orthogonal matrix is invertible, by Theorem 6.9 its columns make a basis for $\mathbb{R}^n$, implying that any complete orthonormal system is a basis for $\mathbb{R}^n$.

There is a surprising corollary.

Corollary 6.4 *If the rows of an $n \times n$ matrix $\mathbf{A}$ all have length one and are pairwise orthogonal, then the columns of $\mathbf{A}$ are pairwise orthogonal and have length one too.*

Proof. By the assumption of the corollary $\mathbf{AA}^T = \mathbf{I}$ implying that $\mathbf{A}^T$ is the inverse matrix for $\mathbf{A}$. Then by the Theorem 3.1 we obtain that $\mathbf{A}^T\mathbf{A} = \mathbf{I}$ implying the conclusion stated in the corollary. □

Definition 6.11 A square matrix $\mathbf{A}$ is called to be **orthogonally diagonalisable** if there is an **orthogonal** matrix $\mathbf{P}$ and a diagonal matrix $\mathbf{D}$ such that

$$\mathbf{P}^T\mathbf{AP} = \mathbf{D}.$$

Notice that $\mathbf{P}^T = \mathbf{P}^{-1}$.

Theorem 6.10 *An $n \times n$ matrix $\mathbf{A}$ is orthogonally diagonalizable if and only if there is an orthonormal basis of $\mathbb{R}^n$ consisting of eigenvectors of $\mathbf{A}$.*

Proof. An $n \times n$ matrix $\mathbf{A}$ is diagonalisable if and only if there is an invertible matrix

$$\mathbf{P} = \begin{pmatrix} \mathbf{v}_1 \ \mathbf{v}_2 \ \cdots \ \mathbf{v}_n \end{pmatrix}$$

such that $\mathbf{AP} = \mathbf{PD}$. The latter means that the vectors $\mathbf{v}_j$ are eigenvectors of $\mathbf{A}$. The additional requirement of orthogonality means that $\{\mathbf{v}_1, \mathbf{v}_2, \ldots, \mathbf{v}_n\}$ is an orthogonal basis for $\mathbb{R}^n$ as stated.   □

Lemma 6.3 *If $\mathbf{A}$ is an $n \times n$ real symmetric matrix, then its eigenvalues and eigenvectors are real.*

Proof. We consider in $\mathbb{C}^n$ the hermitian inner product:

$$\langle \mathbf{x}, \mathbf{y} \rangle = \mathbf{x}^T \overline{\mathbf{y}} = x_1 \overline{y}_1 + x_2 \overline{y}_2 + \cdots + x_n \overline{y}_n.$$

Suppose that $\mathbf{Ax} = \lambda \mathbf{x}$ for some $\mathbf{0} \neq \mathbf{x} \in \mathbb{C}^n$. Then

$$\lambda \|\mathbf{x}\|^2 = \langle \mathbf{Ax}, \mathbf{x} \rangle = \langle \mathbf{x}, \mathbf{Ax} \rangle = \overline{\lambda} \|\mathbf{x}\|^2 \Rightarrow \boxed{\lambda = \overline{\lambda}}.$$

If λ is real, then $\mathbf{x} \in \mathrm{N}(\mathbf{A} - \lambda \mathbf{I})$ is also real.   □

Lemma 6.4 *If a real $n \times n$ matrix $\mathbf{A}$ is symmetric, then eigenvectors corresponding to distinct eigenvalues are orthogonal.*

Proof. Let $\mathbf{Av}_1 = \lambda_1 \mathbf{v}_1$, $\mathbf{Av}_2 = \lambda_2 \mathbf{v}_2$, where $\lambda_1 \neq \lambda_2$, then both λ_1 and λ_2 are real and

$$\lambda_1 \langle \mathbf{v}_1, \mathbf{v}_2 \rangle = \langle \mathbf{Av}_1, \mathbf{v}_2 \rangle = \langle \mathbf{v}_1, \mathbf{Av}_2 \rangle = \lambda_2 \langle \mathbf{v}_1, \mathbf{v}_2 \rangle \Rightarrow \boxed{\langle \mathbf{v}_1, \mathbf{v}_2 \rangle = 0}.$$

□

> **Theorem 6.11** *A square matrix* $\mathbf{A}$ *with real entries is orthogonally diagonalisable if and only if* $\mathbf{A}$ *is symmetric.*

Proof. If $\mathbf{A}$ is orthogonally diagonalisable, then $\mathbf{P}^T\mathbf{A}\mathbf{P} = \mathbf{D} \Leftrightarrow \mathbf{A} = \mathbf{P}\mathbf{D}\mathbf{P}^T$. It follows that

$$\mathbf{A}^T = \left(\mathbf{P}\mathbf{D}\mathbf{P}^T\right)^T = \mathbf{P}\mathbf{D}\mathbf{P}^T = \mathbf{A}.$$

If $\mathbf{A}$ is an $n \times n$ symmetric matrix, then its characteristic polynomial $p_\mathbf{A}(\lambda)$ of degree n has a complex root λ_1 by Theorem 4.11. By Lemma 6.3, the number λ_1 is real. Let $\mathbf{v}_1 \in N(\mathbf{A} - \lambda_1\mathbf{I})$ be the eigenvector corresponding to the eigenvalue λ_1 and let $V_1 = \text{Lin}\{\mathbf{v}_1\}^\perp$. If $V_1 = \{\mathbf{0}\}$, then $n = 1$ and the proof is finished. Otherwise, $n > 1$ and for every vector $\mathbf{v} \in V_1$:

$$(\mathbf{A}\mathbf{v}, \mathbf{v}_1) = (\mathbf{v}, \mathbf{A}\mathbf{v}_1) = \lambda_1(\mathbf{v}, \mathbf{v}_1) = 0,$$

which implies that $\mathbf{A}\mathbf{v} \in V_1$. It follows that $\mathbf{A}$ defines a linear operator on V_1, $\dim(V_1) = n - 1$. Then the characteristic polynomial of this operator, see (4.30), has degree $n - 1 = \dim(V_1)$. Since $n > 1$, it has at least one complex root λ_2, which is an eigenvalue for $\mathbf{A}$. By Lemma 6.3, the number λ_2 is real. Let $\mathbf{v}_2 \in N(\mathbf{A} - \lambda_2\mathbf{I})$ be the eigenvector corresponding to the eigenvalue λ_2 and let $V_2 = \text{Lin}\{\mathbf{v}_1, \mathbf{v}_2\}^\perp$. If $V_2 = \{\mathbf{0}\}$, then $n = 2$ and the proof is finished. Otherwise, $n > 2$ and the process can be continued by induction. It is clear that it will produce the required orthogonal basis of eigenvectors in n steps. The proof is completed by an application of Theorem 6.10. $\qquad\qquad\square$

Since the proof of Theorem 6.11 looks a little bit tricky, it looks reasonable to run it on a concrete matrix.

Problem 6.8 Diagonalise orthogonally the symmetric matrix

$$\mathbf{A} = \begin{pmatrix} 2 & 1 & 1 \\ 1 & 2 & 1 \\ 1 & 1 & 2 \end{pmatrix}.$$

Solution: We evaluate the characteristic polynomial of $\mathbf{A}$:

$$p_\mathbf{A}(\lambda) = \begin{vmatrix} 2-\lambda & 1 & 1 \\ 1 & 2-\lambda & 1 \\ 1 & 1 & 2-\lambda \end{vmatrix} = (1-\lambda)(\lambda-1)(\lambda-3) + (1-\lambda)^2 = -(\lambda-1)^2(\lambda-4).$$

We find an eigenvector corresponding to $\lambda_1 = 4$:

$$\begin{pmatrix} -2 & 1 & 1 \\ 1 & -2 & 1 \\ 1 & 1 & -2 \end{pmatrix} \sim \begin{pmatrix} 1 & 0 & -1 \\ 0 & 1 & -1 \\ 0 & 0 & 0 \end{pmatrix} \Rightarrow \mathbf{v}_1 = \begin{pmatrix} 1 \\ 1 \\ 1 \end{pmatrix}.$$

Then $V_1 = \text{Lin}\{\mathbf{v}_1\}^{\perp}$ is the plane $x + y + z = 0$ with $\mathbf{v}_1$ as a normal vector. Since

$$\begin{pmatrix} 2 & 1 & 1 \\ 1 & 2 & 1 \\ 1 & 1 & 2 \end{pmatrix} \begin{pmatrix} x \\ y \\ z \end{pmatrix} = \begin{pmatrix} 2x + y + z \\ x + 2y + z \\ x + y + 2z \end{pmatrix},$$

we see that the matrix $\mathbf{A}$ keeps the plane V_1 invariant and defines a linear operator on V_1. Following Definition 4.17, we find the Gauss basis for the subspace V_1:

$$\mathbf{w}_1 = \begin{pmatrix} -1 \\ 1 \\ 0 \end{pmatrix}, \quad \mathbf{w}_2 = \begin{pmatrix} -1 \\ 0 \\ 1 \end{pmatrix}.$$

Since $\mathbf{A}\mathbf{w}_1 = \mathbf{w}_1$, $\mathbf{A}\mathbf{w}_2 = \mathbf{w}_2$, we see that the matrix $\mathbf{A}$ is the diagonal matrix in the basis $\{\mathbf{w}_1, \mathbf{w}_2\}$ with entries equal 1 on the main diagonal. It follows that the characteristic polynomial of this matrix is $(\lambda - 1)^2$. We can put $\mathbf{v}_2 = \mathbf{w}_1$. Finally,

$$\mathbf{v}_3 = \mathbf{w}_2 - \frac{(\mathbf{w}_2, \mathbf{w}_1)}{(\mathbf{w}_1, \mathbf{w}_1)} \mathbf{w}_1 = \begin{pmatrix} -1 \\ 0 \\ 1 \end{pmatrix} - \frac{1}{2} \begin{pmatrix} -1 \\ 1 \\ 0 \end{pmatrix} = \begin{pmatrix} -1/2 \\ -1/2 \\ 1 \end{pmatrix}$$

Thus, we obtained the orthonormal system

$$\mathbf{u}_1 = \frac{1}{\sqrt{3}} \begin{pmatrix} 1 \\ 1 \\ 1 \end{pmatrix}, \quad \mathbf{u}_2 = \frac{1}{\sqrt{2}} \begin{pmatrix} -1 \\ 1 \\ 0 \end{pmatrix}, \quad \mathbf{u}_3 = \sqrt{\frac{2}{3}} \begin{pmatrix} -1/2 \\ -1/2 \\ 1 \end{pmatrix}. \tag{6.17}$$

We have

$$\mathbf{P} = \begin{pmatrix} \mathbf{u}_1 & \mathbf{u}_2 & \mathbf{u}_3 \end{pmatrix} = \begin{pmatrix} \frac{1}{\sqrt{3}} & \frac{-1}{\sqrt{2}} & \frac{-1}{\sqrt{6}} \\ \frac{1}{\sqrt{3}} & \frac{1}{\sqrt{2}} & \frac{-1}{\sqrt{6}} \\ \frac{1}{\sqrt{3}} & 0 & \sqrt{\frac{2}{3}} \end{pmatrix}, \quad \mathbf{D} = \begin{pmatrix} 4 & 0 & 0 \\ 0 & 1 & 0 \\ 0 & 0 & 1 \end{pmatrix}. \quad \square$$

In general, it is not necessary to repeat the proof of Theorem 6.11 as we did in the above solution to Problem 6.8. Theorem 6.11 guarantees the orthogonality of eigenvectors with different eigenvalues. In case $\text{geo}_{\mathbf{A}}(\lambda_k) > 1$, we apply the Gram-Schmidt orthogonalization process, as we did in the solution to Problem 6.8. Notice, that in fact Lemma 6.4 was not used in the proof of Theorem 6.11, and can be obtained as its corollary.

Theorem 6.12 (Spectral Theorem) *Let $\mathbf{A}$ be a symmetric $n \times n$ matrix and let $\{\lambda_1, \cdots, \lambda_n\}$ be a complete list of its eigenvalues (not necessarily different) and $\{\mathbf{u}_1, \ldots, \mathbf{u}_n\}$ the corresponding list of its eigenvectors of unit lengths. Then for every $\mathbf{x}$*

$$\mathbf{A}\mathbf{x} = \sum_{k=1}^{n} \lambda_k (\mathbf{x}, \mathbf{u}_k) \mathbf{u}_k. \tag{6.18}$$

Proof. Since $\mathbf{A}$ is orthogonally diagonalizable, we see that $\text{Lin}\{\mathbf{u}_1, \cdots, \mathbf{u}_n\} = \mathbb{R}^n$. Hence, (6.18) holds for every $\mathbf{x} \in \mathbb{R}^n$ if and only if it holds for the vectors $\mathbf{x} = \mathbf{u}_j$, $j = 1, 2, \ldots, n$. Then

$$\mathbf{A}\mathbf{u}_j = \lambda_k \mathbf{u}_j = \sum_{k=1}^{n} \lambda_k (\mathbf{u}_j, \mathbf{u}_k) \mathbf{u}_k.$$

since $\mathbf{u}_j \perp \mathbf{u}_k$ for $k \neq j$ and $(\mathbf{u}_j, \mathbf{u}_j) = 1$. $\qquad \square$

Formula (6.18) represents the action of the linear transformation defined by the matrix $\mathbf{A}$ as a linear combination with coefficients λ_k of linear operators $E_k(\mathbf{x}) = (\mathbf{x}, \mathbf{u}_k)\mathbf{u}_k$. Each operator E_k is the orthogonal projection onto $\text{Lin}\{\mathbf{v}_k\}$. Thus, by formula (6.18) the action of a symmetric matrix $\mathbf{A}$ on columns $\mathbf{x}$ in $\mathbb{R}^n$ coincides with the action on $\mathbf{x}$ of the linear combination of orthogonal projectors

$$\sum_{k=1}^{n} \lambda_k E_k. \tag{6.19}$$

This decomposition is called the **spectral decomposition** of a symmetric matrix $\mathbf{A}$. The term 'spectral' reflects the presence of the numbers λ_k in formula (6.19). The set of all such numbers λ_k is called the **spectrum** of $\mathbf{A}$. If we put $\lambda_k = 1$ for each k, then $\mathbf{A} = \mathbf{I}$ is the identity matrix. According to (6.19), then the identity mapping I can be decomposed into the sum of orthogonal projections:

$$I = \sum_{k=1}^{n} E_k. \tag{6.20}$$

The matrix $\mathbf{E}_k$ of the orthogonal projection E_k can be easily calculated by formula (4.22):

$$\mathbf{E}_k = \left(E_k(\mathbf{e}_1) \cdots E_k(\mathbf{e}_n) \right) = \left((\mathbf{e}_1, \mathbf{u}_k)\mathbf{u}_k \cdots (\mathbf{e}_n, \mathbf{u}_k)\mathbf{u}_k \right). \tag{6.21}$$

Problem 6.9 Find the matrix of the orthogonal projector E_1 in the spectral decomposition of $\mathbf{A}$ corresponding to the eigenvalue 4 of the symmetric matrix $\mathbf{A}$ in Problem 6.8.

Solution: By (6.17)

$$\mathbf{u}_1 = \frac{1}{\sqrt{3}}\begin{pmatrix} 1 \\ 1 \\ 1 \end{pmatrix} \Rightarrow (\mathbf{e}_1, \mathbf{u}_1)\mathbf{u}_1 = (\mathbf{e}_2, \mathbf{u}_1)\mathbf{u}_1 = (\mathbf{e}_3, \mathbf{u}_1)\mathbf{u}_1 = \frac{1}{\sqrt{3}}\mathbf{u}_1.$$

It follows that

$$\mathbf{E}_1 = \frac{1}{3}\begin{pmatrix} 1 & 1 & 1 \\ 1 & 1 & 1 \\ 1 & 1 & 1 \end{pmatrix}. \quad \square$$

Problems

Prob. 189 — If $\mathbf{A}$ is an $n \times n$ symmetric real matrix, then $N(\mathbf{A}) = N(\mathbf{A}^k)$ for every positive integer k.

Prob. 190 — If λ is an eigenvalue of an $n \times n$ real symmetric matrix $\mathbf{A}$ then every generalised eigenvector is an eigenvector: $N(\mathbf{A} - \lambda\mathbf{I}) = G(\mathbf{A} - \lambda\mathbf{I})$.

Prob. 191 — Let $\mathbf{A}$ be an $n \times n$ matrix. Is it row equivalent to the orthogonal projection onto its row space?

Prob. 192 — Let
$$\mathbf{A} = \begin{pmatrix} 1 & 2 & 0 & -1 \\ 0 & 0 & 1 & 1 \\ 0 & 0 & 0 & 0 \end{pmatrix}.$$

Find the orthogonal projection onto $\mathbf{row}(\mathbf{A})^T$.

Prob. 193 — Let
$$\mathbf{A} = \begin{pmatrix} -10 & 7 & -7 \\ 7 & -2 & 5 \\ -7 & 5 & 8 \end{pmatrix}.$$

(a) Prove that $\{-17, 2, 11\}$ is the complete list of all eigenvalues for $\mathbf{A}$.

(b) Find the matrix $\mathbf{E}_3$ of the orthogonal projection E_3 in the spectral decomposition of $\mathbf{A}$ corresponding to $\lambda_3 = 11$.

(c) Consider the polynomial
$$f(x) = \frac{(x + 17)(x - 2)}{252}$$

and prove that $f(\mathbf{A}) = \mathbf{E}_3$. Observe that $f(11) = 1$ and $f(-17) = f(2) = 0$.

Prob. 194 — Determine whether the matrix is orthogonal, and if so find its inverse.

$$\mathbf{A} = \begin{pmatrix} 1/2 & 1/2 & 1/2 & 1/2 \\ 1/2 & -5/6 & 1/6 & 1/6 \\ 1/2 & 1/6 & 1/6 & -5/6 \\ 1/2 & 1/6 & -5/6 & 1/6 \end{pmatrix}$$

Prob. 195 — Prove that if $\mathbf{x}$ is an $n \times 1$ matrix, then the matrix

$$\mathbf{A} = \mathbf{I}_n - \frac{2}{\mathbf{x}^T\mathbf{x}}\mathbf{x}\mathbf{x}^T$$

is both orthogonal and symmetric.

Prob. 196 — Check wether it is possible to orthogonally diagonalize the matrix

$$A = \begin{pmatrix} 3 & -2 & 1 \\ -2 & 6 & -2 \\ 1 & -2 & 3 \end{pmatrix}.$$

Write down this diagonalization explicitly if it is possible.

Prob. 197 — Prove that if A is a symmetric orthogonal matrix, then 1 and -1 are the only possible eigenvalues.

Prob. 198 — Prove that if A is any $m \times n$ matrix, then $A^T A$ has an orthonormal set of n eigenvectors.

Prob. 199 — If $\{u_1, u_2, \ldots, u_n\}$ is an orthonormal basis for $\mathbb{R}^n$, and if A can be expressed as

$$A = c_1 u_1 u_1^T + c_2 u_2 u_2^T + \cdots + c_n u_n u_n^T,$$

with real $c_1, \ldots, c_n$, then A is symmetric and has eigenvalues $c_1, \ldots, c_n$.

Prob. 200 — Let $v = \begin{pmatrix} 1 & 0 & 1 \end{pmatrix}^T$. Find a matrix P that orthogonally diagonalizes the matrix $I - vv^T$.

Prob. 201 — For the matrix A in **Prob. 98** find the orthogonal projection on its range $R(A)$.

6.5 Quadratic Forms

Definition 6.12 A quadratic form in two variables x and y is defined by the formula

$$q(x, y) = ax^2 + 2cx + by^2 = x^T A x,$$

where

$$x = \begin{pmatrix} x \\ y \end{pmatrix}, \quad A = \begin{pmatrix} a & c \\ c & b \end{pmatrix}.$$

Definition 6.13 A quadratic form in n variables, $n \geq 2$ is an expression of the form

$$q(x, y) = x^T A x,$$

where A is a symmetric matrix of size $n \times n$ and $x \in \mathbb{R}^n$.

Problem 6.10 Write the quadratic form

$$q(x_1, x_2, x_3) = 5x_1^2 + 10x_2^2 + 2x_1^2 + 4x_1x_2 + 2x_1x_3 - 6x_2x_3$$

in the matrix form $\mathbf{x}^T \mathbf{A} \mathbf{x}$.

Solution: Notice that in the matrix $\mathbf{A}$ the entries a_{ij} with $i \neq j$ are equal halves of the coefficients at $x_i x_j$ just because the product $x_i x_j$ enters the formula for q ones, whereas $a_{ij} = a_{ji}$ are shown in $\mathbf{A}$ twice.

$$\mathbf{A} = \begin{pmatrix} 5 & 2 & 1 \\ 2 & 10 & -3 \\ 1 & -3 & 2 \end{pmatrix} \Rightarrow \mathbf{x}^T \mathbf{A} \mathbf{x} = \begin{pmatrix} x_1 & x_2 & x_3 \end{pmatrix} \begin{pmatrix} 5 & 2 & 1 \\ 2 & 10 & -3 \\ 1 & -3 & 2 \end{pmatrix} \begin{pmatrix} x_1 \\ x_2 \\ x_3 \end{pmatrix}$$

$$= \begin{pmatrix} x_1 & x_2 & x_3 \end{pmatrix} \begin{pmatrix} 5x_1 + 2x_2 + x_3 \\ 2x_1 + 10x_2 - 3x_3 \\ x_1 - 3x_2 + 2x_3 \end{pmatrix} =$$

$$x_1(5x_1 + 2x_2 + x_3) + x_2(2x_1 + 10x_2 - 3x_3) + x_3(x_1 - 3x_2 + 2x_3) =$$
$$5x_1^2 + 10x_2^2 + 2x_1^2 + 4x_1x_2 + 2x_1x_3 - 6x_2x_3. \quad \square$$

Lemma 6.5 *For every quadratic form $q(\mathbf{x})$ there is only one symmetric matrix $\mathbf{A}$ such that $q(\mathbf{x}) = \mathbf{x}^T \mathbf{A} \mathbf{x}$ for every $\mathbf{x} \in \mathbb{R}^n$.*

Proof. A construction of a symmetric matrix $\mathbf{A}$ by a given quadratic form q is shown in Problem 6.10. For every $\mathbf{x}$ and $\mathbf{y}$ in $\mathbb{R}^n$ and a symmetric matrix $\mathbf{A}$ corresponding to a quadratic form q, we have

$$2\mathbf{y}^T \mathbf{A} \mathbf{x} = (\mathbf{x} + \mathbf{y})^T \mathbf{A} (\mathbf{x} + \mathbf{y}) - \mathbf{y}^T \mathbf{A} \mathbf{y} - \mathbf{x}^T \mathbf{A} \mathbf{x} = q(\mathbf{y} + \mathbf{x}) - q(\mathbf{y}) - q(\mathbf{x}). \quad (6.22)$$

Substituting $\mathbf{y} = \mathbf{e}_i$, $\mathbf{x} = \mathbf{e}_j$ in (6.22), we obtain that

$$2a_{ij} = q(\mathbf{e}_i + \mathbf{e}_j) - q(\mathbf{e}_j) - q(\mathbf{e}_i). \quad \square$$

Definition 6.14 Suppose that $q(\mathbf{x})$ is a quadratic form. Then

- $q(\mathbf{x})$ is **positive definite** if $q(\mathbf{x}) \geq 0$ for all $\mathbf{x}$, and $q(\mathbf{x}) = 0$ only when $\mathbf{x} = \mathbf{0}$,
- $q(\mathbf{x})$ is **positive semi-definite** if $q(\mathbf{x}) \geq 0$ for all $\mathbf{x}$,
- $q(\mathbf{x})$ is **negative definite** if $q(\mathbf{x}) \leq 0$ for all $\mathbf{x}$, and $q(\mathbf{x}) = 0$ only when $\mathbf{x} = \mathbf{0}$,
- $q(\mathbf{x}$ is **negative semi-definite** if $q(\mathbf{x}) \leq 0$ for all $\mathbf{x}$,
- $q(\mathbf{x})$ is **indefinite** if it is neither positive definite, nor positive semi-definite, nor negative definite, nor negative semi-definite.

This classification of quadratic forms has a clear geometrical meaning if the orthogonal diagonalization of the symmetric matrix associated with the quadratic form q is available. We have $q(\mathbf{x}) = \mathbf{x}^T \mathbf{A} \mathbf{x}$, where $\mathbf{A}$ is a real symmetric matrix. The matrix $\mathbf{A}$

can be orthogonally diagonalised $\mathbf{A} = \mathbf{PDP}^T$, where $\mathbf{P}$ is a **real orthogonal** matrix and $\mathbf{D} = \text{Diag}(\lambda_1, \lambda_2, \ldots, \lambda_n)$, λ_j being **real eigenvalues** of $\mathbf{A}$. It follows that

$$q(\mathbf{x}) = \mathbf{x}^T \mathbf{A} \mathbf{x} = \mathbf{x}^T \mathbf{PDP}^T \mathbf{x} = \left(\mathbf{P}^T \mathbf{x}\right)^T \mathbf{D} \left(\mathbf{P}^T \mathbf{x}\right) = \mathbf{z}^T \mathbf{D} \mathbf{z} = \lambda_1 z_1^2 + \lambda_2 z_2^2 + \cdots + \lambda_n z_n^2,$$

where $\mathbf{z} = \mathbf{P}^T \mathbf{x} \Leftrightarrow \mathbf{x} = \mathbf{Pz}$. These calculations immediately imply the following theorem.

Theorem 6.13 *Suppose that the quadratic form $q(\mathbf{x})$ has matrix representation $q(\mathbf{x}) = \mathbf{x}^T \mathbf{A} \mathbf{x}$. Then:*

- *$q(\mathbf{x})$ is **positive definite** if and only if all eigenvalues of $\mathbf{A}$ are positive,*
- *$q(\mathbf{x})$ is **positive semi-definite** if and only if all eigenvalues of $\mathbf{A}$ are non-negative,*
- *$q(\mathbf{x})$ is **negative definite** if and only if all eigenvalues of $\mathbf{A}$ are negative,*
- *$q(\mathbf{x}$ is **negative semi-definite** if and only if all eigenvalues of $\mathbf{A}$ are non-positive,*
- *$q(\mathbf{x})$ is **indefinite** if and only if some eigenvalues of $\mathbf{A}$ are negative, and some are positive.*

Theorem 6.14 *Let $q(\mathbf{x}) = \mathbf{x}^T \mathbf{A} \mathbf{x}$, where $\mathbf{x} \in \mathbb{R}^2$, and*

$$\mathbf{A} = \begin{pmatrix} a & c \\ c & b \end{pmatrix}.$$

Then

- *If $|\mathbf{A}| > 0$ and $a > 0$, then $\lambda_1 > 0$, $\lambda_2 > 0$ and $\mathbf{A}$ is positive definite.*
- *If $|\mathbf{A}| > 0$ and $a < 0$, then $\lambda_1 < 0$, $\lambda_2 < 0$ and $\mathbf{A}$ is negative definite.*
- *If $|\mathbf{A}| < 0$, then λ_1 and λ_2 have opposite signs and $\mathbf{A}$ is indefinite.*

Proof. If $\mathbf{x} = \begin{pmatrix} x & y \end{pmatrix}^T$ and $a \neq 0$, then

$$q(\mathbf{x}) = ax^2 + 2cxy + by^2 = a\left(x + \frac{c}{a}y\right)^2 + \frac{ab - c^2}{a}y^2 = a\left(x + \frac{c}{a}y\right)^2 + \frac{|\mathbf{A}|}{a}y^2.$$

So, if $a > 0$ and $|\mathbf{A}| > 0$, then $q(\mathbf{x}) > 0$ for $\mathbf{x} < 0$. The discriminant $d_\mathbf{A}$ of the characteristic polynomial $p_\mathbf{A}(\lambda) = \lambda^2 - (a + b)\lambda + |\mathbf{A}|$ is

$$d_\mathbf{A} = (a + b)^2 - 4(ab - c^2) = (a - b)^2 + 4c^2 > 0.$$

If $a > 0$ and $|\mathbf{A}| = ab - c^2 > 0$, then $b > 0$ and

$$(a + b)^2 - d_{\mathbf{A}} = 4(ab - c^2) = 4|\mathbf{A}| > 0,$$

implying that all roots of the characteristic polynomial are positive. If they are positive, then by Vieta's theorem $|\mathbf{A}| > 0$ and $a + b > 0$. The condition $ab - c^2 > 0$ implies that both a and b have the same sign. Since $a + b > 0$ this shows that $a > 0$. This proves the first statement. Other statements are proved similarly. $\square$

Lemma 6.6 *If $q(\mathbf{x}) = \mathbf{x}^T \mathbf{A} \mathbf{x}$ is positive definite, then $a_{ii} > 0$ for every i.*

Proof. $a_{ii} = \mathbf{e}_i^T \mathbf{A} \mathbf{e}_i = q(\mathbf{e}_i) > 0$. $\square$

Definition 6.15 Let $\mathbf{A}$ be an $n \times n$ matrix. Then the determinant

$$\mathbf{A}\begin{pmatrix} i_1 & i_2 & \dots & i_p \\ k_1 & k_2 & \dots & k_p \end{pmatrix} = \begin{vmatrix} a_{i_1 k_1} & a_{i_1 k_2} & \cdots & a_{i_1 k_p} \\ a_{i_2 k_1} & a_{i_2 k_2} & \cdots & a_{i_2 k_p} \\ \vdots & \vdots & \ddots & \vdots \\ a_{i_p k_1} & a_{i_p k_2} & \cdots & a_{i_p k_p} \end{vmatrix}$$

is called a **minor** of the matrix $\mathbf{A}$

Definition 6.16 Let $\mathbf{A}$ be an $n \times n$ matrix. Then the **principal minors** of $\mathbf{A}$ are the n determinants formed from the first r rows and the first r columns of $\mathbf{A}$, $1 \leq r \leq n$:

$$a_{11}, \begin{vmatrix} a_{11} & a_{12} \\ a_{21} & a_{22} \end{vmatrix}, \begin{vmatrix} a_{11} & a_{12} & a_{13} \\ a_{21} & a_{22} & a_{23} \\ a_{31} & a_{32} & a_{33} \end{vmatrix}, \cdots, \begin{vmatrix} a_{11} & a_{12} & \cdots & a_{1n} \\ a_{21} & a_{22} & \cdots & a_{2n} \\ \vdots & \vdots & \ddots & \vdots \\ a_{n1} & a_{n2} & \cdots & a_{nn} \end{vmatrix}.$$

Theorem 6.15 (Silvester's Criterion) *Let $\mathbf{A}$ be an $n \times n$ symmetric matrix. Then $\mathbf{A}$ is positive definite if and only if all its principal minors are positive.*

The proof of the necessity in Theorem 6.15 follows from Lemma 6.7.

Lemma 6.7 *Let $\mathbf{A}_{r \times r}$ be the matrix $(a_{ij})_{1 \leq i, j \leq r}$. If $\mathbf{A}$ is positive definite, then $\mathbf{A}_{r \times r}$ is positive definite for every r, $1 \leq r \leq n$.*

Proof. If $0 \neq \mathbf{x} = \left(x_1 \ x_2 \ \cdots \ x_r\right)^T$ and $\mathbf{x}_r = \left(x_1 \ x_2 \ \cdots \ x_r \ 0 \ \cdots \ 0\right)^T$, then

$$\mathbf{x}^T \mathbf{A}_{r \times r} \mathbf{x} = \mathbf{x}_r^T \mathbf{A} \mathbf{x}_r > 0.$$

Since all eigenvalues of a positive definite matrix are positive and the determinant of a matrix equals the product of its eigenvalues, we conclude that

$$|\mathbf{A}_{r \times r}| > 0. \qquad \qquad \square$$

Proof of sufficiency in Theorem 6.15. This part of the proof is more complicated and is based on **Lagrange's method of diagonalization**. The idea of the proof is to reduce the quadratic form of the symmetric matrix $\mathbf{A}$ to a diagonal form by a sequence of non-orthogonal transformations soi that the quadratic form obtained be positive. Then we may apply Theorem 6.13. We begin with the case $n = 2$ and first complete the square for the quadratic form:

$$q(\mathbf{x}) = a_{11}x_1^2 + 2a_{12}x_1x_2 + a_{22}x_2^2 = a_{11}\left(x_1 + \frac{a_{12}}{a_{11}}x_2\right)^2 + a_{11}\left(\frac{a_{22}}{a_{11}} - \frac{a_{12}^2}{a_{11}^2}\right)x_2^2$$

$$\begin{aligned} y_1 &= x_1 + \frac{a_{12}}{a_{11}}x_2 \\ y_2 &= x_2 \end{aligned} \Leftrightarrow \mathbf{y} = \mathbf{S}\mathbf{x} = \begin{pmatrix} 1 & \frac{a_{12}}{a_{11}} \\ 0 & 1 \end{pmatrix}\mathbf{x} \Rightarrow q(\mathbf{x}) = |\mathbf{A}_{1\times1}|y_1^2 + \frac{|\mathbf{A}_{2\times2}|}{|\mathbf{A}_{1\times1}|}y_2^2. \quad (6.23)$$

Let us consider the diagonal matrix $\mathbf{D}(d_1, d_2)$, where $d_1 = |\mathbf{A}_{1\times1}|$ and $d_2 = |\mathbf{A}_{2\times2}|/|\mathbf{A}_{1\times1}|$. Then formulas in (6.23) can be summarized as follows:

$$(\mathbf{S}^{-1}\mathbf{y})^T \mathbf{A}(\mathbf{S}^{-1}\mathbf{y}) = \mathbf{y}^T \mathbf{D}(d_1, d_2)\mathbf{y}.$$

By Lemma 6.5, we conclude that

$$(\mathbf{S}^{-1})^T \mathbf{A}(\mathbf{S}^{-1}) = \mathbf{D}(d_1, d_2).$$

If $|\mathbf{A}_{1\times1}| > 0$ and $|\mathbf{A}_{2\times2}| > 0$, then $q(\mathbf{y}) > 0$ for any $\mathbf{y} = \mathbf{S}\mathbf{x} \neq \mathbf{0}$. Since $\mathbf{S}$ is an invertible matrix, this implies that $q(\mathbf{x}) > 0$ for every $\mathbf{x} \neq \mathbf{0}$. Then all eigenvalues of $\mathbf{A}$ are positive by Theorem 6.13.

If $n > 2$ then the terms in the quadratic form involving x_1 are all listed in the formula

$$a_{11}x_1^2 + 2a_{12}x_1x_2 + \cdots + 2a_{1n}x_1x_n.$$

Since the form is positive definite, it must be positive when $x_2 = \cdots = x_n = 0$ and $x_1 \neq 0$, which implies that $a_{11} > 0$. Hence

$$a_{11}\left(x_1 + \sum_{k=2}^{n} \frac{a_{1k}}{a_{11}}x_k\right)^2 - a_{11}\left(\sum_{k=2}^{n} \frac{a_{1k}}{a_{11}}x_k\right)^2 =$$

$$a_{11}\left\{x_1 + \sum_{k=2}^{n}\frac{a_{1k}}{a_{11}}x_k - \sum_{k=2}^{n}\frac{a_{1k}}{a_{11}}x_k\right\} \cdot \left\{x_1 + 2\sum_{k=2}^{n}\frac{a_{1k}}{a_{11}}x_k\right\} =$$

$$a_{11}x_1^2 + 2a_{12}x_1x_2 + \cdots + 2a_{1n}x_1x_n.$$

This formula naturally lends itself to the following change of variables:

$$y_1 = x_1 + \sum_{k=2}^{n}\frac{a_{1k}}{a_{11}}x_k, \quad y_2 = x_2, \quad \ldots, \quad y_n = x_n.$$

$$y = S_1 x = \begin{pmatrix} 1 & \frac{a_{12}}{a_{11}} & \cdots & \frac{a_{1n}}{a_{11}} \\ 0 & 1 & \cdots & 0 \\ \vdots & \vdots & \ddots & \vdots \\ 0 & 0 & \cdots & 1 \end{pmatrix} x, \quad |S_1| = 1.$$

The nonsingular transformation of variables S_1 reduces the quadratic form q to

$$q_1(y) = a_{11}y_1^2 + \sum_{i,j=2}^{n} b_{ij}y_iy_j, \quad q(x) = q_1(y) = q_1(S_1x).$$

Since q_1 is positive definite, the form $\sum_{i,j=2}^{n} b_{ij}y_iy_j$ is positive definite as well, and we may continue the algorithm of completing square. As result, we obtain a sequence S_i matrices with $\det(S_i) = 1$

It follows that

$$(S_1^{-1}S_2^{-1}\cdots S_{r-1}^{-1})^T A_{r\times r}(S_1^{-1}S_2^{-1}\cdots S_{r-1}^{-1}) = \text{Diag}(d_1, d_2, \ldots, d_r), \quad (6.24)$$

where $\det(S_j) = 1$ for $j = 1, \ldots, r-1$. Using the multiplicative property of determinants (see Theorem 3.29), we obtain **Jacobi's formulas** by evaluating determinants of the both sides of formula (6.24):

$$d_1 d_2 \cdots d_r = |A_{r\times r}| \Rightarrow \boxed{d_1 = a_{11} > 0, \ d_r = \frac{|A_{r\times r}|}{|A_{(r-1)\times(r-1)}|}, \ r > 1}. \quad (6.25)$$

Since all principal minors are positive, we obtain by (6.25) that all numbers d_j are positive, which implies that the quadratic form of the symmetric matrix A is positive for non-zero vectors x. Theorem 6.13 completes the proof of sufficiency. $\square$

Problem 6.11 Using the above methods investigate the quadratic form

$$q(x, y, z) = 4x^2 + 2y^2 + 4z^2 - 2xy + 4yz - 6xz. \quad (6.26)$$

Solution by Sivester's Criterion.

$$|\mathbf{A}_{1\times1}| = 4, \ |\mathbf{A}_{2\times2}| = \begin{vmatrix} 4 & -1 \\ -1 & 2 \end{vmatrix} = 7, \ |\mathbf{A}_{3\times3}| = \begin{vmatrix} 4 & -1 & -3 \\ -1 & 2 & 2 \\ -3 & 2 & 4 \end{vmatrix} = 6.$$

The quadratic form is positive definite.

Solution by Lagrange's Method

$$q(x,y,z) = 4x^2 + 2y^2 + 4z^2 - 2xy + 4yz - 6xz =$$

$$\left(4x^2 + 2\cdot(-1)xy + 2\cdot(-3)xz\right) + 2y^2 + 4z^2 + 4yz =$$

$$4\left(x - \frac{1}{4}y - \frac{3}{4}z\right)^2 - 4\left(\frac{1}{4}y + \frac{3}{4}z\right)^2 + 2y^2 + 4z^2 + 4yz =$$

$$4u^2 + \frac{7}{4}y^2 + \frac{7}{4}z^2 + \frac{5}{2}yz = 4u^2 + \left(\frac{7}{4}y^2 + 2\frac{5}{4}yz\right) + \frac{7}{4}z^2 =$$

$$4u^2 + \frac{7}{4}\left(y + \frac{5}{7}z\right)^2 - \frac{7}{4}\cdot\frac{25}{49}z^2 + \frac{7}{4}z^2 =$$

$$4u^2 + \frac{7}{4}v^2 + \frac{6}{7}w^2 = |\mathbf{A}_{1\times1}|u^2 + \frac{|\mathbf{A}_{2\times2}|}{|\mathbf{A}_{1\times1}|}v^2 + \frac{|\mathbf{A}_{3\times3}|}{|\mathbf{A}_{2\times2}|}w^2. \quad \square \quad (6.27)$$

Lagrange's Method applied to the symmetric matrix $\mathbf{A}$ assotiated with the quadratic form q in 6.26:

$$\mathbf{A} = \begin{pmatrix} 4 & -1 & -3 \\ -1 & 2 & 2 \\ -3 & 2 & 4 \end{pmatrix}$$

can be summarized by the formula

$$\mathbf{u} = \begin{pmatrix} u \\ v \\ w \end{pmatrix} = \begin{pmatrix} 1 & -\frac{1}{4} & -\frac{3}{4} \\ 0 & 1 & \frac{5}{7} \\ 0 & 0 & 1 \end{pmatrix}\begin{pmatrix} x \\ y \\ z \end{pmatrix} = \mathbf{Sx}.$$

Notice that the rows of $\mathbf{S}$ are in one-to-one correspondence with completing squaring in the process shown in (6.27). Then

$$q_1(\mathbf{u}) = \mathbf{u}^T \begin{pmatrix} 4 & 0 & 0 \\ 0 & \frac{7}{4} & 0 \\ 0 & 0 & \frac{6}{7} \end{pmatrix}\mathbf{u} = \mathbf{x}^T\mathbf{S}^T\begin{pmatrix} 4 & 0 & 0 \\ 0 & \frac{7}{4} & 0 \\ 0 & 0 & \frac{6}{7} \end{pmatrix}\mathbf{Sx} \Rightarrow \mathbf{A} = \mathbf{U}^T\mathbf{U}, \ \text{ where}$$

$$\mathbf{U} = \begin{pmatrix} 2 & 0 & 0 \\ 0 & \frac{\sqrt{7}}{2} & 0 \\ 0 & 0 & \frac{\sqrt{6}}{\sqrt{7}} \end{pmatrix}\begin{pmatrix} 1 & -\frac{1}{4} & -\frac{3}{4} \\ 0 & 1 & \frac{5}{7} \\ 0 & 0 & 1 \end{pmatrix} = \begin{pmatrix} 2 & -\frac{1}{2} & -\frac{3}{2} \\ 0 & \frac{\sqrt{7}}{2} & \frac{5\sqrt{7}}{14} \\ 0 & 0 & \frac{\sqrt{6}}{\sqrt{7}} \end{pmatrix}.$$

The following theorem is a consequence of the Jacobi's formulas (6.25), since for nagative defined forms all numbers $d_1, d_2, \cdots, d_n$ are negative.

Theorem 6.16 *Let* **A** *be an* $n \times n$ *symmetric matrix. Then* **A** *is negative definite if and only if* $a_{11} < 0$ *and the sequence of all its principal minors alternates in signs.*

Theorem 6.17 *If there is an invertible matrix* **S** *such that*

$$\mathbf{S}^T \mathbf{A} \mathbf{S} = \text{Diag}\,(d_1, d_n, \ldots, d_n)\,, \quad d_i > 0,$$

then the matrix **A** *is positive definite.*

Proof. Since **S** is invertible, the equation $\mathbf{x} = \mathbf{S}\mathbf{y}$ has a unique solution. Then

$$\mathbf{x}^T \mathbf{A} \mathbf{x} = (\mathbf{S}\mathbf{y})^T \mathbf{A}(\mathbf{S}\mathbf{y}) = \mathbf{y}^T \text{Diag}\,(d_1, d_n, \ldots, d_n)\,\mathbf{y} > 0.$$

$\square$

Theorem 6.18 *A symmetric matrix* **A** *is positive definite if and only if there exists an invertible upper triangular matrix* **U** *such that*

$$\mathbf{A} = \mathbf{U}^T \mathbf{U}.$$

Proof. If $\mathbf{A} = \mathbf{U}^T \mathbf{U}$, then

$$\mathbf{x}^T \mathbf{A} \mathbf{x} = \mathbf{x}^T \mathbf{U}^T \mathbf{U} \mathbf{x} = \langle \mathbf{U}\mathbf{x}, \mathbf{U}\mathbf{x} \rangle > 0$$

for $\mathbf{x} \neq 0$ since **U** is invertible. If **A** is positive definite, then by Lagrange's method

$$q(\mathbf{x}) = \mathbf{x}^T \mathbf{A} \mathbf{x} = \mathbf{u}^T \text{Diag}(d_1, \ldots, d_n)\mathbf{u} =$$
$$\mathbf{x}^T \mathbf{S}^T \text{Diag}(d_1, \ldots, d_n)\mathbf{S} \Rightarrow \mathbf{U} = \text{Diag}(\sqrt{d_1}, \ldots, \sqrt{d_n})\mathbf{S}. \quad \square$$

Problem 6.12 Write the quadratic form $q(x, y) = 7x^2 - 4xy + 4y^2$ as $\mathbf{x}^T \mathbf{A} \mathbf{x}$, where **A** is a symmetric matrix and **x** is a column vector. Find an orthogonal matrix **P** and a diagonal matrix **D** such that $\mathbf{P}^T \mathbf{A} \mathbf{P} = \mathbf{D}$. Express the quadratic form as

$$q(x, y) = \lambda_1 u^2 + \lambda_2 v^2$$

for appropriate constants λ_1, λ_2 and coordinates u, v. Use this information to sketch a graph of the curve

$$7x^2 - 4xy + 4y^2 = 24$$

in the xy-plane.

Solution:

$$q(x,y) = 7x^2 - 2 \cdot 2xy + 4y^2 = \begin{pmatrix} x & y \end{pmatrix} \begin{pmatrix} 7 & -2 \\ -2 & 4 \end{pmatrix} \begin{pmatrix} x \\ y \end{pmatrix} \Rightarrow$$

$$\mathbf{A} = \begin{pmatrix} 7 & -2 \\ -2 & 4 \end{pmatrix} \Rightarrow |\mathbf{A}_{1 \times 1}| = 7 > 0, \; |\mathbf{A}_{2 \times 2}| = 24 > 0.$$

Hence $\mathbf{A}$ is positive definite by Silvester's Criterion.

$$p_{\mathbf{A}}(\lambda) = \begin{vmatrix} 7 - \lambda & -2 \\ -2 & 4 - \lambda \end{vmatrix} = (\lambda - 8)(\lambda - 3).$$

$$\begin{cases} \lambda_1 = 8: & \mathbf{A} - 8\mathbf{I} = \begin{pmatrix} -1 & -2 \\ -2 & -4 \end{pmatrix} \sim \begin{pmatrix} 1 & 2 \\ 0 & 0 \end{pmatrix} \Rightarrow \mathbf{v}_1 = \begin{pmatrix} -2 \\ 1 \end{pmatrix}, \\ \lambda_2 = 3: & \mathbf{A} - 3\mathbf{I} = \begin{pmatrix} 4 & -2 \\ -2 & 1 \end{pmatrix} \sim \begin{pmatrix} 2 & -1 \\ 0 & 0 \end{pmatrix} \Rightarrow \mathbf{v}_2 = \begin{pmatrix} 1 \\ 2 \end{pmatrix}. \end{cases}$$

If we follow the standard approach, then the vectors $\mathbf{u}_i$ shown in the left-hand graph below cannot be obtained by the rotation of the coordinate vectors:

$$\mathbf{u}_1 = \frac{1}{\sqrt{5}} \begin{pmatrix} -2 \\ 1 \end{pmatrix}, \; \mathbf{u}_2 = \frac{1}{\sqrt{5}} \begin{pmatrix} 1 \\ 2 \end{pmatrix}, \; \mathbf{P} = \frac{1}{\sqrt{5}} \begin{pmatrix} -2 & 1 \\ 1 & 2 \end{pmatrix}, \; \mathbf{D} = \begin{pmatrix} 8 & 0 \\ 0 & 3 \end{pmatrix}.$$

But if we swap $\mathbf{v}_1$ and $\mathbf{v}_2$, then the orthonormal system $\{\mathbf{u}_1, \mathbf{u}_2\}$

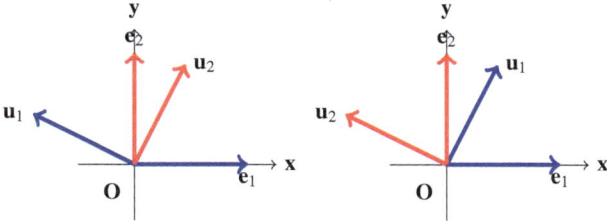

Fig. 6.1: Two transformations.

$$\mathbf{u}_1 = \frac{1}{\sqrt{5}} \begin{pmatrix} 1 \\ 2 \end{pmatrix}, \; \mathbf{u}_2 = \frac{1}{\sqrt{5}} \begin{pmatrix} -2 \\ 1 \end{pmatrix}, \; \mathbf{P} = \frac{1}{\sqrt{5}} \begin{pmatrix} 1 & -2 \\ 2 & 1 \end{pmatrix}, \; \mathbf{D} = \begin{pmatrix} 3 & 0 \\ 0 & 8 \end{pmatrix}$$

is obtained from the standard coordinate system as a rotation anti-clockwise by the angle $\theta = \arccos 1/\sqrt{5}$. Then

$$\mathbf{x}^T \mathbf{A} \mathbf{x} = \mathbf{x}^T \mathbf{P} \mathbf{D} \mathbf{P}^T \mathbf{x} \overset{\mathbf{z} = \mathbf{P}^T \mathbf{x}}{=} \mathbf{z}^T \mathbf{D} \mathbf{z} \overset{\mathbf{z} = (u, v)^T}{=} 3u^2 + 8v^2;$$

$$\begin{pmatrix} u \\ v \end{pmatrix} = \frac{1}{\sqrt{5}} \begin{pmatrix} 1 & 2 \\ -2 & 1 \end{pmatrix} \begin{pmatrix} x \\ y \end{pmatrix}$$

Notice that Lagrange's method gives

$$7x^2 - 4xy + 4y^2 = 7 \left(x - \frac{2}{7} y \right)^2 + \frac{24}{7} y^2 \Rightarrow \begin{cases} u = x - \frac{2}{7} y \\ v = y \end{cases} \Rightarrow \mathbf{S} = \begin{pmatrix} 1 & -\frac{2}{7} \\ 0 & 1 \end{pmatrix}.$$

The matrix $\mathbf{S}$ being not orthogonal, does not allow us to draw the picture of the curve.   □

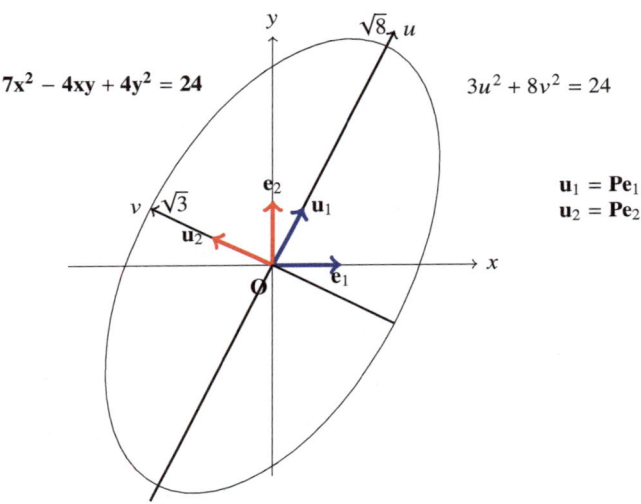

Fig. 6.2: The ellipse.

Problem 6.13 A grocery store carries two brands of cat food, a local brand that it obtains at the cost of 30 cents per can and a well-known national brand it obtains at the cost of 40 cents per can. The grocer' management estimates that if the local brand is sold for x cents per can and the national brand for y cents per can, then approximately $70 - 5x + 4y$ cans of the local brand and $80 + 6x - 7y$ cans of the national brand will be sold each day. How should the grocer price each brand to maximize total daily profit from the sale of cat food?

Comments In this problem, variable x denotes the selling price p_l of a local can and variable y denotes the selling price p_n of a national brand can. Then

$$\begin{aligned} q_l &= 70 - 5p_l + 4p_n, \\ q_n &= 80 + 6p_l - 7p_n. \end{aligned} \tag{6.28}$$

The choice of $\pm$ signs at p_l and p_n in both demand functions has economic meaning. Let us investigate the first demand function $q_l = 70 - 5p_l + 4p_n$. If the price p_l for a local brand can decreases, then obviously the sold number q_l increases. If the price p_n increases, then people would prefer to buy cans of the local brand which increases q_l. The second equation can be analysed similarly.

Solution: The total profit Π is calculated by the formula

$$\Pi(x, y) = \underbrace{(70 - 5x + 4y)}_{\text{items sold}} \cdot \underbrace{(x - 30)}_{\text{profit per item}} + \underbrace{(80 + 6x - 7y)}_{\text{items sold}} \cdot \underbrace{(y - 40)}_{\text{profit per item}}$$

$$= -5x^2 + 10xy - 7y^2 - 20x + 240y - 5300.$$

The critical points of $\Pi(x, y)$ are the points at which its gradient is zero.

$$\nabla\Pi = \Pi_x \mathbf{i} + \Pi_y \mathbf{j} = \mathbf{0}_2 \Leftrightarrow \begin{cases} \Pi_x &= -10x + 10y - 20 = 0 \\ \Pi_y &= 10x - 14y + 240 = 0 \end{cases} \Leftrightarrow \begin{cases} x &= 53 \\ y &= 55 \end{cases} \quad (6.29)$$

At this point,

$$\Delta\Pi \approx \begin{pmatrix} \Delta x & \Delta y \end{pmatrix} \begin{pmatrix} \Pi_{xx} & \Pi_{xy} \\ \Pi_{yx} & \Pi_{yy} \end{pmatrix} \begin{pmatrix} \Delta x \\ \Delta y \end{pmatrix}.$$

Since

$$\Pi_{xx} = -10, \ \Pi_{yy} = -14, \ \Pi_{xy} = \Pi_{yx} = 10,$$

we obtain that

$$\begin{pmatrix} \Pi_{xx} & \Pi_{xy} \\ \Pi_{yx} & \Pi_{yy} \end{pmatrix} = \begin{pmatrix} -10 & 10 \\ 10 & -14 \end{pmatrix} = \mathbf{A}$$

We see that the characteristic polynomial $p_{\mathbf{A}}(\lambda) = \lambda^2 + 24\lambda + 40$ has two negative roots $\lambda_{1,2} = -12 \pm \sqrt{104}$. Since the matrix $\mathbf{A}$ is symmetric, it is orthogonally diagonalizable, and in new coordinates $\mathbf{x} = \mathbf{Pu}$ the quadratic form corresponding to $\mathbf{A}$ is negative: $\lambda_1 u_1^2 + \lambda_2 u_2^2$. It follows that $\Delta\Pi \leq 0$, implying that $(53, 55)$ is the point of maximum for the profit Π. □

Using matrices, we can simplify the evaluation of $\nabla\Pi = \Pi_x \mathbf{i} + \Pi_y \mathbf{j}$. We denote by $\mathbf{x} = \begin{pmatrix} x & y \end{pmatrix}^T$ the vector of cans selling prices $x = p_l$, $y = p_n$, and by $\mathbf{c} = \begin{pmatrix} 30 & 40 \end{pmatrix}^T$ the vector of the wholesale prices. Finally, we put

$$\mathbf{A} = \begin{pmatrix} -5 & 4 \\ 6 & -7 \end{pmatrix}, \ \mathbf{b} = \begin{pmatrix} 70 \\ 80 \end{pmatrix}$$

to transform (6.28) into a matrix form

$$\mathbf{q} = \begin{pmatrix} q_l \\ q_n \end{pmatrix} = \mathbf{b} + \mathbf{Ax}.$$

Then the profit Π is the dot product

$$\Pi = (\mathbf{x} - \mathbf{c})^T \mathbf{q} = (\mathbf{x}^T - \mathbf{c}^T)(\mathbf{b} + \mathbf{Ax}). \quad (6.30)$$

We apply the Gradient Product Rule in a vector form to (6.30):

$$\nabla\Pi = \nabla(\mathbf{x}^T - \mathbf{c}^T)(\mathbf{b} + A\mathbf{x}) + (\mathbf{x} - \mathbf{c})\nabla(\mathbf{b} + A\mathbf{x})$$

Since $\nabla\mathbf{x}^T = \begin{pmatrix} \mathbf{i} & \mathbf{j} \end{pmatrix}$, we obtain the following formula for the gradient of the profit:

$$\nabla\Pi = \begin{pmatrix} \mathbf{i} & \mathbf{j} \end{pmatrix}(\mathbf{b} + A\mathbf{x}) + (\mathbf{x}^T - \mathbf{c}^T)A\begin{pmatrix} \mathbf{i} \\ \mathbf{j} \end{pmatrix}. \tag{6.31}$$

In our case, formula (6.31) yields

$$\nabla\Pi = \begin{pmatrix} \mathbf{i} & \mathbf{j} \end{pmatrix}\begin{pmatrix} 70 - 5x + 4y \\ 80 + 6x - 7y \end{pmatrix} + \begin{pmatrix} x - 30 & y - 40 \end{pmatrix}\begin{pmatrix} -5\mathbf{i} + 4\mathbf{j} \\ 6\mathbf{i} - 7\mathbf{j} \end{pmatrix} =$$
$$(70 - 5x + 4y)\mathbf{i} + (80 + 6x - 7y)\mathbf{j} + (x - 30)(-5\mathbf{i} + 4\mathbf{j}) + (y - 40)(6\mathbf{i} - 7\mathbf{j}) =$$
$$(-10x + 10y - 20)\mathbf{i} + (10x - 14y + 240)\mathbf{j},$$

which coincides with calculations shown in (6.29).

Problem 6.14 A supermarket sells three brands of coffee: Brand A (local), Brand B (national, and Brand C (premium). The supermarket buys these at different costs (Brand A (local), \$3 per can, Brand B (national) \$4 per can, and Brand C (premium) \$5 per can) and aims to set prices to maximize profit. The supermarket management established statistically that the demand functions q_A, q_B, q_C for each of these products are defined by the following system of equations

$$\begin{cases} q_A &= -20 - p_A + p_B + p_C \\ q_B &= -11 + 2p_A - p_B + p_C \\ q_C &= 3 + p_A + p_B - 2p_C \end{cases}$$

where p_A, p_B, p_C are the selling prices of each can in three brands. How should the supermarket price each brand to maximize total daily profit from the coffee?

Solution: To simplify notations, we put $x = p_A$, $y = p_B$, $z = p_C$. Then

$$A = \begin{pmatrix} -1 & 1 & 1 \\ 2 & -1 & 1 \\ 1 & 1 & -2 \end{pmatrix}, \quad \mathbf{b} = \begin{pmatrix} -20 \\ -11 \\ 3 \end{pmatrix}, \quad \mathbf{c} = \begin{pmatrix} 3 \\ 4 \\ 5 \end{pmatrix}$$

and by (6.30) the gradient of daily profit Π is calculated by the formula

$$\nabla\Pi = \begin{pmatrix} \mathbf{i} & \mathbf{j} & \mathbf{k} \end{pmatrix}(\mathbf{b} + A\mathbf{x}) + (\mathbf{x}^T - \mathbf{c}^T)A\begin{pmatrix} \mathbf{i} \\ \mathbf{j} \\ \mathbf{k} \end{pmatrix}. \tag{6.32}$$

In our case (6.32) yields

$$\nabla \Pi = \begin{pmatrix} \mathbf{i} & \mathbf{j} & \mathbf{k} \end{pmatrix} \begin{pmatrix} -20 - x + y + z \\ -11 + 2x - y + z \\ 3 + x + y - 2z \end{pmatrix} + \begin{pmatrix} x - 3 & y - 4 & z - 5 \end{pmatrix} \begin{pmatrix} -\mathbf{i} + \mathbf{j} + \mathbf{k} \\ 2\mathbf{i} - \mathbf{j} + \mathbf{k} \\ \mathbf{i} + \mathbf{j} - 2\mathbf{k} \end{pmatrix} =$$

$$(-20 - x + y + z)\mathbf{i} + (-11 + 2x - y + z)\mathbf{j} + (3 + x + y - 2z)\mathbf{k} +$$
$$(x - 3)(-\mathbf{i} + \mathbf{j} + \mathbf{k}) + (y - 4)(2\mathbf{i} - \mathbf{j} + \mathbf{k}) + (z - 5)(\mathbf{i} + \mathbf{j} - 2\mathbf{k}) =$$
$$(-30 - 2x + 3y + 2z)\mathbf{i} + (-15 + 3x - 2y + 2z)\mathbf{j} + (-6 + 2x + 2y - 4z)\mathbf{k}.$$

It follows that $\nabla \Pi = \mathbf{0}$ if and only if x, y, z satisfy the system of linear equations:

$$\begin{cases} -2x + 3y + 2z & = 30 \\ 3x - 2y + 2z & = 15 \\ 2x + 2y - 4z & = 6 \end{cases}$$

The reduced row echelon form of the extended matrix of the coefficient matrix is given by

$$\text{rref} \begin{pmatrix} -2 & 3 & 2 & | & 1 & 0 & 0 \\ 3 & -2 & 2 & | & 0 & 1 & 0 \\ 2 & 2 & -4 & | & 0 & 0 & 1 \end{pmatrix} = \begin{pmatrix} 1 & 0 & 0 & | & 1/15 & 4/15 & 1/6 \\ 0 & 1 & 0 & | & 4/15 & 1/15 & 1/6 \\ 0 & 0 & 1 & | & 1/6 & 1/6 & -1/12 \end{pmatrix}$$

It follows that

$$\begin{pmatrix} x \\ y \\ z \end{pmatrix} = \begin{pmatrix} 1/15 & 4/15 & 1/6 \\ 4/15 & 1/15 & 1/6 \\ 1/6 & 1/6 & -1/12 \end{pmatrix} \begin{pmatrix} 30 \\ 15 \\ 6 \end{pmatrix} = \begin{pmatrix} 7 \\ 10 \\ 7 \end{pmatrix}.$$

Since

$$\begin{pmatrix} \Pi_{xx} & \Pi_{xy} & \Pi_{xz} \\ \Pi_{yx} & \Pi_{yy} & \Pi_{yz} \\ \Pi_{zx} & \Pi_{zy} & \Pi_{zz} \end{pmatrix} = \begin{pmatrix} -2 & 3 & 2 \\ 3 & -2 & 2 \\ 2 & 2 & -4 \end{pmatrix} = \mathbf{B},$$

we find that the characteristic polynomial

$$p_\mathbf{B}(\lambda) = -\lambda^3 - 8\lambda^2 - 3\lambda + 60 = -(\lambda + 5)(\lambda^2 + 3\lambda - 12)$$

of the symmetric matrix $\mathbf{B}$ has one positive root and two negative

$$\lambda_1 = -5, \; \lambda_2 = \frac{-3 - \sqrt{57}}{2}, \; \lambda_3 = \frac{\sqrt{57} - 3}{2}.$$

This means that the point $x = 7$, $y = 10$, $z = 7$ is not a point of local maximum, implying that such a point does not exist. □

Problems

Prob. 202 — Let

$$\mathbf{A} = \begin{pmatrix} 7 & -2 \\ -2 & 4 \end{pmatrix}$$

be a symmetric matrix.

- Using Theorem 6.14, check that the matrix $\mathbf{A}$ is positive definite.
- Using Lagrange's diagonalization, find an upper triangular matrix $\mathbf{U}$ such that $\mathbf{A} = \mathbf{U}^T \mathbf{U}$.
- Orthogonally diagonalize the matrix $\mathbf{A}$.
- Find the spectral decomposition (6.19) for the symmetric matrix $\mathbf{A}$. Find explicit formulas for the spectral projectors in (6.19).
- Using previous calculations, show that the spectral projectors satisfy (6.20).

Prob. 203 — Let

$$\mathbf{A} = \begin{pmatrix} 1 & 1 & 0 \\ 1 & 5 & 1 \\ 0 & 1 & 1 \end{pmatrix}$$

be a symmetric matrix.

- Using Theorem 6.15, check that the matrix $\mathbf{A}$ is positive definite.
- Using Lagrange's diagonalization, find an upper triangular matrix $\mathbf{U}$ such that $\mathbf{A} = \mathbf{U}^T \mathbf{U}$.
- Orthogonally diagonalize the matrix $\mathbf{A}$.
- Find the spectral decomposition (6.19) for the symmetric matrix $\mathbf{A}$. Find explicit formulas for the spectral projectors in (6.19).
- Using previous calculations, show that the spectral projectors satisfy (6.20).

Prob. 204 — Suppose that $\mathbf{A}$ is a symmetric transition matrix of a regular Markov chain. Using (6.19), show that

$$\lim_{k \to +\infty} \mathbf{A}^k = E_1,$$

where E_1 is the spectral projector corresponding to the eigenvalue 1. Check if this statement remains true if the regularity condition is omitted.

Chapter 7
Regression

7.1 Chebyshev's Regression

In 1864, P.L.Tchebysheff published a pioneering paper on polynomial regressions, see Tchebysheff (1854). By Lagrange's formula (see also Theorem 3.35) the polynomial y of degree n satisfying $y(x_i) = y_i$, $i = 0, 1, \ldots, n$ is given by

$$y(x) = \sum_{i=0}^{n} \frac{L(x)}{(x - x_i)L'(x_i)} y_i, \quad L(x) = (x - x_0)(x - x_1) \cdots (x - x_n).$$

The following identity is central for Chebyshev's method:

$$\frac{1}{x - x_0} + \frac{1}{x - x_1} + \cdots + \frac{1}{x - x_n} = \frac{L'(x)}{L(x)}$$

Using the Long Division of Polynomials, we can write:

$$\frac{L'(x)}{L(x)} = \cfrac{1}{(A_1 x + B_1) + \cfrac{1}{(A_2 x + B_2) + \cfrac{1}{(A_3 x + B_3) + \cfrac{1}{\ddots}}}}$$

We can stop the algorithm of continued fraction at any step k and transform the expression obtained to the ratio of two polynomials. Then $\psi_k(x)$ denotes the denominator of this fraction (called the k-th convergent of the continued fraction).

S. Khrushchev, *Linear Algebra with Applications to Economics*, Classroom Companion: Economics, https://doi.org/10.1007/978-3-031-68682-5_7

Theorem 7.1 *The Lagrange interpolation polynomials $y(x)$ can be represented as partial sums of the following series*

$$A_1 \sum_{i=0}^{n} y_i - A_2 \psi_1(x) \sum_{i=0}^{n} \psi_1(x_i) y_i + A_3 \psi_2(x) \sum_{i=0}^{n} \psi_2(x_i) y_i - \cdots .$$

*The first k summands of this series give the polynomial of degree $k - 1$ which is the **least squares** polynomial for our data $y_0, y_1, \ldots, y_n$.*

Problem 7.1 Apply Chebyshev's Method to find the least squares quadratic polynomial for the data points:

t_i	1	1.5	2	2.5	3	3.5	4	4.5	5
y_i	−0.15	0.24	0.68	1.04	1.21	1.15	0.86	0.41	−0.08

$$L(x) = -7087.5 + 27343.1x - 44951.1x^2 + 41496.3x^3 -$$
$$23786.4x^4 + 8805.56x^5 - 2110.5x^6 + 316.5x^7 - 27x^8 + x^9$$
$$L'(x) = 27343.1 - 89902.1x + 124489.x^2 - 95145.8x^3 +$$
$$44027.8x^4 - 12663.x^5 + 2215.5x^6 - 216x^7 + 9x^8$$

$$L(x) = \left(\frac{x}{9} - \frac{1}{3}\right) L'(x) + P(x)$$
$$P(x) = 2026.88 - 5662.38x + 6534.38x^2 - 4051.04x^3 +$$
$$1461.25x^4 - 307.417x^5 + 35.x^6 - 1.66667x^7$$

$$\boxed{A_1 = \frac{1}{9}, \quad \psi_1(x) = \frac{x}{9} - \frac{1}{3}} .$$

By the Long Division of polynomials:

$$L'(x) = (-5.4x + 16.2) P(x) + Q(x)$$
$$Q(x) = -5492.25 + 12773.5x - 11944.8x^2 + 5766.75x^3 -$$
$$1520.06x^4 + 207.9x^5 - 11.55x^6$$

$$\boxed{A_2 = -5.4, \quad \psi_2(x) = -5.4x + 16.2} .$$

$$P(x) = (0.1443x - 0.4329)\, Q(x) + R(x)$$

$$R(x) = -350.722 + 659.8x - 479.732x^2 + 169.018x^3 - \;\Rightarrow\; \boxed{A_3 = 0.1443}$$

$$28.9286x^4 + 1.92857x^5$$

Finally,

$$\sum_{i=1}^{9} y_i = 5.36, \quad \sum_{i=1}^{9} \psi_1(x_i)y_i = 0.070002, \quad \sum_{i=1}^{9} \psi_2(x_i)y_i = 3.782.$$

$$A_1 \sum_{i=0}^{n} y_i - A_2\psi_1(x) \sum_{i=0}^{n} \psi_1(x_i)y_i + A_3\psi_2(x) \sum_{i=0}^{n} \psi_2(x_i)y_i =$$

$$\frac{1}{9} \cdot 5.36 + 5.4 \cdot \left(\frac{x}{9} - \frac{1}{3}\right) + 0.1443(-0.6x^2 + 3.6x - 4.39999) \cdot 3.782 =$$

$$\boxed{-0.327445x^2 + 2.00667x - 1.93171}$$

1. A great advantage of Chebyshev's Method is that it allows one to calculate the least squares approximations consecutively gradually increasing the degrees of the polynomials of the best approximation.

2. A disadvantage is that the described algorithm sometimes gives zero residues. This can be eliminated if one writes the continued fraction in a slightly different form.

3. Nowadays, Chebyshev's Method is applied by making use of Orthogonal Polynomials instead of Chebyshev's Continued Fraction. This allows one to construct the polynomials $\psi_k(x)$ via the so-called three-tern recurrence relation.

7.2 Least Squares and Regression

In practical applications, especially if $R(A) \neq \mathbb{R}^m$ even small mistakes in data $\mathbf{b}$ result in the system inconsistency. To overcome these difficulties a method of the normal equations was developed.

> **Definition 7.1** Given a linear system $\mathbf{Ax} = \mathbf{b}$ of m equations in n unknowns, find a vector $\mathbf{x}$ in $\mathbb{R}^n$ that minimizes $\|\mathbf{b} - \mathbf{Ax}\|$ with respect to the Euclidean inner product in $\mathbb{R}^n$. Such a vector, if it exists, is called a **least squares solution**

of $\mathbf{Ax} = \mathbf{b}$. The vector $\mathbf{b} - \mathbf{Ax}$ is called the **least squares error vector**, and the number $\|\mathbf{b} - \mathbf{Ax}\|$ is called the **least squares error**.

Lemma 7.1 *Let W be a subspace of a finite-dimensional inner product space V, and $\mathbf{b}$ be a vector in V. Then $\mathrm{proj}_W \mathbf{b}$ is the unique best approximation to $\mathbf{b}$ in W:*

$$\|\mathbf{b} - \mathrm{proj}_W \mathbf{b}\| < \|\mathbf{b} - \mathbf{w}\|$$

for every $\mathbf{w} \in W$, $w \neq \mathrm{proj}_W \mathbf{b}$.

Proof. If $\mathbf{w} \in W$, then

$$\mathbf{b} - \mathbf{w} = (\mathbf{b} - \mathrm{proj}_W \mathbf{b}) + (\mathrm{proj}_W \mathbf{b} - \mathbf{w}) \Rightarrow \|\mathbf{b} - \mathbf{w}\|^2 = \|\mathbf{b} - \mathrm{proj}_W \mathbf{b}\|^2 + \|\mathrm{proj}_W \mathbf{b} - \mathbf{w}\|^2 \Rightarrow$$
$$\|\mathbf{b} - \mathbf{w}\|^2 > \|\mathbf{b} - \mathrm{proj}_W \mathbf{b}\|^2 \quad \text{if} \quad \mathbf{w} \neq \mathrm{proj}_W \mathbf{b}. \quad \square$$

Definition 7.2 The equation $\mathbf{A}^T \mathbf{A} \mathbf{x} = \mathbf{A}^T \mathbf{b}$ is called the **normal equation** or the **normal system** associated with $\mathbf{Ax} = \mathbf{b}$.

Theorem 7.2 *For every linear system $\mathbf{Ax} = \mathbf{b}$ of m equations in n unknowns, the associated normal system of n equations in n unknowns*

$$\mathbf{A}^T \mathbf{A} \mathbf{x} = \mathbf{A}^T \mathbf{b} \tag{7.1}$$

is consistent, and all its solutions are least squares solutions of the initial system $\mathbf{Ax} = \mathbf{b}$. Moreover, if W is the column space of $\mathbf{A}$, and $\mathbf{x}$ is any least squares solution of $\mathbf{Ax} = \mathbf{b}$, then the orthogonal projection of $\mathbf{b}$ on W is given by the formula

$$\mathrm{proj}_W \mathbf{b} = \mathbf{Ax}. \tag{7.2}$$

Proof. If $W = \mathrm{CS}(\mathbf{A})$, then $\mathrm{N}(\mathbf{A}^T) = W^\perp$. It follows that

$$\mathbf{Ax} = \mathrm{proj}_W \mathbf{b} \Leftrightarrow \mathbf{b} - \mathbf{Ax} = \mathbf{b} - \mathrm{proj}_W \mathbf{b} \Leftrightarrow \mathbf{A}^T (\mathbf{b} - \mathbf{Ax}) =$$
$$\mathbf{A}^T (\mathbf{b} - \mathrm{proj}_W \mathbf{b}) = \mathbf{0} \Leftrightarrow \mathbf{A}^T (\mathbf{b} - \mathbf{Ax}) = \mathbf{0} \Leftrightarrow \mathbf{A}^T \mathbf{A} \mathbf{x} = \mathbf{A}^T \mathbf{b}. \quad \square$$

Lemma 7.2 *For any $m \times n$ matrix $\mathbf{A}$ the following are equivalent:*

(**a**) *The column vectors of $\mathbf{A}$ are linearly independent.*
(**b**) $\mathbf{A}^T \mathbf{A}$ *is invertible.*

Proof. Applying (2.28) to $\mathbf{A} := \mathbf{A}^T$, we obtain:

$$\mathrm{CS}(\mathbf{A}) \oplus \mathrm{N}(\mathbf{A}^T) = \mathbb{R}^m.$$

(a) $\Rightarrow$ (b) If a vector $\mathbf{x}$ is in the null space of an $n \times n$ matrix $\mathbf{A}^T \mathbf{A}$, then

$$\mathbf{A}\mathbf{x} \in \mathrm{N}(\mathbf{A}^T) \cap \mathrm{CS}(\mathbf{A}) \Rightarrow \mathbf{A}\mathbf{x} = \mathbf{0} \Rightarrow \mathbf{x} = \mathbf{0},$$

since the columns of $\mathbf{A}$ are linearly independent.

(b) $\Rightarrow$ (a) Suppose to the contrary that the columns of $\mathbf{A}$ are linearly dependent. Then there exists a nonzero vector $\mathbf{x} \in \mathbb{R}^n$ such that $\mathbf{A}\mathbf{x} = \mathbf{0}$. It follows that $\mathbf{A}^T \mathbf{A}\mathbf{x} = \mathbf{0}$, implying that $\mathbf{A}^T \mathbf{A}$ is not invertible. $\qquad\square$

Theorem 7.3 *If $\mathbf{A}$ is an $m \times n$ matrix with linearly independent column vectors, then for every $m \times 1$ matrix $\mathbf{b}$, the linear system $\mathbf{A}\mathbf{x} = \mathbf{b}$ has a unique least squares solution. This solution is given by*

$$\mathbf{x} = (\mathbf{A}^T \mathbf{A})^{-1} \mathbf{A}^T \mathbf{b}. \tag{7.3}$$

Moreover, if W is the column space of $\mathbf{A}$, then the orthogonal projection of $\mathbf{b}$ on W is

$$\mathrm{proj}_W \mathbf{b} = \mathbf{A}\mathbf{x} = \mathbf{A}(\mathbf{A}^T \mathbf{A})^{-1} \mathbf{A}^T \mathbf{b}. \tag{7.4}$$

Proof. By Lemma 7.2 the matrix $(\mathbf{A}^T \mathbf{A})^{-1}$ exists. Applying $(\mathbf{A}^T \mathbf{A})^{-1}$ to (7.1), we obtain (7.3). Substituting the formula for $\mathbf{x}$ obtained in (7.3) in the formula (7.2), we obtain (7.4). $\qquad\square$

Problem 7.2 A sales organization obtains the following data relating the number of salespersons to annual sales

Number of salespersons	5	6	7	8	9	10
Annual sales	2.3	3.2	4.1	5.0	6.1	7.2

Let x denote the number of salespersons and y denote the annual sales (in millions of dollars).

(a) Find the least square line relating x and y.
(b) Use the equation obtained in (a) to estimate the annual sales, when there are 14 salespersons.

Solution by Calculus: The Linear Regression Problem is to find a linear function $\mathbf{y} = \mathbf{a}\mathbf{x} + \mathbf{b}$ such that the expression

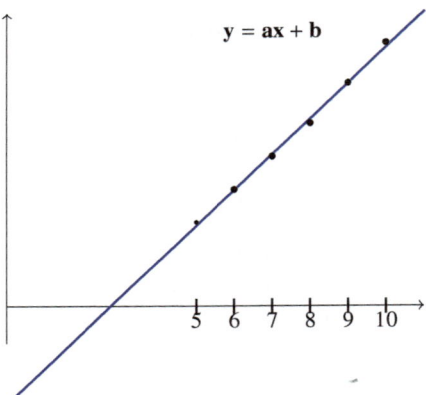

Fig. 7.1: The regression line

$$D = \sum_{i=1}^{n} (y(x_i) - y_i)^2$$

is as small as possible. Elementary Algebra shows that

$$D = a^2 \sum_{i=1}^{n} x_i^2 + 2ab \sum_{i=1}^{n} x_i + nb^2 - 2a \sum_{i=1}^{n} x_i y_i - 2b \sum_{i=1}^{n} y_i + \sum_{i=1}^{n} y_i^2.$$

By the gradient test the minimum is attained at the point with zero gradient:

$$\frac{\partial D}{\partial a} = 2a \sum_{i=1}^{n} x_i^2 + 2b \sum_{i=1}^{n} x_i - 2 \sum_{i=1}^{n} x_i y_i = 0$$

$$\frac{\partial D}{\partial b} = 2nb + 2a \sum_{i=1}^{n} x_i - 2 \sum_{i=1}^{n} y_i = 0$$

$$\Rightarrow \begin{cases} a \sum_{i=1}^{n} x_i^2 + b \sum_{i=1}^{n} x_i &= \sum_{i=1}^{n} x_i y_i \\ a \sum_{i=1}^{n} x_i + bn &= \sum_{i=1}^{n} y_i. \end{cases}$$

Applying statistical notation, we can write

$$\frac{z_1 + z_2 + \cdots + z_n}{n} = E(z).$$

Lemma 7.3 *The determinant of the linear system in two unknowns a and b*

$$\begin{cases} aE(x^2) + bE(x) &= E(xy) \\ aE(x) + b &= E(y) \end{cases}$$

is positive:

$$\Delta = \begin{vmatrix} E(x^2) & E(x) \\ E(x) & 1 \end{vmatrix} = E(x^2) - (E(x))^2 > 0.$$

Proof. By the Cauchy-Schwarz inequality

$$\left| \sum_{i=1}^{n} x_i \right| = \left| \sum_{i=1}^{n} x_i \cdot 1 \right| < \left(\sum_{i=1}^{n} x_i^2 \right)^{1/2} \left(\sum_{i=1}^{n} 1 \right)^{1/2} \Rightarrow$$

$$\left| \frac{1}{n} \sum_{i=1}^{n} x_i \right| = \left| \sum_{i=1}^{n} x_i \cdot 1 \right| < \left(\frac{1}{n} \sum_{i=1}^{n} x_i^2 \right)^{1/2} \Rightarrow E(x^2) - (E(x))^2 > 0.$$

Notice the equality in the Cauchy-Schwarz inequality may occur only if $\mathbf{x} = (x_1, x_2, \ldots, x_n)$ is proportional to $(1, 1, \ldots, 1)$, which is impossible since x_i is an increasing sequence.   □

By Cramer's Rules, see Theorem 3.33,

$$a = \frac{\Delta_a}{\Delta} = \frac{E(xy) - E(x)E(y)}{E(x^2) - (E(x))^2}, \quad b = \frac{\Delta_b}{\Delta} = \frac{E(x^2)E(y) - E(x)E(xy)}{E(x^2) - (E(x))^2}.$$

In Problem 7.2, we have $n = 6$ and

$$E(x) = \frac{5 + 6 + \cdots + 10}{6} = \frac{45}{6}, \quad E(x^2) = \frac{5^2 + 6^2 + \cdots + 10^2}{6} = \frac{355}{6}$$

$$E(y) = \frac{2.3 + 3.2 + 4.1 + 5 + 6.1 + 7.2}{6} = \frac{27.9}{6} = 4.65$$

$$E(xy) = \frac{5 \cdot 2.3 + 6 \cdot 3.2 + 7 \cdot 4.1 + 8 \cdot 5 + 9 \cdot 6.1 + 10 \cdot 7.2}{6} = \frac{226.3}{6}$$

$$a = \frac{E(xy) - E(x)E(y)}{E(x^2) - (E(x))^2} = 0.974286$$

$$\Rightarrow \boxed{y = 0.974x - 2.65714}. \quad □$$

$$b = \frac{E(x^2)E(y) - E(x)E(xy)}{E(x^2) - (E(x))^2} = -2.65714$$

Solution by Gram-Schmidt orthogonalization process: The vector

$$\mathbf{y}_0 = \begin{pmatrix} 2.3 & 3.2 & 4.1 & 5 & 6.1 & 7.2 \end{pmatrix}$$

is a given vector in $\mathbb{R}^6$. We consider the subspace

$$W = \{ \mathbf{z} : z_k = ax_k + b, \ k = 1, \ldots, 6 \text{ for some } a, b \in \mathbb{R} \},$$

where $x_k = k + 4$ are the elements of the set $E = \{5, 6, 7, 8, 9, 10\}$. It is clear that $W = \text{Lin}\{1|E, x|E\}$. Then the **least squares line** is given by $\text{proj}_W \mathbf{y}_0$.

The vectors

$$\mathbf{v}_1 = \begin{pmatrix} 1 & 1 & 1 & 1 & 1 & 1 \end{pmatrix}, \quad \mathbf{v}_1 = \begin{pmatrix} 5 & 6 & 7 & 8 & 9 & 10 \end{pmatrix}$$

can be orthogonalised: $\mathbf{u}_1 = \mathbf{v}_1$,

$$\mathbf{u}_2 = \mathbf{v}_2 - \frac{(\mathbf{v}_2, \mathbf{v}_1)}{(\mathbf{v}_1, \mathbf{v}_1)} \mathbf{v}_1.$$

We have

$$(\mathbf{v}_1, \mathbf{v}_1) = 6, \quad (\mathbf{v}_2, \mathbf{v}_1) = 5 + 6 + 7 + 8 + 9 + 10 = 45$$

$$\mathbf{u}_2 = \mathbf{v}_2 - \frac{2}{15}\mathbf{v}_1 = (-2.5, -1.5, -0.5, 0.5, 1.5, 2.5).$$

$$(\mathbf{u}_2, \mathbf{u}_2) = (25 + 9 + 1 + 1 + 9 + 25)/4 = 17.5$$
$$(\mathbf{y}_0, \mathbf{u}_1) = 2.3 + 3.2 + 4.1 + 5 + 6.1 + 7.2 = 27.9$$
$$(\mathbf{y}_0, \mathbf{u}_2) = (-5 \cdot 2.3 - 3 \cdot 3.2 - 4.1 + 5 + 3 \cdot 6.1 + 5 \cdot 7.2)/2 = 17.05$$

Taking into account that $\mathbf{u}_2 = (x - 7.5)|E$, we obtain that

$$\text{proj}_W \mathbf{y}_0 = \frac{(\mathbf{y}_0, \mathbf{u}_1)}{(\mathbf{u}_1, \mathbf{u}_1)}\mathbf{u}_1 + \frac{(\mathbf{y}_0, \mathbf{u}_2)}{(\mathbf{u}_2, \mathbf{u}_2)}\mathbf{u}_2 = \frac{27.9}{6}\mathbf{u}_1 + \frac{34.1}{35}\mathbf{u}_2 =$$

$$\frac{34.1}{35}(x - 7.5) + \frac{27.9}{6} \Rightarrow \boxed{\mathbf{y(x) = 0.974x - 2.657}}. \quad \square$$

Solution by Theorem 7.3. We consider the matrix with two linearly independent rows

$$\mathbf{A}^T = \begin{pmatrix} 1 & 1 & 1 & 1 & 1 & 1 \\ 5 & 6 & 7 & 8 & 9 & 10 \end{pmatrix}.$$

Then

$$(\mathbf{A}^T\mathbf{A})^{-1}\mathbf{A}^T\mathbf{y}_0 = \begin{pmatrix} -2.65714 \\ 0.974286 \end{pmatrix} \Rightarrow \boxed{\mathbf{y(x) = 0.974x - 2.657}}. \quad \square$$

A general problem of approximation data plots with polynomial lemniscates can be stated as follows.

Problem 7.3 Suppose that we are given n data points

$$(x_1, y_1), \ (x_2, y_2), \ \ldots, (x_n, y_n),$$

with $x_1, x_2 < \cdots < x_n$. Find the polynomial

$$y(x) = a_m x^m + a_{m-1} x^{m-1} + \cdots + a_1 x + a_0, \ m \leq n - 1,$$

that fits the data given (in the sense of the least quadratic deviation) the best possible way.

The system of equations $y(x_i) = y_i$, $i = 1, \ldots, n$ can be solved in the set of polynomials y of degree m if and only if the system of linear equations

$$\mathbf{Ax} = \begin{pmatrix} x_1^m & x_1^{m-1} & \cdots & x_1 & 1 \\ x_2^m & x_2^{m-1} & \cdots & x_2 & 1 \\ \vdots & \vdots & \ddots & \vdots & \vdots \\ x_n^m & x_n^{m-1} & \cdots & x_n & 1 \end{pmatrix} \begin{pmatrix} a_m \\ a_{m-1} \\ \vdots \\ a_0 \end{pmatrix} = \begin{pmatrix} y_1 \\ y_2 \\ \vdots \\ y_n \end{pmatrix} = \mathbf{b}$$

has a solution. Notice that the columns of $\mathbf{A}$ are linearly independent. Otherwise we could have a polynomial of degree $m < n$ vanishing at n points. It follows that the system $\mathbf{Ax} = \mathbf{b}$ has a unique least squares solution.

Problem 7.4 The following data show atmospheric pollutants y_i at half-hour intervals t_i

t_i	1	1.5	2	2.5	3	3.5	4	4.5	5
y_i	−0.15	0.24	0.68	1.04	1.21	1.15	0.86	0.41	−0.08

Find a polynomial least square approximation (the least square lemniscate).

$$y = -0.327446x^2 + 2.00668x - 1.93171$$

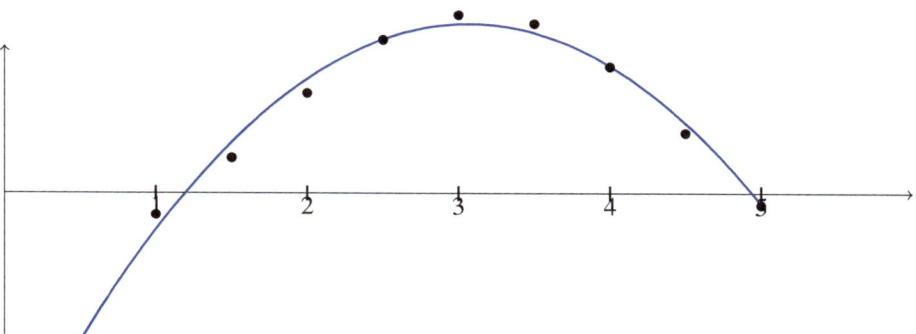

Solution: A plot of these data points suggests that a quadratic polynomial $y = a_2 t^2 + a_1 t + a_0$. may produce a good model for these data. Following the theory presented, we have: $\mathbf{Ax} = \mathbf{b}$, where

$$A = \begin{pmatrix} 1 & 1 & 1 \\ 2.25 & 1.5 & 1 \\ 4 & 2 & 1 \\ 6.25 & 2.5 & 1 \\ 9 & 3 & 1 \\ 12.25 & 3.5 & 1 \\ 16 & 4 & 1 \\ 20.25 & 4.5 & 1 \\ 25 & 5 & 1 \end{pmatrix}, \quad x = \begin{pmatrix} a_2 \\ a_1 \\ a_0 \end{pmatrix}, \quad b = \begin{pmatrix} -0.15 \\ 0.24 \\ 0.68 \\ 1.04 \\ 1.21 \\ 1.15 \\ 0.86 \\ 0.41 \\ -0.08 \end{pmatrix} \qquad (7.5)$$

The normal equation (system): $\boxed{A^T A x = A^T b}$. The matrix product $A^T A$ is

$$\begin{pmatrix} 1 & 2.25 & 4 & 6.25 & 9 & 12.25 & 16 & 20.25 & 25 \\ 1 & 1.5 & 2 & 2.5 & 3 & 3.5 & 4 & 4.5 & 5 \\ 1 & 1 & 1 & 1 & 1 & 1 & 1 & 1 & 1 \end{pmatrix} \begin{pmatrix} 1 & 1 & 1 \\ 2.25 & 1.5 & 1 \\ 4 & 2 & 1 \\ 6.25 & 2.5 & 1 \\ 9 & 3 & 1 \\ 12.25 & 3.5 & 1 \\ 16 & 4 & 1 \\ 20.25 & 4.5 & 1 \\ 25 & 5 & 1 \end{pmatrix}$$

$$A^T A = \begin{pmatrix} 1583.25 & 378 & 96 \\ 378 & 96 & 27 \\ 96 & 27 & 9 \end{pmatrix}, \quad Ab = \begin{pmatrix} 54.65 \\ 16.71 \\ 5.36 \end{pmatrix} \Rightarrow$$

$$\begin{pmatrix} 1583.25 & 378 & 96 \\ 378 & 96 & 27 \\ 96 & 27 & 9 \end{pmatrix} \begin{pmatrix} a_2 \\ a_1 \\ a_0 \end{pmatrix} = \begin{pmatrix} 54.65 \\ 16.71 \\ 5.36 \end{pmatrix} \Rightarrow$$

$$\begin{pmatrix} a_2 \\ a_1 \\ a_0 \end{pmatrix} = \begin{pmatrix} -0.327 \\ 2.0067 \\ -1.9317 \end{pmatrix} \Rightarrow \boxed{y = -0.327446x^2 + 2.00668x - 1.93171}.$$

Notice that we obtained exactly the same result as in Problem 7.1 by Chebyshev's method.

7.3 Weak Generalized Inverse

> **Definition 7.3** An $n \times m$ matrix $\mathbf{A}^g$ is called a **weak generalised inverse** (WGI) for an $m \times n$ matrix $\mathbf{A}$ if
>
> $$\mathbf{A}\mathbf{A}^g\mathbf{A} = \mathbf{A}. \qquad (7.6)$$

The extended matrix $(\mathbf{A}|\mathbf{I}_m)$ is row equivalent to the matrix $(\mathrm{rref}(\mathbf{A})|\mathbf{B})$. Let J be the leading list $1 \leq j_1 < \cdots < j_r \leq n$ and $\mathbf{E}_J$ be the leading list matrix for $\mathbf{A}$, see (3.21). We recall that $r = \mathrm{rank}(\mathbf{A})$. By Theorem 3.17, the vector $\mathbf{p} = \mathbf{E}_J\mathbf{B}\mathbf{y}$ is the Gauss particular solution to the system $\mathbf{A}\mathbf{x} = \mathbf{y}$ for every $\mathbf{y} \in R(\mathbf{A})$. If $\mathbf{y} = \mathbf{A}\mathbf{x}$ then $\mathbf{p} = \mathbf{E}_J\mathbf{B}\mathbf{A}\mathbf{x}$ satisfies $\mathbf{A}\mathbf{x} = \mathbf{A}\mathbf{p} = \mathbf{A}\mathbf{E}_J\mathbf{B}\mathbf{A}\mathbf{x}$ for every $\mathbf{x}$ in $\mathbf{R}^n$, implying that

$$\mathbf{A} = \mathbf{A}\,(\mathbf{E}_J\mathbf{B})\,\mathbf{A}.$$

Hence the matrix $\mathbf{E}_J\mathbf{B}$ is a weak generalized inverse for the matrix $\mathbf{A}$. It is called the **Gauss generalised inverse**.

Lemma 7.4 *If $\mathbf{A}^g$ is a WGI of a matrix $\mathbf{A}$ and the system $\mathbf{A}\mathbf{x} = \mathbf{b}$ has a solution $\mathbf{x}$, then $\mathbf{p} = \mathbf{A}^g\mathbf{b}$ is a particular solution for the system $\mathbf{A}\mathbf{x} = \mathbf{b}$.*

Proof.
$$\mathbf{A}\mathbf{p} = \mathbf{A}\mathbf{A}^g\mathbf{b} = \mathbf{A}\mathbf{A}^g\mathbf{A}\mathbf{x} = \mathbf{A}\mathbf{x} = \mathbf{b}. \qquad \square$$

> **Theorem 7.4** *Let $\mathbf{A}$ and $\mathbf{G}$ be matrices of order $m \times n$ and $n \times m$ respectively. Then the following conditions are equivalent:*
>
> **(i)** *The matrix $\mathbf{G}$ is a weak generalized inverse for $\mathbf{A}$.*
> **(ii)** *For any $\mathbf{y} \in R(\mathbf{A})$ the vector $\mathbf{x} = \mathbf{G}\mathbf{y}$ is a particular solution to the system $\mathbf{A}\mathbf{x} = \mathbf{y}$.*

> *Proof.* **(i)** $\Rightarrow$ **(ii)** by Lemma 7.4.
>   **(ii)** $\Rightarrow$ **(i)** If $\mathbf{A}\mathbf{G}\mathbf{y} = \mathbf{y}$ for any $\mathbf{y} \in R(\mathbf{A})$, we obtain that $\mathbf{A}\mathbf{G}\mathbf{A}\mathbf{z} = \mathbf{A}\mathbf{z}$ for every $\mathbf{z}$ implying that $\mathbf{A}\mathbf{G}\mathbf{A} = \mathbf{A}$. $\square$

The following theorem describes all WGI matrices for any matrix $\mathbf{A}$.

Theorem 7.5 *Let $\mathbf{A}$ be an $m \times n$ matrix of rank $r = \operatorname{rank}(\mathbf{A})$. Then all weak generalized inverses for $\mathbf{A}$ are described by the formula*

$$\mathbf{A}^g = \mathbf{Q}^{-1} \left(\begin{array}{c|c} \mathbf{I}_r & \mathbf{U} \\ \hline \mathbf{V} & \mathbf{W} \end{array} \right) \mathbf{P}^{-1}, \tag{7.7}$$

where $\mathbf{P}$ and $\mathbf{Q}$ are invertible matrices in the factorization (3.29), $\mathbf{U}$ is an arbitrary $r \times (m - r)$ matrix, $\mathbf{V}$ arbitrary $(n - r) \times r$ matrix, and $\mathbf{W}$ arbitrary $(n - r) \times (m - r)$ matrix.

Proof. By Theorem 3.19, for every $m \times n$ matrix $\mathbf{A}$ there are an $m \times m$ invertible matrix $\mathbf{P}$ and an $n \times n$ invertible matrix $\mathbf{Q}$ such that

$$\mathbf{A} = \mathbf{P} \begin{pmatrix} \mathbf{I}_r & \mathbf{0} \\ \mathbf{0} & \mathbf{0} \end{pmatrix} \mathbf{Q}, \tag{7.8}$$

where $r = \operatorname{rank}(\mathbf{A})$. If $\mathbf{A}^g$ is defined by the formula:

$$\mathbf{A}^g = \mathbf{Q}^{-1} \begin{pmatrix} \mathbf{I}_r & \mathbf{U} \\ \mathbf{V} & \mathbf{W} \end{pmatrix} \mathbf{P}^{-1}, \tag{7.9}$$

where $\mathbf{U}$, $\mathbf{V}$, and $\mathbf{W}$ any matrices of the corresponding dimensions, then

$$\mathbf{A}\mathbf{A}^g\mathbf{A} = \mathbf{P} \begin{pmatrix} \mathbf{I}_r & \mathbf{0} \\ \mathbf{0} & \mathbf{0} \end{pmatrix} \mathbf{Q}\mathbf{Q}^{-1} \begin{pmatrix} \mathbf{I}_r & \mathbf{U} \\ \mathbf{V} & \mathbf{W} \end{pmatrix} \mathbf{P}^{-1}\mathbf{P} \begin{pmatrix} \mathbf{I}_r & \mathbf{0} \\ \mathbf{0} & \mathbf{0} \end{pmatrix} \mathbf{Q} =$$

$$\mathbf{P} \begin{pmatrix} \mathbf{I}_r & \mathbf{0} \\ \mathbf{0} & \mathbf{0} \end{pmatrix} \begin{pmatrix} \mathbf{I}_r & \mathbf{U} \\ \mathbf{V} & \mathbf{W} \end{pmatrix} \begin{pmatrix} \mathbf{I}_r & \mathbf{0} \\ \mathbf{0} & \mathbf{0} \end{pmatrix} \mathbf{Q} = \mathbf{P} \begin{pmatrix} \mathbf{I}_r & \mathbf{U} \\ \mathbf{0} & \mathbf{0} \end{pmatrix} \begin{pmatrix} \mathbf{I}_r & \mathbf{0} \\ \mathbf{0} & \mathbf{0} \end{pmatrix} \mathbf{Q} = \mathbf{P} \begin{pmatrix} \mathbf{I}_r & \mathbf{0} \\ \mathbf{0} & \mathbf{0} \end{pmatrix} \mathbf{Q} = \mathbf{A}.$$

If $\mathbf{A}\mathbf{A}^g\mathbf{A} = \mathbf{A}$, then we substitute the expression (7.8) for $\mathbf{A}$ in this formula and obtain a matrix equality:

$$\begin{pmatrix} \mathbf{I}_r & \mathbf{0} \\ \mathbf{0} & \mathbf{0} \end{pmatrix} \mathbf{Q}\mathbf{A}^g\mathbf{P} \begin{pmatrix} \mathbf{I}_r & \mathbf{0} \\ \mathbf{0} & \mathbf{0} \end{pmatrix} = \begin{pmatrix} \mathbf{I}_r & \mathbf{0} \\ \mathbf{0} & \mathbf{0} \end{pmatrix}. \tag{7.10}$$

If we represent $\mathbf{Q}\mathbf{A}^g\mathbf{P}$ as a block matrix of the form

$$\mathbf{Q}\mathbf{A}^g\mathbf{P} = \begin{pmatrix} \mathbf{X} & \mathbf{U} \\ \mathbf{V} & \mathbf{W} \end{pmatrix},$$

then (7.10) implies that $\mathbf{X} = \mathbf{I}_r$, which results in (7.9). $\qquad \square$

Definition 7.4 An $n \times m$ matrix $\mathbf{L}$ is called the **left inverse matrix** for an $m \times n$ matrix $\mathbf{A}$ if $\mathbf{LA} = \mathbf{I}_n$.

Theorem 7.6 *An $m \times n$ matrix $\mathbf{A}$ has a left inverse matrix $\mathbf{L}$ if and only if* rank$(\mathbf{A}) = n$.

Proof. If $\mathbf{LA} = \mathbf{I}_n$ and $\mathbf{Ax} = \mathbf{0}_m$, then $\mathbf{x} = \mathbf{I}_n\mathbf{x} = \mathbf{LAx} = \mathbf{L0}_m = \mathbf{0}_n$. implying that $N(\mathbf{A}) = \{\mathbf{0}_n\}$. By the Rank-Nullity Theorem rank$(\mathbf{A}) = n$.

If rank$(\mathbf{A}) = n$ then $N(\mathbf{A}) = \{\mathbf{0}_n\}$ by the Rank-Nullity Theorem. By Theorem 1.26, $m \geq n$. It follows that $\mathbf{E}_J = \begin{pmatrix} \mathbf{I}_n & \mathbf{0}_{n \times (m-n)} \end{pmatrix}$. Given $\mathbf{x}$ in $\mathbb{R}^n$ the vector $\mathbf{p} = \mathbf{E}_J\mathbf{BAx}$ is the Gauss particular solution to the equation $\mathbf{Ap} = \mathbf{Ax}$. Since $N(\mathbf{A}) = \{\mathbf{0}_n\}$, we conclude that $\mathbf{p} = \mathbf{x}$. It follows that $\mathbf{E}_J\mathbf{B}$ is a left inverse matrix for $\mathbf{A}$. $\qquad\square$

Theorem 7.7 *If an $m \times n$ matrix $\mathbf{A}$ has a left inverse $\mathbf{L}$ then every weak generalized inverse $\mathbf{A}^g$ is its left inverse as well.*

Proof. Multiplying the identity (7.6) by a given left inverse matrix $\mathbf{L}$ we obtain that

$$\mathbf{I}_n = \mathbf{LA} = \mathbf{LAA}^g\mathbf{A} = (\mathbf{LA})\,\mathbf{A}^g\mathbf{A} = \mathbf{I}_n\mathbf{A}^g\mathbf{A} = \mathbf{A}^g\mathbf{A},$$

which shows that $\mathbf{A}^g$ is a left inverse matrix for $\mathbf{A}$. $\qquad\square$

Corollary 7.1 *If an $m \times n$ matrix $\mathbf{A}$ has a left inverse then the Gauss generalized inverse is its left inverse as well.*

Corollary 7.2 *If an $m \times n$ matrix $\mathbf{A}$ has a left inverse then the matrix*

$$\left(\mathbf{A}^T\mathbf{A}\right)^{-1}\mathbf{A}^T \tag{7.11}$$

is the left inverse for $\mathbf{A}$.

Proof. If $\mathbf{A}$ is left invertible then $N(\mathbf{A}) = \{\mathbf{0}_n\}$. It follows that the $n \times n$ matrix $\mathbf{A}^T\mathbf{A}$ is non-singular. Hence, $\left(\mathbf{A}^T\mathbf{A}\right)^{-1}$ exists. Finally,

$$\left(\mathbf{A}^T\mathbf{A}\right)^{-1}\mathbf{A}^T\mathbf{A} = \left(\mathbf{A}^T\mathbf{A}\right)^{-1}\left(\mathbf{A}^T\mathbf{A}\right) = \mathbf{I}_n. \qquad\square$$

Problem 7.5 Find the Gauss generalized inverse for

$$\mathbf{A} = \begin{pmatrix} 1 & -1 & 1 \\ 0 & 1 & 0 \\ 1 & -1 & 1 \\ 1 & 1 & 0 \end{pmatrix}.$$

Find the left inverse for $\mathbf{A}$ by (7.11). Show that the Gauss inverse for $\mathbf{A}$ is its left inverse.

Solution: We have

$$\text{rref}(\mathbf{A}) = \begin{pmatrix} 1 & 0 & 0 \\ 0 & 1 & 0 \\ 0 & 0 & 1 \\ 0 & 0 & 0 \end{pmatrix} \Rightarrow J = \{1,2,3\}, \quad \mathbf{E}_J = \begin{pmatrix} 1 & 0 & 0 & 0 \\ 0 & 1 & 0 & 0 \\ 0 & 0 & 1 & 0 \end{pmatrix}.$$

Elementary calculations with the extended matrix for $\mathbf{A}$ shows that

$$\text{rref}\,(\mathbf{A}|\mathbf{I}_4) = \begin{pmatrix} 1 & 0 & 0 & 0 & -1 & 0 & 1 \\ 0 & 1 & 0 & 0 & 1 & 0 & 0 \\ 0 & 0 & 1 & 0 & 2 & 1 & -1 \\ 0 & 0 & 0 & 1 & 0 & -1 & 0 \end{pmatrix}.$$

It follows that the matrix

$$\mathbf{E}_J\mathbf{B} = \begin{pmatrix} 1 & 0 & 0 & 0 \\ 0 & 1 & 0 & 0 \\ 0 & 0 & 1 & 0 \end{pmatrix} \begin{pmatrix} 0 & -1 & 0 & 1 \\ 0 & 1 & 0 & 0 \\ 0 & 2 & 1 & -1 \\ 1 & 0 & -1 & 0 \end{pmatrix} = \begin{pmatrix} 0 & -1 & 0 & 1 \\ 0 & 1 & 0 & 0 \\ 0 & 2 & 1 & -1 \end{pmatrix}$$

is the Gauss generalized inverse for $\mathbf{A}$. The matrix

$$\left(\mathbf{A}^T\mathbf{A}\right)^{-1}\mathbf{A}^T = \begin{pmatrix} 0 & -1 & 0 & 1 \\ 0 & 1 & 0 & 0 \\ 1/2 & 2 & 1/2 & -1 \end{pmatrix}$$

is another generalised inverse for $\mathbf{A}$ which is also a left inverse. Finally,

$$\mathbf{E}_J\mathbf{B}\mathbf{A} = \begin{pmatrix} 0 & -1 & 0 & 1 \\ 0 & 1 & 0 & 0 \\ 0 & 2 & 1 & -1 \end{pmatrix} \begin{pmatrix} 1 & -1 & 1 \\ 0 & 1 & 0 \\ 1 & -1 & 1 \\ 1 & 1 & 0 \end{pmatrix} = \mathbf{I}_3,$$

implying that the Gauss generalized inverse is the left inverse for $\mathbf{A}$. $\qquad\square$

If $\mathbf{A}^g = \left(\mathbf{A}^T\mathbf{A}\right)^{-1}\mathbf{A}^T$, then by Theorem 6.8

$$\mathbf{A}^g\mathbf{A} = \mathbf{I}_n, \quad \mathbf{A}\mathbf{A}^g = \mathbf{A}\left(\mathbf{A}^T\mathbf{A}\right)^{-1}\mathbf{A}^T = \mathbf{P}_{R(\mathbf{A})}.$$

Definition 7.5 An $n \times m$ matrix $\mathbf{R}$ is called the **right inverse matrix** for an $m \times n$ matrix $\mathbf{A}$ if $\mathbf{AR} = \mathbf{I}_m$.

Theorem 7.8 *An $m \times n$ matrix $\mathbf{A}$ has a right inverse matrix $\mathbf{R}$ if and only if* rank($\mathbf{A}$) = m.

Proof. By Theorem 1.5,

$$\mathbf{AR} = \mathbf{I}_m \Leftrightarrow \mathbf{R}^T \mathbf{A}^T = \mathbf{I}_m.$$

It follows that $\mathbf{R}$ is a right inverse for $\mathbf{A}$ if and only if $\mathbf{R}^T$ is a left inverse for $\mathbf{A}^T$. The result follows by Theorem 7.6.   □

Corollary 7.3 *If an $m \times n$ matrix $\mathbf{A}$ has a right inverse then the matrix*

$$\mathbf{A}^T \left(\mathbf{AA}^T \right)^{-1}$$

is the right inverse for $\mathbf{A}$.

Theorem 7.9 *If $\mathbf{A}^g$ is a WGI of an $m \times n$ matrix $\mathbf{A}$, then*

1. *$\mathbf{AA}^g$ is a projection of $\mathbb{R}^m$ onto $R(\mathbf{A})$.*
2. *$\mathbf{A}^g \mathbf{A}$ projects $\mathbb{R}^n$ parallel to $N(\mathbf{A})$.*
3. *$\mathbf{I} - \mathbf{A}^g \mathbf{A}$ is a projection onto $N(\mathbf{A})$.*

Proof. **1.** Let $\mathbf{P} = \mathbf{AA}^g$. Then $\mathbf{P}^2 = \mathbf{AA}^g \mathbf{AA}^g = \mathbf{AA}^g = \mathbf{P}$, implying that $\mathbf{P}$ is a projection. Next,

$$R(\mathbf{A}) = R(\mathbf{AA}^g \mathbf{A}) \subset R(\mathbf{AA}^g) = R(\mathbf{P}) \subset R(\mathbf{A}),$$

which implies that $\mathbf{P} = \mathbf{AA}^g$ is a projection onto $R(\mathbf{A})$.
     2. It is clear that

$$N(\mathbf{A}) = N(\mathbf{AA}^g \mathbf{A}) \supset N(\mathbf{A}^g \mathbf{A}) \supset N(\mathbf{A}) \Rightarrow N(\mathbf{A}) = N(\mathbf{A}^g \mathbf{A}).$$

Finally, $\mathbf{AA}^g \mathbf{AA}^g = (\mathbf{AA}^g \mathbf{A})\mathbf{A}^g = \mathbf{AA}^g$.

3. We have already proved in **2** that $A^g A$ is a projection. Then $(I - A^g A)^2 = 1 - A^g A - A^g A + (A^g A)^2 = I - A^g A$ is a projection as stated. Since $A(I - A^g A) = 0$, we conclude that $Q = I - A^g A$ is a projection onto a subspace of $N(A) = N(A^g A)$. In fact, this projection is onto. Indeed, if $x \in N(A^g A)$, then

$$A^g A x = 0 \Rightarrow x = x - A^g A x = Qx.$$

Theorem 7.10 *For any $m \times n$ matrix A, the matrix equation $Ax = b$ is consistent if and only if*

$$AA^g b = b.$$

Further, when it is consistent, its solutions are given by the formula

$$x = A^g b + (I - A^g A) w,$$

where w is any vector in $\mathbb{R}^n$.

Proof. The matrix equation $Ax = b$ is consistent if and only if $b \in R(A) = R(AA^g)$, see Theorem 6.3. Since $P = AA^g$ is a projection onto $R(A)$ by Theorem 6.3, the inclusion $b \in R(A)$ is equivalent to $AA^g b = b$.

We have

$$Ax = AA^g b + A (I - A^g A) w = b + (A - AA^g A) w = b.$$

By 3 of Theorem 6.3 every $z \in N(A)$ can be represented as $Qw = w - A^g A$. Since any solution is a sum of a particular solution $A^g b$ and the zero solution $z \in N(A)$, the theorem is proved. $\qquad\square$

Problem 7.6 Find a WGI for the matrix

$$A = \begin{pmatrix} 1 & -1 & 2 \\ 0 & 2 & -2 \\ 1 & 1 & 0 \end{pmatrix}.$$

Solution: We solve the system $Ax = b$ using the method of the extended matrix: $Ax = I_3 b$. We have

$$\text{rref}\,(\mathbf{A}|\mathbf{I}_3) = \begin{pmatrix} 1 & 0 & 1 & 0 & -1/2 & 1 \\ 0 & 1 & -1 & 0 & 1/2 & 0 \\ 0 & 0 & 0 & 1 & 1 & -1 \end{pmatrix} = (\text{rref}(\mathbf{A})|\mathbf{B}) \Rightarrow \mathbf{A}\mathbf{x} = \mathbf{b} \Leftrightarrow$$

$$\begin{pmatrix} 1 & 0 & 1 \\ 0 & 1 & -1 \\ 0 & 0 & 0 \end{pmatrix}\begin{pmatrix} x_1 \\ x_2 \\ x_3 \end{pmatrix} = \begin{pmatrix} 0 & -1/2 & 1 \\ 0 & 1/2 & 0 \\ 1 & 1 & -1 \end{pmatrix}\begin{pmatrix} b_1 \\ b_2 \\ b_3 \end{pmatrix} = \begin{pmatrix} -b_2/2 + b_3 \\ b_2/2 \\ b_1 + b_2 - b_3 \end{pmatrix}.$$

By (3.21),

$$\mathbf{E}_J\mathbf{B} = \begin{pmatrix} 1 & 0 & 0 \\ 0 & 1 & 0 \\ 0 & 0 & 0 \end{pmatrix}\begin{pmatrix} 0 & -1/2 & 1 \\ 0 & 1/2 & 0 \\ 1 & 1 & -1 \end{pmatrix} = \begin{pmatrix} 0 & -1/2 & 1 \\ 0 & 1/2 & 0 \\ 0 & 0 & 0 \end{pmatrix}.$$

Then $\mathbf{A}(\mathbf{E}_J\mathbf{B})\mathbf{A} = \mathbf{A}$, implying that $\mathbf{E}_J\mathbf{B}$ is the Gauss generalized inverse.

By Kronecker-Capelli Theorem, see Theorem 1.29, the system $\mathbf{A}\mathbf{x} = \mathbf{b}$ is consistent if and only if $b_3 = b_1 + b_2$. The non-leading variable for this system is $x_3 = t$. Then observing that x_1 and x_2 are the only leading variables, we can represent the general solution to $\mathbf{A}\mathbf{x} = \mathbf{b}$ as follows ($x_3 = t$):

$$\mathbf{x} = \begin{pmatrix} x_1 \\ x_2 \\ x_3 \end{pmatrix} = \begin{pmatrix} -b_2/2 + b_3 - t \\ b_2/2 + t \\ t \end{pmatrix} = \begin{pmatrix} b_1 + b_2/2 \\ b_2/2 \\ 0 \end{pmatrix} + t\begin{pmatrix} -1 \\ 1 \\ 1 \end{pmatrix} \Rightarrow \begin{pmatrix} 1 & 1/2 & 0 \\ 0 & 1/2 & 0 \\ 0 & 0 & 0 \end{pmatrix}\begin{pmatrix} b_1 \\ b_2 \\ b_3 \end{pmatrix} = \mathbf{x}_p.$$

$$\mathbf{C} = \begin{pmatrix} 1 & 1/2 & 0 \\ 0 & 1/2 & 0 \\ 0 & 0 & 0 \end{pmatrix} \Rightarrow \mathbf{C}\mathbf{b} = \begin{pmatrix} 1 & 1/2 & 0 \\ 0 & 1/2 & 0 \\ 0 & 0 & 0 \end{pmatrix}\begin{pmatrix} b_1 \\ b_2 \\ b_3 \end{pmatrix} = \mathbf{x}_p$$

is a particular solution. It is easy to see that $\mathbf{C}$ is a WGI for the matrix $\mathbf{A}$ and $\mathbf{A}$ is a WGI for the matrix $\mathbf{C}$. The matrices

$$\mathbf{A}\mathbf{C} = \begin{pmatrix} 1 & 0 & 0 \\ 0 & 1 & 0 \\ 1 & 1 & 0 \end{pmatrix}, \quad \mathbf{C}\mathbf{A} = \begin{pmatrix} 1 & 0 & 1 \\ 0 & 1 & -1 \\ 0 & 0 & 0 \end{pmatrix}$$

are idempotent. Since both matrices are not symmetric, they represent skew projections.

Problem 7.7 Find all WGI for the matrix

$$\mathbf{A} = \begin{pmatrix} 1 & -1 & 2 \\ 0 & 2 & -2 \\ 1 & 1 & 0 \end{pmatrix}.$$

Solution: We write elementary row operations above $\sim$, and their inverses in the boxes below:

$$\mathbf{A} = \begin{pmatrix} 1 & -1 & 2 \\ 0 & 2 & -2 \\ 1 & 1 & 0 \end{pmatrix} \quad \begin{array}{c} r_3 := \underset{\sim}{r_3} - r_1 \\ \hline r_3 := r_3 + r_1 \end{array} \quad \begin{pmatrix} 1 & -1 & 2 \\ 0 & 2 & -2 \\ 0 & 2 & -2 \end{pmatrix} \quad \begin{array}{c} r_3 := \underset{\sim}{r_3} - r_2 \\ \hline r_3 := r_3 + r_2 \end{array}$$

$$\begin{pmatrix} 1 & -1 & 2 \\ 0 & 2 & -2 \\ 0 & 0 & 0 \end{pmatrix} \quad \begin{array}{c} r_2 := \underset{\sim}{r_2}/2 \\ \hline r_2 := 2r_2 \end{array} \quad \begin{pmatrix} 1 & -1 & 2 \\ 0 & 1 & -1 \\ 0 & 0 & 0 \end{pmatrix} \quad \begin{array}{c} r_1 := \underset{\sim}{r_1} + r_2 \\ \hline r_1 := r_1 - r_2 \end{array} \quad \begin{pmatrix} 1 & 0 & 1 \\ 0 & 1 & -1 \\ 0 & 0 & 0 \end{pmatrix} = \mathrm{rref}(\mathbf{A}).$$

Now we write the product of elementary matrices corresponding to the boxed operations from the left to the right:

$$\mathbf{A} = \underbrace{\begin{pmatrix} 1 & 0 & 0 \\ 0 & 1 & 0 \\ 1 & 0 & 1 \end{pmatrix}}_{r_3 := r_3 + r_1} \underbrace{\begin{pmatrix} 1 & 0 & 0 \\ 0 & 1 & 0 \\ 0 & 1 & 1 \end{pmatrix}}_{r_3 := r_3 + r_2} \underbrace{\begin{pmatrix} 1 & 0 & 0 \\ 0 & 2 & 0 \\ 0 & 0 & 1 \end{pmatrix}}_{r_2 := 2r_2} \underbrace{\begin{pmatrix} 1 & -1 & 0 \\ 0 & 1 & 0 \\ 0 & 0 & 1 \end{pmatrix}}_{r_1 := r_1 - r_2} \begin{pmatrix} 1 & 0 & 1 \\ 0 & 1 & -1 \\ 0 & 0 & 0 \end{pmatrix}$$

$$\Rightarrow \mathbf{A} = \mathbf{P} \cdot \mathrm{rref}(\mathbf{A}), \quad \mathbf{P} = \begin{pmatrix} 1 & -1 & 0 \\ 0 & 2 & 0 \\ 1 & 1 & 1 \end{pmatrix}.$$

We write elementary **column** operations above $\sim$, and their inverses in the boxes below:

$$\mathrm{rref}(\mathbf{A}) = \begin{pmatrix} 1 & 0 & 1 \\ 0 & 1 & -1 \\ 0 & 0 & 0 \end{pmatrix} \quad \begin{array}{c} c_3 := \underset{\sim}{c_3} + c_2 \\ \hline c_3 := c_3 - c_2 \end{array} \quad \begin{pmatrix} 1 & 0 & 1 \\ 0 & 1 & 0 \\ 0 & 0 & 0 \end{pmatrix} \quad \begin{array}{c} c_3 := \underset{\sim}{c_3} - c_1 \\ \hline c_3 := c_3 + c_1 \end{array}$$

$$\begin{pmatrix} 1 & 0 & 0 \\ 0 & 1 & 0 \\ 0 & 0 & 0 \end{pmatrix} \Rightarrow \mathrm{rref}(\mathbf{A}) = \begin{pmatrix} 1 & 0 & 0 \\ 0 & 1 & 0 \\ 0 & 0 & 0 \end{pmatrix} \underbrace{\begin{pmatrix} 1 & 0 & 0 \\ 0 & 1 & -1 \\ 0 & 0 & 1 \end{pmatrix}}_{c_3 := c_3 - c_2} \underbrace{\begin{pmatrix} 1 & 0 & 1 \\ 0 & 1 & 0 \\ 0 & 0 & 1 \end{pmatrix}}_{c_3 := c_3 + c_1} \Rightarrow$$

$$\mathrm{rref}(\mathbf{A}) = \begin{pmatrix} 1 & 0 & 0 \\ 0 & 1 & 0 \\ 0 & 0 & 0 \end{pmatrix} \mathbf{Q}, \quad \mathbf{Q} = \begin{pmatrix} 1 & 0 & 1 \\ 0 & 1 & -1 \\ 0 & 0 & 1 \end{pmatrix}.$$

$$\mathbf{A}^g = \mathbf{Q}^{-1} \begin{pmatrix} 1 & 0 & a \\ 0 & 1 & b \\ c & d & e \end{pmatrix} \mathbf{P}^{-1} = \begin{pmatrix} 1 & 0 & -1 \\ 0 & 1 & 1 \\ 0 & 0 & 1 \end{pmatrix} \begin{pmatrix} 1 & 0 & a \\ 0 & 1 & b \\ c & d & e \end{pmatrix} \begin{pmatrix} 1 & 1/2 & 0 \\ 0 & 1/2 & 0 \\ -1 & -1 & 1 \end{pmatrix} =$$

$$\frac{1}{2} \begin{pmatrix} 2 - 2a - 2c + 2e & 1 - 2a - c - d + 2e & 2a - 2e \\ -2b + 2c - 2e & 1 - 2b + c + d - 2e & 2b + 2e \\ 2c - 2e & c + d - 2e & 2e \end{pmatrix}$$

In particular, if $a = b = c = d = e = 0$, then

$$C = \begin{pmatrix} 1 & 1/2 & 0 \\ 0 & 1/2 & 0 \\ 0 & 0 & 0 \end{pmatrix}$$

is a weak generalized inverse for $\mathbf{A}$.

Solution by Theorem 3.9. By Theorem 3.9, we have $\mathbf{A} = \mathbf{P}\mathrm{rref}(\mathbf{A})$, where the first columns of $\mathbf{P}$ are the columns of the leading list J for the matrix $\mathbf{A}$. The leading list for the matrix $\mathbf{A}$ is $J = \{1,2\}$. It follows that

$$\mathbf{P} = \begin{pmatrix} 1 & -1 & a \\ 0 & 2 & b \\ 1 & 1 & c \end{pmatrix},$$

where a,b,c are arbitrary real numbers such that $\mathbf{P}^{-1}$ exists. To be consistent with the previous solution, we may put $a = b = 0, c = 1$. We have

$$\mathrm{rref}\left(\mathrm{rref}(\mathbf{A})^T\right) = \mathrm{rref}\begin{pmatrix} 1 & 0 & 0 \\ 0 & 1 & 0 \\ 1 & -1 & 0 \end{pmatrix} = \begin{pmatrix} 1 & 0 & 0 \\ 0 & 1 & 0 \\ 0 & 0 & 0 \end{pmatrix}.$$

By Theorem 3.9,

$$\mathrm{rref}(\mathbf{A})^T = \mathbf{R}\begin{pmatrix} 1 & 0 & 0 \\ 0 & 1 & 0 \\ 0 & 0 & 0 \end{pmatrix} = \begin{pmatrix} 1 & 0 & a \\ 0 & 1 & b \\ 1 & -1 & c \end{pmatrix}\begin{pmatrix} 1 & 0 & 0 \\ 0 & 1 & 0 \\ 0 & 0 & 0 \end{pmatrix},$$

where we again put $a = b = 0, c = 1$. Then $\det(\mathbf{R}) = 1$, implying that $\mathbf{R}$ is invertible, and passing to the transpose matrices we get

$$\mathrm{rref}(\mathbf{A}) = \begin{pmatrix} 1 & 0 & 0 \\ 0 & 1 & 0 \\ 0 & 0 & 0 \end{pmatrix}\begin{pmatrix} 1 & 0 & 1 \\ 0 & 1 & -1 \\ 0 & 0 & 1 \end{pmatrix} = \begin{pmatrix} 1 & 0 & 0 \\ 0 & 1 & 0 \\ 0 & 0 & 0 \end{pmatrix}\mathbf{Q}.$$

We got the required factorization with less efforts that in the first solution. □

One More Method to find a WGI Let $\mathrm{rank}(\mathbf{A}) = r$. Chose any $r \times r$ nonsingular submatrix of $\mathbf{A}$. For convenience let

$$\mathbf{A} = \begin{pmatrix} \mathbf{A}_{11} & \mathbf{A}_{12} \\ \mathbf{A}_{21} & \mathbf{A}_{22} \end{pmatrix},$$

where $\mathbf{A}_{11}$ is a nonsingular $r \times r$ matrix. All leading columns of $\mathbf{A}$ pass through $\mathbf{A}_{11}$. This means that the rest columns are linear combinations of the first r columns. In other words there exists a matrix $\mathbf{X}$ such that

$$\mathbf{A}_{12} = \mathbf{A}_{11}\mathbf{X}, \quad \mathbf{A}_{22} = \mathbf{A}_{21}\mathbf{X}.$$

Then

$$\mathbf{A}^g = \begin{pmatrix} \mathbf{A}_{11}^{-1} & \mathbf{0} \\ \mathbf{0} & \mathbf{0} \end{pmatrix}.$$

Indeed:

$$\mathbf{A}\mathbf{A}^g\mathbf{A} = \begin{pmatrix} \mathbf{A}_{11} & \mathbf{A}_{11}\mathbf{X} \\ \mathbf{A}_{21} & \mathbf{A}_{21}\mathbf{X} \end{pmatrix} \begin{pmatrix} \mathbf{A}_{11}^{-1} & \mathbf{0} \\ \mathbf{0} & \mathbf{0} \end{pmatrix} \begin{pmatrix} \mathbf{A}_{11} & \mathbf{A}_{11}\mathbf{X} \\ \mathbf{A}_{21} & \mathbf{A}_{21}\mathbf{X} \end{pmatrix} =$$

$$\begin{pmatrix} \mathbf{A}_{11} & \mathbf{A}_{11}\mathbf{X} \\ \mathbf{A}_{21} & \mathbf{A}_{21}\mathbf{X} \end{pmatrix} \begin{pmatrix} \mathbf{I}_r & \mathbf{X} \\ \mathbf{0} & \mathbf{0} \end{pmatrix} = \begin{pmatrix} \mathbf{A}_{11} & \mathbf{A}_{11}\mathbf{X} \\ \mathbf{A}_{21} & \mathbf{A}_{21}\mathbf{X} \end{pmatrix} = \mathbf{A}.$$

For

$$\mathbf{A} = \begin{pmatrix} 1 & -1 & 2 \\ 0 & 2 & -2 \\ 1 & 1 & 0 \end{pmatrix} \quad \text{we have} \quad \mathbf{A}_{11} = \begin{pmatrix} 1 & -1 \\ 0 & 2 \end{pmatrix}.$$

Hence,

$$\mathbf{A}^g = \begin{pmatrix} \mathbf{A}_{11}^{-1} & \mathbf{0} \\ \mathbf{0} & \mathbf{0} \end{pmatrix} = \begin{pmatrix} 1 & 1/2 & 0 \\ 0 & 1/2 & 0 \\ 0 & 0 & 0 \end{pmatrix}.$$

Problems

Prob. 205 — Show that all WGI for $\mathbf{A}$ are given by the formula

$$\mathbf{A}^g + \mathbf{B}, \quad \text{where} \quad \mathbf{A}\mathbf{B}\mathbf{A} = \mathbf{0}.$$

Prob. 206 — In some cases it is convenient to represent matrices in a block form:

$$\mathbf{A} = \begin{pmatrix} 1 & 2 & 5 \\ 3 & 4 & 6 \\ 7 & 8 & 9 \end{pmatrix} = \left(\begin{array}{cc|c} 1 & 2 & 5 \\ 3 & 4 & 6 \\ \hline 7 & 8 & 9 \end{array} \right) = \begin{pmatrix} \mathbf{A}_{11} & \mathbf{A}_{12} \\ \mathbf{A}_{21} & \mathbf{A}_{22} \end{pmatrix}.$$

Check the following calculations:

$$\mathbf{A}^2 = \left(\begin{array}{c|c} \mathbf{A}_{11}^2 + \mathbf{A}_{12}\mathbf{A}_{21} & \mathbf{A}_{11}\mathbf{A}_{12} + \mathbf{A}_{12}\mathbf{A}_{22} \\ \hline \mathbf{A}_{21}\mathbf{A}_{11} + \mathbf{A}_{22}\mathbf{A}_{21} & \mathbf{A}_{21}\mathbf{A}_{12} + \mathbf{A}_{22}^2 \end{array} \right) = \left(\begin{array}{cc|c} 42 & 50 & 62 \\ 57 & 70 & 93 \\ \hline 94 & 118 & 164 \end{array} \right).$$

Prob. 207 — Suppose that $\text{rank}(\mathbf{A}) = \text{rank}(\mathbf{A}_{11}) = r$ and that

$$\mathbf{A} = \begin{pmatrix} \mathbf{A}_{11} & \mathbf{A}_{12} \\ \mathbf{A}_{21} & \mathbf{A}_{22} \end{pmatrix} = \begin{pmatrix} \mathbf{A}_{11} & \mathbf{A}_{11}\mathbf{X} \\ \mathbf{A}_{21} & \mathbf{A}_{21}\mathbf{X} \end{pmatrix}.$$

Then

$$\mathbf{A} = \begin{pmatrix} \mathbf{A}_{11} \\ \mathbf{A}_{21} \end{pmatrix} \begin{pmatrix} \mathbf{I}_r & \mathbf{X} \end{pmatrix}$$

is a rank factorization of **A**.

7.4 Strong Generalised Inverses

> **Definition 7.6** Let **A** be an $m \times n$ matrix. A **strong generalised inverse** (SGI) of **A** denoted by $\mathbf{A}^G$ is any $n \times m$ matrix such that
>
> (i) $\mathbf{A}\mathbf{A}^G\mathbf{A} = \mathbf{A}$.
> (ii) $\mathbf{A}^G\mathbf{A}\mathbf{A}^G = \mathbf{A}^G$.
> (iii) $\mathbf{A}\mathbf{A}^G$ is the orthogonal projection of $\mathbb{R}^m$ onto $R(\mathbf{A})$.
> (iv) $\mathbf{A}^G\mathbf{A}$ is the orthogonal projection of $\mathbb{R}^n$ onto
>
> $$N(\mathbf{A})^\perp = \text{Row}(\mathbf{A}).$$

By Theorem 6.3 both $\mathbf{A}^G\mathbf{A}$ and $\mathbf{A}\mathbf{A}^G$ are projections. This implies that ($\mathbf{i} - \mathbf{iv}$) are equivalent to the **Penrose equations**:

(1) $\mathbf{A}\mathbf{A}^G\mathbf{A} = \mathbf{A}$.
(2) $\mathbf{A}^G\mathbf{A}\mathbf{A}^G = \mathbf{A}^G$.
(3) $(\mathbf{A}\mathbf{A}^G)^T = \mathbf{A}\mathbf{A}^G$.
(4) $(\mathbf{A}^G\mathbf{A})^T = \mathbf{A}^G\mathbf{A}$.

Lemma 7.5 *The SGI is unique if exists.*

Proof. Suppose that **X** and **Y** are SGI for a matrix **A**. Then by the Penrose equations we have:

$$\mathbf{X} \overset{(2)}{=} \mathbf{X}\,(\mathbf{AX}) \overset{(3)}{=} \mathbf{X}\,(\mathbf{AX})^T \overset{(iii)}{=} \mathbf{X}\,(\mathbf{AYAX})^T = \mathbf{X}(\mathbf{AX})^T(\mathbf{AY})^T \overset{(iii)}{=}$$

$$\mathbf{X}(\mathbf{AX})(\mathbf{AY}) = \mathbf{XAY} \overset{(iii)}{=} (\mathbf{XA})^T\,\mathbf{Y} \overset{(ii)}{=} (\mathbf{XA})^T\,\mathbf{YAY} \overset{(iii)}{=} (\mathbf{XA})^T\,(\mathbf{YA})^T\,\mathbf{Y} =$$

$$\mathbf{A}^T\mathbf{X}^T\mathbf{A}^T\mathbf{Y}^T\mathbf{Y} = (\mathbf{AXA})^T\,\mathbf{Y}^T\mathbf{Y} = \mathbf{A}^T\mathbf{Y}^T\mathbf{Y} = (\mathbf{YA})\,\mathbf{Y} = \mathbf{Y}. \quad \square$$

To obtain a **Formula for SGI**, we apply Theorem 3.12, which allows one to constructively write a rank factorization of the given matrix $\mathbf{A} = \mathbf{BC}$, $\text{rank}(\mathbf{A}) = r$. By definition $\text{rank}(\mathbf{B}) = \text{rank}(\mathbf{C}) = r$ and the sizes of **B** and **C** are $m \times r$ and $r \times n$ correspondingly.

$$
\begin{cases}
\mathbf{B}_{\text{left}}^{-1} = \left(\underbrace{\mathbf{B}^T}_{r \times m} \underbrace{\mathbf{B}}_{m \times r} \right)^{-1} \underbrace{\mathbf{B}^T}_{r \times m} \Rightarrow \text{size} = r \times m \\[2em]
\mathbf{C}_{\text{right}}^{-1} = \underbrace{\mathbf{C}^T}_{n \times r} \left(\underbrace{\mathbf{C}}_{r \times n} \underbrace{\mathbf{C}^T}_{n \times r} \right)^{-1} \Rightarrow \text{size} = n \times r
\end{cases}
$$

Theorem 7.11 *Let* $\mathbf{A}$ *be an* $m \times n$ *matrix of rank* r *and* $\mathbf{A} = \mathbf{BC}$ *be its rank factorization. Then*

$$
\mathbf{A}^G = \mathbf{C}_{\text{right}}^{-1} \mathbf{B}_{\text{left}}^{-1}, \quad \text{size}(\mathbf{A}^G) = n \times m.
$$

Proof.

1. $\mathbf{A}\mathbf{A}^G\mathbf{A} = (\mathbf{BC})\,\mathbf{C}_{\text{right}}^{-1}\mathbf{B}_{\text{left}}^{-1}\,(\mathbf{BC}) = \mathbf{B}\mathbf{I}_r\mathbf{I}_r\mathbf{C} = \mathbf{BC} = \mathbf{A}$

2. $\mathbf{A}^G\mathbf{A}\mathbf{A}^G = \mathbf{C}_{\text{right}}^{-1}\mathbf{B}_{\text{left}}^{-1}\,(\mathbf{BC})\,\mathbf{C}_{\text{right}}^{-1}\mathbf{B}_{\text{left}}^{-1} = \mathbf{C}_{\text{right}}^{-1}\mathbf{B}_{\text{left}}^{-1} = \mathbf{A}^G$

3. $\mathbf{A}\mathbf{A}^G = (\mathbf{BC})\,\mathbf{C}_{\text{right}}^{-1}\mathbf{B}_{\text{left}}^{-1} = (\mathbf{BC})\left[\mathbf{C}^T \left(\mathbf{CC}^T\right)^{-1} \left(\mathbf{B}^T\mathbf{B}\right)^{-1} \mathbf{B}^T \right]$

   $= \mathbf{B}\left(\mathbf{B}^T\mathbf{B}\right)^{-1}\mathbf{B}^T$ is symmetric

4. $\mathbf{A}^G\mathbf{A} = \left[\mathbf{C}^T \left(\mathbf{CC}^T\right)^{-1} \left(\mathbf{B}^T\mathbf{B}\right)^{-1} \mathbf{B}^T \right](\mathbf{BC}) = \mathbf{C}^T\left(\mathbf{CC}^T\right)^{-1}\mathbf{C}$

   is symmetric

The Penrose equations are satisfied. Hence $\mathbf{C}_{\text{right}}^{-1}\mathbf{B}_{\text{left}}^{-1} = \mathbf{A}^G$. $\qquad \square \qquad\qquad \square$

Problem 7.8 Find the SGI for the matrix

$$
\mathbf{A} = \begin{pmatrix} 1 & -1 & 2 \\ 0 & 2 & -2 \\ 1 & 1 & 0 \end{pmatrix}.
$$

Solution: We have already proved that

$$
\mathbf{A} = \begin{pmatrix} 1 & -1 & 2 \\ 0 & 2 & -2 \\ 1 & 1 & 0 \end{pmatrix} = \mathbf{BC} = \begin{pmatrix} 1 & -1 \\ 0 & 2 \\ 1 & 1 \end{pmatrix} \begin{pmatrix} 1 & 0 & 1 \\ 0 & 1 & -1 \end{pmatrix}
$$

$$\mathbf{C}_{\text{right}}^{-1} = \mathbf{C}^T \left(\mathbf{C}\mathbf{C}^T\right)^{-1} = \begin{pmatrix} 2/3 & 1/3 \\ 1/3 & 2/3 \\ 1/3 & -1/3 \end{pmatrix}$$

$$\mathbf{B}_{\text{left}}^{-1} = \left(\mathbf{B}^T\mathbf{B}\right)^{-1}\mathbf{B}^T = \begin{pmatrix} 1/2 & 0 & 1/2 \\ -1/6 & 1/3 & 1/6 \end{pmatrix}$$

$$\mathbf{A}^G = \mathbf{C}_{\text{right}}^{-1}\mathbf{B}_{\text{left}}^{-1} = \frac{1}{3}\begin{pmatrix} 2 & 1 \\ 1 & 2 \\ 1 & -1 \end{pmatrix} \cdot \frac{1}{6}\begin{pmatrix} 3 & 0 & 3 \\ -1 & 2 & 1 \end{pmatrix} = \boxed{\frac{1}{18}\begin{pmatrix} 5 & 2 & 7 \\ 1 & 4 & 5 \\ 4 & -2 & 2 \end{pmatrix}}.$$

Problem 7.9 Find the strong generalized inverse $\mathbf{A}^G$ of the matrix

$$\mathbf{A} = \begin{pmatrix} 1 & 0 & 1 & 1 \\ 0 & 1 & 1 & -1 \\ 1 & 1 & 2 & 0 \end{pmatrix}.$$

Solution: **Step 1.** We evaluate $\text{rref}(\mathbf{A})$:

$$\text{rref}(\mathbf{A}) = \begin{pmatrix} 1 & 0 & 1 & 1 \\ 0 & 1 & 1 & -1 \\ 0 & 0 & 0 & 0 \end{pmatrix}.$$

Step 2. $\text{rank}(\mathbf{A}) = 2$, $J = \{1, 2\}$.

Step 3.

$$\mathbf{B} = \left(\text{Col}_1(\mathbf{A})\ \text{Col}_2(\mathbf{A})\right) = \begin{pmatrix} 1 & 0 \\ 0 & 1 \\ 1 & 1 \end{pmatrix}, \quad \mathbf{C} = \text{rref}_p(\mathbf{A}) = \begin{pmatrix} 1 & 0 & 1 & 1 \\ 0 & 1 & 1 & -1 \end{pmatrix}.$$

Step 4. A rank factorization of $\mathbf{A}$ is given by:

$$\mathbf{A} = \begin{pmatrix} 1 & 0 & 1 & 1 \\ 0 & 1 & 1 & -1 \\ 1 & 1 & 2 & 0 \end{pmatrix} = \begin{pmatrix} 1 & 0 \\ 0 & 1 \\ 1 & 1 \end{pmatrix}\begin{pmatrix} 1 & 0 & 1 & 1 \\ 0 & 1 & 1 & -1 \end{pmatrix} = \mathbf{BC}.$$

Step 5.

$$\mathbf{C}_{\text{right}}^{-1} = \mathbf{C}^T \left(\mathbf{C}\mathbf{C}^T\right)^{-1} = \frac{1}{3}\begin{pmatrix} 1 & 0 \\ 0 & 1 \\ 1 & 1 \\ 1 & -1 \end{pmatrix}$$

Step 6.

$$\mathbf{B}_{\text{left}}^{-1} = \left(\mathbf{B}^T\mathbf{B}\right)^{-1}\mathbf{B}^T = \begin{pmatrix} 2/3 & -1/3 & 1/3 \\ -1/3 & 2/3 & 1/3 \end{pmatrix}$$

Step 7.

$$\mathbf{A}^G = \mathbf{C}_{\text{right}}^{-1}\mathbf{B}_{\text{left}}^{-1} = \frac{1}{3}\begin{pmatrix} 1 & 0 \\ 0 & 1 \\ 1 & 1 \\ 1 & -1 \end{pmatrix}\begin{pmatrix} 2/3 & -1/3 & 1/3 \\ -1/3 & 2/3 & 1/3 \end{pmatrix} = \frac{1}{9}\begin{pmatrix} 2 & -1 & 1 \\ -1 & 2 & 1 \\ 1 & 1 & 2 \\ 3 & -3 & 0 \end{pmatrix}.$$

Problem 7.10 Find a strong generalized inverse $\mathbf{A}^G$ of the matrix

$$\mathbf{A} = \begin{pmatrix} 1 & 1 & 0 \\ 0 & 1 & 1 \\ -1 & 1 & 2 \\ 1 & 0 & -1 \end{pmatrix}.$$

Solution:
Step 1. We evaluate $\text{rref}(\mathbf{A})$:

$$\text{rref}(\mathbf{A}) = \begin{pmatrix} 1 & 0 & -1 \\ 0 & 1 & 1 \\ 0 & 0 & 0 \\ 0 & 0 & 0 \end{pmatrix}.$$

Step 2. $\text{rank}(\mathbf{A}) = 2$, $J = \{1,2\}$.
Step 3.

$$\mathbf{B} = \left(\text{Col}_{j_1}(\mathbf{A})\ \text{Col}_{j_2}(\mathbf{A})\right) = \begin{pmatrix} 1 & 1 \\ 0 & 1 \\ -1 & 1 \\ 1 & 0 \end{pmatrix}, \quad \mathbf{C} = \text{rref}_p(\mathbf{A}) = \begin{pmatrix} 1 & 0 & -1 \\ 0 & 1 & 1 \end{pmatrix}.$$

Step 4. We find a rank factorization of the matrix $\mathbf{A}$:

$$\mathbf{A} = \begin{pmatrix} 1 & 1 & 0 \\ 0 & 1 & 1 \\ -1 & 1 & 2 \\ 1 & 0 & -1 \end{pmatrix} = \begin{pmatrix} 1 & 1 \\ 0 & 1 \\ -1 & 1 \\ 1 & 0 \end{pmatrix}\begin{pmatrix} 1 & 0 & -1 \\ 0 & 1 & 1 \end{pmatrix} = \mathbf{BC}.$$

Step 5.

$$\mathbf{C}_{\text{right}}^{-1} = \mathbf{C}^T\left(\mathbf{CC}^T\right)^{-1} = \frac{1}{3}\begin{pmatrix} 2 & 1 \\ 1 & 2 \\ -1 & 1 \end{pmatrix}$$

Step 6.

$$\mathbf{B}_{\text{left}}^{-1} = \left(\mathbf{B}^T \mathbf{B}\right)^{-1} \mathbf{B}^T = \frac{1}{3} \begin{pmatrix} 1 & 0 & -1 & 1 \\ 1 & 1 & 1 & 0 \end{pmatrix}$$

Step 7.

$$\mathbf{A}^G = \mathbf{C}_{\text{right}}^{-1} \mathbf{B}_{\text{left}}^{-1} = \frac{1}{9} \begin{pmatrix} 2 & 1 \\ 1 & 2 \\ -1 & 1 \end{pmatrix} \begin{pmatrix} 1 & 0 & -1 & 1 \\ 1 & 1 & 1 & 0 \end{pmatrix} = \frac{1}{9} \begin{pmatrix} 3 & 1 & -1 & 2 \\ 3 & 2 & 1 & 1 \\ 0 & 1 & 2 & -1 \end{pmatrix}.$$

Problem 7.11 State what is meant by a left inverse of an $m \times n$ matrix. Then consider the matrix

$$\mathbf{A} = \begin{pmatrix} -1 & 1 \\ 1 & 0 \\ 1 & 2 \end{pmatrix}.$$

(**i**) State a property of $\mathbf{A}$ that ensures that $\mathbf{A}$ has a left inverse.

(**ii**) Find all of the left inverses of the matrix $\mathbf{A}$.

(**iii**) Describe the range of the matrix $\mathbf{A}$ geometrically, and find its Cartesian equation. Find a basis of $R(\mathbf{A})^\perp$.

Solution: A left inverse of an $m \times n$ matrix is an $n \times m$ matrix $\mathbf{L}$ such that $\mathbf{LA} = \mathbf{I}_n$, where $\mathbf{I}_n$ is the $n \times n$ identity matrix.

(**i**) An $m \times n$ matrix $\mathbf{A}$ has a left inverse if and only if rank$(\mathbf{A}) = n$, i.e. $\dim R(\mathbf{A}) = n$. In case of the matrix $\mathbf{A}$ given, we see that its columns are not proportional. Hence rank$(\mathbf{A}) = 2$ and a left inverse exists.

(**ii**) Left inverses are determined by the equation:

$$\begin{pmatrix} 1 & 0 \\ 0 & 1 \end{pmatrix} = \begin{pmatrix} a & b & c \\ d & e & f \end{pmatrix} \begin{pmatrix} -1 & 1 \\ 1 & 0 \\ 1 & 2 \end{pmatrix} = \begin{pmatrix} -a+b+c & a+2c \\ -d+e+f & d+2f \end{pmatrix}.$$

So, we obtain the following systems:

$$\begin{cases} -a+b+c &= 1 \\ a+2c &= 0 \end{cases} \text{ and } \begin{cases} -d+e+f &= 0 \\ d+2f &= 1 \end{cases} \Leftrightarrow$$

$$\begin{cases} a &= -2c \\ b &= 1-3c \end{cases} \text{ and } \begin{cases} d &= 1-2f \\ e &= 1-3f \end{cases} \overset{c=s,\, f=t}{\Rightarrow}$$

$$\mathbf{L} = \begin{pmatrix} -2s & 1-3s & s \\ 1-2t & 1-3t & t \end{pmatrix}, \ s,t \in \mathbb{R}.$$

(**iii**) The range of $\mathbf{A}$ is the column space of $\mathbf{A}$, i.e. the linear span of two columns which is nothing but the plane passing through these vectors. A normal vector to this plane is determined by the formula:

$$\begin{vmatrix} \mathbf{i} & \mathbf{j} & \mathbf{k} \\ -1 & 1 & 1 \\ 1 & 0 & 2 \end{vmatrix} = 2\mathbf{i} + 3\mathbf{j} - \mathbf{k}.$$

The equation of the plane is $2x + 3y - z = 0$. The normal vector $\mathbf{n} = (2, 3, -1)^T$ to this plane makes a basis $\{\mathbf{n}\}$ for $R(\mathbf{A})^\perp$.   □

Problem 7.12 Consider the matrix

$$\mathbf{A} = \begin{pmatrix} -1 & 1 \\ 1 & 0 \\ 1 & 2 \end{pmatrix}.$$

(**iv**) Find the left inverse of $\mathbf{A}$ which is also the strong generalised inverse of $\mathbf{A}$. For the purpose of this question, call this left inverse $\mathbf{L}^G$.

(**v**) What linear transformation is represented by the matrix $\mathbf{AL}^G$ where $\mathbf{L}^G$ is the matrix in part (**iv**)? Give a full description of this linear transformation in terms of subspaces of $\mathbb{R}^3$ that you should identify.

Solution:

(**iv**) Since $\text{rank}(\mathbf{A}) = 2$ the rank factorization of $\mathbf{A}$ is the product $\mathbf{A} \cdot \mathbf{I}_2$. Then

$$\mathbf{A}^G = \mathbf{A}_{\text{left}}^{-1} = (\mathbf{A}^T \mathbf{A})^{-1} \mathbf{A}^T =$$

$$\left[\begin{pmatrix} -1 & 1 & 1 \\ 1 & 0 & 2 \end{pmatrix} \begin{pmatrix} -1 & 1 \\ 1 & 0 \\ 1 & 2 \end{pmatrix} \right]^{-1} \begin{pmatrix} -1 & 1 & 1 \\ 1 & 0 & 2 \end{pmatrix} =$$

$$\left[\begin{pmatrix} 3 & 1 \\ 1 & 5 \end{pmatrix} \right]^{-1} \begin{pmatrix} -1 & 1 & 1 \\ 1 & 0 & 2 \end{pmatrix} = \frac{1}{14} \begin{pmatrix} 5 & -1 \\ -1 & 3 \end{pmatrix} \begin{pmatrix} -1 & 1 & 1 \\ 1 & 0 & 2 \end{pmatrix} =$$

$$\boxed{\frac{1}{14} \begin{pmatrix} -6 & 5 & 3 \\ 4 & -1 & 5 \end{pmatrix} = \mathbf{L}^G}.$$

(**v**) The linear transformation $\mathbf{AL}^G$ is the orthogonal projection onto the subspace $R(\mathbf{A})$, which is the plane $\{(x, y, z)^T : 2x + 3y - z = 0\}$. The null space of this orthogonal projection is the linear span $\text{Lin}\{(2, 3, -1)^T\}$.   □

Theorem 7.12 *If* $\text{hef}(\mathbf{A}) = \mathbf{H}$, *then there is an invertible matrix* $\mathbf{S}$ *such that* $\mathbf{SA} = \mathbf{H}$. *Then* $\mathbf{S}$ *is a WGI for* $\mathbf{A}$.

Proof. Since $\mathbf{H}$ is obtained from $\mathbf{A}$ by elementary row operations, $\mathbf{H} = \mathbf{SA}$, where $\mathbf{S}$ is the product of the corresponding elementary matrices. Hence $\mathbf{S}$ is invertible. Since $\mathbf{H}^2 = \mathbf{H}$ and $\mathbf{S}$ is invertible, we have

$$\mathbf{SASA} = \mathbf{SA} \Rightarrow \mathbf{ASA} = \mathbf{A} \Rightarrow \mathbf{S} \in \mathrm{WGI}(\mathbf{A}).$$

□

Theorem 7.3 provides a formula for a least square solution if the columns of an $m \times n$ matrix are linearly independent. The SGI gives an exact formula in any case.

Theorem 7.13 *Let $\mathbf{A}$ be an $m \times n$ matrix. Then the general least square solution of the matrix equation $\mathbf{Ax} = \mathbf{b}$ is given by*

$$\mathbf{x} = \mathbf{A}^G\mathbf{b} + \left(\mathbf{I} - \mathbf{A}^G\mathbf{A}\right)\mathbf{w}, \ \mathbf{w} \in \mathbb{R}^n.$$

Proof. A least squares solution to the matrix equation $\mathbf{Ax} = \mathbf{b}$ is a vector $\mathbf{x}$ which minimises the quantity $||\mathbf{Ax} - \mathbf{b}||$. It follows that $\mathbf{x}^*$ is the least square solution if and only if $\mathbf{Ax}^*$ equals the orthogonal projection of $\mathbf{b}$ onto $\mathrm{R}(\mathbf{A})$, i.e.

$$\mathbf{Ax}^* = \mathbf{AA}^G\mathbf{b}.$$

It is clear that $\mathbf{x}^* = \mathbf{A}^G\mathbf{b}$ is a particular solution to this system. Hence, the general solution is given by

$$\mathbf{x}^* = \mathbf{A}^G\mathbf{b} + \mathbf{z}, \ \mathbf{z} \in \mathrm{N}(\mathbf{A}).$$

Since $\mathbf{I} - \mathbf{A}^G\mathbf{A}$ is a projection onto $\mathrm{N}(\mathbf{A})$, we obtain the required formula for the solution. □

Problem 7.13 Find all of the possible solutions to the least squares problem given by the matrix equation $\mathbf{Ax} = \mathbf{b}$, where

$$\mathbf{A} = \begin{pmatrix} 1 & -1 & 2 \\ 0 & 2 & -2 \\ 1 & 1 & 0 \end{pmatrix}, \ \mathbf{b} = \begin{pmatrix} -1 \\ 0 \\ 1 \end{pmatrix}.$$

Solution: All solutions are given by the formula

$$\mathbf{x} = \mathbf{A}^G\mathbf{b} + (\mathbf{I} - \mathbf{A}^G\mathbf{A})\mathbf{w} = \mathbf{A}^G\mathbf{b} + \mathbf{z}, \ \mathbf{w} \in \mathbb{R}^3, \ \mathbf{z} \in \mathrm{N}(\mathbf{A}).$$

We know that

$$\mathbf{A}^G = \frac{1}{18}\begin{pmatrix} 5 & 2 & 7 \\ 1 & 4 & 5 \\ 4 & -2 & 2 \end{pmatrix}, \mathrm{N}(\mathbf{A}) = \mathrm{Lin}\left\{\begin{pmatrix} -1 \\ 1 \\ 1 \end{pmatrix}\right\}.$$

Hence

$$\mathbf{x}^* = \frac{1}{18}\begin{pmatrix} 5 & 2 & 7 \\ 1 & 4 & 5 \\ 4 & -2 & 2 \end{pmatrix}\begin{pmatrix} -1 \\ 0 \\ 1 \end{pmatrix} + t\begin{pmatrix} -1 \\ 1 \\ 1 \end{pmatrix} = \frac{1}{9}\begin{pmatrix} 1 \\ 2 \\ -1 \end{pmatrix} + t\begin{pmatrix} -1 \\ 1 \\ 1 \end{pmatrix}, \ t \in \mathbb{R}.$$

Notice that

$$\begin{pmatrix} 1 & 2 & -1 \end{pmatrix}\begin{pmatrix} -1 \\ 1 \\ 1 \end{pmatrix} = 0,$$

which implies that

$$\|\mathbf{x}^*\|^2 = \frac{1}{81}\left\|\begin{pmatrix} 1 \\ 2 \\ -1 \end{pmatrix}\right\|^2 + t^2\left\|\begin{pmatrix} -1 \\ 1 \\ 1 \end{pmatrix}\right\|^2.$$

This formula shows that $\mathbf{A}^G\mathbf{b}$ is the least square solution which is the closest to zero. □

Problems

Prob. 208 — Find the Hermite echelon form hef($\mathbf{A}$), see Definition 1.24, for the matrix

$$\mathbf{A} = \begin{pmatrix} 3 & 6 & 9 \\ 1 & 2 & 5 \\ 2 & 4 & 10 \end{pmatrix}.$$

Prob. 209 — Show that if hef($\mathbf{H}$) = $\mathbf{H}$, then $\mathbf{H}^2 = \mathbf{H}$.

Prob. 210 — If hef($\mathbf{A}$) = $\mathbf{H}$, then there is an invertible matrix $\mathbf{S}$ such that $\mathbf{SA} = \mathbf{H}$. Then $\mathbf{S}$ is a WGI for $\mathbf{A}$.

Prob. 211 — Find the generalized inverse $\mathbf{A}^G$ for the matrix

$$\mathbf{A} = \begin{pmatrix} 1 & 2 & 1 \\ 2 & 1 & 0 \end{pmatrix}.$$

Prob. 212 — Prove that for any $m \times n$ matrix $\mathbf{A}$ the generalized inverse matrix $\mathbf{A}^G$ satisfies the equations:

$$\text{rank}(\mathbf{A}) = \text{rank}(\mathbf{A}^G) = \text{rank}(\mathbf{AA}^G) = \text{rank}(\mathbf{A}^G\mathbf{A}).$$

Prob. 213 — Find the strong generalised inverse matrix for

$$\mathbf{A} = \begin{pmatrix} 1 & 0 & 1/2 \\ 1 & 0 & 1/2 \\ 0 & -1 & -1/2 \\ 0 & -1 & -1/2 \end{pmatrix}.$$

Prob. 214 — Find all of the possible solutions to the least squares problem given by the matrix equation $\mathbf{Ax} = \mathbf{b}$ where

$$\mathbf{A} = \begin{pmatrix} 1 & 1 & 2 & 3 \\ 1 & -1 & 0 & 1 \\ 1 & 1 & 2 & 3 \end{pmatrix}, \quad \mathbf{b} = \begin{pmatrix} -1 \\ 0 \\ 1 \end{pmatrix}$$

Prob. 215 — Let $\left(\mathbf{A}^T \mathbf{A}\right)^g$ be any generalized inverse of $\mathbf{A}^T \mathbf{A}$, where $\mathbf{A}$ is an $m \times n$ matrix. Show that then

(a) $\left[\left(\mathbf{A}^T \mathbf{A}\right)^g\right]^T$ is a generalized inverse of $\mathbf{A}^T \mathbf{A}$,
(b) using the formula $\mathbf{A} = \mathbf{A}\mathbf{A}^G \mathbf{A}$, show that

$$\mathbf{A}\left(\mathbf{A}^T \mathbf{A}\right)^g \mathbf{A}^T \mathbf{A} = \mathbf{A},$$

(c) the matrix $\mathbf{A}\left(\mathbf{A}^T \mathbf{A}\right)^g \mathbf{A}^T$ does not depend on the choice of the generalized inverse $\left(\mathbf{A}^T \mathbf{A}\right)^g$,
(d) the matrix $\mathbf{A}\left(\mathbf{A}^T \mathbf{A}\right)^g \mathbf{A}^T$ is symmetric even if $\left(\mathbf{A}^T \mathbf{A}\right)^g$ is not symmetric.

Prob. 216 — Show that $\mathrm{tr}\,(\mathbf{A}) = \mathrm{rank}\,(\mathbf{A})$ for any idempotent matrix $\mathbf{A}$.

Prob. 217 — Find the parabola that gives the best least squares approximation to the points

$$(-1,1), \ (0,-1), \ (1,0), \ (2,2).$$

Prob. 218 — (a) Find the strong generalized inverse $\mathbf{A}^G$ for the matrix

$$\mathbf{A} = \begin{pmatrix} 4 & -3 & -3 \\ 5 & -4 & -5 \\ -1 & 1 & 2 \end{pmatrix}.$$

(b) Describe all least square solutions to the equation $\mathbf{Ax} = \mathbf{b}$, where $\mathbf{b} = (1,0,0)^T$.
(c) Determine the least square solution of $\mathbf{Ax} = \mathbf{b}$ which is the closest to the origin.
(d) Using the strong generalized inverse $\mathbf{A}^G$ find the orthogonal projection of the vector $\mathbf{b} = (1,0,0)^T$ onto the column space of the matrix $\mathbf{A}$.

Prob. 219 — Find the least squares solutions, the least squares error vector, and the least squares error of the linear system

$$\begin{cases} 3x_1 + 2x_2 - x_3 & = 2 \\ x_1 - 4x_2 + 3x_3 & = -2 \ . \\ x_1 + 10x_2 - 7x_3 & = 1 \end{cases}$$

Prob. 220 — Find the linear function $y = ax + b$, which fits the following data best

$$(1,1), \quad (2,3), \quad (3,2).$$

Prob. 221 — Find the strong genaralized inverses for the following matrices:

$$\mathbf{A} = \begin{pmatrix} 9 & 18 & 2 & 11 & 2 \\ 3 & 6 & 1 & 4 & 0 \\ 5 & 10 & 1 & 6 & 1 \\ 4 & 8 & 1 & 5 & 1 \end{pmatrix}, \quad \mathbf{B} = \begin{pmatrix} 1 & 1 & 5 & -4 & 2 \\ 8 & 4 & 28 & -19 & 14 \\ 3 & 1 & 9 & -9 & 4 \\ 3 & 2 & 12 & -9 & 5 \end{pmatrix}.$$

References

Anthony, M. and Biggs, N. 2024. Mathematics for economics and finance (2nd edition). Cambridge University Press.

Anthony, M., Harvey, M. 2012. Linear Algebra: Concepts and Methods. Cambridge University Press.

Axler, S. 2015. Linear Algebra Done Right. Springer.

Kennedy, J.G., Snell, J.L. 1976. Finite Markov Chains. Springer.

Khrushchev, S. 2023. Automatic Digital Workplace of a Teacher in Microsoft 365 Education. Amazon.

Kolman, B., Hill, D. 2007. Elementary Linear Algebra with Applications (9th Edition) Pearson.

Meier, J. 2008. Groups, Graphs and Trees. Cambridge University Press.

Narici, L., Beckenstein, E. 2011. Topological Vector Spaces. CRC.

Rockafellar, R.T. 1990. Convex Analysis. Princeton.

Swetz, F.J. 2008. Legacy of the LUOSHU. A.K. Peters.

Tchebysheff, S. 1854. On a formula in Analysis. *Bulletin de la Classe phys,-mathem. de l'Acad. Imp. des Science de St.Petersburg*, **XIII**, 210.

Vanderbei, R.J. 2014. Linear Programming. Foundations and Extensions. Fourth Edition. Springer.

Index

© The Author(s), under exclusive license to Springer Nature Switzerland AG 2024
S. Khrushchev, *Linear Algebra with Applications to Economics*, Classroom Companion:
Economics, https://doi.org/10.1007/978-3-031-68682-5